# International Conference on Chemical Engineering and Applied Sciences (ICChEAS 2019)



## Strengthening Innovative Chemical Engineering and Applied Sciences in the Digital Age

IOP Conference Series: Materials Science and Engineering
Volume 845

Banda Aceh, Indonesia
23 - 24 October 2019

ISBN: 978-1-7138-3612-4
ISSN: 1757-8981

**Printed from e-media with permission by:**

Curran Associates, Inc.
57 Morehouse Lane
Red Hook, NY 12571

**Some format issues inherent in the e-media version may also appear in this print version.**



Institute of Physics
Dirac House, Temple Back
Bristol BS1 6BE UK

Phone: 44 1 17 929 7481
Fax: 44 1 17 920 0979

techtracking@iop.org

**Additional copies of this publication are available from:**

Curran Associates, Inc.
57 Morehouse Lane
Red Hook, NY 12571 USA
Phone: 845-758-0400
Fax: 845-758-2633
Email: curran@proceedings.com
Web: www.proceedings.com

# TABLE OF CONTENTS

IOP Conf. Series: Materials Science and Engineering **845** (2020) 011001 doi:10.1088/1757-899X/845/1/011001

# PREFACE

We are very pleased to present to you the Proceedings of the International Conference on Chemical Engineering and Applied Sciences (ICChEAS) 2019. With the theme "Strengthening Innovative Chemical Engineering and Applied Sciences in the Digital Age", this served as a single platform for academicians, researchers and practitioners to exchange information and knowledge, present new developments and research findings, and develop new networking for future collaborations. The conference taking place in Banda Aceh, the Province of Aceh, Indonesia during 23-24 October 2019 was organized by the Chemical Engineering Department, Syiah Kuala University, Indonesia and co-organized by to the Faculty of Chemical Engineering, Universiti Teknologi MARA (UiTM) Malaysia, the Department of Chemical Engineering of Riau University (UNRI) Indonesia, and Vocational School, Diponegoro University (UNDIP) Indonesia.

The event was attended by more than 152 participants consisting 68 presentations coming from chemical engineering and its related fields as well as from applied sciences background, either from Indonesia and overseas. The keynote speeches were delivered by internationally acclaimed professors while plenary speakers are distinguished professors from leading local and international institutions. The technical presenters are well-respected researchers from universities in Indonesia and ASEAN.

The conference proceedings consisting selected papers submitted to the conference coming from various universities, research institutes, and industries. After a rigorous review process, where each paper was reviewed by at least two experts, high quality papers were accepted for presentation at the conference and for publication in the proceedings. Through this online Scopus-listed proceedings, we are sharing with you the papers from the ICChEAS 2019.

We would like to thank all the authors for their contribution to the conference as well as to the proceedings and also to the keynote and plenary speakers, organizing committee members, reviewers, chairpersons, volunteers, sponsors and all participants who have made the ICChEAS) 2019 a success.

Prof. Dr. Abrar Muslim, ST. M.Eng.
Conference Chair
Chemical Engineering Department
Syiah Kuala University, Banda Aceh, Indonesia

IOP Conf. Series: Materials Science and Engineering **845** (2020) 011002 doi:10.1088/1757-899X/845/1/011002

# Peer review statement

All papers published in this volume of *IOP Conference Series: Materials Science and Engineering* have been peer reviewed through processes administered by the proceedings Editors. Reviews were conducted by expert referees to the professional and scientific standards expected of a proceedings journal published by IOP Publishing.

IOP Conf. Series: Materials Science and Engineering **845** (2020) 012001 doi:10.1088/1757-899X/845/1/012001

# Potassium enrichment of treated palm oil mill effluent in a fixed-bed contactor

**S H Pranolo[1*], Margono[1], Paryanto[1], A Muizuddin[1] and K T Ayyubbiru[1]**
[1]Study Program of Chemical Engineering, Faculty of Engineering, Universitas Sebelas Maret, Jl. Ir. Sutami 36A, Jebres, Surakarta, Indonesia 57126

E-mail: sunu_pranolo@staff.uns.ac.id

**Abstract.** Treated Palm Oil Mill Effluent (POME) is the effluent of POME-based biogas production process. This liquid is potential as a raw material of liquid organic fertilizer, but the potassium content should be upgraded. Ash of empty palm fruit bunches (EPFB-a) which is locally available in huge amount is an attractive source of potassium. Potassium enrichment of treated POME was performed by contacting this liquid with EPFB-a in a fixed bed contactor. This research studied the effects of treated POME flow rate and height of EPFB-a bed on potassium mass transfer coefficient from solid to liquid phase in the contactor. Treated POME and EPFB-a were obtained from a crude palm oil mill in Rantau Sakti Village, Tambusai Utara Sub-district, Rokan Hulu Regency of Riau Province. The treated POME flow rates were set at 3.72; 3.36; and 2.52 L/min. EPFB-a bed height were fixed at 0.03, 0.06, and 0.09 m. The performance of this fixed bed contactor was evaluated based on extracted potassium content.

## 1. Introduction

In recent years, Indonesia was the top producer of worldwide crude palm oil (CPO). It is also predicted that the country may provide 39% - 60% of the international CPO demand. This demand would be supplied from 565 to 880 Mton fresh fruit bunch [1]. Water specific consumption of CPO production is about 5.0 - 7.5 tons/ton and more than 50% of water ends up as POME. It means that every ton of CPO production generates 2.5 - 3.75 ton of POME. This abundant quantity of POME is a potential pollutant to the water sources in the nearby of palm oil mills as its elevated content of BOD and COD [2]. However, its content depends on the production process and fresh fruit bunch (FFB) utilized from as low as 16 g $O_2$/mL when using advanced technology process at capacity 150 MT of FFB/hour and as high as 100 g $O_2$/mL when using more primitive one at capacity 2.5 MT of FFB/hour. Many studies focused on treatment of converting POME to biogas in an anaerobic digester using both traditional and advance methods. Microbiological approach [3]–[5], photocatalytic degradation [6], photo-oxidation method [7], adsorption using several types of bio-adsorbents [8], [9], , and electrocoagulation method [10] successfully demonstrated COD removal.

Beside POME, crude palm oil mills also generate a large amount of solid waste which is dominated by empty palm fruit bunch (EPFB) of about 23% for every ton of FFB. Moisture content was reported as high as 60% - 70%. The chemical characteristics of EFB was reported as 43.3% of cellulose, 26.2% of hemicellulose, and 30.5% of lignin[11]. Recently, this enormous amount of empty fruit bunch appears as problematic solution to its utilization. Traditionally, EPFB is dumped in palm oil plantation where

* To whom any correspondence should be addressed.

decomposes naturally which releases methane gas to the atmosphere. This action may contribute greenhouse gas significantly while more environmentally friendly methods are now available. The EPFB is possible to be a feedstock for bio-oil production[12], converted to syngas via gasification [13], pelleted for pyrolysis process [14], and source of cellulose and nanocrystalline cellulose for heavy metal sorption [15], [16]. Production of methane by anaerobic co-digestion of POME and EPFB mixture was also studied. Co-digestion has positively affected to methane production at proper ratio and size of EFB [11].

In this mean time, effluent of anaerobic biodigester based on POME is scarcely reused and is not an area of interest for research among environmentalists, industrialists and scientists. Reusing the effluent for composting and cleaning process of EPFB was proposed in order to save the operating costs of a palm oil mill. But, most of this biodigester effluent is just discharged to the waterbody though its COD and BOD values are sometimes not low enough to comply with the regulation limit. It is advisable to be combined with subsequent aerobic treatment or for other application such as composting [2].

Based on mass transfer phenomena, potassium in EPFB-a can be extracted with the effluent of biodigester to enrich the effluent with potassium. A solid-liquid mass transfer can be processed in several types of contactors, such as fixed beds reactor [17], [18], fluidized beds reactor [17], [19], trickle bed reactor [20], rotating tubular packed bed of spheres reactor [21], and agitated vessel [22]. Most studies showed that the mass transfer operation depends on solid-liquid mass transfer coefficient ($k_La$) which is affected by liquid flow characteristic and easiness of inter phase contact. The flow characteristic is represented with Reynolds number, and the contact easiness is described with Schmidt number. Construction of contactor especially fixed bed contactor were observed affecting the solid-liquid mass transfer coefficient. This is represented as the ratio of bed thickness and particle diameter [21] or the ratio of particle diameter and reactor diameter [18].

This research is intended for reusing both effluent of biodigester based on POME and EPFB-a as raw material of liquid organic fertilizer. As a fertilizer, potassium content of the effluent is low enough about 2270 mg/L [6] that addition of potassium from other sources is necessary. While, EPFB ash content may attain 13.65% [12]. Typically, biomass ash contains about 24.2% - 27.7% of potassium [23]–[25]. Thus, EPFB-a is potential source of potassium for producing liquid organic fertilizer from the effluent. In this research, extracting potassium of EPFB-a with treated POME was processed in a fixed bed contactor.

## 2. Experimental technique

### 2.1. Calculation method

A model of potassium balance of EPFB-a in a fixed bed contactor is written as equation (1) with $X_A$ is potassium mass fraction in EPFB-a, $\rho_B$ is density of EPFB-a (g/cm$^3$), $C_A$ is potassium concentration in treated POME (g/cm$^3$), and $C_A^*$ is saturated potassium in treated POME. The saturated potassium is approximated with equilibrium concentration of potassium in the water at 25°C (0.6274 g/cm$^3$) as treated POME was an aqueous solution (Haynes, 2011). The analysis showed that at initial condition, mass fraction of potassium in EPFB-a was 1.07% and potassium in treated POME was 3.37x10$^{-4}$ g/cm$^3$. By analyzing the treated POME after extracting for potassium content, equation (1) was solved numerically for calculating the $k_La$ values of every experiment run.

$$\frac{dX_A}{dt} = \frac{k_La}{\rho_B}(C_A - C_A^*) \quad (1)$$

The variables that affected solid-liquid mass transfer coefficient ($k_La$) in the contactor are density ($\rho$), viscosity ($\mu$) and flow rate ($v$) of treated POME, column diameter ($D$), and EPFB-a bed height ($z$). A mathematical correlation among solid-liquid mass transfer coefficient and the related variables was developed according to equation (2) with using dimensional analysis. A quantitative model should be formulated as equation (3) with $C_1$, $C_2$ and $C_3$ are constant parameters which are specific to this case.

$$k_la = f(\rho, \mu, v, D, z) \quad (2)$$

$$\frac{k_L a \ D^2 \ \rho}{\mu} = C_1 \ \left(\frac{\rho v D}{\mu}\right)^{C2} \left(\frac{z}{D}\right)^{C3} \tag{3}$$

Matrices technique was applied to calculate those parameters which were applied to predict the solid-liquid mass transfer using equation (3). The relative difference between prediction and experimental values of the coefficient was expressed as average error.

### *2.2. Experimental setup*

A fixed bed contactor experimental set consists of a fixed bed column 0.15 m in inner diameter and 2.50 m in height, 5 L effluent tank made from stainless steel, 5 L treated POME plastic storage tank, 0.25" valves, 3/8" of inner pipe diameter and 25 W electric pump (see figure 1). Treated POME and EPFB-a were collected from a palm oil mill in Rantau Sakti Village, Tambusai Utara Sub-district, Rokan Hulu Regency of Riau Province.

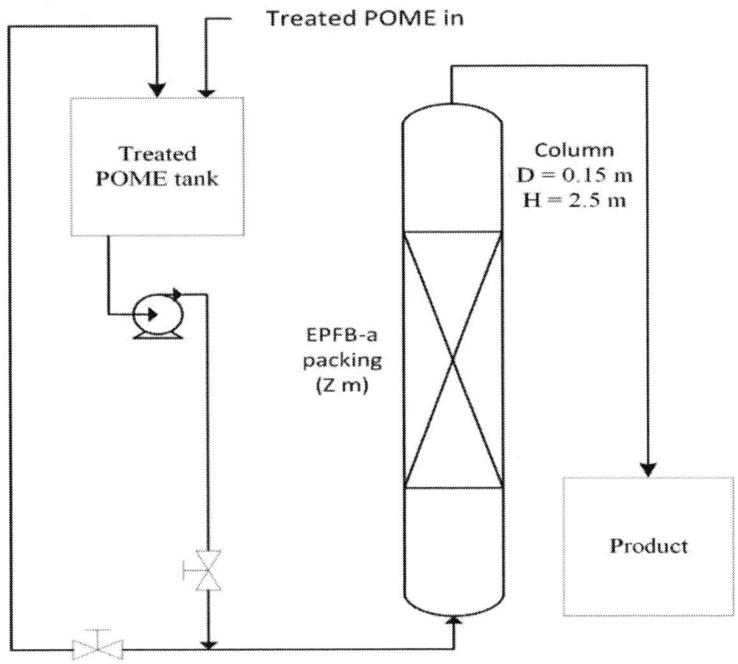


**Figure 1.** Schematic drawing of the fixed bed contactor for the experiments.

Before extracting, EPFB-a was stacked in the column using 300 mesh filter sheets at a certain height for each experiment: 0.03, 0.06, and 0.09 m. The EPFB-a bed was installed in the distance of 1.50 m from the bottom to avoid inlet flow effect to the bed. Treated POME from the tank was introduced to the column by pumping it to the bottom inlet at various flow rates (2.52, 3.36, and 3.72 L/min) for each EPFB-a stack height by adjusting the valves. The contact between treated POME and EPFB-a took place when the liquid passed through the bed. After contacting, the liquid flowed out from the top of the column to product tank. The first sample of outlet liquid was taken from the tank after 10 seconds of processed for potassium analysis.

The following samples were taken every 5 seconds after that. Potassium content in the samples were analyzed according to Indonesian National Standard (SNI 19-7030-2005) using AA-6200 Atomic Absorption Flame Emission Spectrophotometer (AAS) from Shimadzu. Prior to the use, the instrument

IOP Conf. Series: Materials Science and Engineering **845** (2020) 012001 doi:10.1088/1757-899X/845/1/012001

was standardized with potassium solution 1000 ppm in 2% $HNO_3$then a standard curve was constructed. The AAS analysis showed potassium content in a sample as absorbance and by using the standard curve potassium concentration was expressed as mass fraction.

## 3. Results and discussion

### *3.1. Sample analysis*

Figure 2 describes an increasing of potassium in treated POME after flowing through the EPFB-a bed in the column at different bed height. The *x*-axis of Figure 1 indicates relative difference between potassium content in outlet treated POME and potassium content of fresh treated POME. Obviously, the potassium content in EPFB-a is the highest at the beginning of the process which causes more potassium transferred to the treated POME. It conforms with mass transfer law (Fick's law) and previous studies [19], [20], [22] that the concentration driving force would be the highest at the initial condition and this encourages more potassium to move from the solid to liquid phase.

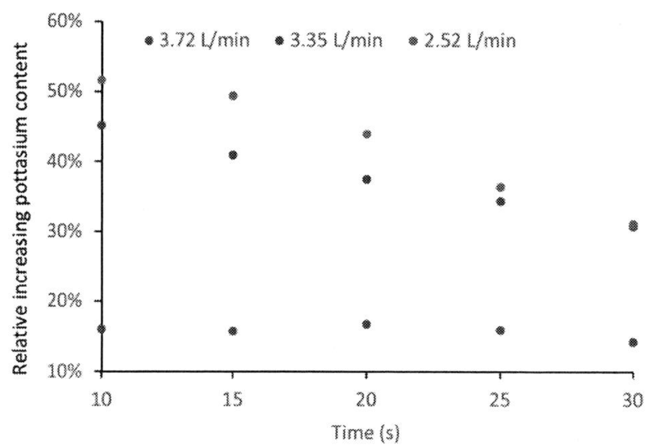


(a) $z = 0.03$ m

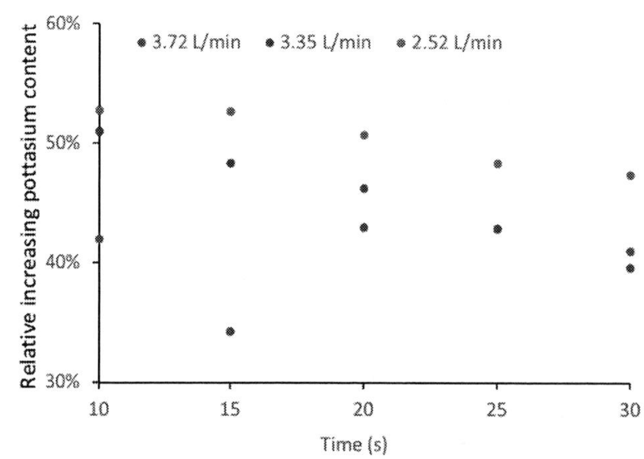


(b) $z = 0.06$ m

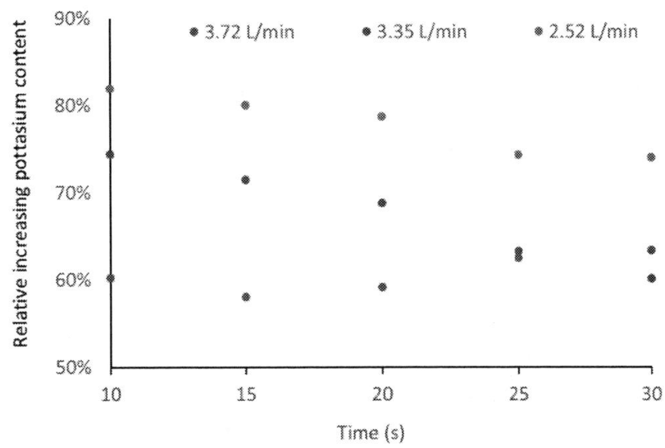


(c) $z = 0.09$ m

**Figure 2.** Relative increasing potassium content in treated POME as function of treated POME volumetric rate and EPFB-a bed height: (a) z=0.03 m; (b) z=0.06 m; and (c) z=0.09 m.

Table 1 shows all the experimental results when one liter of treated POME extracts potassium of EPFB-a bed in one second. According to the results the highest the EPFB-a bed the more potassium in the bed, thus the more potassium would be extracted from the EPFB-a bed. The results also indicated that the longer the treated POME residence time the more potassium extracted. In this study, the residence time was less and after 30 seconds of process it achieved 74.09%. This was equivalent to 0.0516 gram of potassium extracted with one liter of treated POME every second. Only 14.21% increasing of potassium was obtained when a bed height of 0.03 m and treated POME flow rate of 3.72 L/min were applied to the process. At this condition, treated POME was capable to extract potassium from EPFB-a bed at a rate of 0.0105 gram/(L sec).

**Table 1.** Extraction rate of potassium of EPFB-a (gram) with one liter of treated POME in one second.

| Volumetric Rate of treated POME (L/min) | EPFB-a bed height (m) 0.03 | 0.06 | 0.09 |
|---|---|---|---|
| 2.52 | 0.0269 | 0.0334 | 0.0516 |
| 3.35 | 0.0241 | 0.0297 | 0.0443 |
| 3.72 | 0.0105 | 0.0270 | 0.0409 |

*3.2. Solid-liquid mass transfer coefficient*

Based on sample analysis results, potassium content in EPFB-a and treated POME, solid-liquid mass transfer coefficient of potassium from EPFB-a to treated POME was predicted at every single experiment. Firstly, the $k_La$ was calculated with equation (1) numerically. Figure 3 and figure 4 present the effect of flow characteristic of treated POME which is expressed as Reynolds Number (*Re*) and geometrical factor which is expressed as ratio of EPFB-a bed height and inner column diameter (*z*/*D*) on solid-liquid mass transfer coefficient which is expressed as a non-dimensional number of $(k_L a\, D^2 \rho)/\mu$. It is found that effect of flow characteristic on solid-liquid mass transfer coefficient is inversely proportional to the ratio of EPFB-a bed height and inner column diameter. Some previous studies also presented similar effect of Reynolds number on the coefficient, although the studies used different solid-liquid phase and contacting method. The geometrical factor was also expressed with different ways [19].

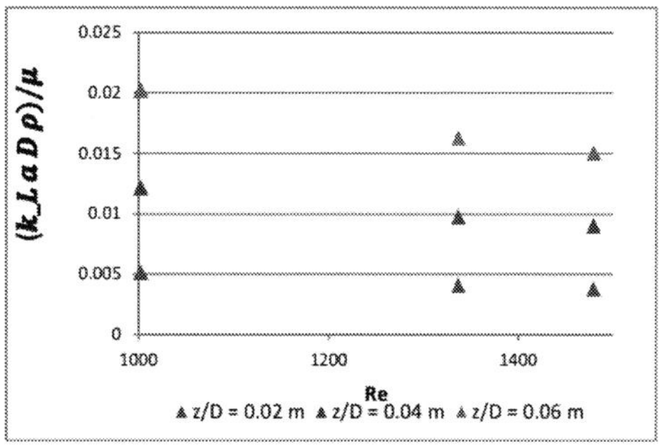


**Figure 3.** Effect of Reynolds Number on $\left[\frac{k_L a\ D^2\ \rho}{\mu}\right]$

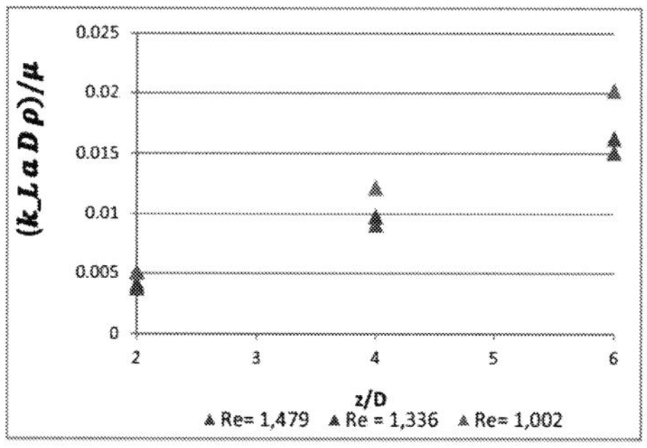


**Figure 4.** Effect of geometrical factor z/D on $\left[\frac{k_L a\ D^2\ \rho}{\mu}\right]$

The value of solid-liquid mass transfer coefficient ($k_La$) varies with Reynolds Number and geometrical factor ranges in this study. It was calculated using equation (1) and the sample analysis results. Liquid flows of all the experiments were in the range of laminar regime, and the geometrical factor were set at 2.00, 4.00 and 6.00. Calculation the parameters on equation (2) needs a set of three equation (2) which relate $(k_L a\ D^2\rho)/\mu$ with Reynolds Number and geometrical value as formulated with equation (2). Every equation in the set should consist different Reynolds Number and geometrical value. By using matrices method, the parameters were calculated. The results were applied on equation (2) for calculating the solid-liquid mass transfer equation ($k_La$). Table 2 shows all the calculated coefficients in the unit of 1/second. It is found that the coefficient decreases with the increasing of Reynolds Number, but increasing with the increasing of geometrical factor. It is formulated in a mathematical model of dimensionless numbers as:

$$\frac{k_L a\ D^2\ \rho}{\mu} = 0.3969\ Re^{-0.7549}\left(\frac{z}{D}\right)^{1.2515} \quad (4)$$

**Table 2.** Calculated solid-liquid mass transfer coefficient ($k_La$) at various Reynolds Numbers and geometrical factors using equation (2).

| Reynolds Number | Geometrical Factor ($z/D$) 2.00 | 4.00 | 6.00 |
|---|---|---|---|
| 1001 | 2.03 x $10^{-5}$ | 4.83 x $10^{-5}$ | 8.02 x $10^{-5}$ |
| 1335 | 1.63 x $10^{-5}$ | 3.89 x $10^{-5}$ | 6.46 x $10^{-5}$ |
| 1479 | 1.51 x $10^{-5}$ | 3.60 x $10^{-5}$ | 5.98 x $10^{-5}$ |

The parameter values of $C_1$, $C_2$ and $C_3$ of equation (3) are 0.3969, -0.7549, and 1.2515 respectively. The average error was determined as ratio of the absolute difference between experimental value and calculated value of $k_La$ to experimental value of $k_La$. Thus, the correlation of three non-dimensional numbers as in (2.3) can be implemented to predict the solid-liquid mass transfer coefficient of potassium from EPFB-a and treated POME in a fixed bed column. The correlation should be applied in laminar flow regime and geometrical factor 2.00 – 6.00.

## 4. Conclusion

The potassium content in treated POME as raw material of liquid organic fertilizer is potentially to be upgraded by extracting potassium content of EPFB-a in a fixed bed column. This process is represented with solid-liquid mass transfer coefficient which depends on chemical and physical characteristics of treated POME, and ratio of bed height and bed diameter of EPFB-a in the column. The formulated mathematical model of dimensionless numbers is applicable to laminar flow and range of $\left(\frac{Z}{D}\right)$ from 2.00 to 6.00 with an average error of about 14.55%.

## References

[1] Afriyati D, Kroeze K and Saad A 2016 *Sci. Total Environ.* **557** 562
[2] Ji C M, Eong P P, Ti T B, Seng C E and Ling C K 2013 *Renew. Sust. Energ. Rev.* **26** 717
[3] Poh P E and Chong M F 2010 *Chem. Eng. Trans.* **21** 811
[4] Tan S P, Kong H F, Bashir M J K, Lo P K, Ho C D and Ng C A 2017 *Bioresour. Technol.***245** 916
[5] Islam M A, Yousuf A, Karim A, Pirozzi D, Khan M R and Wahid Z A 2018 *J. Clean. Prod.* **172** 1779
[6] Alhaji M H, Sanaullah K, Lim S, Khan A, Hipolito C N, Abdullah M O, Bhawani S A and Jamil T 2016 *Process Saf. Environ.* **102** 673
[7] Alhaji MH, Sanaullah K, Salleh SF, Baini R, Lim SF, Rigit ARH, Said KAM and Khan A 2018 *Process Saf. Environ.* **117** 180
[8] Ibrahim I, Hassan M A, Abd-Azis S, Shirai Y, Andou Y, Othman M R, Ali A A M and Zakari M R 2017 *J. Clean. Prod.* **141** 122
[9] Lam S S, Liew R K, Cheng C K, Rasit N, Ooi C K, Ma N L, Ng J, Lam W H, Chong C T and Chase H A 2018 *J. Environ. Manage.* **213** 400
[10] Nasrullah M, Lakhveer S, Mohamad Z, Norsita S, Krishnan S, Wahida N and Zularisam A W 2017 *Water Res. Ind.* **17** 7
[11] Saelor S, Kongjan P and O-Thong S 2017 *Energy Procedia* **138** 717
[12] Chang S H 2014 *Biomass Bioenergy* **62** 174
[13] Li Y H and Chen H S 2018 *Int. J. Hydrogen Energ.* **43** 667
[14] Nyakuma B B, Ahmad A, Johari A, Abdullah T A T, Oladokun O and Aminu D Y 2015 *Chem. Eng. Trans.* **45** 1327
[15] Daneshfozoun S, Nazir M S, Abdullah B and Abdullah M A 2015 *Chem. Eng. Trans.* **37** 679
[16] Shanmugarajah B, Kiew PL, Chew I M L, Choong T S Y and Tan K W 2015 *Chem. Eng. Trans.* **45** 1705
[17] Derksen JJ 2014 *Chem. Eng. J.* **255** 233
[18] Harraz AA, El Gheriany IA, Abdel Azis MH, Zewail TM and Konsowa AH 2015 *Biochem. Eng. J.* **103** 1
[19] Kalaga D V, Dhar A, Dalvi S V and Joshi J B 2014 *Chem. Eng. J.* **245** 323
[20] Nicol W and Joubert R 2013 *Chem. Eng. J.***230** 361
[21] Abdel-Aziz MH, Nirdosh I and Sedahmed GH 2018*Int. J. Heat Mass Tran.***126** 1129
[22] Bong E Y, Eshtiaghi N, Wu J and Parthasarathy R 2015 *Chem. Eng. Res. Des.***100** 148
[23] Imaduddin M, Yoeswono, Wijaya K and Tahir I 2008 *Bull. Chem. Reac. Catal.***3** 14
[24] Lahijani P, Zainal Z A, Mohamed A R and Mohammadi M 2013 *Bioresour. Technol.***132** 351
[25] Gbaraneh L D and Chu G C S 2016 *Int. J. Sci. Res. Edu.* **4(08)** 5674

# Characterization of Modified Polyurethane Foam Adsorbents for Mercury Adsorption Applications

**Darmadi[1], M R Lubis[1], R Saputra[1], M D Alfarabi[1], S Sarah[1] and Adisalamun[1*]**
Chemical Engineering Department, Syiah Kuala University.

E-mail: adisalamun@che.unsyiah.ac.id

**Abstract.** The use of modified polyurethane foam adsorbents for mercury adsorption applications can reduce the impact of mercury pollution. This study aims to study the characteristics of adsorbents made from castor oil, MDI (Methylene diphenyl diisocyanate), glycerol, distilled water and modified by the addition of zeolite and bentonite. Polyurethane foam which has optimum performance is further characterized to determine physical and chemical properties, including: functional groups (FTIR), morphology (SEM) and area (BET). The results of FTIR Characterization showed that in polyurethane foams without modification there was a C = O group which is a typical group in the polyurethane group shown in wavenumber (1654-1516 $cm^{-1}$, 1680-1514 $cm^{-1}$, and 1593-1516 $cm^{-1}$) Whereas in zeolite modified polyurethane foam there is a C = CH group with a wavenumber of 3008 $cm^{-1}$ which is a typical group for zeolite whereas in bentonite there is a C = CH group with a wavenumber of 3007 $cm^{-1}$ which is a typical group for bentonite. The results of SEM characterization show that zeolite modified polyurethane foam has a smooth and flat structure with small pores and polyurethane foam with bentonite has irregular or irregular particle shape. The average BET area obtained in the PUF modification is 1.3778 $m^2\ g^{-1}$. Whereas for the smallest BET area obtained at synthetic PUF 19.1 gr + Zeolite 40% for 0.773 $m^2\ g^{-1}$, the area is influenced by the comparison of composition from polyurethane foam synthesis.

## 1. Introduction

Polyurethane is a synthetic polymer formed by the reaction between monomers containing at least two functional groups of isocyanates and other monomers containing at least two hydroxy groups in the presence of a catalyst. The basic raw material in producing polyurethanes is polyols and isocyanates. The molecular structure and size of this compound influences the polymerization reaction and polyurethane structure obtained [1].

Polyurethane foam can be stable in acidic and basic solutions and will not change the structure when heated to a temperature of 180°C [2,3]. Thus, polyurethane foam material is very suitable to be used as an adsorbent for the separation process with immersion techniques and column methods.

Polyurethane foam has been widely applied to various separation processes or pre-concretion methods. The ability of polyurethane foam as an adsorbent has been investigated by Vidoti et al. [4] and Yassin [5]. Lately, several researchers have also modified polyurethane foam to expand its application. Moawed et al. [6] modified polyurethane foam with Rhodamin B to separate trivalent metal ions Bi (III), Sb (III) and Fe (III). In addition, Li et al. [7] modified polyurethane foam with oleophilic monomers as absorbent materials for diesel and kerosene and water. In this study, the polyurethane foam that will be

[*] To whom any correspondence should be addressed.

produced is modified using natural fillers both inorganic and organic. The inorganic fillers that will be used are bentonite and zeolite, while the organic fillers used are chitosan and cellulose. The resulting polyurethane foam is expected to have good mechanical and absorbent properties and to be characterized according to polyurethane foam.

Characterization of polyurethane is a very important part of a polyurethane industry to determine the application of polyurethane. This polyurethane application is determined by the properties possessed by the polyurethane material. To find out the properties of polyurethane that have been synthesized, characterization of physical and chemical properties is carried out including morphology and molecular structure. In this study used several characterization methods, among others: molecular structure analysis with Fourier Transform Infrared Spectroscopy (FTIR), analysis of surface with scanning electrons microscopys (SEM).

Analysis of the data used in this study are: functional group analysis with FTIR. This analysis aims to ensure the formation of polyurethane foams from functional groups displayed by FTIR spectra and also structural analysis of the surface of polyurethane foam with SEM. The analysis aims to characterize the morphology of polyurethane foams [8]. The use of polyurethane foam as an absorbent material is very rare to use because the procedure for making flexible polyurethane foam has not been widely published. With the scarcity of this information, the product form of polyurethane foam produced is often not as expected. According to the information obtained, the economic loss due to the process of making the process of making polyurethane foam is quite large. This is due to the lack of knowledge regarding the nature of the polyurethane flexible foam during its formation reaction.

In the process of making polyurethane foams it can be seen the nature of polyurethane foam during the polymerization reaction takes place. The advantages possessed by polyurethane foam which are modified by zeolite and bentonite are good chemical, mechanical and absorbency properties.As for the consideration in the selection of polyurethane foam making materials are the factors of material characteristics, economic value, and environment. Characteristics of materials are factors that provide special advantages related to the physical-chemical properties of materials, in this case include the properties of absorption to water, thermal conductivity, and so on. These physical chemical properties are generally used as the main reference in determining the selection of materials in the manufacture of polyurethane foams [9].

Based on the description above, in this study the polyurethane foam adsorbent was characterized to see the properties of polyurethane produced as an adsorbent, so as to provide information and benefits in the industrial field for the application of mercury adsorption.

## 2. Methodology

### *2.1. Materials and tools*

The material used in this study is castor oil as polyol, methylene diphenyl diisocyanate (MDI) (chemically pure, Fuchen LLC China), distilled water as blowing agent, glycerol (Merck, Steinheim, Germany), Hg Standard Solution (Merck, Steinheim, Germany), zeolite, bentonite. The tools used are glassware, Hot Plate, ball mill, digital scales, ovens, furnaces, 100 mesh sieves, Scanning Electron Microscopy With Energy-Dispersive X-Ray (SEM-EDX, Quanta F250, FEI, Oregon, USA), Atomic Absorption Spectroscopy (AAS, Analyst 800, Perkin Elmer Co, Norwalk, CT, USA), Fourier Transform Infrared Spectroscopy (FTIR, Nicolet 6700 FTIR, Thermo, MA, USA).

### *2.2 Preparation of polyurethane foam adsorbents*

Preparation of the modified zeolite is done by one-shot method by trial and error. At first the mixture that will be made in a beaker glass using a hot plate is made by mixing 10 g of castor oil, 1 g of distilled water, with a variation of glycerol mass (0.2 g; 0.4 g; and 0.6 g), with zeolite (4 g; 5 g; and 6 g) and variations in MDI mass (4 g; 5 g; and 60 g). The mixture is stirred with a speed of 250 rpm for 10 minutes and a temperature of 70°C to form a bubble. Then poured into $6 \times 6 \times 3$ cm$^3$ aluminum mold. The foam formed in $6 \times 6 \times 3$ cm$^3$ aluminum mold is placed in an oven with a temperature of 70°C for 2 hours. Each polyurethane foam that has been formed and hardened, then diced 0.5 cm$^3$ [10].

Polyurethane mixture made in beaker glass and using a hot plate, the mixture is made by varying the mass of the material which is castor oil 10 g, 1 g of distilled water, with a composition of glycerol (4 g; 5 g; and 6 g), variation of bentonite (4 g; 5 g; and 6 g) and MDI (4 g; 5 g; and 6 g). The mixture is stirred at a speed of 250 rpm for 10 minutes at a temperature of 70°C until the foam is formed or for about 20 seconds, then poured into $6 \times 6 \times 3$ cm$^3$ aluminum mold. The foam formed is cured at 70°C for 2 hours. After being removed from the oven and cooled, bentonite modified polyurethane foam formed was then cut into 0.5 cm$^3$ cubes.

### *2.3. Characterization*
Modified polyurethane foam adsorbent was characterized using FTIR, SEM, and visual observations. FTIR analysis was carried out to determine the functional groups of absorbent polyurethane foam, while SEM analysis was carried out to see the morphological structure of the polyurethane foam adsorbent.

## 3. Results and Discussion
### *3.1. FTIR analysis*
The functional groups found on the surface of polyurethane foam (PU) which are modified by zeolite and bentonite are strongly influenced by the process of making, heating and activating. Castor oil, MDI, glycerol, $H_2O$, zeolite and bentonite are the main elements in the formation of pore surfaces of polyurethane foam modified by zeolite and bentonite. This element can have a significant influence on the application of the adsorbent, ion surface during the adsorption process and on the surface morphology of polyurethane foam. Figure 1 shows the FTIR spectrum for polyurethane foam and modified polyurethane foam.

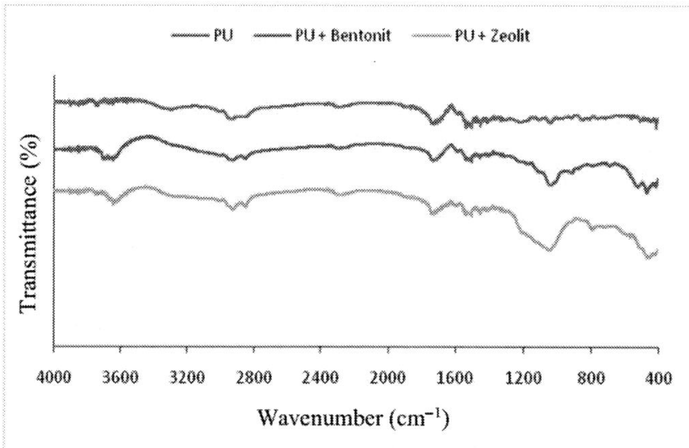


**Figure 1.** The IR spectra of PU, PU Bentonite and PU + Zeolite.

Table 1 shows that the N-H group in each foam is shown at 3700 cm$^{-1}$ absorption number. Addition of zeolite and bentonite in the process of foam synthesis can increase the hydroxyl group, as evidenced by the wave band that extends around the absorption number of 3700 cm$^{-1}$. The C = C-H group was not readable on polyurethane foam without modification, while the type of foam was read at a wavenumber of 3007-3008 cm$^{-1}$. In polyurethane foam without modification the C≡N group appears at 2690 cm$^{-1}$ absorption number, which indicates that there is excess MDI. FTIR spectrum from PU + zeolite foam and bentonite + foam did not show C≡N because MDI had reacted. Cluster C = O is a typical group in the polyurethane group shown in all foams (1654-1516 cm$^{-1}$, 1680-1514 cm$^{-1}$, and 1593-1516 cm$^{-1}$). Shifting the absorption number to the C = O group to the lower direction shows the characteristics of the modified material from zeolite or bentonite. Meanwhile, C = C is a group found in polyurethane, zeolite and bentonite, indicated by all types of foam at absorption numbers of 1413 $cm^{-1}$. The C-OH group can be seen from the wavenumbers of 1041-1047 $cm^{-1}$ which are general groups in castor oil, zeolite and bentonite.

**Table 1.** Results of FTIR PU Foam spectrum, PU Foam + Zeolite, PU + Bentonite Foam.

| Functional Group | Wavenumber ($cm^{-1}$) | | |
|---|---|---|---|
| | PU Foam | PU Form + Zeolite | PU Form + Bentonite |
| N-H | 3699 | 3799 | 3741 |
| C-C-H | 2935 - 2856 | 2929 - 2854 | 2931 – 2854 |
| C=C-H | - | 3008 | 3007 |
| C≡N | 2690 | - | - |
| C=O | 1654 – 1516 | 1680 - 1514 | 1593 – 1516 |
| $CH_2$ | 1448 | 1448 | 1449 |
| $CH_3$ | 1448-1331 | 1456-1311 | 1449-1309 |
| C=C | 1413 | 1413 | 1413 |
| C-O-C | 1219 | - | 1221 |
| C-OH | 1041 | 1141-1043 | 1047 |

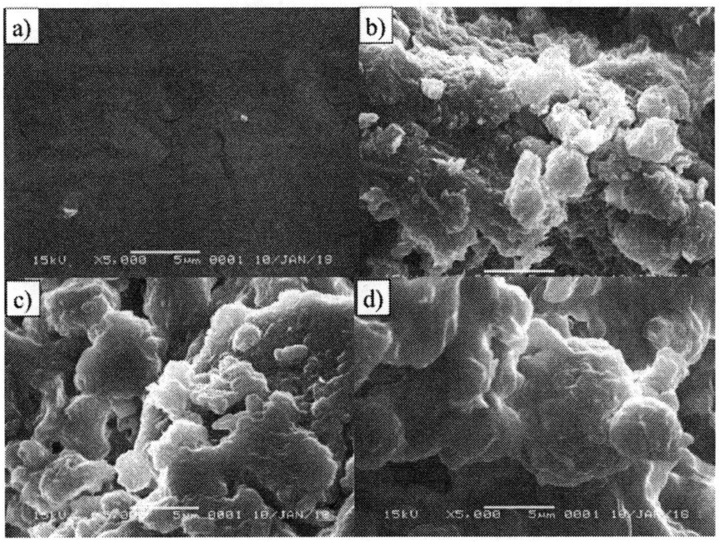


**Figure 2.** The Scanning Electron Microscopy (SEM) result of (a) PU foam, (b) PU foam + 4 g zeolite, (c) PU foam + 5 g zeolite, and (d) PU foam + 6 g zeolite.

*3.2. Surface morphology*
Scanning electron microscopy (SEM) is used to characterize the morphology of polyurethane foams because mechanical properties are mainly influenced by clear open cell density, thickness, and size. This technique is used to observe the open cell structure, which is typical of flexible polyurethane foam.

The surface morphology of polyurethane foam was analyzed using SEM at 5000 times magnification. The results of observations using SEM are shown in Figure 2 and Figure 3. Figure 2 shows that the polyurethane foam without modification has a smooth and flat surface structure with small pores. There is empty space between one particle and another particle. Whereas in polyurethane foam modified with zeolite it has a rough surface structure, the presence of empty spaces and the formation of zeolite particles in the form of spherulite. These particles have irregular size and distribution. Zeolite particles dominate the surface structure of polyurethane foams and if the zeolite concentration given is higher, the size of the spherulite structure is greater. Figure 3 shows that polyurethane foam modified with bentonite has irregular or irregular particles. The formed particles also have different sizes and are not distributed throughout the empty spaces. Similar to zeolite bentonite addition also dominates the structure surface of polyurethane foam, the higher the concentration of bentonite given, the larger particle size will be formed and more evenly distributed to the entire surface of polyurethane foam.

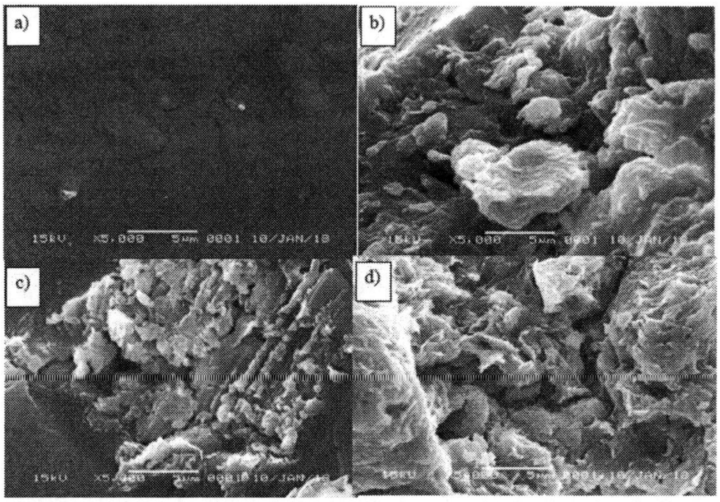


**Figure 3.** The Scanning Electron Microscopy (SEM) result of (a) PU foam, (b) PU foam + 4 g bentonite, (c) PU foam + 5 g bentonite, and (d) PU Foam + bentonite 6 g.

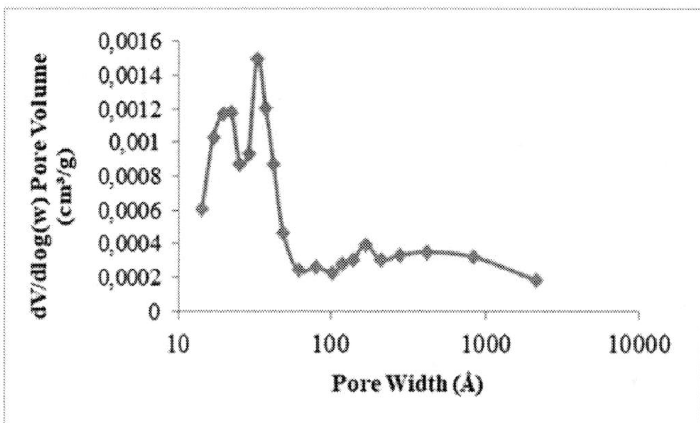


**Figure 4.** BET analysis results Polyurethane foam (PU), synthesis PUF 19.1 g and Zeolite 40%.

### *3.3. Surface Area*

The characterization of the area of polyurethane foam with the BET method (Brunauer-Emmett-Teller) aims to determine the material surface area, pore distribution of material and isotherm adsorption of a gas in the material. The BET method is used for flat material surfaces and does not have plates on each layer that can be used to explain the area of the sample. Table 2 shows the result of BET testing taken based on the characteristics of zeolite and bentonite polyurethane (PUF) foam.

The area of BET for synthetic PUF modified with zeolite obtained a varied pore surface area for each sample code. The average BET area obtained in the PUF modification is 1.3778 $m^2$ $g^{-1}$. This small surface area is caused by a comparison of the composition of synthetic PUF when making synthetic PUF. So that it will lead to the ability of synthetic PUF modified with zeolite to absorb Hg metal contained in water.

**Table 2.** Result of BET analysis.

| Sample Code | Sample Mass (g) | BET Surface Area A / ($m^2$ $g^{-1}$) |
|---|---|---|
| **PUF** | 0.2868 | 1.7596 |
| **PUF 18.6 g Zeolite 50%** | 0.2991 | 1.3733 |
| **PUF 19.5 g Zeolite 60%** | 0.2787 | 0.8561 |
| **PUF 16.9 g Zeolite 40%** | 0.2867 | 1.0221 |
| **PUF 20 g.5 g Zeolite 50%** | 0.2828 | 1.0349 |
| **PUF 19.1 g Zeolite 40%** | 0.2624 | 0.773 |
| **PUF 18.7 g Zeolite 50%** | 0.2614 | 2.8257 |

## 4. Conclusion

The conclusions that can be drawn from this study are as follows:

1. The results of FTIR characterization show that there are several chemical groups present in the polyurethane foam without modification and also on polyurethane foams modified by zeolite and bentonite.
2. The results of SEM characterization prove that polyurethane foam without modification has a smooth and flat surface with small nodule and no empty space, whereas in zeolite modified polyurethane foam the particles are spherulite, zeolite dominates the surface structure whereas in polyurethane foam modified bentonite is the same with zeolite, bentonite also dominates the structure of the surface of polyurethane foam.
3. The largest BET surface area obtained in PUF modification of 18.7 g + 50% Zeolite is 2.8257 $m^2$ $g^{-1}$. Whereas the smallest BET area was obtained at synthetic PUF 19.1 g + Zeolite 40% at 0.773 $m^2$ $g^{-1}$.

## Reference

[1] Triwulandari E, Prihastuti H, Haryono A and Susilo E 2008 *J. Sains Mat. Indo.* ed. December 31
[2] El-Shahat M F, Moawed E A and Burham N 2008 *J. Hazard Mater.* **160** 629
[3] Moawed E A 2006 *Ana. Chim. Acta.* **580** 263
[4] Vidotti E C, Cancino J C, Olibeira C C, Rollemberg M and do Carmo E 2005 *Ana. Sci.* **21** 149
[5] Yassin S A 2012 *J. Sci.* **23** 5

[6] Moawed E A, Farag A B and El-Shahat M F 2011 *J. Saudi Chem. Soc.* **17** 47
[7] Li H, Liu L and Yang F 2013 *Procedia Environ. Sci.* **18** 528
[8] Rohaeti E 2005 *Prosiding Seminar Nasional Penelitian, Pendidikan, dan Penerapan MIPA* K-1
[9] Yeon K H, Lee J W, Lee J S, and Moon S H 2002 *J. Apl. Polym. Sci.* **86**(7) 173
[10] Kumari S, Chauhan G S and Ahn J H 2016 *Chem. Eng.* **304** 728

# Residence Time Distribution Study in Laboratory Scale Struvite Crystallization Reactor with Online Conductivity Measurement

**R Ulfa[1], I Machdar [1*], Y Yunardi [1], W Rinaldi [1] and J Jauharlina [2]**
[1]Chemical Engineering Department, Faculty of Engineering, Syiah Kuala University
Jalan Tengku Syech Abdur Rauf No. 7, Darussalam, Banda Aceh 23111, Indonesia
[2]Plant Protection Department, Faculty of Agriculture, Syiah Kuala University
Jaan Tgk. Hasan Krueng Kalee No. 3, Darussalam, Banda Aceh 23111, Indonesia

E-mail: machdar@unsyiah.ac.id

**Abstract.** Elimination Residence time distribution (RTD) analysis is necessary to explain the hydrodynamic behavior of a continuous stirred tank reactor. In this study, RTD was applied in a laboratory scale struvite crystallization reactor to estimate the degree on non-ideality and its effect on the reactor performance i.e. effective volume, actual hydraulic residence time (HRT), and material dispersion. An impulse-response tracer (NaCl saturated solution) approach and a conductivity meter to detect of the NaCl at the effluent of the reactor were applied. To evaluate effects on the reactor performance, variation of impeller locations, impeller speeds, and flowrates were conducted. The main findings from this study showed that the difference between actual and theoretical HRT was found around as around 10%-34% and 30%-37% for condition of the impeller location at the upper and lower part, respectively. These percentages were corresponding to the reactor's effective volume, lower percentage means higher effective volume. This finding correlated to the value of dispersion number. It was found to be about 1.612 and 0.49 for the impeller location at the upper and lower part, respectively. Finally, it was observed that the tracer was being almost 100% recovered in the reactor due to a good flow pattern and its impeller location affects the reactor's performance.

## 1. Introduction

Content of ammonia in the urea fertilizer plant wastewater is often discharged to natural water bodies without any control over the generation of eutrophication and other problems in water cycle [1]. The removal and recovery of ammonia from wastewater can be treated by biological, physical, or chemical methods, either alone or in combination [2-5]. It is of great challenge to find an appropriate technology that promotes ammonia recovery from the wastewater to produce valuable material and simultaneously polish the quality of the wastewater effluent.

Currently, ammonia content in the urea fertilizer plant wastewater can be reduced by recovering it in the forms of crystaline product called struvite [6]. Struvite is relatively easy to dry and handle and gives potential as a slow-release fertilizer [7]. The solubility of struvite ($MgNH_4PO_4.6H_2O$, magnesium phosphate hexahydrate), determined in deionized water was found to be 169.2 mg/L at 25°C, with the maximum value of 212.7 mg/L at 35°C [8]. Shu et al. [9] reviewed that 365 kg struvite application in

* To whom any correspondence should be addressed.

agriculture field is enough to fertilize 2.6 ha land. Potency of struvite as a fertilizer has been studied [10]. They found that struvite released less N during the first 3 to 6 weeks and more N during the last 9 weeks than that of ammonium phosphate.

The formation of struvite crystal develops in two chemical steps: nucleation or crystal birth and crystal growth which a combination of several factors such as the crystal state of initial compounds, thermodynamic of liquid-solid equilibrium, phenomena of mass transfer between solid and liquid phases, and kinetic of reaction [11-12]. In struvite crystallization process mixing shows a strong effect on the number of crystals formed and consequently, on their size [13].

The method of operation of crystallization process has an important aspect of the performance of struvite production. Many researchers have conducted experiments to get highest removal of ammonia through struvite precipitation from different types of reactors. Hutnik et al. [14] designed a continuous flow type cylindrical tank equipped with a heating/cooling jacket. The tank was also equipped with a draft type circular tube with a three-paddle propeller. Matynia et al. [15] developed jet-pump crystallizer which operates continuous reaction crystallization process. Rahman et al. [16] designed an simple reactor where feeds were continuously added. Ali and Schneider [17] designed a batch reactor with auto control of feed and temperature. Considering the above views, especially in the mixing effect in the crystallization reactor, the current study was conducted to test the effects of impeller location, impeller speed, and flow rate on reactor's effective volume and to estimate its behavior concerning actual hydraulic retention time (HRT) and material dispersion number.

## 2. Methodology

### *2.1. Experimental setup of reactor crystallizer*

The crystallizer reactor was a cylindrical tank made of Plexiglas. The schematic diagram and photograph are presented in Figure 1. The reactor was designed in two distinct portions, i.e. reaction and crystal growth section (upper part), and sedimentation section (lower part). The upper part dimension of the reactor was 40 cm in height, and 40 cm in diameter. The lower part dimension of the reactor was 40 cm in height, and 30 cm in diameter. The reactor had a total liquid volume of 90 liters. The liquid was stirred by a stainless steel propeller-type impeller (number of blades 4; blade length 15 cm; blade width 3 cm). Dosing (JCMB 45-5 AILIPU Solenoid Dosing) were used to deliver fluid to the reactor. The reactor was continuously operated at room condition. There is an internal circulator which pushed the reactants in solution downwards in a draft tube. The draft tube was installed in the upper part of the reactor. The approach of installing the draft tube in the upper part of the reactor was intended for eliminating crystal product loss at the reactor effluent. Larger-crystal products are accumulated at the bottom of the reactor, and they were discharged by opening the bottom valve.

### *2.2. Tracer study*

The experimental study was conducted in a continuous flow mode. A schematic diagram of the RTD experiments is presented in Figure 1. In the RTD studies saturated NaCl solution was used as a representative tracer in order to describe the liquid flow behavior in the reactor. A simple impulse-response method was used in this study. A conductivity probe connected to a conductivity meter (Corning Pinnacle 541, Cole-Palmer, IL, USA) was used to detect the concentration of the NaCl solution at the effluent of the reactor. Three sets of studies were conducted varying the impeller locations, impeller speeds, and flowrates to examine their effect on the liquid flow behavior. The impeller location was set at the lower (20 cm from the reactor bottom) and the upper parts (55 cm from the liquid surface). Each variation of the impeller locations was carried out by modifying the impeller speed (80, 130, and 180 rpm). Liquid flowrates used were 30, 40, and 50 L/h equivalent to HRT of 3, 2.25, and 1.8 h, respectively, corresponding to its impeller speed variations.

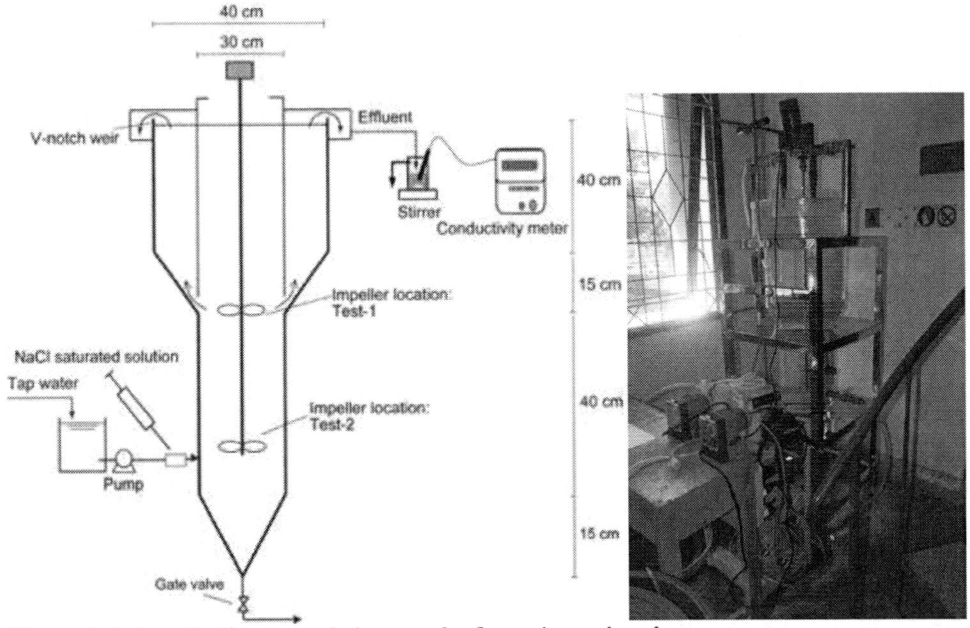


**Figure 1.** Schematic diagram and photograph of experimental study.

### *2.3. RTD Analysis*

The residence time distributions of tracer in the reactor ~~is~~ explained by the *E(t)* curve that gives the variation of tracer concentration at the effluent of the reactor. The curve is given by equation (1) [18].

$$E(t) = \frac{C(t)}{\int_0^\infty C(t)\,dt} \tag{1}$$

It explains insignificant appearance how much time fluid element spent in the reactor and describes a fluid distribution (hydraulic behavior) in the reactor. The mean residence time ($t_{ave}$) can be calculated with the uses of equation (2). It is a function of $C_i$, the measured tracer concentration at time $t_i$, and $\Delta t_i$ which is the interval between two measurements. The variance of the RTD, σ, and the normalized variance, $\sigma_\theta^2$, can be evaluated as equations (3) and (4), respectively. A dispersion number (D/uL) has been formulated by Levenspiel [18] and given in equation (5).

$$t_{ave} = \frac{\sum_i t_i C_i \Delta t_i}{\sum_i C_i \Delta t_i} \tag{2}$$

$$\sigma^2 = \frac{\int_0^\infty t^2 C\,dt}{\int_0^\infty C\,dt} - t_{ave} \tag{3}$$

$$\sigma_\theta^2 = \frac{\sigma^2}{t_{ave}^2} \tag{4}$$

$$\sigma_\theta^2 = 2\left(\frac{D}{uL}\right) - 2\left(\frac{D}{uL}\right)^2\left(1 - e^{-\frac{uL}{D}}\right) \tag{5}$$

Index distribution of the tracer or effective volume of the reactor is assumed by ratio of actual HRT to theoretical one. The calculated dispersion number (D/uL), is a parameter that measures the extent of axial dispersion [18]. The degree of fluid mixing in the reactor was estimated by the Morril dispersion index (MDI) formula as also applied in a plug flow reactor by Machdar et al. [19]. The MDI was calculated from the $T_{10}$ and $T_{90}$ parameters, which were estimated from the RTD curve by calculating the time interval related to 10% and 90% of the area under the curve, respectively.

## 3. Results and discussions

### *3.1. Residence time distributions (RTDs)*

The RTD was enumerated using the tracer results from Figure 2 and equation (2). A tracer result summary including data analysis is presented in Table 1. Figure 2 presents the tracer concentration (conductivity values) at the effluent of the reactor versus time for the two experiments of each study. The variability between each curve for the given tracer was relatively narrow, which explained that the conducted tracer study was consistence.

All curves in Figure 2 demonstrated that two essential evidences i.e., a delay time tracer was firstly detected and tailing or extend in tracer tracking. The first condition was related to the minimum time required for the tracer reaching conductivity detector (in the outlet). The second condition likely indicated the presence of a stagnant volume of the reactor. The tailing phenomenon was probably caused by the tracer passes a stagnant (dead) volume of the reactor and slowly released by local mixing, and finally generating long tracer tracking.

As shown in Table 1, the average value of MDI under impeller location at upper and lower part were 8.10 and 6.35, respectively. A larger MDI indicates more mixing in the reactor [20].

### *3.2. Effect of impeller location and speed*

It was observed that, the actual HRT was mainly shorter than that of the theoretical one for all being tested. The general investigation from Figure 2 is that the impeller location at the upper part is related to higher effective volume of the reactor. In other words, the impeller location affects the reactor's performance. The effective volume of the reactor was found about 81% and 67% in average for the impeller location at the upper and lower part, respectively. As shown in Table 1, this investigation is more noticeable from the higher dispersion numbers calculated at the upper part than that of at the lower part of the impeller location. The average value of dispersion number was found to be 1.6 and 0.7 for the impeller location at the upper and lower part, respectively. As the impeller speed increases, the dispersion number rises for all flow rates tested.

### *3.3. Effect of flowrate or HRT*

Figure 2 also presents the experimental RTD curves at flow or HRT. After each RTD experiment, the tracer concentration in the effluent stream became zero. This means that the tracer was being almost 100% recovered in the reactor due to a good flow pattern. An increase of flowrate or reduce of HRT from 30 L/h (HRT = 3 h) to 50 L/h (HRT = 1.8 h) results in a increasing effective volume, but it was not significant difference in impeller speed of 80 rpm.

From the tracer data analysis, at condition of the impeller location at the upper part, the difference between the actual and the theoretical HRT was found as 34%, 14%, and 10% in average for the flowrate of 30 L/h, 40 L/h, and 50 L/h, respectively. Concurrently, at condition of the impeller location at the lower part, the difference was found as 37%, 33%, and 30% on average.

Percentage of the difference between the actual and the theoretical HRT were comparable to the stagnant (dead) volume in the reactor. In this study, lower flowrate or higher HRT decreased reactor's effective volume. This effect was likely due to the lower water turbulence, leading to fluid short-circuiting or channeling in the internal body of the reactor. This phenomen of short-circuiting in the stirred tank reactor were also reported by other researchers as a well [21, 22]. These drawbacks can primary be improved through a better agitation.

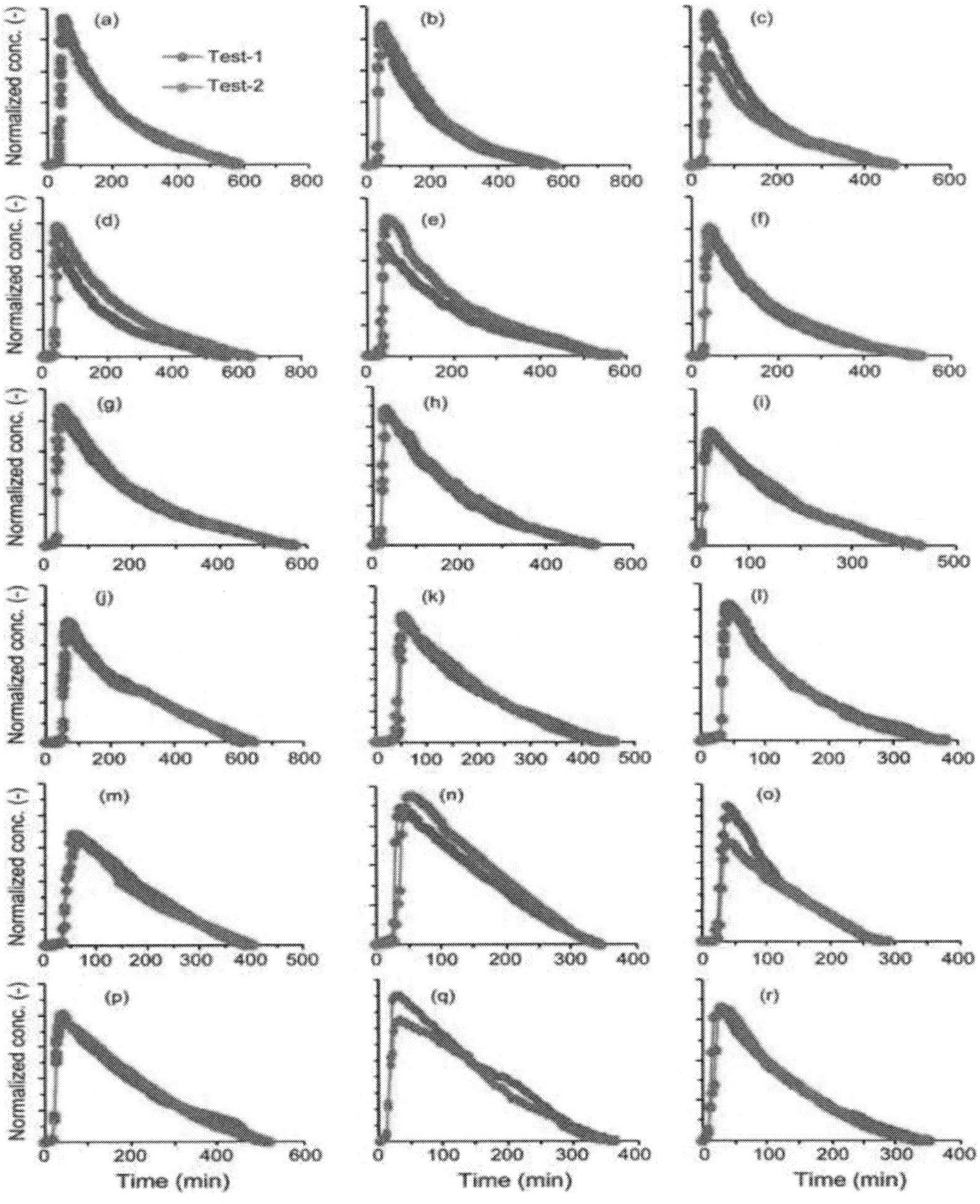


**Figure 2.** RTD curves.a to i: impeller location at the upper part. j to r: impeller location at the lower part. a to c: impeller speed 80 rpm, flowrate 30, 40, 50 L/h. d to f: impeller speed 130 rpm, flowrate 30, 40, 50 L/h. g to i: impeller speed 180 rpm, flowrate 30, 40, 50 L/h. j to l: impeller speed 80 rpm, flowrate 30, 40, 50 L/h. m to o: impeller speed 130 rpm, flowrate 30, 40, 50 L/h. p to r: impeller speed 180 rpm, flowrate 30, 40, 50 L/h.

**Table 1**. Summary calculation of RTD analysis.

| Impeller location | Impeller speed (rpm) | Flowrate (L/h) – (HRT) | Average Actual HRT (h) | Average Effective volume (%) | Average Dispersion number (D/uL) | Average MDI (-) |
|---|---|---|---|---|---|---|
| The upper part | 80 | 30 (3) | 1.92 | 64 | 1.6 | 7.22 |
| | | 40 (2.25) | 1.94 | 86 | 1.2 | 6.75 |
| | | 50 (1.8) | 1.59 | 88 | 1.5 | 7.04 |
| | 130 | 30 (3) | 2.03 | 68 | 1.9 | 8.14 |
| | | 40 (2.25) | 2.18 | 97 | 1.1 | 8.90 |
| | | 50 (1.8) | 1.72 | 96 | 2.1 | 7.67 |
| | 180 | 30 (3) | 2.02 | 67 | 2.1 | 10.16 |
| | | 40 (2.25) | 1.70 | 75 | 1.3 | 5.88 |
| | | 50 (1.8) | 1.55 | 86 | 1.4 | 11.11 |
| | Average | | | 81 | 1.6 | 8.10 |
| The lower part | 80 | 30 (3) | 2.64 | 82 | 0.7 | 6.03 |
| | | 40 (2.25) | 1.80 | 80 | 0.5 | 5.25 |
| | | 50 (1.8) | 1.43 | 79 | 0.4 | 5.07 |
| | 130 | 30 (3) | 1.49 | 50 | 0.3 | 4.71 |
| | | 40 (2.25) | 1.39 | 62 | 0.3 | 5.12 |
| | | 50 (1.8) | 1.12 | 62 | 0.3 | 5.00 |
| | 180 | 30 (3) | 1.74 | 58 | 2.6 | 9.38 |
| | | 40 (2.25) | 1.37 | 61 | 0.8 | 7.96 |
| | | 50 (1.8) | 1.24 | 69 | 0.8 | 8.66 |
| | Average | | | 67 | 0.7 | 6.35 |

## 4. Conclusions

The experimental results led to three main conclusions: (1) impeller location and impeller speed have significant effect on actual HRT and effective volume of the reactor; (2) material tracer do not travel completely inside of the reactor due to fluid short-circuiting or channeling; and (3) the results further indicated that in order to increase reactor's effective volume, mixing system and impeller speed must be considered.

**Acknowledgment**
This study was supported by the Ministry of Research, Technology, and Higher Education of the Republic of Indonesia under the national competitive research scheme.

**References**

[1] Yetilmezsoy K, Ilhan F, Kocak E and Akbin H M 2017 *J. Clean. Prod.* **152** 88
[2] Yu R, Geng J, Ren H,Wang Y and Ke Xu 2012 *Bioresrc. Tech.* **124** 292
[3] Tan X S P, Tan W K, Teo, Li T K 2006 *J. Membrane. Sci.* **271** 59
[4] Calli B, Mert B and Inanc B 2005. *Chemosphere* **59** 819
[5] Jeong Y K and Hwang S J 2005 *Acta Crystallographica* **96** 1
[6] Machdar I, Depari S D, Ulfa R, Muhammad S, Hisbullah A B and Safrul W 2018 *IOP Conf. Ser.: Mater. Sci. Eng.* 358
[7] Westerman P W, Bowers K E, Zering K D and Ad-cock M E 2008 *ASABE Paper No. 083892.* Presented at ASABE Annual International Meeting, June 29-July 2

[8] Bhuiyan M I, Mavinic D S, and Beckie R D 2007 *Environ. Technol.* **28** 1015
[9] Shu L, Schneider P, Jegatheesan V and Johnson J 2006 *Bioresour. Technol.* **97** 2211
[10] Johnston A E and Richards I R 2004 *Phosphorus Research Bulletin* **15** 52
[11] Rahman Md, Saleh M A M, Rashid U, Ahsan A, Hossain M M and Ra C S 2014 *Arabian Journal of Chemistry* **7** 139
[12] Ohlinger K N, Young T M and Schroeder E D 1999 *J. Environ. Eng.* **125** 730
[13] Regy S, Mangin D, Klein J P and Lieto J 2002 *Lagep Report*. available at/ http://www.nhm.ac.uk/research-curation/departments/mineralogy/ researchgroups/phosphaterecovery/LagepReportS
[14] Hutnik N, Wierzbowska B, Piotrowski K and Matynia A 2013 *J. Sci. Technol.* **3** 58
[15] Matynia A, Wierzbowska B, Hutnik N, Mazienczuk A, Kozik A and Piotrowski K 2013 *Proc. Environ. Sci.* **18** 766
[16] Rahman M M, Liu Y H, Kwag J H, Ra C S 2011 *J. Hazard. Mater.* **186** 2026
[17] Ali M I and Schneider P A 2008 *Chem. Eng. Sci.* **63** 3514
[18] Levenspiel O 1999 *Chemical Reaction Engineering*, 3rd ed., Wiley, New York.
[19] Machdar I, Muhammad S, Onodera T and Syutsubo K 2018 *Environ. Eng. Res.* **23** 195
[20] Hart F L 1979 *Water Pollut. Control Fed.* **51** 2868
[21] Obonukut M E, Inyang U E, Bassey P G and Etuk B R 2017 *J. Multidisciplinary Eng. Science Studies* (JMESS) **3** 5
[22] Gitis V and Asraf-snir M 2011 *Chem. Eng. J.* **166** 79

# The treatment effect of tensile strength on polypropylene filled with silica from rice husk ash

**F Y Guswara[1], F Ihsan[1], S Aprilia[1*], Zuhra[1] and Fitriani[1]**
[1]Chemical Engineering Department, Syiah Kuala University, Darussalam, Banda Aceh, Aceh, 2311 Indonesia

E-mail: sriaprilia@unsyiah.ac.id

**Abstract.** Polypropylene (PP) is a plastic that can be utilized in many sectors such as food wrappers and eyeglass lenses. Polypropylene which is filled with silica from rice husk ash can improve it is mechanical properties. The purpose of this study was to examine the effect of the addition rice husk ash on the mechanical properties of polypropylene composite. The extracted silica filtered with 325 mesh sieve. Silica was analyzed using Scanning Electron Microscopy (SEM), Scanning Electron Microscopy-Energy Dispersive X-ray (SEM-EDX), Fourier Transform Infra-Red (FTIR), X-ray Diffraction (XRD), and X-ray Fluorescence (XRF). Polypropylene composite is made using a simple mixing method. Polypropylene composite film filled with silica from rice husk ash was analyzed using tensile strength and elongation at break as well as functional group tests (FTIR) and composite structure (SEM). The result of this study was (1) Silica extracted from rice husk ash produced higher silica purity which is 21.93 % weight, (2) The extracted silica was in crystalline form, (3) The silica in polypropylene film was in siloxane form and (4) Polypropylene composite film with the addition of 2 % silica has the highest tensile strength and elongation at break (5.36 kgf/mm$^2$ and 5.00 %).

## 1. Introduction

One of the plastics type commonly used to make plastic bags and easily find on the market is polypropylene (PP) [1]. PP is a type of saturated aliphatic polymer which is a semi-crystalline polymer consisting of two parts which is crystalline and amorphous phase also has a high melting point, transparent, and good impermeability. PP can develop into a matrix of composite materials to obtain higher tensile strength and elongation [2].

The disadvantages of PP polymers are that they have a low density compared to other types of polymers such as polyethylene and polystyrene which cause poor mechanical properties [3]. To improve the mechanical properties, addition of filler is needed to add the density of the PP polymer. Polymers that arranged composites as matrix will produce composites with final properties that are highly dependent on the characteristics of the polymer. Addition of filler to the composite will also increase the hardness, and the degradation process also improves the mechanical properties of a polymer such as tensile strength and percent elongation [4]. Most of researches utilized different recycled wastes to improve PP composites properties with the environmental friendly filler. Among various filler used for composites, one of them is rice husk ash (RHA). RHA is an agricultural residue by product abundantly available in Indonesia [5, 6].

[*] To whom any correspondence should be addressed.

RHA is a by-product produced from the process of rice milling and often used as a growing media and a fuel source for burning. Improper handling of RHA will result in environmental pollution [7]. RHA contains 94% silica of total weight that can be used as a natural filler for composites [8]. To isolate silica in RHA can be used in a very easy process and relatively inexpensive cost using the extraction method. The extraction method is easier to use in alkaline and acidic solutions. Alkaline compounds such as KOH, NaOH, and $Na_2CO_3$ can dissolve silica in the amorphous phase, where acid compounds can precipitate more dissolved silica and for the acid that commonly used for the extraction of silica are acetic acid, oxalic acid, citric acid, and hydrochloric acid [9]. Many researchers have conducted silica research using extraction processes with a variety of variations of raw materials and solvents. Rafiee et al. [10] used rice husk (RH) and acetic acid solution, the resulting silica was 6 nm in the form of 99.90% amorphous phase. Amalia et al. [11] also conducted a similar study with the same raw material and using NaOH solution for 12 hours, obtained 82.61% silica yield. Amalia et al. [11] and Diharjo et al. [12] states that the use of silica as a filler in PP composites can increase the tensile strength, impact toughness, and composite density.

In this study the effect of the process of isolating silica from RHA using acetic acid extraction treatment, characteristics of the resulting silica and the use of silica as a filer to enhance the properties of PP composite film on the effect toward morphology, physical and mechanical properties were studied.

## 2. Experimental

### *2.1. Preparation of polypropylene filled with silica*

The rice husk ash (RHA) used in this study for extracting silica was obtained from a rice mill and ground to powder using blender. RHA was soaked in 3 M glacial acetic acid solution for 2 hours at room temperature. The solution is then washed and filtered until the pH is neutral and dried at 105°C for 2 hours. The resultant silica is filtered with a 325 mesh sieve and were stored for further analysis. The process of preparing polypropyene (PP) composites film by dissolving PP using xylene until homogeneous with constant temperature at 190°C and then filler from RHA was loading with various concentrations of 0-5 % wt into the solution. The composites solution was cast on flat stainless steel and were stored for further analysis.

### *2.2. Characterization*

#### *2.2.1. Crystallinity analysis*

Crystallinity and armphous phase was determined using XRD Shimadzu XDR 7000 at an operating voltage 40 kV and applied current was 30 mA.

#### *2.2.2. Morphology analysis*

To determine the structure formed on silica and polypropylene composite film analyze by scanning electron microscopy (SEM). The sample is attached to a holder then in thin layer of platinum in a vacuum state with enlargement is 1000x.

#### *2.2.3. Composition analysis*

The particle composition of silica from RHA was analyzed using X-ray fluorescence (XRF).

#### *2.2.4 Functional group analysis*

The structure of functional group sample silica and polypropylene composite film has analysis using IRAffinity-1S FTIR spectrometer with scanning ranges of 4000 $cm^{-1}$ to 400 $cm^{-1}$ at room temperature.

#### *2.2.5. Mechanical analysis*

Tensile strength and elongation analysis were performed using the HT-8503 Universal Testing Machines device. The polyproylene composites film was cut following ASTM D638.

## 3. Results and discussion

### *3.1. Silica particle from RHA analysis*

#### *3.1.1. Crystallinty of Silica particle analysis*

Silica has crystalline structure contrary to component from RHA, which are amorphous in nature. The evaluation of the amount of phase form of silica is very important. For this purpose, there are some specific methods, one of them is estimated by the silica activity index [10].

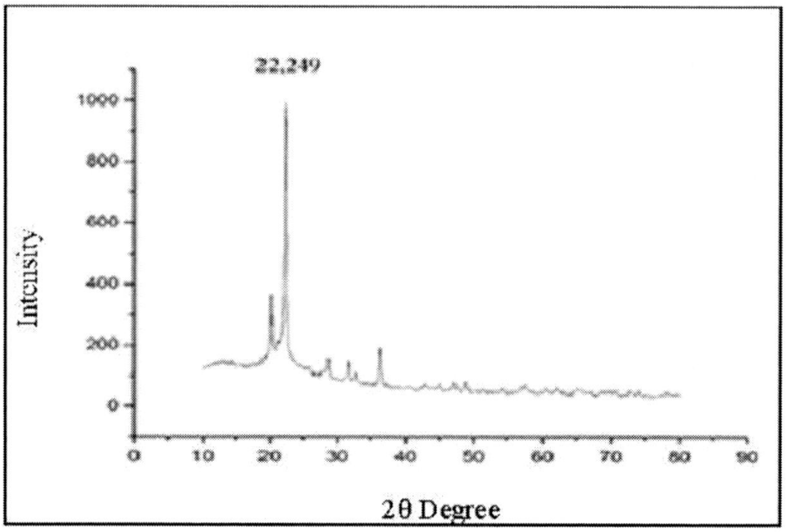


**Figure 1.** X-ray diffraction patterns of silica from RHA.

From Figure 1, it can be seen that the silica particles obtained through the combustion of rice husks ash with chemical treatment using extraction by acetic acid at room temperature was crystalline. The diffractogram also have a high intensity level which is $2\theta = 22.249^{o}$ is a prove that silica was in a crystalline form. Similar result obtained by olawale et al. [13] states that the particles of rice husk ash are crystalline which they have proven using Raman Spectra. The diffraction peak at $2\theta$ of $20^{o}$, $22^{o}$ and $36^{o}$ is a typical silica characteristic which attributed to the presence of an ordered crystalline structure and the acid treatment of RHA did not affect the silica structure which consistent with the present study [14,15].

#### *3.1.2. Morphology analysis of silica particles*

The morphology structure of silica particle from RHA was studied using SEM, as shown in Figure 2. The structure of sample was a polygonic and irregular shapes. It is also seen that agglomeration of several silica particles forms larger particles. This might be due to the large surface area of the silica and forming a strong electric dipole that can agglomerate [13]. The average particle size of silica is 100μm and with an irregular morphology is seen.

These particle are homogenous and agglomerate, indicate that parts of micropores exist in silica particle from RHA. A similar result observed that precipitation of acetic acid produce the solid and the aggregation rate of silica particle occurs faster when pH decreases because of acetic acid [16]. These solid disappear when is washed and filtered until pH neutral, indicating the removal of impurities [14].

*3.1.3. Composition analysis of RHA*

To evaluate the effect of purification parameter and to confirm the presence of silica particle, SEM-EDX was carried out on silica from RHA with acetic acid treatment. There are four main elements found in the rice husk ash which has been extracted by acetic acid, namely carbon, oxygen, silica, and potassium. The result of composition analysis of RHA are tabulated in Table 1.

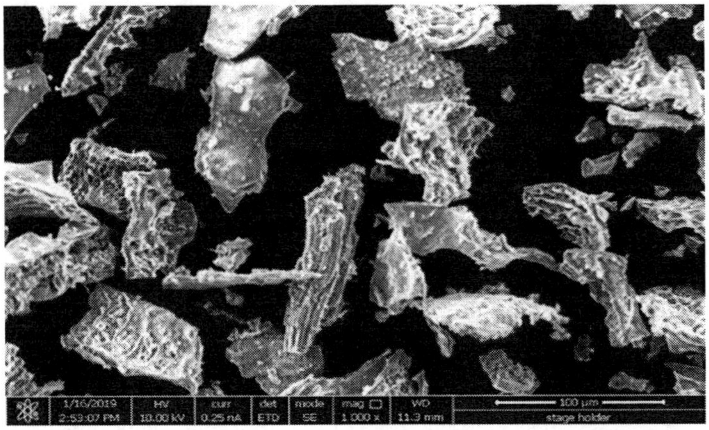


**Figure 2.** Silica analysis with SEM at 1000 x magnification.

**Table 1.** Analysis of RHA content.

| Element | Total Particles | Weight (%) | Atom (%) |
|---|---|---|---|
| C | 7.797 | 34.38 | 45.10 |
| O | 6.460 | 42.95 | 42.30 |
| Si | 6.230 | 21.93 | 12.30 |
| K | 94 | 0.75 | 0.30 |
| Total | 20.581 | 100.00 | 100.00 |

The composition was reported based on weight percentage shows that RHA treated with acetic acid containing 21.93 wt% of silica and 34.38 wt% of oxygen. This occurred because there is a siloxane functional groups (Si-O-Si) in RHA treated with acetic acid. Very high carbon content indicates that silica which is mostly contained in the outer surface of rice husk ash was not all soluble by the solvent of acetic acid and causing carbon (C) particles trapped in the pores of RHA. This can cause the carbon as impurity that obtained in the particles still high [17]. Based on the results, other impurities except carbon (C) and potassium (K) has been dissolved during the extraction process using acetic acid. The result obtained by Arifin et al. [5] shows that RHA without treatment using acetic acid contained of 89.83 wt% $SiO_2$ but there were still impurities in RHA such as $Fe_2O_3$, CaO, MnO, $K_2O$, and MgO. This shows that in this study acetic acid was able to reduce impurities so that only the remaining potassium (K), Oxygen (O), and carbon (C) elements were left.

*3.1.4. Functional group analysis of silica particle from RHA*

FTIR spectra of silica particle are presented in Figure 3 showed all types of $SiO_2$ are strong bonds at the peak of 1110-1080 $cm^{-1}$. Apart from these strong bonds, there are other peaks that can help in identifying the types of $SiO_2$ such as α-Quartz ($SiO_2$ crystalline forms), spoted at the peak of 1105, 796, and 621

$cm^{-1}$ [18]. At wavelengths of 1105 $cm^{-1}$ and 796 $cm^{-1}$, the siloxane - (Si-O) n- functional group is caused by the presence of stretching vibrations in the symmetrical and asymmetrical Si-O-Si group.

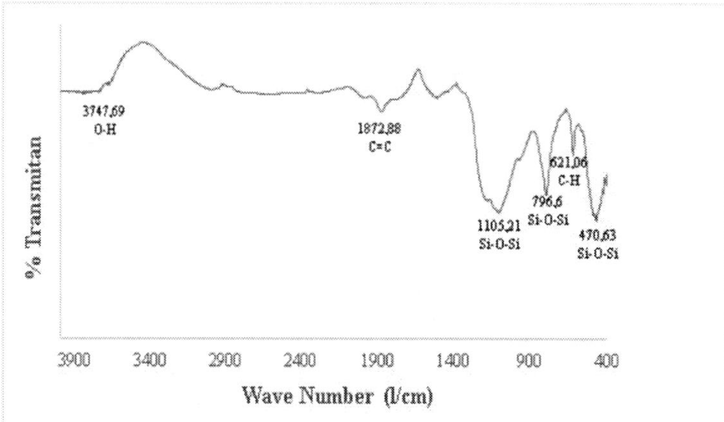


**Figure 3.** FTIR analysis of silica particles from RHA.

Three main peaks are shows strong absorption at 470, 796, 1105 $cm^{-1}$, indicating the presence of silica particle. The peak also shows the vibration of oxygen atom joined with another atoms in the asymmetric stretching due to vibrations of Si-O-Si band [19]. The O–H stretching vibrations appear in the region of 3500–3800 $cm^{-1}$. The band resulted from silanol OH groups and water bound to the silica surface by hydrogen bonding [14]. Research study by Premaratne et al. [28] also indicated that O-H stretching can happen due to physically and chemically absorbed water molecules on the $SiO_2$ surface.

The FTIR peaks in this study were similar result to those in the silica particle spectra obtained from RHA [14,20,19]. Futhermore, the FTIR found changes in the chemical component of silica from RHA. In comparison with RHA sample spectra from Liou et al [14] study and RHA treated with acetic acid show different absorption. The breadth and height of O–H and Si–O bands in the RHA sample are much smaller than extracted silica samples. The finding is due to the compact agglomerate of silica particles produced by a gelation reaction that leads to increased strength of intermolecular hydrogen bonding.

### *3.2. Polypropylene/silica composites film analysis*

#### *3.2.1. Morphology analysis of composites film*

The SEM images of polypropylene (PP) composites film with various silica particle from RHA content were shown in Figure 4. The surface structure in Figure 3(a) is PP without silica and Figure 3(b)-(f) is a PP filled with 1, 2, 3, 4, and 5 wt% silica.

Figure 4 shows that the PP composites film matrix forms a layer that produces a void. In Figure 4 (a) there is no void seen in the matrix. This because there is no addition of silica to the sample. In Figure 4 (b), (c); and (d) indicate improvement in bonding between silica from RHA and polypropylene. it appears that the silica spread and fill the PP composites matrix and form a void [21]. The number of void which slightly increases where the filler particles are fill the PP matrix.

In Figure 4 (e) and (f) the distance between the silica looks far apart and the PP composites matrix looks rigid and solid. This occurs due to uneven stirring when mixing silica into PP composites film. Increasing concentration of filler to the PP matrix can affect the microstructure of the polymer which causes uneven distribution of silica and reduced the tensile strength of composites film [22]. The adhesion of the filler-polymer matrix increase is improved with increasing filler. However increasing more filler concentration can reduce the properties of film due to non-covalent bonding and hydrogen bonding with the OH group and oxygen of each of group carboxil bonding between the filler and the polymer matrix which in line with studies Ginting et al. [23], Kord et al. [24] and Bhat et al. [25].

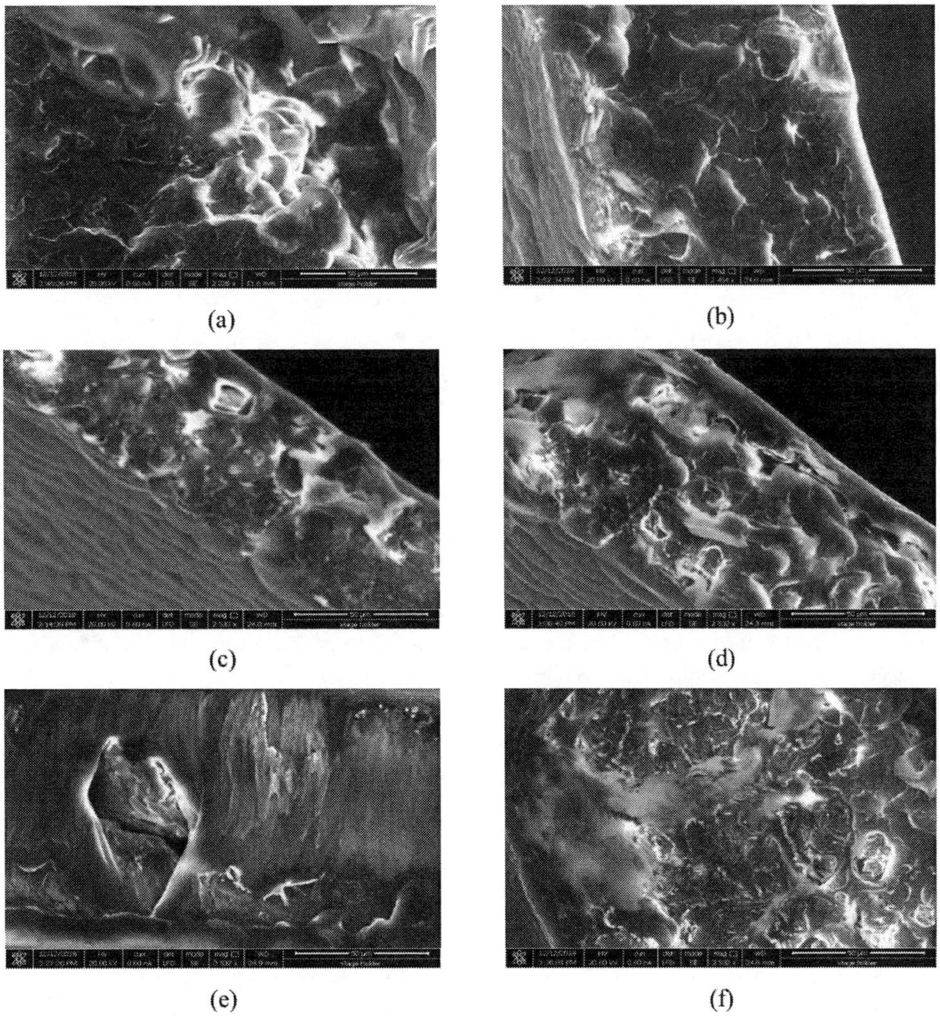


**Figure 4.** SEM images with 2000x magnification for polypropylene composites film (a) without silica (b) 1 wt% silica (c) 2 wt% silica (d) 3 wt% silica (e) 4 wt% silica and (f) 5 wt% silica.

*3.2.2. Functional group of PP/silica composites film*

The obtained FTIR spectra for PP composites filled with silica and without silica were performed in Figure 1. The overall result shows that structures of PP composites filled with 5 wt% silica are almost the same with pure PP polymers. However, when viewed in greater detail, there is a difference that is in the PP composites film with silica show the new peak (blue circle) at 621.08 $cm^{-1}$ which is a functional group of $CH_2$ whereas in pure PP composites is nowhere to be found [26,27]. Spectrum composites film at 1076 $cm^{-1}$ and spectra rice husk ash at 1078 $cm^{-1}$, it indicates the presence of a silica functional group in the form of Si-O-Si. The existence of such bonds may state that the composite contains silica because of the filler [28].

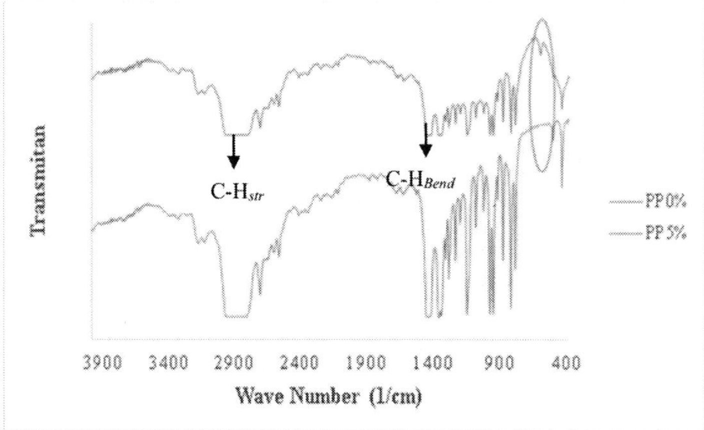


**Figure 5.** FTIR analysis results comparison between composite PP without silica and with silica.

From the FTIR spectra also shows that the composite PP without silica is made from polypropylene polymer at peak 2.875 $cm^{-1}$,which is a methyl functional group ($-CH_3$) showing the $C\text{-}H_{str}$ bond which is a typical area of the PP polymer. At peak 1.442 $cm^{-1}$ and 1.373 $cm^{-1}$ indicate to the methyl $C\text{-}H_{bend}$ functional group. The peaks detected in the range of 1200-950 $cm^{-1}$ are attributed to stretching of C-O-C, C-O, and C-OH functional groups [26,28].

### *3.2.3. Tensile strength analysis*

In this study, tensile strength are measured in order to evaluate the PP/silica composites film. Figure 6 shows the result of tensile strength for all sample composites filled with silica and without silica. The tensile strength of samples were filled with 0, 1, 2, 3, 4, 5 wt% silica were found to be 4.29; 3.69; 5,36; 4.40; 4.02; and 5.18 kgf / $mm^2$ respectively.

From Figure 6 it can be seen that the highest tensile strength value is PP composites film with 2 wt% silica fillers. It was caused by the addition of fillers with an amount of < 10 % resulting in a closer distance between particles in the composites film which causes the composites to become stronger [27]. The best silica composition to be added to PP composites is 0.1 gram. This is similar with what was done in this study that PP samples with a filler addition of 2 wt % were proportional to 0.08 gr of silica added to the PP composite. The decrease in tensile strength occurred in PP composites of 3 and 4 wt%. This is because agglomeration occurs in PP composites [23].

### *3.2.4. Elongation analysis*

Figure 7 shows the result of elongation for all sample composites filled with silica and without silica. The elongation of samples were filled with 0, 1, 2, 3, 4, 5 wt% silica were found to be 3,72; 3,12; 5,00; 2,18; 4,10; dan 4,28 % respectively.

This can occur because the addition of filler into the composites film pores results in a closer distance between fillers due to the tight arrangement, this can cause the value of stiffness of composites film increase while the flexibility decreases [29]. Sample with the addition of 2 wt% silica has the highest elongation strength value at 5 % of elongation. This happens because the most optimum addition of fillers is 0.1 gram [30]. In addition, this can occur because of the influence of many things such as irregular inter-filler arrangement, different thicknesses, and pore formation in composite samples that have an impact on the forces acting on the composite, resulting in poorly produced elongation, which in line with studies Aprilia et al. [27].

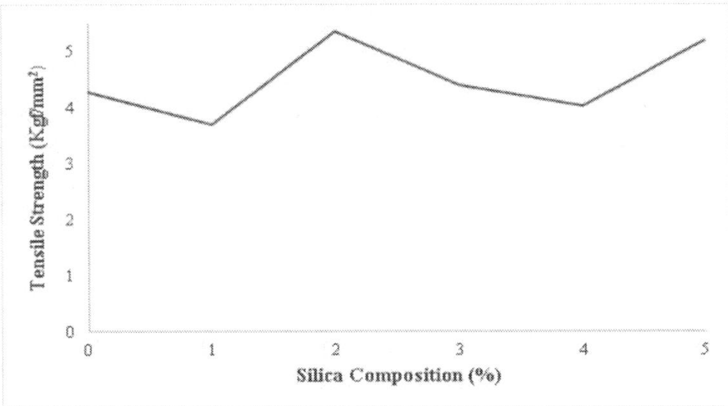


**Figure 6.** Silica filler effect on tensile strength of composite polymers.

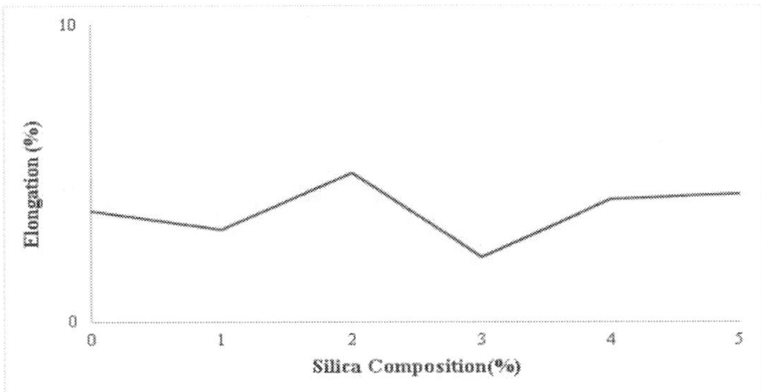


**Figure 7.** Silica filler effect on elongation at break of composite polymers.

## 4. Conclusions

In this study, the effect of acetic acid treatment on RHA produced a higher silica purity level which is 21.93% by weight. Chemical treatment of rice husk ash using acetic acid produces silica in a crystalline state. Analysis of functional groups shows the presence of silica indicated by the emergence of a peak at the peak of $808 cm^{-1}$ and $623 cm^{-1}$ which is a siloxane (Si-O-Si) functional group. PP composites film with the addition of 2 wt% silica has the highest tensile strength and elongation at 5.36 kgf / $mm^2$ and 5.00%.

**Acknowledgements**

The authors would like to thank The Research Institutes and Community Service University of Syiah Kuala, Ministry of Research, Technology and Higher Education who have funded this research in accordance with the Research Contract No. 30/UN11.2/PP/SP3/2019.

**References**

[1] Pudjiastuti W, Listyarini A and Sudirman 2012 *J. Indust. Res.* **6** 51
[2] Mardiyati, Steven, Rizkiansyah R R and Purnomo I 2016 *J. Mach.* **25** 73
[3] Abubakar 2013 *Logaritma* **1** 98
[4] Deswita D, Handanayani A and Yulianti E 2014 *J. Chem. Packag.* **37** 45
[5] Arifin B, Aprilia S, Alam P N, Mulana F, Anaska D M and Putri D E 2017 *MATEC Web of Conferences* 1
[6] Saravanan S D and Kumar M S 2013 *Int. Conf. Des. Manufact.* **64** 1505
[7] Chandra A, Miryanti Y I P A, Widjadja L B and Pramuika A 2012 *Eng. J.* **2** 1
[8] Shukla S K, Nidhi, Sudha, Pooja, Namrata, Charu, Akshay, Silvi, Manisha, Rizwana, Bharadvaja A, Dubey G C and Tiwari A 2013 *Adv. Mater. Lett.* **4** 714
[9] Handanayani P A, Nurjanah E and Rengga W D P 2015 *Renew. Natur. Mater. J.* **4** 55
[10] Rafiee E, Shahebrahimi L, Feyzi M and Shaterzadeh M 2012 *Int. Nano Lett.* **2** 29
[11] Amalia S R, Fajarwati M, Fitriawan M, Aji M P and Yulianto A 2016 *Physic National Conference* 1
[12] Diharjo K, Elharomy I and Purwanto A 2014 *J. Mech. Eng.* **5** 27
[13] Olawale O, Akinmoladun A I, Oyawale F A and Rejoice A 2013 *Int. J. Sci. Eng. Res.* **4** 1
[14] Liou T H and Yang C C *Mater. Sci. Eng.* **176** 521
[15] Hasan A F, Abdelghny A M, Elhadidy H and Youssef A M 2014 *J. Sol-Gel Sci. Technol.* **69** 465
[16] Affandi S, Setyawan H, Winardi S, Purwanto A and Balgis R 2009 *Adv. Powd. Technol.* **20** 468
[17] Solihudin, Noviyanti A R and Rukiah 2015 *Chem. Natur. Acta* **3** 11
[18] Launer P J 2013 *Infrared Analysis Of Organosilicon Compounds: Spectra Structure Correlations* (Morrisville: Galest Inc) 175
[19] Amutha K, Ravibaskar R and Sivakumar G 2010 *J. Nanotech. Appl.* **4** 61
[20] Premaratne W A P J, Priyadarshan W M G I, Gunawardena S H P and De Alwis A A P 2013 *J. Sci. univ. Kelaniya* **8** 33
[21] Rahman A, Farid M and Ardhyananta H 2016 *ITS Eng. J.* **5** 9
[22] Pustak A, Pucic I, Denac M, Svab I, Pohleven J, Musil V and Smit I 2012 *J. Appl. Polym. Sci.* 30
[23] Ginting E M, Wirjosentono B, Bukit N and Agusmar H 2014 *Chem. Mater. Res.* **6** 14
[24] Kord B 2011 *Bio Resour.* **6** 1351
[25] Bhat A H, Abdul and Khalil H P S 2011 *Bio Resour.* **6** 1288
[26] Coates J 2000 *Interpretation of Infrared Spectra A Practical Approach* (Chichester: John Wiley & Sons Ltd) 1
[27] Aprilia S, Arifin B, Arahman N, Abubakar, Amin A, Wicaksono A V and Bakhtiar D 2019 *Rasayan J. Chem.* **12** 994
[28] Nizamuddin S, Jadhav S, Qureshi S S, Baloch H A, Siddiqu M H, Mubarak N M, Mubarak, Grifin G, Madapusi S, Tanksale A and Ahamed M I 2019 *Sci. Rep. J.* **5445** 1
[29] Kada D, Migneault, S Tabak G and Koubaa A 2016 *J. Mater. Sci.* **36** 1243
[30] Moosa A and Awad A Y 2016 *Int. J. Curr. Eng. Technol.* **6** 804

# Pulp and paper industry: An overview on pulping technologies, factors, and challenges

**K F Rullifank[1], M E Roefinal[1], M Kostanti[1], L Sartika[1] and Evelyn[1*]**
[1]Department of Chemical Engineering, Faculty of Engineering
Universitas Riau, Pekanbaru 28293, Riau, Indonesia

E-mail: evelyn@eng.unri.ac.id

**Abstract.** The pulp and paper sector currently plays important part in the world's economy. In this paper, we present a review of pulping technologies to convert logs or wood chips and non-woody materials into pulp for use in papermaking which consist of mechanical, chemical and semi-mechanical. It also talks about the history of papermaking and various important factors involved in producing high quality pulp and paper such as H-factor (time and temperature), alkaline charge, liquor to wood ratio, and sulphidity. This paper also discusses about challenges facing by the pulp and paper industry or processes, namely high use of energy and chemicals, as well as generation of liquid and solid wastes of large quantities. This paper will be handier for those who are new to the field of pulping research for paper production.

## 1. Introduction

Modern paper making evolved from an ancient art in Egyptian civilization which used papyrus as writing media at around 3000 BC. The characteristics of the papers in this era are not the same as the papers used today. A more skillful papermaking later was reported from a Chinese named T'sai Lun in 100 AD, utilizing a wide variety of materials such as bamboo or mulberry fibers. This method of papermaking was then spread throughout Middle East and Rome in the Mediterranean Sea after several centuries until it reached Europe, England, and America at the beginning of 15th century. In 1799, the first continuous papermaking machine called Fourdrinier machines was patented by a Nicholas Louis Robert from France [1]. The milestones in the pulp and paper industry continued further with the invention of cylinder paper machine by John Dickinson in 1809, leading to a rise in Fourdrinier machines usage for making thin paper. Seventeen years later, steam cylinder was introduced for drying the papers. The United States began to use the Fourdrinier engine only in 1927. Mechanical process of making pulp from wood of low quality was discovered in 1814 by Friedrich Gottlob Keller, whereas soda process was just introduced in 1854 by Charles Watt and Hugh Burgess. Sulphite process was found three years later by Benjamin Chew Tilgh, an American chemist, through his British Patent. This process was able to produce a good and ready bleached pulp. Finally, Kraft or sulphate process was discovered from a basic experiment of Carl Dahl in Danzig in 1884 which is a dominant method used today [2]. Those milestones provided the basis for modern pulp and paper industry.

[*] To whom any correspondence should be addressed.

### *1.1. Wood Composition and cooking liquor*

There are two types of raw materials for paper making, namely wood (Acacia, Eucalyptus, etc.) and non-wood (banana, kenaf, bamboo, etc.) materials. Of these two types, the pulp and paper industry uses wood raw materials. Wood is part of the stem or branches and twigs of plants that is harden due to lignification. Wood is available in large quantities, easily cultivated, cheap, and contains a lot of cellulose. Wood as raw materials are divided into long fiber (softwood) and short fiber wood (hardwood) (Table 1). Each type of wood produces different types of pulp [2]. Cooking liquor is another component in pulp making and divided into three categories: (i) white liquor, which is the main cooking chemical consists of caustic (NaOH), sodium sulphite ($Na_2S$), and sodium carbonate ($Na_2CO_3$); (ii) black liquor, which is a washing liquid containing lignin and dissolved chemicals and is used to meet the needs of cooking fluid in the cooking process, and; (iii) white chemicals, which contain chlorine dioxide ($ClO_2$) for increasing the whiteness of pulp.

**Table 1.** Chemical composition of hardwood and softwood [3].

| Component | Hardwood (%) | Softwood (%) |
|---|---|---|
| Cellulose | 45 | 42 |
| Hemicellulose | 30 | 28 |
| Lignin | 20 | 27 |
| Extractive | 5 | 3 |

### *1.2. Manufacturing of pulp and paper*

Pulp manufacturing starts with transporting wood coming from the forest to the wood piling place called wood yard. This is then followed by wood skin peeling using debarking drum, which is a paring tool drum containing a lifting bar by turning motion. Peeled skin will fall through holes in the wall slot. After exfoliation, the bark is converted into chips called the chipping process. Chips that have met the standard will proceed to the next process, which is cooking process. Aim of this process is to dissolve the lignin component in wood using chemical and heat. The cooking process takes place in a large pressurized vessel called a digester by addition of white liquor. Subsequently, washing is carried out to separate fibers from water-soluble pliers, which consists of organic compounds i.e. lignin and also organic compounds as leftovers from cooking chemicals. Screening using a filter followed to separate fibers or impurities based on size. These components can reduce the quality of pulp and can cause waste of chemicals in the bleaching processes. Bleaching is the next step which is applied by the pulp and paper industry in four sequential stages, namely Initial Dioxide Stage (D0), Oxidation-Resistant Extraction (EOP) stage, Chlorine Dioxide stage 1, and Chlorine Dioxide stage 2. D0 is gradation and separation of lignin, whereas EOP is elimination of lignin. Chlorine Dioxide stage 1 is aimed to increase brightness and Chlorine Dioxide stage 2 is the final stage to perfect the level of bleaching stage. Finally, the pulp goes to paper production stage in which the products can be traded for use as stationery, office supplies, and others. In addition to paper, pulp can be directly sold to investors who need to change it into new products such as cardboard, tissue, etc.

## 2. Overview of pulping technology

Pulping process is the separation of lignin to obtain fiber or cellulose from fibrous material. Thus, cellulose must be clean from lignin so that the quality of the paper obtained does not change colour during use. *Mechanical, Chemical, and Semi-Mechanical Pulping* are three general classes of turning wood into pulp [3]. Mechanical pulping is entirely dependent on physical action whereas chemical pulping needs chemicals to release the fibers. *Semi-Mechanical Pulping* combined mechanical and chemical actions to obtain the fibers. Results from this process is called pulp, which will then be processed into paper in the paper machine [4].

IOP Conf. Series: Materials Science and Engineering **845** (2020) 012005 doi:10.1088/1757-899X/845/1/012005

### *2.1. Mechanical Pulping*

A tool such as grinding is used for scraping in mechanical pulping. First, the wood is skinned and then cut and crushed to produce chips. Then, the raw materials are milled to release fibers, followed by filtering these materials so that cellulose is separated from the others [3]. The yield from this process is around 90-98%. Mechanical slurries contain high lignin and short fibers which make it fragile. There are several methods of mechanical pulping namely *Mechanical Pulp Refiner (RMP), Pressure Groundwood (PGW), Stone Groundwood (SGW), Thermomechanical Pulp (TMP)* and *Chemi-Thermomechanical Pulping (CTMP)*, in which the main difference is whether the pulp produced from a refiner or stone grinder [5]. Other advantages of mechanical pulping are low production costs and suitable for printing paper due to good ink absorption, compressibility, opacity, high bulk density, and more environmentally friendly. Nonetheless, drawbacks arisen from this process are: high loads of shieve and knots; yellow color of paper when exposed to bright lights due to high lignin content, and; high fiber damage and low pulp strength due to shorter fiber. Figure 1 illustrates the CTMP process.

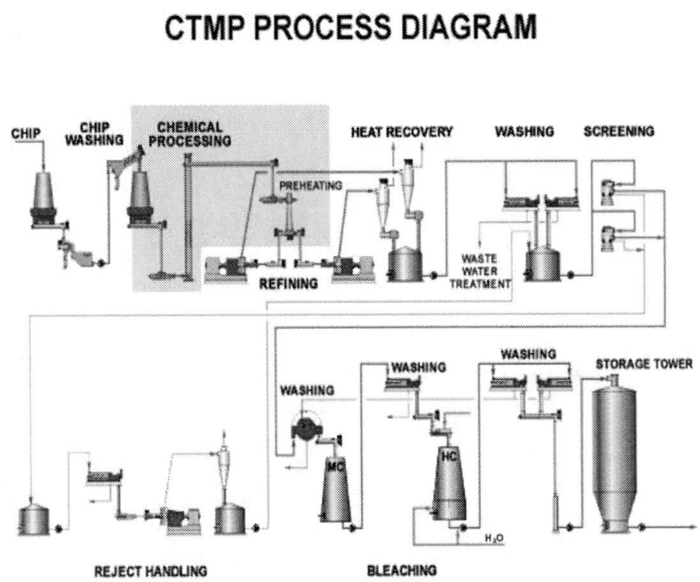


**Figure 1.** Illustration of the *Chemi-Thermomechanical Pulping* process [5].

### *2.2. Chemical Pulping*

Pulping with a chemical process aims to damage and dissolve the fiber binding agent consisting of lignin and pentose using chemicals. This process of damaging and dissolving is called cooking. The process of cooking raw materials with chemical solutions is carried out in a reactor called digester. During cooking, lignin reacts with chemical solutions and forms dissolved compounds that are easily washed. Some cellulose also reacts, producing a low pulp yield. Chemical pulping can be divided into three types, namely soda, sulfite and Kraft pulp making [5]. Soda and sulfate processes use alkali chemicals in cooking liquids, so composting pulp follows. Compared to mechanical method, chemical process usually produces higher pulp strength and paper quality, reuse chemicals and energy more efficiently, and products can be used for making viscose [3]. However, several drawbacks arisen from this process are a lower yield and greater production cost (40-52%), a pungent odor from the pulp, and more pollution to the environment due to chemicals used.

*2.2.1. Soda pulping*

The cooking solution used in soda pulping is sodium hydroxide (NaOH). The soda solution function in the process is for hydrolysing lignin and other fiber binding agents so that the fiber contained in the raw material will be released. Lacking of cooking chemicals will result in darken pulp and difficulty in bleaching. However, excessive cooking chemicals can reduce yield by degradation of cellulose fibers. Cellulose degradation with concentrated sodium hydroxide solution can occur at temperatures above 1000 °C. If the temperature is too high, amount of degraded carbohydrate will be greater than dissolved lignin, which will reduce the yield and thickness of the pulp. Nonetheless, it is easy to recover sodium hydroxide from liquor as an advantage from this process [6]. Soda-Anthraquinone (Soda-AQ) has also been used to make pulp made of mostly non-woody plants, in which the anthraquinone acts as a redox catalyst resulting in increased yield [7].

*2.2.2. Sulphite pulping*

This process uses active chemicals as follows: sulphite acid, calcium bisulphite, sulphur dioxide expressed in solution as Ca $(HSO_3)_2$ with excess of $H_2SO_3$. Softwood is the common raw material and $SO_2$ and Ca $(HCO_3)_2$ are the cooking solution (equations 1 and 2) [6]. Lignin which is bound to cellulose will react with a solution of Ca $(HSO_3)_2$ (equations 3-5). The process produces a high yield in certain kappa numbers, higher white degree of pulp that is not bleached, and pollution of less than 1. However, this process is rarely used in industry due to the high costs.

$$S + O_2 \rightarrow SO_2 \quad (1)$$

$$2SO_2 + H_2O + CaCO_3 \rightarrow Ca(HSO_3)_2 + CO_2 \quad (2)$$

$$Ca(HSO_3)_2 \rightarrow Ca^{2+} + 2HSO_3^- \quad (3)$$

$$Lignin + HSO_3^- \rightarrow SO_2 + LigninOH^- \quad (4)$$

$$LigninOH^- + HSO_3 \rightarrow LigninSO_3^- + H_2O \quad (5)$$

*2.2.3. Kraft pulping*

Kraft technology found by Carl Ferdinand Dahl is chief and superior in pulp production [6]. This process combines sodium hydroxide and sodium sulphide at a temperature of 160-180 °C (320-356 °F) and pH>12, producing fibrous material resulted from destruction of bonds of lignin macromolecules. The following cooking reaction can be shown for Kraft process (equation 6).

$$\text{NaOH}+NaS_2 + Wood \rightarrow Na - org. + S - org. + NaHS \quad (6)$$

*2.2.4. Semi-mechanical pulping*

Semi-mechanical pulping is a method of combining chemical process with mechanical process. The chemical used in this process is sodium sulphate to remove cellulose lignin bonds and get perfect fiber separation. Included in this process is *Chemi-Thermo Mechanical Pullping (CTMP)* which uses heating to degrade lignin so that the pulp has a lower yield but with better quality than pulp with the mechanical process [4]. Following are the advantages of semi-mechanical pulping: a fairly bright color pulp, a longer pulp resistance, and 65-85% of yield. On the contrary, several disadvantages noted are a more complicated process and large production costs.

*2.2.5. Other chemical pulping methods*

Other methods in pulp production are organosolv and hydrotropic methods. Organosolv is a method of separating fibers using organic chemicals such as methanol, ethanol, acetone, acetic acid and others. In this process, decomposition of lignin is caused by breaking of ether bond. This process does not use sulphur so that it is safer for the environment and recycling black liquor can be done easily [4]. It also requires relatively low capital infestation due to less waste and easy refining of alcohol [8]. Printing paper and dissolving pulp are examples products of organosolv process which have lower strength property than the Kraft process. With respect to hydrotropic method, fibers separation is carried out using water-soluble and environmentally friendly reagents known since the middle of last century [9]. Hydrotropic is an aqueous salt solution that at the same temperature, could provide greater volubility for substances that are slightly soluble (such as lignin) than water. Example of commonly used hydrotropic salt is sodium xylene sulfonate. Relatively pure lignin can be recovered by deposition and used for conversion to other chemical products [10].

*2.2.6. Biopulping*
Biopulping is the process of separating fibers (cellulose) with other substances to produce pulp using certain microorganisms. There are two biopulping processes namely lignin depolymerisation and esterification of oxalate. These processes are resulted from fungal penetration on wood chip's lumen and cells, increasing saturation of water fiber as one of the results [4]. Fungi from *Basidiomycetes* class have been known able to degrade lignin in the wood weathering process. Lignin degradation involving the activity of the ligninolytic enzyme produced by the white root fungi, namely lignin peroxidase, manganese peroxidase, and laccase [11]. The third enzyme is a multi-extracellular enzyme that involves in the lignin depolymerization process. Some research results show that biopulping is relatively superior to the Kraft process because it is environmentally friendly [12]. Nonetheless, this process is slow compared to chemicals and requires biological control in a complicated process.

## 3. Cooking equipments, parameters, and pulp analysis
The pulp cooking process is usually carried out using a tool called digester. Digester is a cook container that must withstand high temperature, alkaline or chemicals with high concentration and large pressure. Digester is an upright cylinder-shaped chips/chip-in-shape cookware that is designed for high pressure and temperature. There are two filter strainers in the digester, one is located at the top of the digester called relief strainer and the other one is located in the middle of the digester called middle strainer. The function of strainer is to keep the wood fibers that are being cooked not out of the digester at the time to circulate the cooked fluid and at the time of wasting the existing gases in the digester. There two types of digesters, namely batch and continuous digesters.

Batch digester is a large digester with a total filled volume of chip and cooking liquor range between 70 and 350 $m^3$. Usually a factory has six to eight batch digesters so while some cook, others can fill up, go into the blow tank, and so on. Steam provides heating of digester and dilution of cooking liquid, which can be directly or indirectly. When steam passing the inside of the digester tubes, steam recycling and more uniform heating can be obtained. A continuous digester is a tubular digester in which the chip enters and exits the digester continuously, passes over a stream containing five major stages (presteaming-impregnation-heating-cooking-washing). This type of digester has been considered more efficient in terms of space, control, energy, and chemical usage. A special feeder is often used to allow chips entering the commonly pressurized digester with the help of a rotary valve. After passing through the high-pressure feeding, the cooker will take the chips to the digester. Conditions for impregnation are usually 45 minutes at 130 °C, allowing uniform cooking of chips at these process conditions [4]. Below are influencing parameters in Kraft pulping.

*3.1. Alkali Strength*
Alkali strength for softwoods is normally about 12 to 14% effective alkali on dry woods, whereas the requirement is typically lower (8 to 10%) for hardwoods with active alkali about 21%. Providing sufficient composition of alkali is very important for completion of cooking and adjustment of reaction rate. The common employed method is to utilize the minimum practicable alkali charge and vary the cooking temperature to achieve the desired rate. Higher alkali charge could cause a slight reduction in hemicellulose retention at given kappa number and changes the composition of retained hemicellulose [6]. Alkali in the cooking liquor is basically consumed for three different reactions: lignin reaction, neutralisation of organic acids, and resin reaction in the wood [13].

### *3.2. H-factor*

The H-factor is the cooking time-temperature needed to achieve a certain kappa number, which is described by the area under the Kappa number (K) curve versus time (in hr) (Fig. 2). For instance, a value of the H-factor of 925 is proportional to a one-hour isothermal cooking at 170 °C, considering other cooking conditions (effective alkali concentration, liquor-to-wood-ratio, etc.) are maintained constant. This will result in a sufficient precision of the predicted kappa number [5]. However, the cooking process can not eliminate the lignin perfectly without damaging the pulp. Therefore, it will be followed by an advanced process that is bleaching to prevent reject pulp.

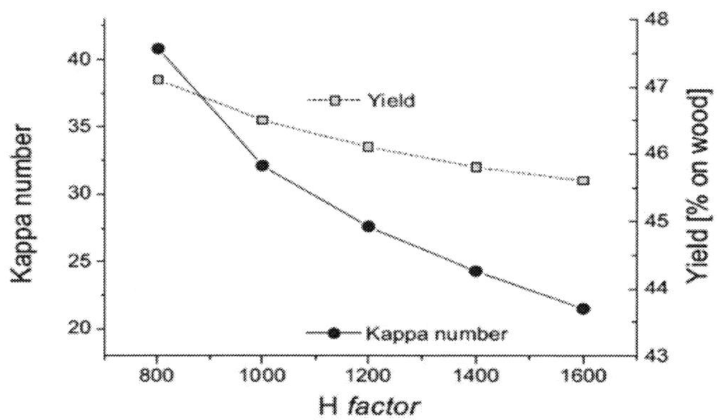


**Figure 2.** Effect of H-factor on kappa number and yield (Spruce/Pine = 1:1; liquor to wood ratio = 3.8:1; maximum temperature = 170 °C; effective alkali charge = 19%; sulphidity = 38%) [5].

### *3.3. Liquor to wood ratio*

For adequate penetration, it is essential to supply sufficient liquor for wetting chips' surfaces. For batch cooking, three-quarter filled digester is normally employed at the beginning. Gradual rising in the liquid level due to chip moisture and lignin in the liquid phase during cooking [6]. Sufficient white liquor is supplied in the digester to provide the specified alkali charge. Balance of the liquid requirement is typically made up with black liquor. Liquor to wood ratio is a ratio between all liquids contained in the digester with a dry wood chip, usually at ratio of 3:1 to 5:1. A greater dilution would decrease the concentration of active chemical and thereby reducing the reaction rate.

### *3.4. Sulphidity*

Delignification selectivity (yield vs. kappa number; viscosity vs. kappa number) is also primary and important parameter. High selectivity requires concentration of sulphide as high as possible, especially during transition from initial to bulk delignification [14]. A carbon–carbon bond cleavage of the b-c-linkage yielding formaldehyde and styryl aryl structures could present during insufficient supply of sulphide ions, reducing the selectivity. Effects of $Na_2S$ in Kraft pulping of softwoods are prone until 15% sulphidity, resulting in higher yield and a stronger pulp, but the benefits are less dramatic at higher sulphidities [6].

### *3.5. Impregnation*

In each process of chemical pulping, it is very important to ensure alkali distributes uniformly in all parts of the wood. Impregnation process goal is to optimize penetration of chemical substances into wood so that the resulting pulp is uniform, high quality and does not leave the wood that does not react with liquor. Ideally, the cooking should take place with the same concentration in all parts of wood and with a uniform temperature as well. Penetration is defined as transfers of liquid and associated chemicals into air-filled cavities of the wood chips caused by hydrostatic pressurization until saturation of fibers. Liquid diffuses into the pores while entrapped air dissolves into water or diffuse out pores [5]. Impregnation process is usually done by inserting hot steam or hot black liquor into wood chip.

### *3.6. Pulp Analysis*

Kappa number, viscosity, and alpha cellulose are three important components in pulp analysis. Kappa number is often used in mill control work to achieve degree of delignification and chemical requirement for bleaching, in which the percentage lignin approximately equals to K x 0.15. Kappa number is commonly determined from reactions of lignin and other non-cellulosic constituents of pulp fibers with acidic permanganate. Kleppe [15] reported that total pulp yield in Kraft pulping of southern pine and southern mixed hardwoods is influenced by kappa number (Fig. 3). A good indication of cellulose degree of polymerization can be obtained by measuring the viscosity of cellulose solution of known concentration. Alpha cellulose (the long-chain molecular fraction of the holocellulose) is defined chemically as that portion which resists solubilization of the shorter-chain cellulose and hemicelluloses in strong caustic solution. The chemical determination is carried out through pulp extraction under specified conditions. The soluble portion is then measured by oxidation with potassium dichromate, and the alpha cellulose fraction is calculated by its difference [6].

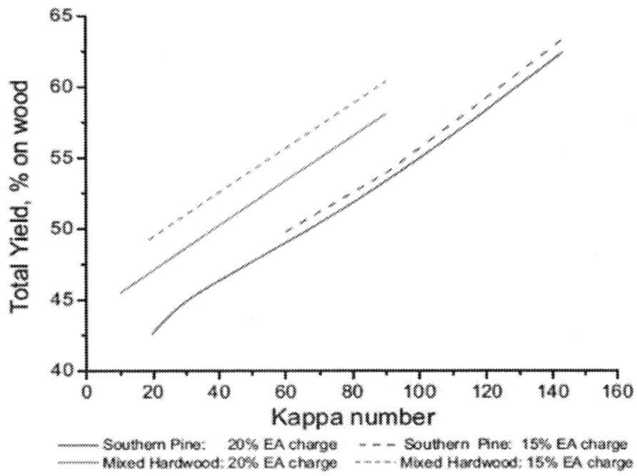


**Figure 3.** Total pulp yield versus kappa number in Kraft pulping of southern pine and southern mixed hardwoods [14].

## 4. Challenges to face

### *4.1. Energy usage in pulp and paper mills*

The characteristics of energy technology used for the pulp and paper industry depend on the type of process used. Each process requires certain energy that may use fuels such as coal, gas and oil, or electricity. The fuel is mainly used for steam generation and the main electricity user in an industry is an electric motor. Regarding mechanical pulping, the process requires considerable electrical energy. In contrast, chemical pulping requires large amounts of steam and most of the energy consumed is for pulping and paper drying. However, this process produces a by-product in the form of black liquor. This by-product can be used to generate steam and electricity in an incinerator. Black liquor can produce energy about 22 GJ per tonne of pulp produced. Thus, it can result in energy surplus depending on the configuration of process in the pulp mill. Gasification technology has also an opportunity to increase efficiency in the use of black liquor. In the gasification process, hydrocarbons are converted into synthetic gas (syngas) in the form of carbon monoxide and hydrogen mixture. This synthetic gas can be used as raw materials in chemical industry or for driving gas turbine used in the power plant. This technology is often called black liquor gasification combined cycle (BLGCC) which fuelling can also use biomass such as leather and wood chip. With respect to biopulping, reducing the environmental impact and production costs due to small use of chemicals and energy are the main goals of this process. However, the process is slower compared to the mechanical and chemical processes and requires proper control when working with biological material.

### *4.2. Waste generation and management*

The wastes produced from the pulp and paper production can be divided into three categories, namely liquid, solid and air or gas. Liquid waste is generally processed using a Waste Water Management Installation (WWTP). The WWTP system classifies and combines its operating units (physical, chemical, and biological). Operation unit associated with physical is screening. Screening is used as an efficient and inexpensive way to set aside large sized suspended materials (e.g. fiber) and weak black liquor filtration. Fibers and weak black liquor are not desired in cooking and the concentration unit, respectively. Regarding to chemical process, chemicals addition to waste water is usually carried out to remove particles that are difficult to settle. Some examples of wastes or processes that use chemicals are remaining liquid from the bleaching results using chemical chlorine dioxide, caustic soda extraction, and chemical solution from cooking. With respect to biological process, the main objective of waste water treatment is to collect and eliminate or decompose dissolved organic matter which is biodegradable by microorganism activity.

The pulp and paper industry generally produces solid waste in the form of stone from lime and contains soda. This must be disposed of in a safe and comfortable environment. Solid waste must be disposed of at the place of final disposal that is secure. There are two other types of solid wastes in the industry processed using bark boilers and lime kilns. Bark boilers are used to burn bark, while lime kilns are used for processing lime sludge. The air emission produced from the pulp production process is usually recycled using a tool such as blow gas treatment. Electrostatic dust precipitator and wet scrubber are two examples of air emission units available in recovery boiler and recaustizing unit, respectively. Some wastes or processes that produce air emissions along with their handling are polluted condensate from the digester process, non-condensable gas, high pressure steam produced from burning organic matter, evaporation of the rest of chemicals due to the heat from washing units, and wastes containing solid particles from boiler chimney.

## 5. Conclusion

The history of making pulp and paper starts from ancient method using bamboo as the main ingredients and now in a modern way using Acacia and Eucalyptus along with technological advances in paper products of higher quality. There are several technologies known in papermaking process, namely mechanical, chemical, and biological pulping, with chemical Kraft pulping as the most dominant method. In the Kraft process, high quality pulp is produced using temperature of 160 to 180 °C with 120 minutes of cooking and 60 minutes of heating time, liquor to wood ratio of 3:1 to 5:1, and active alkali of 21% of dry chip. The chemical process unavoidably has drawbacks to the energy usage and environment. Therefore, more advanced technologies and continuous improvement are needed pertinent to environment and energy controls for the benefit of industries and society.

**Acknowledgement**
The travel grant from Tanoto Foundation is appreciated.

## Reference

[1] Shreve R H 1984 *The chemical process industries* (Tokyo: Mc GrawBook Company, Ltd)
[2] Mertawijaya A W 1981 *Atlas kayu Indonesia* (Bogor: Penerbit Badan Penelitian dan Pengembangan Pertanian)
[3] Casey J P 1980 *Pulp and paper chemistry and chemical* (New York: Wild Interscience Publication)

[4] Biermann C J 1996 *Hand book of pulping and papermaking* (California, USA: Academic Press)
[5] Sixta H 2006 *Handbook of pulp* (Lenzig: Willey-VCH Verlog GmbH and Co.) Vol 1
[6] Smook G A 1982 *Handbook for pulp and paper technologists* (Canada: Canadian Pulp and Paper Association)
[7] Omer S H, Khider T O, Elzaki O T, Mohieldin S D and Shomeina S K 2019 *BMC Chem. Eng.* **1** 1
[8] Sawyer C N, Mc Carty P and Parkin G F 2003 *Chemistry for environmental engineering and science* (Boston: Mc Graw-Hill)
[9] Devendra L P, Kumar K and Pandey M A 2016 *Biores. Technol.* **213** 350
[10] Brauns F E 1952 *The chemistry of lignin* (New York: Academic Press
[11] Setliff E C, Marton R and Granzow S G 1990 *TAPPI J.* **73** 141
[12] Blanchette R A, Burnes T A and Leatham G F 1988 *Biomass* **15** 93
[13] Sarkanen K V and Ludwig C H 1971 *Lignins: occurrence, formation, structure and reactions* (New York: Wiley-Interscience)
[14] Jensen W 1977 *Puukemia: Suomen paperi-insinöörien Yhdistys Ry.* (Helsinki: Suomen paperi-insinöörien yhdistys)
[15] Kleppe P J 1970 *TAPPI J.* **53** 35

# Kinetics of geopolymer's solidification: Effect of concentrations and curing temperatures

**Aman[1*], A Awaluddin[2], A Ahmad[1], M Olivia[3], Herman S[1] and MI Fermi[1]**
[1]Department of Chemical Engineering,Faculty of Technology University of Riau Panam 24293,Pekanbaru Indonesia
[2]Department of Chemistry, Mathematics and Natural Sciences University of Riau Panam 24293,Pekanbaru Indonesia
[3]Department of Civil Engineering, Faculty of Technology University of Riau Panam 24293,Pekanbaru Indonesia

E-mail: aman@lecturer.unri.ac.id

**Abstract**. This paper report the experimental study on the kinetic of geopolymer's solidification by determining setting time by varying activator concentrations of 10 M, 12 M, 14 M and 16 M, and curing temperatures of 30, 60, 80 and 100°C. However prior to the study of the effect the feasiblity of equipment using Vicat Needle. The transformation process of geopolymer crystal formation is analyzed by Avrami's kinetica theory. From the results of the experimental showed the increase activator concentration and curing temperature produce shorter setiing time for geopolymer's solidification. From the Avrami kinetic teory's the growth form of crystal in the geopolymerization showed constant (K) of particle growth is 5.42 x 10-2 and the value of Avrami's component n = 2.310 so that it can be concluded from the perspective of Avrami's kinetic theory, the rate of crystal growth solidification in this geopolymerization process shows a two-dimensional structure.

## 1. Introduction

The solidification method in this study is the geopolymer process produced by reaction of aluminosilicate solids contained in fly ash material with strong base activators producing binders / adhesives such as cement. Factors that influence the properties of geopolymers, namely the type and concentration of activator, activator modulus, curing temperature, curing time, and the ratio of alkaline solution in solids. The results of previous studies indicate that each of the above factors in the right amount becomes a parameter to produce geopolymer concrete/paving blocks with high compressive strength [1,2].

Study the effect of the concentration of alkali solution on the compressive strength in fly ash as a geopolymer concrete base material [3]. It was reported that the higher the concentration of alkali solution affected the increase in compressive strength, the researchers stated that the strength of the geopolymer material is very dependent on the concentration of alkaline solution, in addition to decreasing compressive strength when the ratio of water to solid geopolymers increases. Paver blocks geopolymer from fly ash with variations of NaOH 8 to 12 M activator concentration, the compressive strength increased with increasing the concentration of activator [4].

[*] To whom any correspondence should be addressed.

From the researchers described above there are more physical and chemical properties of geopolymers after the solidification process such as compressive strength, acid resistance, porosity and geopolymer absorption, while the study before the geopolymer solidification process still a few, therefore in this experimental studying the effect of parameters concentration and temperature in the solidification phase in the geopolymerization process and analysis from the perspective of Avrami's kinetic theory. This study also studied the Vicat Needle tool in determining geopolymer solidification and its correlation with Avrami's kinetics theory of crystalline transformation from nucleation phase to crystal growth.

## 2. Materials and methods

### *2.1. Materials*

Materials used in this research, fly ash from PT. RAPP Pelalawan Riau province, Sodium Hydroxide (NaOH) solution from Basic Chemical Labor Department of Chemical Engineering FT-UR, Sodium Silicate ($Na_2SiO_3$) in the form of a solution from PT. Sinar Sakti Kimia Solo, fine / sand aggregate from Rimbo Panjang Quarry Kampar Regency, Riau. The equipments used is the scale, measuring cup, stopwatch, piknometer, oven and Vicat Needle in the Materials Technology laboratory of the Department of Civil Engineering FT-UR.

**Table 1.** Chemical composition of fly ash

| Oxida | (%) by mass |
|---|---|
| SiO2 | 53.616 |
| $Al_2O_3$ | 28.47 |
| $Fe_2O_3$ | 5.525 |
| CaO | 4.974 |
| $P_2O_5$ | 0.569 |
| $SO_3$ | 0.584 |
| $K_2O$ | 0.857 |
| $TiO_2$ | 0.724 |
| LOI | 4,681 |

### *2.2. Preparation*

The kinetics of geopolymer solidification reactions: Preparation of NaOH solutions concentrations of 10,12, 14, and 16 M. Then the activator solutions of NaOH and $Na_2SiO_3$ were mixed with a ratio of 1: 2. Then fly ash : activator solution (NaOH + $Na_2SiO_3$ ) with a ratio of 1: 3 mixed with variations in concentration NaOH, after that the molding process with temperature variations of 30, 60, 80, and 100 °C.

### *2.3. Rate of Test*

The solidification rate test with the Vicat Needl tool and note the solidification rate for each concentration. As addition in this experimental the highest compressive strengh at 14 M NaOH [5]. Explanation and analysis of results obtained using Avrami's Kinetic Theory.

## 3. Result and Discussion

### *3.1. Effect of Temperatures*

From the experimental results, variations in concentration and curing temperatures which results in shorter time for each increase in temperature at a fixed concentration and vice versa each variation in concentration at a fixed temperature against the geopolymer solidification process. Furthermore, from the Kinetic Avrami's Theory, the form of crystal growth in the geopolymerization process shows a two-dimensional structure while certain samples from geopolymers show secondary nucleation in geopolymer growth.

Temperature variation to the rate of solidification. As seen In Figure 1 the relationship between temperature variation on the geopolymer solidification rate at a concentration of 14 M NaOH. Samples with a temperature of 80°C reach the fastest maximum hardening compared to samples with temperatures of 100, 60°C and room temperature. This can be seen from the time it takes the sample temperature of 80 °C for 35 minutes, temperature of 100 °C for 40 minutes, While the temperature of 60 °C and T. room is equal to 60 minutes. This is due to the rate of evaporation of water faster with increasing time. At temperatures of 80 and 100 °C the rate of evaporation of water is faster than the temperature of 60°C and room temperature, but at a temperature of 100 °C the geopolymer product is cracked so that it affects the compressive strength obtained. The best results in this study occurred at an optimum temperature of 80 °C where the geopolymer product did not crack (*cracking*) with the highest compressive strength.

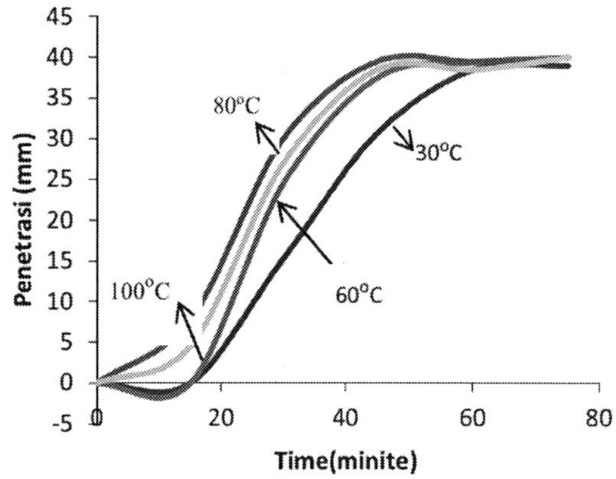


**Figure 1**. Variation in temperature to the rate of solidification in 14 M NaOH.

The increase in curing temperature will accelerate the geopolymerization process due to the reduced water content in the geopolymer material which has an effect on compressive strength [6]. The geopolymer compressive strength is very dependent on the pore size during the geopolymerization process is influenced by the higher temperature the number of pores of the geopolymer material is getting smaller, the optimum temperature is obtained at 80°C [7].

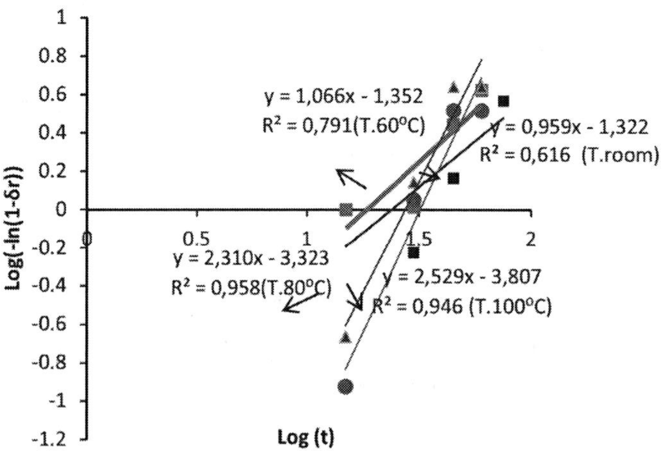


**Figure 2**. Effect variation in temperatur to Avrami's Plot.

As seen Figure 2. The effect of temperature at the optimum concentration NaOH 14 M on geopolymer solidification. From the Avrami plot of the temperature effects of the Avrami exponent value (n) and the growth rate constant (K) formulated and tabulated in Table 2. Based on Table 2. above it can be seen that the value of Avrami's exponents varies in all temperature variations tested. The value (n) starts from 0.959 to 2.529 where the smallest n is at room temperature while the largest (n) value is located at 100 oC. The Avrami's value of this exponent is very important to determine the form of geopolymer growth. Then for the fastest crystal growth rate (K) occurs at a temperature of 80 oC. If the value of n = 1, indicates growth of the nucleus occurs instantly, while the value of n = 2 - 3 growth spherulit and sporadic nucleation forms. The value (n) between 0.959 and 2.529 shows growth in instantaneous nucleation and two-dimensional growth nucleation [8]. However, for the geopolymer growth the expected exponent value starts from 1-2 in two-dimensional shape. These results indicate that the growth of geopolymer forms can be said to be two dimensions.

**Table 2.** Avrami's Exponent for Temperatur 14 M NaOH.

| Temperatur (°C) | Eksponen Avrami (n) | Growth Rate (K),(min$^{-1}$) |
|---|---|---|
| T.room | 0,959 | 4,76 x $10^{-5}$ |
| 60 | 1,066 | 4,45 x $10^{-4}$ |
| 80 | 2,310 | 4,75 x $10^{-2}$ |
| 100 | 2,529 | 1,56 x $10^{-2}$ |

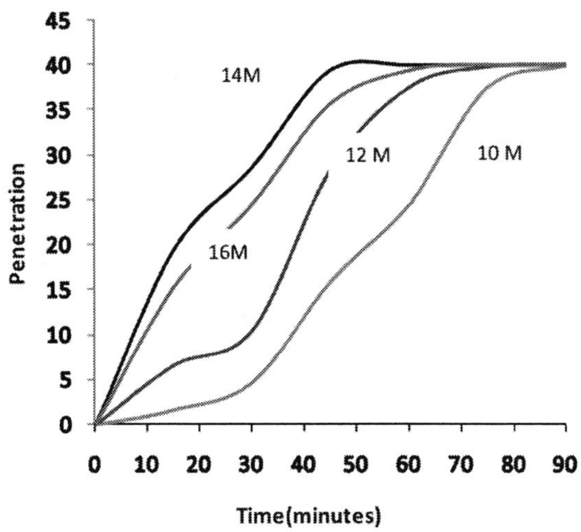


**Figure 3.** Variation in concentration to the rate of solidification at 80°C.

*3.2. Effect of Concentrations*

As seen in figure.3 the relationship between concentration variation on the geopolymer solidification rate at temperature 80°C. Samples with concentration NaOH 14 M reach the fastest maximum hardening compared to samples concentration 16 M, 12 M and 10 M. This can be seen from the time it takes the sample 14 M for 35 minutes, 16 M for 45 minutes, While the others equal to 60 minutes. The amount of dissolution is largely depending on the concentration of NaOH and curing time. Thus, increase the concentrations will increase the number of $OH^-$ ion in the solution for leaching process that lead to acceleration of the geopolymerization process. However based on the literature review, when the concentration of NaOH solution in higher than 15 M, the dissolution supposed to be decreasing due to increase in coagulation of silica [9]. In short the higher molarity of NaOH concentrations lead to geopolymeratization process which the reducing the setting time of geopolymer. As seenFfigure 4 it shown the results for the effect of different activator concentrations on geopolymer compaction. From the Avrami's plot, exponent value (n) and the growth rate constant (K) are formulated and tabulated in Table 3.

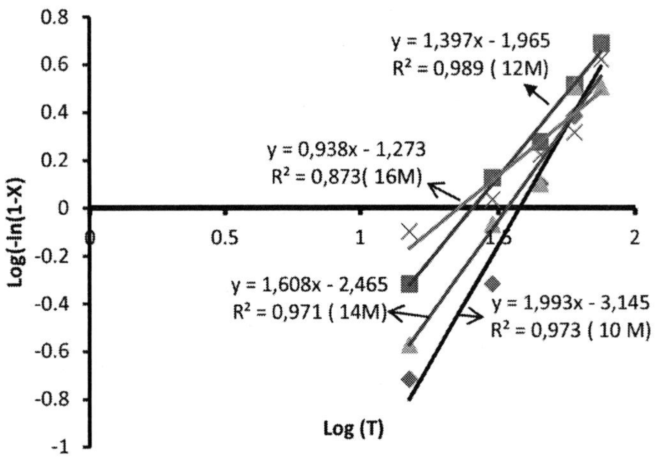


**Figure 4**. Effect variation in concentration to Avrami's Plot.

**Table 3 .** Avrami's Exponent for variation of concentration ,T =80 °C.

| Konsentrasi (M) | Avrami's Exponent **( n )** | Growth Rate **(K)**,(min$^{-1}$) |
|---|---|---|
| 10 | 1,993 | 7,16 x.10$^{-4}$ |
| 12 | 1,397 | 1,08 x 10$^{-2}$ |
| 14 | 1,608 | 5,42 x 10$^{-2}$ |
| 16 | 0,938 | 4,33 x 10$^{-2}$ |

As seen Table 3. that the value of the Avrami's Exponent varies in all variations of the concentration tested. The value (**n**) starts from 0.938 to 1.993 where the smallest is at the concentration of activator 16 M, while the largest (n) value lies in the concentration of activator 10 M. Avrami value of this exponent is very important to determine the form of geopolymer growth. Then for the fastest crystal growth rate (K) occurs at a concentration of activator 14 M is 5.42 x $10^{-2}$. From the results obtained both temperature and concentration variations obtained the same optimum value at 14 M NaOH concentration and 80°C temperature. The value (n) of the variation in concentration between 0.938 and 1.993(2,0) shows growth in instantaneous nucleation and two-dimensional growth nucleation[10,11]. However, for geopolymer growth the expected exponent values start from 1 - 2 in two-dimensional form. These results indicate that the growth of geopolymer forms can be said to be two dimensions.

## 4. Conclusion

From the results and discussion described, it can be concluded that the geopolymer paving blocks produced are: The highest kinetics constant of particle growth (K) reaction with a value of 5.42 x $10^{-2}$ and the value of Avrami's component n = 2,310 obtained at the optimum temperature of 80°C in

14 M NaOH activator concentration. The Avrami's Exponent value range for all samples ranged from 0.959 to 2.529 so it can be concluded that the form of geopolymer growth has 2- dimensional disc structure.

**Acknowledgement**
We would like to express our deepgratitude to LPPM Universitas Riau for research funding Dana DIPA (contract number : 889/UN.19.5.1.3/PT.01.03/ 2019).

**Reference**
[1] Yahya Z and Mahmod K 2015 *Materials* **8**(5) 2227
[2] Olivia M, Putri and Damayanti 2014 Parametric Study on the compressive strength of Palm Oil Fuel Ash (POFA) Geopolymer mortar *The International Conference on Enviromentally Friendly Civil Engineering Construction and Material*, Manado Indonesia
[3] Hardjito D and Rangan B V 2004 Properties of Geopolymer Comcrete with fly ash as Source Material: Effect of Mixture Composition *Seventh CANMET/ACI International Conference on Recent Advance in Concrete Technology* Las Vegas USA
[4] Memon Vaz K 2011 *Int. J. Civil Env. Sci. Construc. Archi. Eng.* **5**(8) 372
[5] Aman, A Awaluddin, A Ahmad and M Olivia 2017 Parametric Study on the compressive strength geopolymer paving block, IOP Conf.Series: Materials Science and Engineering 345 (2018) 012018 doi:10.1088/1757-899X/345/I/012018
[6] Wang and Cheng 2003 Production geopolymer materials by coal fly ash Paper presented at the *7th International Symposium on East Asian Resources Recycling Technology* Tainan,Taiwan
[7] Muniz, Bergna J 2011 *Materials Letters* **65**(6) 995
[8] Mishra A and Choudhary 2000 ARPN *J. Eng. Appl. Sci.* **3**(1) 87
[9] Garcia-Lodeiro L 2010 *Cement and Concrete Research* **40**(1) 27
[10] Alvarez-Ayuso E 2008 *J.Hazard Matter.* **154** 175
[11] Yao X, Zhang Z, Zhu H and Chen Yue 2009 *Thermochimica Acta* **493**(1-2) 49
[12] Ekaputri J Maekawa K and Ishida 2011 The use of geopolymer process for boron fixation in flash Jinan China Proc.7 th *Intl.Symp.Cement and Concrete*
[13] Davidovits J 1994 *World Res. Rev.* **6**(2) 263
[14] Ubolluk and Prinya 2009 *Intl. J. Miner. Proc.* **59**(3) 247
[15] Bergna HE and Robert WO 2005 Colloidal Silica: *Fundamental and Applications* vol.131, Taylor and Prancis
[16] Sathonsaowaphak A, Chindaprasirt and Pimraksa K 2009 *J.Hazard Matter.* **168**(1) 45

# Production of Liquid Bio-Fertilizer from Old Coconut Water and Molasses using Consortium Microbes

**R Darmawan[1*], V G P Dewi[1], M A Rizaldi[1], S R Juliastuti[1], S Gunawan[1], H W Aparamarta[1] and A Wiguno[1]**
[1]Department of Chemical Engineering, Faculty of Industrial Technology, Institut Teknologi Sepuluh Nopember, Surabaya 60111, INDONESIA

E-mail: rdarmawan@chem-eng.its.ac.id

**Abstract.** Increasing issue of returning to nature, the presence of organic fertilizer is one of the things required in agriculture sector. The presence of the liquid organic fertilizer, it is expected to have a positive impact on plants and soil. Moreover, in the manufacturing process, it was utilizing wastes. The liquid fertilizer was made by mixing waste from micro-businesses that produce a porridge (old coconut water) and side product of the sugar factory (molasses). Seven litter solutions from the liquid were mixed using a bioreactor equipped with impeller for 8 incubation days. The isolated microbes were inserted with initial number of $10^{12}$ using counting chamber method. As a phosphate source, the phosphate rock was implemented into reactor. Every 4 days, the samples were taken and analysed for knowing the concentration of NPK and C components. Three consortium microbe variables were carried out for making the fertilizer where the source of microorganisms from enrichment cultures + old coconut water + molasses (EOM); soil + old coconut water + molasses (SOM); and only old coconut water + molasses (OM). The highest (%) of N, P, K and C components was detected at 0.09; 0.04; 10.5 and 1.74, respectively achieved by the EOM after 8 day of incubation period. The bio-fertilizer was applied into the paddy test plants with a hydroponic system.

## 1. Introduction

The presence of waste around us that has not been treated properly yet, environmental issues and economic factors, provides a potential and at once opportunity regarding them. One way is to utilize waste from the production of porridge, in the form of old coconut water and side products from a sugar factory, in the form of molasses. Old coconut water is believed, it's still containing enough nutrients and micro elements that can be used as a medium as well as a source of microorganisms that can be added in the process of making liquid fertilizer. The old coconut water can be found easily and abundantly in the Regency of Ponorogo, East Java Province – Indonesia, because this area is a central industry of porridge, which applies the coconut as its basic material. Referring to the data from BPS – Ponorogo, the number of coconuts that can be produced in this regency is 6.170,09 tons [1] and Indonesia produces coconut about 2.87 million tons with an area of 3.54 million hectares [2].

Likewise, molasses, it's also still having organic elements. Molasses is an energy source that contains about 50% sugar in the form of sucrose (20-30%) and reducing sugars (10-30%). Reducing sugar is a compound that is digested easily and can be directly absorbed by the blood for the process of metabolism in order to obtain energy. The characteristics of molasses have a COD of 65,000-130,000

[*] To whom any correspondence should be addressed.

mg/L and BOD5 of 30,000-96,000 mg/L and have a smell like soy sauce and a dark brown color [3]. Microorganisms contained therein also utilize both of these materials.

Furthermore, utilization of the Lapindo mud in Sidoarjo – Indonesia is as an effort to reduce the negative impact of disasters that have occurred since 2006. For generating bio-fertilizer, it requires culture source as a source of microbes. Lapindo mud is other potential components as a culture source Lapindo mud is composed of 70% water and 30% solids. Salinity of mud is very high (38-40%), so it is salty [4]. Based on the research on the chemical parameters results of Lapindo mud, it is known that the pH value ranges from 6.6 - 7, cation exchange capacity (CEC) of 3.89-35.42 (Me/100g), moisture content of 40.41-60.73 % and total organic carbon content of 54.75-55.47% [5]. Therefore, based on all of the above, this research was conducted. This study aims to make organic liquid fertilizer based on organic materials that utilize consortium microbes sourced from nature.

## 2. Material and Method

### *2.1. Microorganism sources*

The consortium microbes were obtained from some natural environment sources, such as Lapindo Mud, old coconut water and molasses. The mixed cultured from Lapindo mud were enriched for three weeks by following method [6]. The sterile minimum medium was containing some nutrients such as 2 g of $NH_4NO_3$; 2 g of $KH_2PO_4$; 2 g of $K_2HPO_4.3H_2O$; 0.2 g of $MgSO_4.7H_2O$; 0.2 g of $FeCl_3.6H_2O$ and 0.02 g of $CaCl_2$ within 200 mL. For isolation of the mixed cultured, the medium was supplemented using 5 g of mud and 5 g of the Jordan phosphate rock. The phosphate rock composition was described in Table 1. Meanwhile, the other microorganisms were using the old coconut water and molasses directly.

**Table 1.** Jordan phosphate rock test.

| Parameter | **Unit** | **Grade**[b] |
|---|---|---|
| $H_2O$ | % | 2.02 |
| $P_2O_5$ | % adbk[a] | 30.44 |
| CaO | % adbk[a] | 46.71 |
| MgO | % adbk[a] | 0.29 |
| $Na_2O$ | % adbk[a] | 0.64 |
| $K_2O$ | % adbk[a] | 0.03 |
| $Fe_2O_3$ | % adbk[a] | 0.25 |
| $Al_2O_3$ | % adbk[a] | 0.33 |
| $R_2O_3$ | % adbk[a] | 0.57 |
| $SO_3$ | % adbk[a] | 1.06 |

[a] on the basis of dry ingredients
[b] average

### *2.2. Liquid material sources*

The used materials of liquid were collected from coconut water waste of the porridge manufacturer "Jenang Murni" in Ponorogo and side sugar of PT Energi Agro Nusantara (Enero) in Mojokerto – Indonesia.

### *2.3. Fermentation process and manuring paddy hydroponic test*

The liquid medium was prepared by mixing 7 L of old coconut water; 0.01 % v/v of molasses and 0.0024 % w/v of phosphate rock. The fermentation process for 8 days was conducted using the aerobic reactor (Figure 1) inserted by consortium microorganisms as the changed variables. During the experiment, the pH and temperature were maintained at around 7 and 30°C, respectively with aeration supply at a rate of 4 L/min. Three variables microbes used were including the enrichment cultures + old coconut water + molasses (EOM); soil + old coconut water + molasses (SOM); and only old coconut water + molasses (OM). The initial number of microbes were adjusted at $10^{10}$ cells/ mL approximately with counting chamber method [7]. Every 4 days, the liquid bio-fertilizer was checked N, P, K and C-organic contents. The organic contents were analyzed at t = 0 days, t = 4 days, and t = 8 days, then it was tested using paddy hydroponic system for 10 weeks. The manuring test was carried out in the Islamic Boarding Institution (PP) Al-Ahsan, Jombang, East Java – Indonesia.

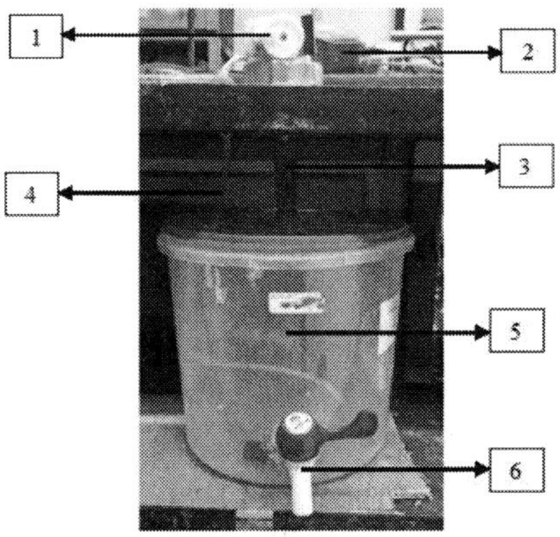


**Figure 1.** Aerobic bioreactor for fermentation process equipped with no 1, 2, 3, 4, 5 and 6 were dynamo agitator, aerator, agitator air hose, reactor, and output valve, respectively

**Figure 2.** Hydroponic system in PP Al- Ahsan, Jombang, East Java – Indonesia.

## 3. Results and Discussion

In the research, it studies the manufacture of liquid fertilizers by utilizing materials and microorganisms that are easily found originating from nature. Three consortium microbe variables were carried out for making the fertilizer where the source of microorganisms from enrichment cultures + old coconut water + molasses (EOM); soil + old coconut water + molasses (SOM); and only old coconut water + molasses (OM). During the fermentation process, bio-fertilizer was analyzed for N, P, K and C content every 4 days. After 8 days, the formed bio-fertilizer was applied to paddy plants with a hydroponic system, in which the system can be seen in the figure 2. One L of organic fertilizer was diluted within 50 L of water. There were three ways for manuring, such as after the paddy test age of 7 days, 30 days, and 60 days. The manuring to the plant test was performed for 10 weeks.

3.1. Observation N, P, K and C elements during the process.

There were some factors or parameters influenced to bio-fertilizer especially amount of N, P, K and C elements during fermentation process, such as the initial element and concentration material sources from coconut water and molasses, microbes used, duration and condition of incubation. The results obtained from the N, P, K and C element analysis can be summarised in Table 2. Based on the analysis component, it was found that the highest number on N, P, and K elements were achieved by EOM at t = 8 days with concentrations of 0.09%, 0.04 % and 10.5 %, respectively and the greatest number on C elements was obtained by OM at t = 8 days with level of 1.78 %.

**Table 2.** Result of analysis N, P, K and C elements.

| Variable | Time | 0 day | 4 days | 8 days |
|---|---|---|---|---|
| | Component (%) | Component (%) | Component (%) | Component (%) |
| EOM | $K_2O$ | 0.26 | 0.51 | **10.5** |
| | $P_2O_5$ | 0.065 | 0.041 | **0.04** |
| | Nitrogen | 0.013 | 0.014 | **0.09** |
| | C-organic | 2.05 | 1.94 | 1.74 |
| SOM | $K_2O$ | 0.26 | 0.37 | 9.61 |
| | $P_2O_5$ | 0.065 | 0.033 | 0.03 |
| | Nitrogen | 0.013 | 0.015 | 0.05 |
| | C-organic | 2.05 | 1.93 | 1.56 |
| OM | $K_2O$ | 0.29 | 0.38 | 5.68 |
| | $P_2O_5$ | 0.058 | 0.038 | 0.03 |
| | Nitrogen | 0.013 | 0.01 | 0.07 |
| | C-organic | 1.77 | 1.79 | **1.78** |

As can be seen from the above table, the utilization of consortium microbes from enrichment cultures + old coconut water + molasses (EOM) reported significantly more N, P, K percentages than soil + old coconut water + molasses (SOM); and only old coconut water + molasses (OM). The result indicated that consortium microorganisms from EOM especially enrichment culture from Lapindo mud can enhance N, P, K compounds well. These culture from Lapindo Mud believed have a critical role for increasing N, P, K components [8] as the nitrogen, phosphor, potassium contributors on the plants. These microbes were believed as a diazotroph bacteria in which have a good role to fix nitrogen [9] due to a nitrogenase enzyme which have capability on for nitrogen and hydrogen combination [10, 11]. Furthermore, the consortium microorganisms (EOM) might be suitable microbe combination to enhance phosphor element and increase potassium component where the condition might be caused by the hydrolysis and proteolysis reactions so that it lead to release $K^+$ from the organic matter [8]

Meanwhile, the highest percentage of C element (1.78%) during process incubation period was obtained by OM. The carbon elements from coconut water and molasses were degraded into a simpler compound with more stable for 8 days incubation period. The three designated variables provide the decreased with enhancing the incubation period where the condition occurs might be caused by absence of the added microorganisms that consume C-organic within coconut water and molasses to enhance the microorganism metabolism as carbon sources [8] and at once as culture sources.

3.2. Observation of height and leaf number on paddy test during manuring

The observation of manuring on paddy test for height and leaf number during 10 weeks can be seen in the Figure 3 and Figure 4, respectively. Figure 3 shows that the paddy plant height increased with increasing manuring time. The largest growth of plant height, it can be achieved by OM. Meanwhile, Figure 4 presents that the number of leaves of rice plants increased until the fourth week then relatively constant with increasing fertilization time. The highest number of rice leaves was also experienced by OM.

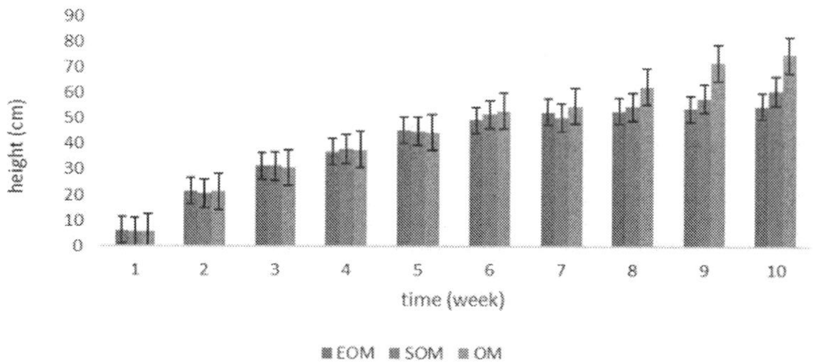


**Figure 3.** Observation of height plant during paddy test.

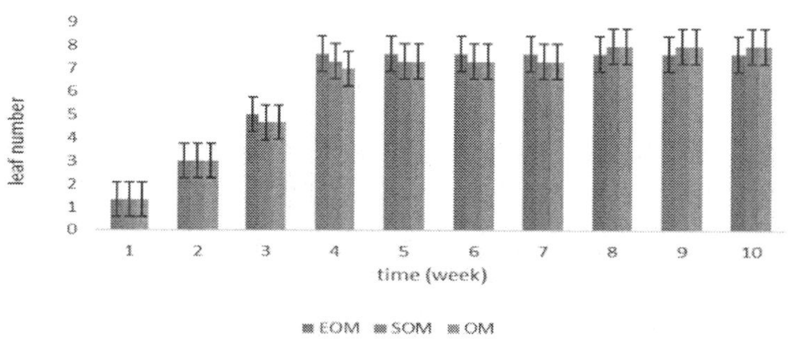


**Figure 4.** Observation of leaf number during paddy test.

The condition of rice plants in the field with the hydroponic system before (0 week) and after fertilizing (5 and 10 weeks) with variables EOM, SOM and OM can be seen in Figures 5, 6, 7, respectively. From the pictures show that the production of liquid bio-fertilizer using natural culture sources with all the variables have a positive response in the hydroponic paddy test, which was indicated by the emergence of paddy fruit.

**Figure 5.** Paddy test was conducted (a); (b) and (c) at 0 week, 5 weeks and 10 weeks, respectively, using enrichment cultures + old coconut water + molasses (EOM).

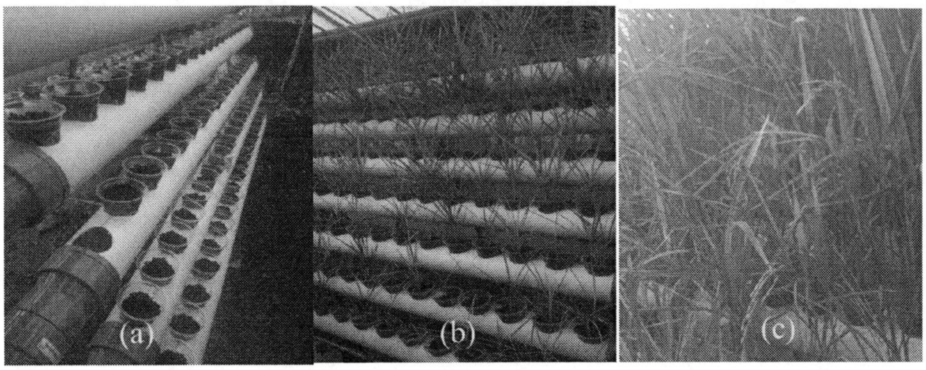


**Figure 6.** Paddy test was conducted (a); (b) and (c) at 0 week, 5 weeks and 10 weeks, respectively, using soil + old coconut water + molasses (SOM).

**Figure 7.** Paddy test was conducted (a); (b) and (c) at 0 week, 5 weeks and 10 weeks, respectively, using only old coconut water + molasses (OM).

## 4. Conclusion

The coconut waste water supplemented by molasses and natural microbes can be used as the potential material and culture sources for generating liquid bio-fertilizer.

**Acknowledgments**

The authors would like to thank the Islamic Boarding Institution (PP) Al-Ahsan, Jombang, East Java – Indonesia for testing of paddy hydroponic system. Furthermore, we also acknowledge to the Institut for Research and Community Services (LPPM) of Institut Teknologi Sepuluh Nopember (ITS) on the Program (*Pengabdian Kepada Masyarakat Berbasis Penelitian*) for financial supporting with agreement document No: 1344/PKS/ITS/2019

## References

[1] BPS Kabupaten Ponorogo, 2015
[2] Directorate General of Estate Crops 2018 *Tree Crop Estate Statistics of Indonesia, Coconut 2015-2017*
[3] Zeng Y F, Liu Z L and Qin Z Z 2009 *J. Hazard Mater*. **162** 682
[4] Arisandi P 2006 *Menebar Bencana Lumpur di Kali Porong* Ecological Observation and Wetlands Conservation
[5] Juniawan A, Rumhayati B dan Ismuyanto B 2013 *Karakteristik Lumpur Lapindo dan Fluktuasi Logam Berat Pb dan Cu pada Sungai Porong dan Aloo* Chemistry Department, Brawijaya University
[6] Darmawan R, Nakata H, Ohta H, Niidome T, Takikawa K and Morimura S 2015 *Bioremed. & Biodegrad.* **6**, 3, doi:10.4172/2155-6199.1000283
[7] Williams R E O 1952 *J. Gen. Microbiol.* **7** 89
[8] Darmawan R, Hidayat F, Juliastuti S R, Hendrianie N, Gunawan S and Aparamarta H W 2019 Liquid Organic Fertilizer from Waste of Coconut Porridge Manufacturer and Molasses using Various Microorganisms, *IOP Conf. Ser.: Mater. Sci. Eng*. 543 012099
[9] Marbun, Posma, Hardy G and Putri J A 2013 Universitas Sumatera Utara
[10] Yin T T, Pin U L and Ghazali A H A 2015 *Trop. Life Scie. Res* . **26** 1
[11] Issar S and Gaur R K 2013 *Intern. J. of Adv. Res* . **1** 8

# Peat water treatment by adsorption using kaolin-based geopolymer

**PS Utama[1], M Olivia[2], BA Prawiranegara[3], ID Agusti[1], JA Pinem[1], L Darmayanti[2] and E Saputra[1*]**
[1]Chemical Engineering Department Universitas Riau, Pekanbaru 28293, Indonesia
[2]Civil Engineering Department Universitas Riau, Pekanbaru 28293, Indonesia
[3]Chemistry Department Universitas Riau, Pekanbaru 28293, Indonesia

E-mail: edysaputra@unri.ac.id

**Abstract**. Geopolymers have zeolite-like chemical compositions, which has some characteristics such as wide surface area and excellent stability so that it can be used for adsorbents for peat water treatment. In this research, the kaolin based geopolymers were applied as an adsorbent for processing the peat water into clean water. The variable used in this research were adsorbent mass (1; 1.5; 2; and 2.5 g), and time (30; 60 and 90 minutes). The results indicated that the kaolin-based geopolymers were very efficient in removing color and organic content. The optimum condition was mass adsorbent of 1 g and time of 30 min, and the result was 100% for removing color and 98.77% for removing organic content.

## 1. Introduction

Peat water is surface water from peat soil with striking features. It has a brownish-red color, contains high organic matter, tastes sour, pH 2-5 and has a low hardness level [1]. Therefore, the quantity of peat water can be used as a water resource, but the quality of peat water cannot yet be used as drinking water. In order for peat water to be utilized, peat water must be treated first so that the parameters on peat water are in accordance with the requirements for clean water quality based on RI Minister of Health Regulation No. 416/MENKES/PER/IX/1990.

One characteristic of peat water is its high color content. The color of peat water is caused by high organic matter (humic acid), where a humic acid is an aromatic group [2]. Adsorption can remove various substances in water such as aromatic compounds, color content, and heavy metal compounds. Adsorption is the process of adsorption of a substance, on an adsorbent surface in the form of solids. There are many adsorbents that have been developed for treating peat water, such as using a composite of $Fe_3O_4$ and chitosan [3], bentonite and montmorillonite nanoparticles [4] and biochar [5].

Geopolymers have zeolite-like chemical compositions. Geopolymers are synthetic polymer materials that can be made from fly ash, kaolin, rice husks, and other materials contained silica and alumina. Geopolymers have been used in the innovation of making concrete to reduce $CO_2$ emissions [6] while Mužek et al. (2013) and Naghsh, M., & Shams, K. (2017) utilized geopolymer as an adsorbent in water treatment [7,8].

Mužek et al. (2013) used coal fly ash from power plants as a geopolymer base material [7]. However, the problem is fly ash from coal power plants is categorized as hazardous and toxic waste

* To whom any correspondence should be addressed.

(PP No. 85 of 1999 concerning Management of Hazardous and Toxic Waste) thus that the use of fly ash is not appropriate for processing peat water into clean water. Therefore, in this study, kaolin was used as the base material for making geopolymer. This study aims to synthesize kaolin-based geopolymer for treating peat water. The effect of adsorbent loading on color and organic substance contents was studied.

## 2. Experimental

*2.1.1. Materials*

The raw material used in this study were peat water from Tapung, Kampar regency, and kaolin from Sigma Aldrich. Other materials used were sodium hydroxide (97 wt. % NaOH Merck), hydrochloric acid (HCl 37 wt. % Merck), sodium silicate ($NaSiO_3$, Merck) and distilled water.

*2.1.2. Methods*

*2.1.2.1. Geopolymer preparation*

The activator solution was made by mixing NaOH of 8 M solution with sodium silicate and stirred continuously until well mixed. Kaolin and activator solution was put in a container and stirred for 10-15 minutes at constant stirring speed. During the stirring process, distilled water was poured slowly until the mixture formed a paste. The paste obtained was put into the mold, then wrapped in aluminium foil, heated for 24 hours in an oven at a temperature of 85 °C. Geopolymer samples were removed from the oven and washed with distilled water until the pH was neutral. The adsorbent particle size of <0.074 mm was used in this study so that the geopolymer was crushed and sieved using No. 200 sieve. The obtained geopolymer then characterized by XRD and SEM.

*2.1.2.2. Adsorption of peat water processes*

Peat water samples were taken from Air Terbit Village, Tapung District, Kampar Regency. Procedure for taking peat water by SNI 6989.57: 2008 concerning the Method of Taking Surface Water. Samples were taken in upstream areas, Desa Air Terbit, Kampar. Then the initial organic content of the sample was analyzed. For maximum results, before the adsorption process, the pH of peat water is lowered by using HCl. The mass variation of the adsorbent used in this study is 1; 1.5; 2; and 2.5 g added to 200 ml of peat water. The mixture was stirred in a test jar with a speed of 240 rpm. Stirring time variations were 30, 60 and 90 seconds. The solution was then filtered and tested for the content of organic matter. Table 1 shows a comparison of several parameters of peat water with clean water quality standards.

**Table 1**. Peat water characteristic

| No | Parameters | Unit | Peat water | Clean Water Requirements [a] | Notes |
|---|---|---|---|---|---|
| 1 | Color | PtCo | 425 | 50 | Not suitable |
| 2 | Ph | - | 4.5 | 6.5-9.0 | Not suitable |
| 3 | Organic matter | mg/l $KMNO_4$ | 51.192 | 10 | Not suitable |

[a]) Based on RI Minister of Health Regulation No. 416 / MENKES / PER / 1990

## 3. Results and discussion

*3.1. Characteristics of Kaolin-Based Geopolymer Adsorbents*

The morphology of the geopolymer adsorbent used in this study was determined by SEM (Scanning Electron Micrographs). Figure 1 shows the SEM micrograph of the geopolymer adsorbent used in this study. It can be seen there are many unreacted particles which varied particle sizes up to 30 μm. It is indicating that not all kaolin used as the raw material polymerized into geopolymer. The geopolymers adsorbent also characterized using XRD. The diffractogram of the geopolymer adsorbent can be seen in Figure 2. The peaks pattern in the diffractogram obtained is similar to the peak patterns of seven days age kaolin based geopolymer from the reference. The peaks patterns in the diffractogram corresponding to a zeolite X patterns [8] and there are kaolinite, quartz, and muscovite. It is indicated that kaolin minerals are still in this geopolymer adsorbent. This is in accordance with the result of SEM characterization that the geopolymer adsorbent made in this study was not perfect. There was part of kaolin not polymerized. Even though the polymerization was incomplete, for catalyst and adsorption application, the incomplete polymerization provides a more active site from the crystalline structure from unreacted kaolin and also zeolite like structure.

**Figure 1.** SEM micrograph of kaolin-based geopolymer with maginifications of 3000x

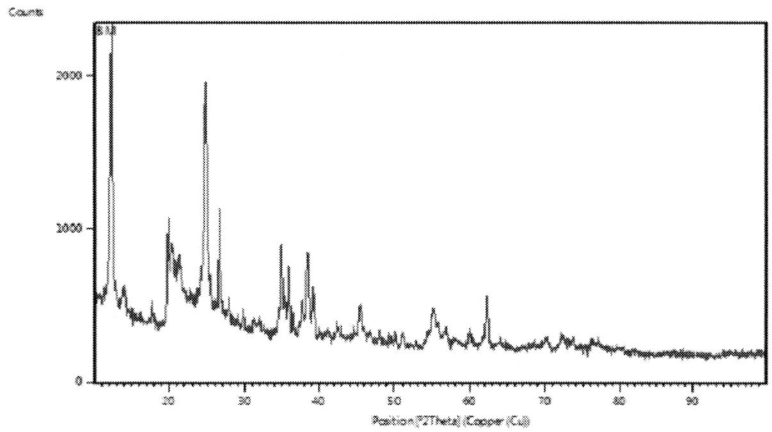


**Figure 2.** The XRD diffraction of solid obtained

*3.2. Peat Water Characteristic*

This research was carried out in a laboratory using peat water with an organic content of 51.19 mg/l, the color was brown with an intensity of 425 PtCo and a pH of 4.5. The three parameters were not yet in accordance with the standards set by the government through the Minister of Health Regulation No.416/MENKES/PER/IX/1990 concerning drinking water and clean water standards.

*3.3. Result of Peat Water Treatment with Kaolin-Based Geopolymer Adsorption*

*3.3.1. Effect of pH on peat water adsorption*

The effect of pH on decreasing the color intensity and organic matter content was determined by carrying out the adsorption at pH 1.5; 3; and 4.5. The pH was lowered by using HCl. Figure 3 shows the effect of pH on decreasing organic matter and color. It can be seen that organic matter and color decreased when pH was lowered from the initial pH of 4.5. The highest removal was 100% for color and 62.96% for the organic matter at pH 1.5. From this finding, it can be concluded that the lower the pH, the higher the removal of organic matter. The higher removal occurs because, at low pH, humic acid has a spherical structure whereas at high pH the structure is rather linear, so if the pH has increased the size of the molecule humic acid will increase, and this causes the adsorption capacity to decrease [9].

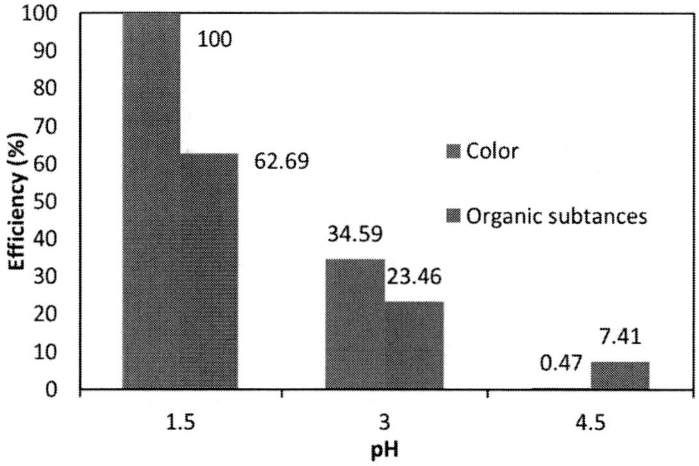


**Figure 3.** The Effect of pH on color reduction efficiency and Organic Substances

*3.3.2. Effect of adsorbent mass and time on peat water adsorption*

The result of color and organic matter reduction by variation of mass of adsorbent are shown in Figures 4 and 5. The sample after processing look clear and this is in accordance with the results of the color test which drops from 425 PtCo to 0 PtCo, the highest color reduction efficiency was obtained at 1 g of adsorbent the result was 100% of color can be reduced. The efficiency of decreasing levels of organic matter was also proportional to the decrease in color. The highest efficiency of decreasing organic matter levels was also at 1 g mass of adsorbent, the organic matter decrease from 51.192 mg/l to 0.63 mg/l it is mean that the geopolymers can adsorb 99.87% of organic matter. From Figures 4 and 5, it can be seen that the mass of the adsorbent is inversely proportional to the efficiency of color and organic matter reduction.

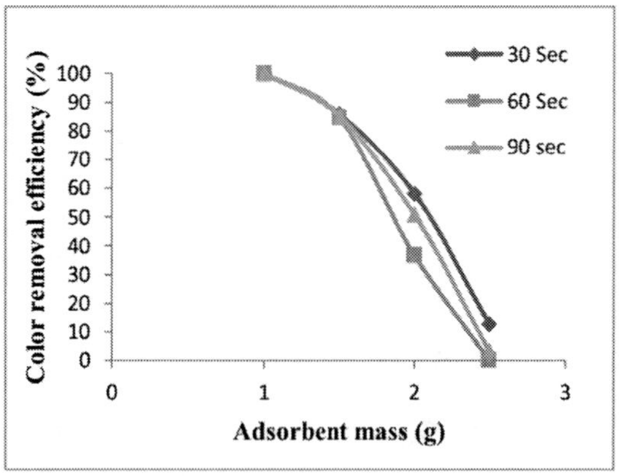


**Figure 4.** Effect of adsorbent mass on color reduction

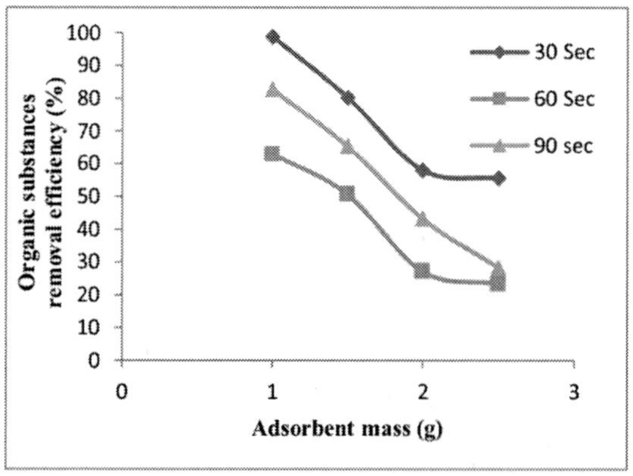


**Figure 5.** Effect of adsorbent mass on organic substances reduction

## 4. Conclusion

The kaolin-based geopolymer can be used effectively to reduce the color and organic matter content in peat water. The highest reduction of organic matter and color of peat water was obtained at pH of 1.5; mass geopolymer of 1 g for 200 $cm^3$ peat water and time of 30 seconds. The results kaolin based-geopolymer obtained can adsorbs 100% of color and 98.77% of organic matter. The high efficiency of color and organic matter adsorption in this study proves that this method is feasible to be used as an alternative for processing peat water.

## References

[1] Fuadi A and Diana S 2018 IOP Conference Series: Materials Science and Engineering. **345**(1) 012.
[2] Sutapa ID and Toruan RL 2015 *Teknologi Indonesia* **37**(3).
[3] Zulfikar MA, Afrita S, Wahyuningrum D and Ledyastuti M 2016 *Envi. Nanotech. Monitor. Manag.* **6** 64.
[4] Derakhshani E and Naghizadeh A 2018 *J. Molecular Liq.* **259** 76.
[5] Jung C, Phal N , Oh J , Chu KH, Jang M and Yoon Y 2015 *J. Hazard. Mater.* **300** 808.
[6] Ekaputri JJ and Triwulan T 2013 *J. Civil Eng.* **20**(1) 1.
[7] Mužek MN, Svilović S, Ugrina M and Zelić J 2016 *Desalination Water Treat.* **57**(23) 10689.
[8] Duxson P, Fernández-Jiménez A, Provis JL. Lukey GC, Palomo A and van Deventer JS 2007 *J. of Mater. Science.* **42**(9) 2917.
[9] Avena MJ, Vermeer AWP and Koopal LK 1999 *Colloids and Surfaces A: Physicochem. Eng. Aspects.* **151**(1-2) 213.

IOP Conf. Series: Materials Science and Engineering **845** (2020) 012009 doi:10.1088/1757-899X/845/1/012009

# Intensification of biolubricant synthesis from waste cooking oil using tetrahydrofuran as co-solvent

**T W Putra[1], R Hardiansyah[1], M R Lubis[1], and M D Supardan[1*]**
[1]Departement of Chemical Engineering, Universitas Syiah Kuala, Darussalam, Banda Aceh-23111, Indonesia

E-mail: m.dani.supardan@unsyiah.ac.id

**Abstarct**. In this study, biolubricant was synthesized through two stages of transesterification. The first transesterification was the reaction between waste cooking oil and ethanol using sodium hydroxide as catalyst to produce fatty acid ethyl esters (FAEE). The second transesterification was the reaction between FAEE and 1-octanol using potassium hydroxide as catalyst with tetrahydrofuran as co-solvent to produce fatty acid octyl esters (biolubricant). This study was focused on the effect of tetrahydrofuran as co-solvent in the second transesterification. Biolubricant yield of 94.42% was obtained at reaction conditions (temperature of 125°C, time of 180 min, reactant molar ratio of 1:4, catalyst amount of 1%-w/w, vacum pressure of 100 mmHg and co-solvent amount of 25%-w/w). The physico-chemical properties of biolubricant were kinematic viscosities of 10.47 $mm^2/s$ and 2.75 $mm^2/s$ at 40°C and 100°C, viscosity index of 103 and acid number of 0.62 mg KOH/g.

## 1. Introduction

The world demand for lubricating oil as protectors of engines caused by friction, heat corrosion has increased in this era [1]. The needs of this lubricating oil are almost entirely fulfilled by petroleum-based lubricating oil (Lubricant). Demand for lubricants continuously increasing furthermore cause problems on petroleum-based energy and environmental pollution. Every year about 55% of residual lubricants have polluted the environment which included soil, water and air [2]. Therefore, renewable and environmentally friendly lubricants are needed to solve these problems.

The development of the synthesis of environmental friendly-based lubricating oil or biolubricant as a substitute for petroleum-based lubricant is still being carried out. Biolubricant is an environmental friendly lubricating oil because it can decompose naturally by microorganisms (biodegradable) and low toxicity [3]. Biolubricant is also renewable because it can be made from vegetable oils such as palm oil, coconut oil, soybean oil, castor oil and sunflower oil [4].

The biolubricant can be synthesized by several methods such as transesterification, selective hydrogenation and epoxidation [5]. Transesterification is the most widely used method and recommended because it can produce yield higher than 90% [4]. Furthermore, the advantage of transesterification is increased oxidative stability so that the biolubricant is not easily damaged by oxidation. However, transesterification requires a high reaction temperature which causes increase energy consumption and production cost [5]. Therefore, the transesterification method still needs to be developed more.

* To whom any correspondence should be addressed.

At this time, the production and application of biolubricant are still limited. In fact, The energy shortage, environmental pollution and technological advances in the automotive and manufacturing in industries are increasingly demanding the use of biolubricant [6]. This problem is caused by the production of biolubricant requires high production cost. This affects biolubricant price which can be more expensive than lubricant. Furthermore, it will be an obstacle in order to produce biolubricant which can be competing for the economic value of commercial lubricant's [7].

The efforts to produce biolubricants with simpler and cheaper processes continue to be developed at this time. The use of cooking oil waste (WCO) as raw material for biolubricants has very important and interesting effects such as to reduce the cost of biolubricant production because its' price is cheaper than other vegetable oils; and to reduce environmental pollution because WCO is waste [8]. Raw material cost takes around 70-80% of total production costs and WCO prices are 30-60% cheaper than other vegetable oils [9]. The study of biolubricant production from WCO using transesterification was carried out by Wang et al. produce biolubricant yields of 85.7%. Studies on biolubricant production from WCO still need to be developed in order to obtain higher yields [9].

The application of co-solvent in the synthesis of biolubricant is still very rare. At this time, the application of co-solvent especially tetrahydrofuran (THF) is still focused on biodiesel synthesis [10]. Co-solvent can accelerate the reaction by increasing the solubility of oil and alcohol so that oil and alcohol can be completed reaction and increase yield [11]. Fransiska et al. have recommended the application of THF as co-solvent in the transesterification. THF has a good ability to dissolve fatty acids (triglycerides) and alcohol. In addition, THF has a low toxicity, boiling point of 67°C which can be separated easily and the price is relatively cheaper than other co-solvents [12].

This study purposes to obtained the intensification of biolubricant synthesis from WCO using THF as co-solvent. This study is expected to help develop the production of biolubricant to be better. Thus, the prospect of biolubricant to substitute lubricant is getting bigger.

## 2. Methodology

### *2.1. Equipments and Chemicals*

The raw material was the WCO from five times frying. The WCO was obtained from palm oil based cooking oil. Cooking oil was used to fry fish, tofu and tempeh for five times. The chemical compounds used were ethanol 96%, sodium hydroxide (NaOH), Potassium hydroxide (KOH), THF ACS reagent 99% (Merck Indonesia), 1-Octanol 99% (Merck Indonesia). Meanwhile, the equipments used were hot plate (Yamato MH-800), vacuum pump (Rocker 3000), Three neck flask 1000 mL (Duran), reflux condenser glassed, magnetic stirrer, rubber tube, thermometer 150°C, beaker glass 500 mL (pyrex), erlenmeyer 250 mL (pyrex).

### *2.2. Biolubricant synthesis*

Two stages of transesterification were carried out for fatty acid octyl esters (biolubricant) synthesis. The first transesterification was the synthesis of fatty acid ethyl esters (FAEE). The second transesterification was the synthesis of biolubricant.

#### *2.2.1. The first transesterification (synthesis of FAEE).*

The first transesterification was the reaction between WCO and ethanol to produce FAEE as below:

$$\underset{\text{Fatty acids (triglycerides)}}{CH_2COOR\text{-}CH_2COOR'\text{-}CH_2COOR} + \underset{\text{Ethanol}}{3\ C_2H_5OH} \xleftrightarrow{NaOH} \underset{\text{FAEE}}{3\ R\text{-}COOC_2H_5} + \underset{\text{Glycerol}}{C_3H_5(OH)_3} \quad (1)$$

The free fatty acid (FFA) of WCO was 1.06% so that could be directly transesterification but if FFA of WCO was 2%, it should be esterification before transesterification [13]. The synthesis of FAEE was done in a three neck flask 1000 mL with relux condenser. WCO was added ethanol with molar ratio (1:6) and NaOH catalyst amount of 0.5%-w/w. The mixture was reacted at temperature of 65°C for 90 min with stirring. The result of reaction was moved into separating funnel and set aside within 18 h so that formed two layers. The upper layer was FAEE and bottom layer was glycerol. The FAEE and glycerol were separated. The FAEE was washed with hot water (50°C) to remove residual ethanol, catalyst and soap. The FAEE was purified at temperature of 110°C to remove residual water and ethanol. Finally, the purified FAEE was obtained to be raw material for the second transesterificaton.

*2.2.2. The second transesterification (synthesis of biolubricant).*
The second transesterification was the reaction between FAEE and 1-octanol to produce biolubricant as below:

$$\underset{\text{FAEE}}{R\text{-}COOC_2H_5} + \underset{\text{1-Octanol}}{C_8H_{17}OH} \overset{KOH}{\longleftrightarrow} \underset{\text{Biolubricant}}{R\text{-}COOC_2H_5C_8H_{17}} \quad (2)$$

The synthesis of biolubricant was done in three neck flask 1000 mL with the reflux condenser which was connected with a vacuum pump. The FAEE was added 1-octanol with molar ratio (1:4) and KOH catalyst amount of 1%-w/w. The mixture was reacted at conditions (temperature = 65, 85, 105, 125 and 135°C, time = 60, 120, 180, 240 and 300 min, THF = 0, 25, and 50%-w/w). The mixture was stirred with 300 rpm at vacuum pressure of 100 mmHg. The result of reaction was moved into separating funnel. The biolubricant was washed with hot water (80°C) to remove residual 1-octanol, catalyst and soap. The biolubricant was then purified at temperature of 200°C to remove residual water and 1-octanol. Finally, the purified biolubricant was collected for analysis.

*2.3. Analysis of biolubricant*
Biolubricant were analyzed by Fourier Transform Infrared Spectroscopy (FTIR) Shimadzu Prestige 6400. This analysis was carried out to determine the functional groups of fatty acid octyl ester based on the wavelength. So the compounds formed in product could be known by comparing them with functional groups in the related references. Basically, the FTIR analysis is carried out to prove that the reaction has been completed to produce the biolubricant.

*2.4. Determination of the physico-chemical properties of biolubricant*
The physico-chemical properties of biolubricant analyzed in this study were density, kinematic viscosity at 40°C and 100°C, viscosity index and acid number. Density was determined by pycnometer 5 mL (IWAKI) according to the ASTM D1217. Kinematic viscosity at 40°C and 100°C was determined by ostwald viscometer according to the ASTM D445 standard method. Viscosity index was calculated according to the ASTM D2270 standard method. Acid number was determined by titration method according to the D974 standard method.

## 3. Results and discussion
*3.1. Functional groups of biolubricant*
The FTIR analysis results obtained spectrum is given in Figure 1. FTIR spectrum compares functional groups of 1-octanol, WCO, FAEE and biolubricant as shown in Table 1. Figure 1 and Table 1 show the difference of functional groups between 1-octanol, WCO, FAEE and biolubricant. The hydroxide group (-OH) as alcohol group appears at 3323.35 $cm^{-1}$ found in 1-octanol. It is not found in WCO, FAEE and biolubricant. The absence of the –OH in biolubricant indicates that the reaction has converted 1-ocanol to biolubricant successfully. It also indicates that biolubricant does not contain the residual 1-octanol KOH catalyst and water. The purification has been successful. The presence of the-OH is not expected in biolubricant. The smaller the –OH in biolubricant indicates that reaction process was close to completion [3].

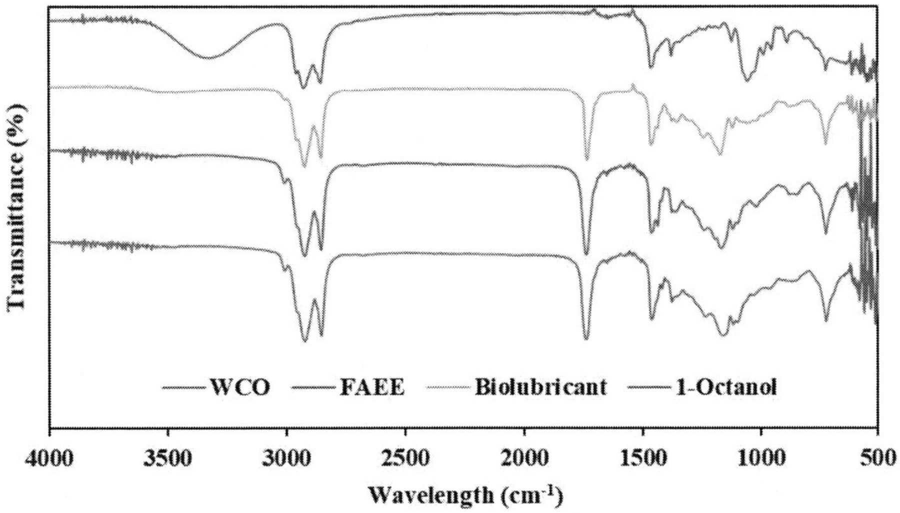


**Figure 1.** FTIR spectrum of waste cooking oil (WCO), fatty acid ethyl esters (FAEE), fatty acid octyl esters (biolubricant) and 1-octanol.

**Table 1.** Function group of waste cooking oil (WCO), fatty acid ethyl esters (FAEE), fatty acid octyl esters (biolubricant) and 1-octanol.

| | Wavelength ($cm^{-1}$) | | | | |
|---|---|---|---|---|---|
| Function group | Reference | WCO | FAEE | Biolubricant | 1-Octanol |
| -OH (alcohol) | 3326[a] | - | - | - | 3323.35 |
| C=O (ester) | 1744[a] | 1743.65 | - | - | - |
| | 1741[b] | - | 1741.72 | - | - |
| | 1738[c] | - | - | 1737.86 | - |
| C-O (ester) | 1168[b] | 1159.22 | 1166.93 | 1170.79 | - |
| | 1056[c] | - | - | 1056.99 | - |
| C-O-H (Alkohol) | 1050[d] | - | - | - | 1053.13 |

[a]Heikal et al. [3]
[b]Ivan-Tan et al. [15]
[c]Chowdhury et al. [8]
[d]Tahari et al. [16]

The ester group (C=O) appears in WCO at 1743.65 $cm^{-1}$, FAEE at 1741.72 $cm^{-1}$ and biolubricant at 1737.86 $cm^{-1}$. The change at 1743.65$cm^{-1}$ to 1741.72 $cm^{-1}$ indicates that the FAEE is obtained from the first transesterification. There is a significant change from 1741.72 $cm^{-1}$ to 1737.86 $cm^{-1}$ indicates that the fatty acid octyl esters (biolubricant) is obtained from the second transesterification. The wavelength of 1737.86 $cm^{-1}$ is the specific wavelength of C=O (ester) group for fatty acid octyl esters because it only appears in 1-octanol based biolubricants [8,14].
The function group of C-O (ester) also appears in WCO at 1159.22 $cm^{-1}$, the FAEE at 1166.93 $cm^{-1}$ and biolubricant at 1170.79 $cm^{-1}$. But, the C-O (ester) at 1056.99 $cm^{-1}$ only appears in biolubricant. The wavelength of 1056.99 $cm^{-1}$ is the specific wavelength of C-O (ester) group for 1-octanol based biolubricants [8,14].

*3.2. Effect of THF and reaction time on yield of biolubricant*
The effect of THF and reaction time on yield of biolubricant at the temperature reaction of 125°C is shown in Figure 2. The result shows that reaction time can increase and decrease yield of biolubricant as well as shown in Figure 2. The reaction time of 60 to 180 min can increase yield of biolubricant because the FAEE and 1-octanol have time to react completely. While at the reaction time of 240 and 300 min, the yield of biolubricant obtained tends to decrease. This decrease is because the reaction has reached the equilibrium, at reaction time of 180 min. Furthermore, this decrease is also caused by the formation of a few soaps at high reaction temperature during the longer of reaction time [15]. In this study, the optimum reaction time obtained is 180 min which the yields of biolubricant obtained are 66.57% (without THF), 94.42% (THF 25%) and 83.25% (THF 50%).

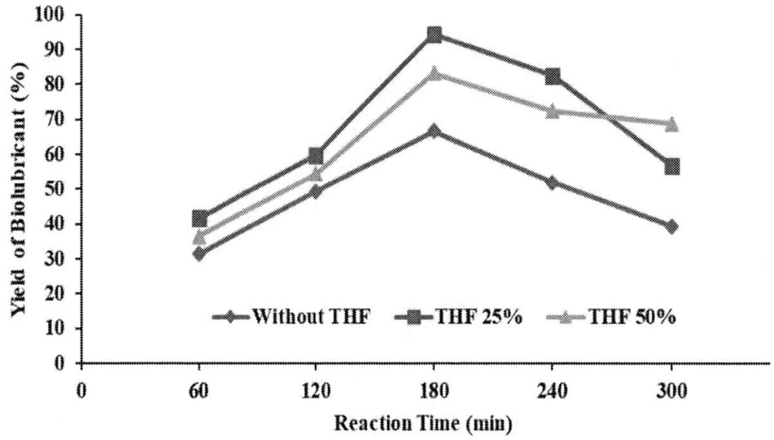


**Figure 2**. Effect of THF and reaction time on yield of biolubricant (molar ratio of FAEE:1-Octanol = 1:4, temperature = 125°C, catalyst 1%-w/w, vacuum pressure 100 mmHg).

The presence of THF can increase yield of biolubricant from 66.57% (without THF) to 94.42% (THF 25%). THF can increase yield of biolubricant by increase the solubility of FAEE and 1-octanol so that the reaction becomes more effective. But, the addition of THF excessively can also decrease yield of biolubricant from 94.42% (THF 25%) to 83.25% (THF 50%). The presence of THF excessively can disturb particle reaction collision between FAEE and 1-octanol. This is due to the effect of excessive dilution of THF [12]. A similar trend was reported by Satriana et al. [17] for glycerolysis reaction. In this study, the optimum addition of THF obtained is 25%-w/w of FAEE.

*3.3. Effect of THF and reaction temperature on yield biolubricant*
The result is obtained that reaction temperature can increase yield of biolubricant at time reaction 180 min as shown Figure 3. The reaction temperature of 65 to 125°C can increase yield of biolubricant from 23.06 to 66.57% (without THF) and 34.94 to 94.42% (THF 25%). This increase is due to the increasing reaction temperature which can increase reaction rate and accelerate reactant particle collisions. At the temperature of 135°C, yield of biolubricant has decreased from 66.57 to 40.36% (without THF) and 94.42 to 27.42% (THF 25%). This is due to 1-octanol has reached the boiling point at temperature of 135°C and vacuum pressure of 100 mmHg [18]. The evaporation of 1-octanol and THF can decrease driving force of reaction so that resulted yield of biolubricant is decreased [15]. In this study, the highest yield of biolubricant is obtained at reaction temperature of 125°C.

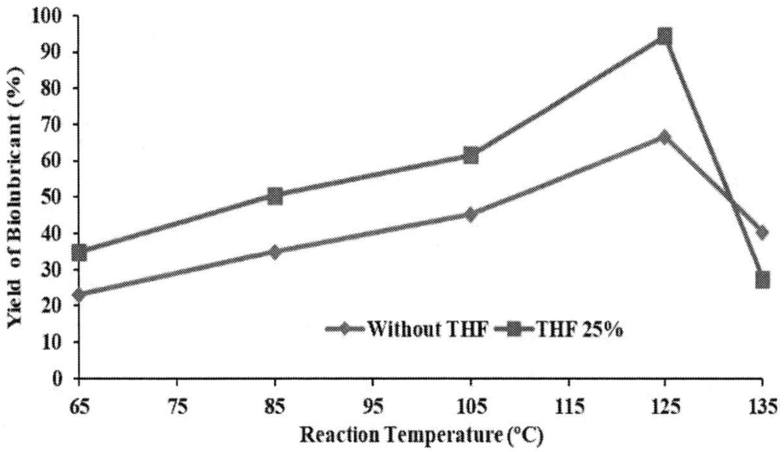


**Figure 3**. Effect of THF and reaction temperature to yield of biolubricant (molar ratio of FAEE:1-Octanol = 1:4, time = 180 min, catalyst 1%-w/w, vacuum pressure 100 mmHg).

### *3.4. Physico-chemical properties of biolubricant*

The pysico-chemical properties of biolubricant include density (40°C), kinematic viscosity (40°C), kinematic viscosity (100°C), viscosity index and acid number as shown in Table 2. Table 2 shows the comparison of biolubricant and lubricant (ISO VG 10). Kinematic viscosity of biolubricant is higher than lubricant. Viscosity index of biolubricant is also higher than lubricant. These results indicate that lubricating biolubricant is better than lubricant. But acid number of biolubricant is higher than lubricant. This is affected by vegetable-based oil sources that contain the high FFA. Another reason is there is no additive added in biolubricant which makes it easily oxidized.

**Table 2**. Physico-chemical properties of biolubricant.

| Physico-chemical properties | Biolubricant | Lubricant (ISO VG 10) |
|---|---|---|
| Density at 40°C (g/mL) | 0.874 | 0.888 |
| Kinematic viscosity at 40°C ($mm^2/s$) | 10.47 | 10 |
| Kinematic viscosity at 100°C ($mm^2/s$) | 2.75 | 2.60 |
| Viscosity index | 103 | 90 |
| Acid number (mg KOH/g) | 0.62 | 0.30 |

## 4. Conclusion

The intensification of biolubricant synthesis from WCO using THF as co-solvent was successfully obtained. The specific function group of C=O (ester) and C-O (ester) from fatty acid octyl esters (biolubricant) was obtained at 1737.86 $cm^{-1}$ and 1056.99 $cm^{-1}$ respectively. The highest yield of biolubricant in this study was 94.42% in reaction conditions (THF 25%, 125°C, 180 min and 100 mmHg). The optimal presence of THF as co-solvent had a significant effect on the quality and quantity of biolubricant. The physico-chemical properties of biolubricant were better than lubricant (ISO VG 10).

## References

[1] Widyawati Y, Suryani A and Romli M 2014 *Int. J. Renew. Energy Dev.* **3** 188-
[2] Madankar C S, Dalai A K and Naik S N 2013 *Ind. Crops Prod.* **44** 139
[3] Heikal E K, Elmelawy M S, Khalil S A and Elbasuny N M 2017 *Egypt J. Pet.* **26** 53
[4] Panchal T M, Patel A, Chauhan D D, Thomas M and Patel J V 2017 *Renew. Sustain. Energy Rev.* **70** 65
[5] Syahir A Z, Zulkifli N W M, Masjuki H H, Kalam M A, Alabdulkarem A, Gulzar M, Khuong L S and Harith M H 2017 *J. Clean. Prod.* **168** 997
[6] Panchal T, Chauhan D, Thomas M and Patel J 2015 *Ind. Crops Prod.* **63** 48
[7] Gryglewicz S, Stankiewicz M, Oko F A and Surawska I 2006 *Tribol. Int.* **39** 560
[8] Chowdhury A, Mitra D and Biswas D 2013 *J. Chem. Technol. Biotechnol.* **88** 139
[9] Wang E, Ma X, Tang S, Yan R, Wang Y, Riley W W and Reaney M J T 2014 *Biomass and Bioenergy* **66** 371
[10] Guan G, Sakurai N and Kusakabe K 2009 *Chem. Eng. J.* **146** 302
[11] Supardan M D, Adisalamun, Lubis Y M, Annisa Y, Satriana and Mustapha W A W 2017 *Pertanika J. Sci. & Technol.* **25**, 1203
[12] Fransiska F, Felicia and Taslim 2014 *Proc. of The 5th Sriwijaya Int. Seminar on Energy and Environmental Science and Technology (Palembang, Indonesia, 10-11 Semptember 2014)* Vol 1 27
[13] Abdullah N H, Hasan S H and Yusoff N R M 2013 *Int. J. Mater. Sci. Eng.* **1** 94
[14] Chowdhury A, Chakraborty R, Mitra D and Biswas D 2014 *Ind. Crops Prod.* **52** 783
[15] Ivan-Tan C T, Islam A, Yunus R and Taufiq-Yap Y H 2017 *J. Clean. Prod.* **148** 441
[16] Tahari M N A, Samidin S, Yarmo M A, Salih N and Salimon J 2015 *Malaysian J. Anal. Sci.* **19** 118
[17] Satriana, Arpi N, Supardan, M D, Gustina, R T and Mustapha W A W 2018 *AIP Conf. Proc.* **1940** 020100
[18] Liley P E, Thomson G H, Daubert T H and Buck E 1997 *Perry's Chemical Engineers' Handbook 7th Edition* ed R H Perry et al. (United States of America: McGraw-Hill) section 2 27

# Palm oil mill fly ash as a low-cost adsorbent for Rhodamine-B removal from industrial wastewater

**J A Pinem[1], P S Utama[1], R S Irianty[1], B A Prawiranegara[2], and Edy Saputra[1*]**
[1]Department of Chemical Engineering Universitas Riau, Pekanbaru 28293, Indonesia
[2]Department of Physics, Faculty of Mathematics and Natural Sciences
Universitas Riau, Pekanbaru 28293, Indonesia

E-mail: edysaputra@unri.ac.id

**Abstract.** In this study, the adsorption properties of Rhodamine-B (RB) on fly ash were investigated. Two different fly ashes conditions, including raw fly ash and activated fly ash, were selected for the adsorption experiment. The adsorption process is conducted by mixing fly ash 0.5; 1; 1,5 and 2g/L with the temperature of 30°C; 35°C; 45°C. The results showed that the HCl activation will enhance the MB adsorption capacity of fly ash, the higher the mass of adsorbent, and the higher mixing temperature, the better the performance of adsorbent is. Then, the great conditions for reducing RB with adsorbent mass 2g/L at temperature of 45°C, with a removal efficiency 90,54%

## 1. Introduction

Textile industries are one of the most thriving industries in Indonesia. Textile is promising export goods that help maintain Indonesian economy[1]. Textile industries always produce waste side products. The most prominent waste from textile industries is colored wastewater. A Dye is the important material used in the textile industry. About 15% of used dyes are not reusable and have to be disposed of as wastewater[2]. Water contamination caused by the dye industry, including food, leather, textile, plastic, cosmetics, paper making, printing, and coloring synthesis has caused more attention. The coloring agent is harmful to humans and the environment [3]. Therefore, coloring agent waste must be treated before disposal. [4] Today, a lot of attention has been paid to removing dyes from industrial wastewater.

There are no general methods to treat and remove dyes from wastewater. It all depends on dyes properties such as chemical composition, concentration, and stability of dyes compound itself. Many methods have been studied for dyes removal from wastewater such as biological, physical and chemical methods. One of those methods, adsorption is gaining more attention due to low energy consumption, highly efficient, and easiness of operation. Adsorption is a phenomenon in which the dissolved dyes molecules are attached to the adsorbent surface[5].

In the adsorption process, an adsorbent is needed which functions to adsorb the adsorbate. One of the prominent materials for adsorbents is POMFA[6]. Indonesia and other southeast Asian countries are world-leading palm oil producers for almost 89% of world production [7]. Hence, POMFA is a potential adsorbent source for treating wastewater because it is inexpensive and abundant. In this study the performance of raw POMFA and activated POMFA and adsorption mechanism on RB removal will be investigated.

* To whom any correspondence should be addressed

## 2. Materials and Methods

### *2.1 Materials*

The materials used in this study included coloring agent Rhodamine-B (RB) ,HCl, NaOH were from Merck. The POMFA was obtained from PTPN V Sei Galuh.

### *2.2. Methods*

POMFA was washed repeatedly with distilled water. After that, it was dried using an oven at 105°C for 6 hours. Dry POMFA is removed from the oven and then mashed using mortar, then sifted using a 200 mesh sieve and stored in airtight plastic. Activation treatment was carried out by mixing 10 g POMFA with 0.1M 100ml HCl. The suspension was stirred using a magnetic stirrer for 24 hours. Then, the POMFA was filtered and washed with distilled water repeatedly until it was clean of impurities and neutral pH. The solid was then dried for 6 hours with an oven at 105°C. Dry POMFA is removed from the oven and directly stored in the desiccator to be used later as an adsorbent. As the adsorbent was ready, adsorption of RB was carried out in a variety of stock solutions (artificial waste) in 1000 $cm^3$ beaker glass. Initially, add the adsorbent to the stock solution. The mixture was stirred at 240 rpm and the temperature varied from 30 to 45 °C. Samples were taken as much as 5 ml through the volume pipette in an interval of 15 minutes for 1.5 hours. Then the sample was centrifuged to segregate the adsorbent with the solution. Furthermore, the concentration of dyes was analyzed by UV-Vis spectrophotometer.

## 3. Results and Discussion

### *3.1. Effect of Adsorbent Mass (Qt)*

This experiment aims to see the effectiveness and ability of the adsorbent to absorb dyes. From an economic efficiency the lower of adsorbent needed better economic efficiency. The mass variation of the adsorbent used is 0.5; 1; 1.5 and 2gr / L. This research was conducted to determine the best conditions for the removal process of artificial textile waste (RB10ppm). The effect of the mass of adsorbent on absorbed dyes (Qt) can be seen in Figure 1.

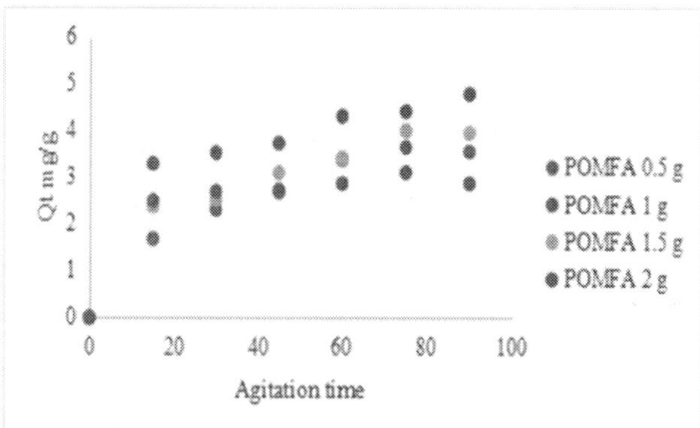


**Figure 1.** Effect of adsorbent mass on the removal of RB

From Figure 1 it can be seen that with the increase in the mass of the adsorbent, the amount of dyes that are absorbed decreases .After certain of time, RB that adsorbed gradually slowed down when it approached equilibrium condition. It might have happened due to the availability of vacant pores surface on the adsorbent. When the initial adsorbent mass is increased from 0.5 gr to 2 gr the adsorption capacity decreased from 4.5 mg/g to 2 mg/g, yet the efficiency increased from 29.18% to 62.06%. From Figure 2, the weight of the adsorbent greatly influences its absorption efficiency. Because of the greater amount of the adsorbent, the more active sites will be available. When there are more active sites available, the amount of dye absorbed will also be greater.

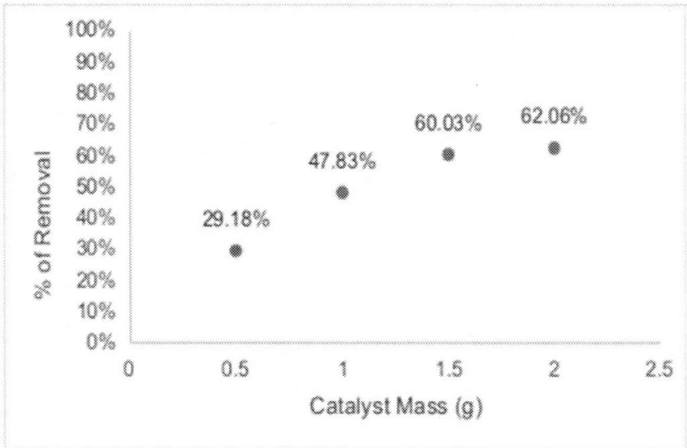


**Figure 2.** Efficiency of RB removal on various adsorbent mass

*3.2. Effect of HCl activation on POMFA adsorben*

Figure 2 shows that POMFA without HCl activation, the adsorption equilibrium occurred at the 75th minute. It was only able to adsorb the dye with the amount of 3.647mg/g. Meanwhile, for activated POMFA, the adsorption equilibrium occurs in the 60th minute with the highest number of absorbed dye is 5.837mg/g. However, the amount of RB absorbed at the time after the equilibrium time (Qe) is almost constant and slowly decreases, this is because the adsorbent has a saturation point in absorbing the adsorbate. The increase in the amount of absorbed dyes in POMFA activated by HCl can be explained as follows. The purpose of activation is to exchange the cations in the adsorbent into $H^+$ ions and release Al, Mg, Fe ions and other impurities from the crystal lattice. During the activation process, impurities found on the surface of the adsorbent which covered the active site of the adsorbent can be removed by dissolving with acid. Dissolution with acid made the adsorbent structures becoming porous so that the active site also increases. Therefore, POMFA which is activated using HCl is better used as an adsorbent compared to non-activated POMFA.

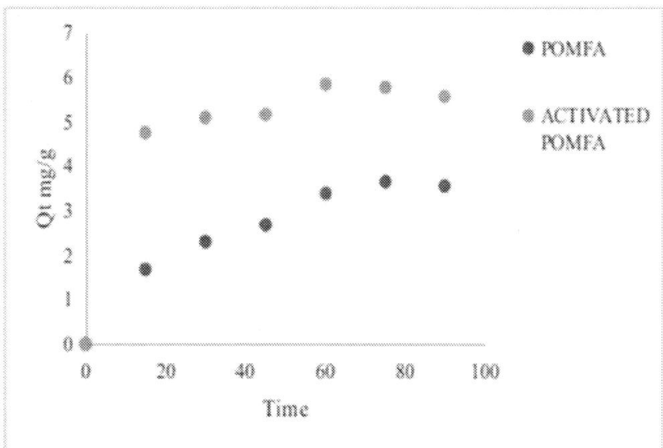


**Figure 3.** Effect of HCl activation on POMFA adsorbent

IOP Conf. Series: Materials Science and Engineering **845** (2020) 012010 doi:10.1088/1757-899X/845/1/012010

### *3.3. Effect of temperature*

From Figure 4 it shows that there is an increase in the amount of RB dyes that are absorbed in each temperature variation over time. At temperature of 30°C the amount of RB that was absorbed at equilibrium was 3.179 mg/g; at temperatures of 35°C was 3.733mg/g, and at temperature of C was 4.477mg/g. An increase in the amount of RB absorbed with increased temperature is caused by increased movement/mobility in the dye molecules. Increased temperature will increase the reactivity of RB so that more substance will interact with active sites on the surface of the adsorbent. Besides that, the greater ion reactivity will also increase ion diffusion into the adsorbent pores [8].

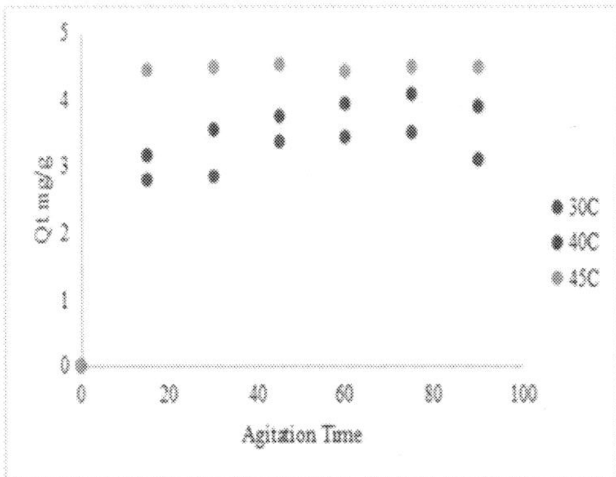

**Figure 4.** Effect of temperature on adsorption process

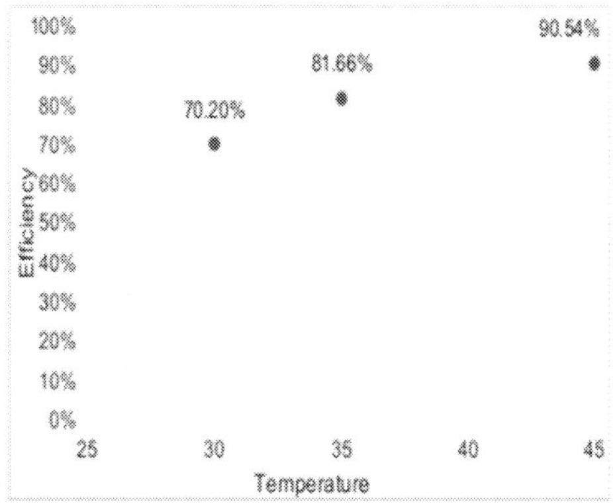

**Figure 5.** Effect of temperature on adsorption efficiency

To determine the adsorption mechanism, the adsorption equilibrium model was tested. The determination of the adsorption mechanism is important to be studied because it will explain how substances interact with the adsorbent and also explain the adsorption capacity of the adsorbent. The data fitting into the model was done using a linear regression method for each temperature variation. The equilibrium model used in this study is the Langmuir and Freundlich equilibrium models [8].The Langmuir model is based on the theory that adsorption occurs on a monolayer system and all adsorption sites at the adsorbent are homogeneous. Whereas Freundlich the isotherm states that the adsorption process occurs on a heterogeneous and multilayer surface. These two models can be described in a linear form as follows[9]:

$$Ce/Qe = 1/q0KL + 1/q0Ce \quad (1)$$

$$\ln qe = \ln KF + 1/n \ln Ce \quad (2)$$

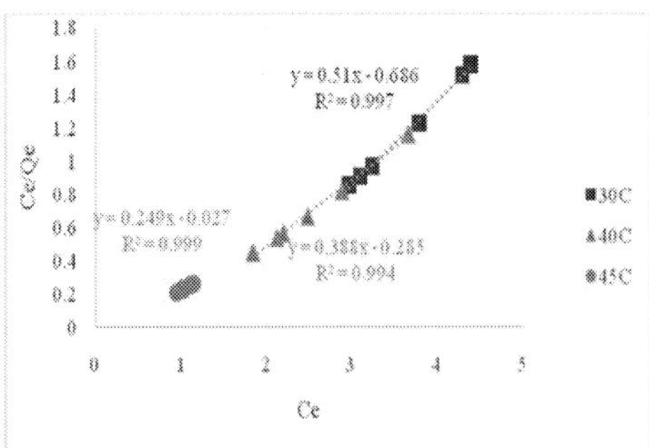

**Figure 6.** Effect of temperature on adsorption process

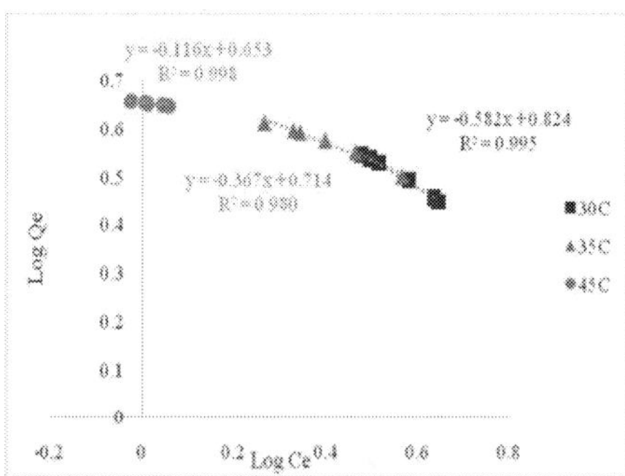

**Figure 7.** Effect of temperature on adsorption efficiency

**Table 1**. Langmuir and Freundlich constant

| Temperature | Langmuir | | | Freundlich | | |
|---|---|---|---|---|---|---|
| (°C) | Qmax (mg.g$^{-1}$) | $K_L$ (Lmg$^{-1}$) | $R^2$ | $K_F$ | n | $R^2$ |
| 30 | 1,961 | -0,743 | 0,999 | 6,677 | -1,716 | 0,995 |
| 35 | 2,576 | -1,357 | 0,9946 | 5,185 | -2,721 | 0,980 |
| 45 | 4,004 | -9,113 | 0,9976 | 4.499 | -8,561 | 0,998 |

Based on the result above, the adsorption process fits the Langmuir equilibrium models. The value of correlation factor ($R^2$) are above 0.99 on all temperatures, admits monolayer adsorption [10].

## 4. Conclusion

The best condition in the removal RB is using FA adsorbent as much as 2 g/L at 45°C with the absorption efficiency of 90.54%. The adsorption mechanism of RB by POMFA is in accordance with the Langmuir isotherm model, the value of $R^2$in the Langmuir isotherm model at temperature 30, 35, and 45C were 0.997; 0.994 and 0.999. According to those statements, the adsorption mechanism of RB on FA is a monolayer adsorption mechanism. From this study, it can be concluded that the activated FA can be used as an adsorbent for RB removal.

## Reference

[1] Arief S and Roza PP 2015 *Jurnal Riset Kimia* **1** 64
[2] Güzel F, Sayğılı H, Sayğılı G A and Koyuncu F 2015 *Journal of Molecular Liquids* **206** 244
[3] Allouche F N, Yassaa, N and Lounici H 2015 Procedia Earth and Planetary Science **15** 596
[4] Fu J, Chen Z, Wang M, Liu S, Zhang J, Zhang J and Xu Q 2015 Chemical Engineering Journal **259** 53
[5] Postai D L, Demarchi C A, Zanatta F, Melo D C C and Rodrigues C A 2016 Alexandria Engineering Journal **55(2)** 1713
[6] Falahiyah F 2015 Doctoral dissertation, Universitas Islam Negeri Maulana Malik Ibrahim
[7] Utama P S, Yamsaengsung, R and Sangwichien, C 2016 Key Engineering Materials **673** 183
[8] Dawood S, Sen T K and Phan C 2014 Water, Air, & Soil Pollution **225**(1) 1
[9] Huang R, Liu Q, Huo J and Yang B 2017 Arabian Journal of Chemistry **10(1)** 24
[10] Banerjee S and Chattopadhyaya M C 2017 Arabian Journal of Chemistry **10** S1629

# Main engine calculation for ferry Ro-Ro 500 DWT ship using electric propulsion

**A. Zulfaidah[1*], Sarwoko[1], Suharto[1], Budhi Santoso[2] and Romadhoni[2]**

[1] Departement of Industry, Constrution and Ship Building Engineering Program, Vocational School, Universitas Diponegoro Semarang, Indonesia
[2] Departement of Naval Architecture Polytecnic of Bengkalis Riau, Indonesia

E-mail: zariany@live.undip.ac.id

**Abstract.** The development of the current design of electrical system to protect the environment in the world. The purpose of this research is to obtain an electric propulsion system and engine power needed by the Ferry RoRo 500 DWT ship. The analysis begins by calculating the total ship resistance using the software. Next, the calculation of determining the generator power is calculated by a mathematical method. Resistance of 500 DWT RORO ferry at 12 knots 67.7 kN service speed. The increase in ship resistance is affected by the volume of the hull below the waterline. The greater the volume the greater the value resistance ship to get. The power play engine generator required to fight the value of ship resistance at the ferry RORO service speed of 984.49 kW.

## 1. Introduction

Ship planning at this time has been developed using installed electrical and electronic devices. The development of the ship's electrical system is getting more detailed, where the electrical system has become part of the propulsion system. While the ship's drive system by using electricity is still limited to alternatives. The electrical system is used on special vessels with the best level of comfort. Cruise ships, warships, unmanned ships, and other special ships are slowly developing electric propulsion systems. The introduction of electric propulsion systems into marine transportation provides the benefits such as the reduction of environmental pollution and improvement of life quality standards in general.

One of the main problems for ship designers is concerning the determination of power requirements for electric power. The use of the main driving force of the ship varies greatly on each engine maker. To reduce power losses, the planner needs to consider the following main factors: optimization of the hull design that results in the least resistance of the ship; choose the type of material that is economical; the shape of the ship's bow that is efficient against waves from the front.

## 2. Method

Efficiency has a function as power to support the performance of the propulsion Should [1]. Based on [2], [4], power can be formulated with effective PE power to move a ship equal to the total thrust force, the resistance of the ship R against the ship's movement and speed v. Effective Horse Power (EHP) is the amount of power needed to overcome the drag force of the hull, so that the ship can move from one place to another with a service speed of Vs.

$$PE = Rtservice \times Vs \quad (1)$$

Where:

* To whom any correspondence should be addressed.

PE = Power Effective (EHP)
Rt = Resistance Total (kN)
Vs = Speed Service (Knots)

Delivery Horse Power (PD) is the power absorbed by the propeller vessel to produce Push Power, or in other words DHP is the power in the channel by the motor motion to the propeller which then ship the conversion into a propelling force. Power in the propeller shaft or DHP tube is calculated from the ratio between the Effective Power or EHP with the Propulsive Coefficient or PC.

$$PD = PE/\eta \quad (2)$$

Where:
PD = Delivery Horse Power (DHP)
PE = Power Effective
η = Effeciency

Shaft Horse Power (SHP) is the measured power to the area in front of the stern tube bearing of the ship's propulsion system. Here the ship has an engine room at the rear, with loss (2-3) %, taken at 2%. So the value of the efficiency of bearings and vane tubes or η.S.η.B is 0.98.

$$SHP = PD/\eta s\eta b \quad (3)$$

Where:
SHP = Shaft Horse Power (SHP)
PD = Delivery Horse Power (DHP)
ηsηb = Efficiency of bearings (0.98)

Brake Horse Power (BHP) is brake power or power received by the axle transmission shaft drive system, which is then operated continuously to move the ship at its service speed (Vs). The amount of power the primary engine or PB needed in planning the propeller and the tube axis of the propeller is not separated by the value of efficiency system gears and transmission or ηG this as planned in relation the power transmission system between the main engine with propeller shaft installed system reduction gear. The gear system on this ship is planned to use a single reduction gear or Suckle Reduction Gear with a loss of 2%. For forwarding and reverse gear or Reversing Gear with 1% loss. From the data, the system can know the value of the efficiency of the transmission gear system or ηg of each system are: ηg = 98% for a single reduction gears.
Where the total coefficient of energy conversion is driving the ratio between the power coming out of the propeller and the electrical energy input used by the ship's propulsion. Total efficiency there are six main components [5]:

$$\eta = \eta C + \eta M + \eta G + \eta S + \eta P + \eta H \quad (4)$$

Where:
ηC = Efficiency of electronic power converter which gives input to the electric drive (0,65 to 0,99)
ηM = Electric motor drive efficiency (0,45 to 0,97)
ηG = Gearbox efficiency (0,95 to 0,97)

$\eta S$ = Shaft line efficiency 0,97 to 0,98)
$\eta P$ = Hull efficiency (0,98 to 1,4)
t = Thrust deduction fraction measured experimentally (0,1 to 0,3)
w = Wake fraction the vessel parameter (0,2 to 0,45)

The decrease in efficiency will affect the speed of the ship. Losing the efficiency value will decrease the speed down by 0.5 from the standard value [6]. Brake Horse Power (SCR) is the output power of the motors in the condition of Continues Service Rating.

$$BHPscr = SHP/\eta g \quad (5)$$

Where:
BHPscr = Brake Horse Power Min (HP)
SHP = Shaft Horse Power (HP)
$\eta g$ = Efficiency gears box

The magnitude of the power of the main driving motor or main motor is the output power on a normal cruise or SCR, where the amount is 80% - 85% of the output power at maximum conditions or MCR. While the output power at the MCR conditions are as follows:

$$BHPmcr = BHPscr/0.85 \quad (6)$$

Where:
BHPmcr = Brake Horse Power Max (HP)
BHPscr = Brake Horse Power Min (HP)

BHP (MCR) has become the da pat is used as a reference in implementing pr OSes elections motor Engine Selection Process. Ship resistance parameters are directly related to the hull when moving. In the design of the ship the main procedures required compatibility between the hull and the ship's propulsion system. Traditionally factors in ship resistance are ship hull obstacles, wave resistance, air resistance, and ship speed. The general formula for ship resistance can be seen in the following [4]:

$$F = F_L + F_d = F_w + F_A + m\, dv/dt \quad (7)$$

Where:
F = Total force requested from the propeller;
$F_L$ = Static hull resistance;
$F_d$ = Dynamic resistance;
$F_w$ = Water resistance;
$F_A$ = Air resistance;
m = Ship displacement;
v = Ship velocity versus water (1 to 50 m/s [3])

In stable ship conditions, just rely on static strength. But when the ship maneuvers, the burden will dramatically increase. Resulting in the sudden fulfillment of electrical power causes disruption to the electrical unit. It could be said to exceed the ship's weight when sailing. The application of propulsion electric system needs to be considered the thrust generated by the propeller. The thrust will be greater to rotate the propeller to reach the desired speed. Comparison with the conventional electric propulsion system is very efficient compared to diesel and turbine in dynamic conditions. The ship's maneuverability is increasing well and in a dynamic ship position. Especially when the ship is at the dock either leaning or reducing the level of collision with other ships. As shown [3] the torque and speed relationship curves are significantly different. Following formula (1)-(4) the condition for power is not proportional to the ideal propeller rotation speed in formula (7).

## 3. Results and Discussion

Case study of electric power propulsion estimation of the main engine ferry Roro 500 DWT. This ferry has an overall length of 45.50 meters; ship width of 12 meters; ship height of 3.2 meters; 2.15 meters ship draft; speed of 12 knots. The ferry is equipped with front and rear ramdoor doors. Ferry passanger capacity is 202 persons; crews 18 persons; car 12 trucks and 7 sedans. The calculation starts from making the ferry model into the MaxSuft education software. Model making is carried out to obtain ferry hydrostatic data. Data from model making can be seen in table 1 below.

**Table 1.** Ferry RORO 500 DWT Hydrostatic data

| Item | Velue | Unit | Holtrop |
|---|---|---|---|
| LWL | 37,802 | m | 37,802 |
| Beam | 11,802 | m | 11,802 |
| Draft | 1,95 | m | 1,95 |
| Displaced volume | 565,06 | m^3 | 565,06 |
| Wetted area | 440,393 | m^2 | 440,393 |
| Prismatic coeff. (Cp) | 0,753 | | 0,753 |
| Waterpl. area coeff. (Cwp) | 0,867 | | 0,867 |
| 1/2 angle of entrance | 36,9 | deg. | 36,9 |
| LCG from midships(+ve for'd) | -1,911 | m | -1,911 |

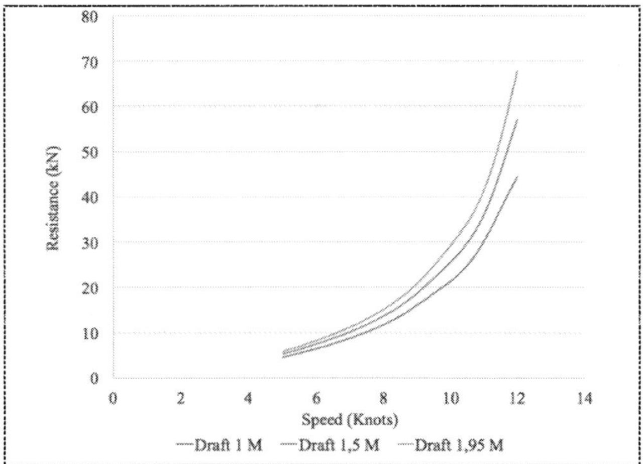


**Figure 1.** Resistance of the curve ship ferry 500 DWT

Determination of engine power is done by using hull speed with 80% hull efficiency. To get a detailed analysis, the division of ship prisoners calculation is done. Calculation of prisoners varies based on the draft draft height. In detail can be seen in the picture below. The prisoners on a 1 meter draft produce a value of 44.5 kN. For a 1,5 meter draft boat, the value of ship resistance is 57.1 kN. Whereas in the draft 1.95 meters the value is 67.7 kN. The value of ship resistance is linear. This shows the influence of the value of ship resistance is affected by the wetted area. Value wetted area of a row of draft 1 meter 330, 747 $m^2$; draft 1.5 meters 388,637 $m^2$ ; draft 1.95 meters 440.393 $m^2$.

The results of power engine calculations using the Holtrop 500 DWT ferry method. Starting from 1 meter draft 343,238 kW; draft 1.5 meters 44 0.358 kW; draft 1.95 meters 522, 427 kW. The addition of the draft fundamentally affects the engine power of the ship. The increase in engine power increases linearly with the speed of the ship. The ship's speed began to increase from 8 knots to 12 knots. The increase in speed between 10 to 12 knots there was a significant increase due to 77.503 kW to 522,427 kW at the full draft. This occurs at such speeds that the engine is in ideal working condition.

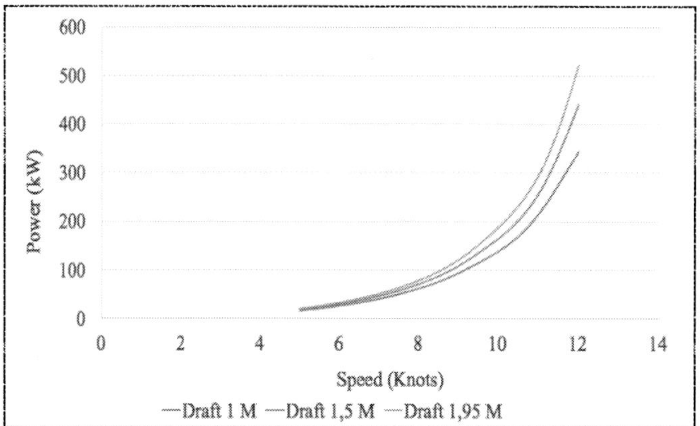


**Figure 2.** Power curve ship ferry 500 DWT

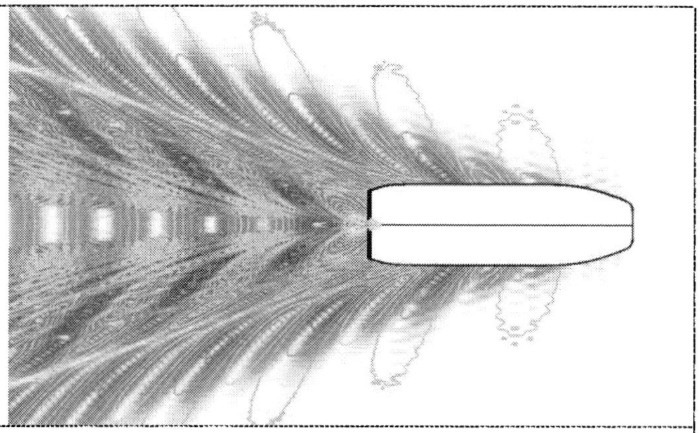

**Figure 3.** Draft wave shape 1,95 meters from ferry 500 DWT

Waveforms at the full draft and at full speed of waves generated by the laminated waveforms. Waves generated due to the shape of the hull below the surface of the water. The laminar waveform of the effect is a linear increase in speed boats. Based on the calculation of engine power using the Holtrop method obtained EHP 522.42 KW (700.58 HP). Efficiency values following formula (4) from the calculations obtained 0.67. Delivery hours power division between EHP with efficiency (2) is 1044.77 Hp. Transmission from the engine to the SHP propeller in formula (3) is 1066.09 Hp. D aya output of the motor peng ger ak on condition Services Continues Rating system transmission efficiency motors with a value of 1122.2 Hp produce. Next, the output power needed by the ship to sail (6) is 1320.23 Hp which is equivalent to 984.49 kW. This value is used as a reference in the Roro 500 DWT ferry engine selection process.

## 4. Conclusion

For energy system analysis, the propulsion must be separated from the propulsion. Analysis of the engine load must be able to convert the ship in static and dynamic conditions. While the analysis of speed, transmission system, and propeller must be able to encourage the use of engine power. Prisoners ship ferry RORO is obtained by variation of the draft vessel to 1 meter, 1.5 meters, and 1.95 meters. The ship's detainees at the full draft condition and full speed of 67.7 kN. Ship resistance on the 1.5-meter draft was 57.1 kN, while the lowest draft obtained a value of 44.5 kN. In theory, the increase in ship resistance due to changes in the volume of the body of the ship under the waterline by 565.06 m 3. The efficiency value has a very significant influence on the determination of the engine power of the ferry RORO 500 DWT. The calculation results show the need for electric generator power to be able to drive the 500 DWT ferry boat BHP mcr power is 1320.23 Hp equivalent to 984.49 kW.

**Acknowledgements**
This research work was supported by the Dean of the Vocational School of Ship building Engineering at Diponegoro University in Semarang.

**References**

[1] J. Medaković, B. Dario and B. Blagojević, 2013, *Int. J. Eng. Sci. Innov.* **2** 155
[2] C. W. Dyson, 1913, *Screw Propellers and Estimation of Power for Propulsion of Ships*, Willey & Sons, London, 60 p.
[3] A. F. Molland, S. R. Turnock and D. A. Hudson, 2011, *Ship Resistance and Propulsion*, Cambridge University Press, p 537 p
[4] V. Vodovozov, 2014, *Electrical Drive: Performance, Design and Control*, Lambert Academic Publishing, Saarbrücken, p 320
[5] Basic Principles of Ship Propulsion. Available at: https://spain.mandieselturbo.com/docs/librariesprovider10/sistemas-propulsivos-marinos/basic-principles-of-ship propulsion.pdf?sfvrsn=2, , August 2019.
[6] Determining Electric Motor Load and Efficiency. A vailable at: https://www.energy.gov /sites/prod/files /2014/04/f15/10097517.pdf, August 2019

# The physical properties, color direction and color difference of stamp ink from gambier (*Uncaria gambir* roxb) with complexing $Al_2(SO_4)_3$ compound

**S. Silfia*, S. Sofyan, and F. Failisnur**

Institution of Research and Standardization of Industry - Padang
Jl. Raya LIK No. 23 Ulu Gadut, Padang, Indonesia

Email: silfiabintiarsul@gmail.com

**Abstract.** The raw materials of stamp ink are usually made from synthetic materials and are toxic that can damage human health. This research uses gambier which is material from the local natural resource. The purpose of this study was to see the influence of $Al_2(SO_4)_3$ on the direction and difference of color, pH, and density of the stamp ink. The study was conducted in several stages, namely the making of gambier extract, the dissolving process of raw gambier with water, drying, reducing the particle size of 100 mesh, and extracting the gambier powder with an alcohol solvent. The complexing compound $Al_2(SO_4)_3$ and other supporting materials were added into the gambier extract to produce stamp ink. The stamp ink was examined the direction and difference of color, pH, and density. The optimum treatment was obtained at a concentration of 30% $Al_2(SO_4)_3$ with the results of test pH 4.053; density 1.16 g/cm$^3$; and color direction L* 41.579; a* 13.023, b* 21.132. The result of the stamp test on white paper showed that the color of the ink is visible, compact, unified and does not expand.

## 1. Introduction

Stamp ink is the liquid used to stamp a document and become an identity of the existence of an institution or organization. This identity is very useful for data validation such as diplomas, decrees, and several other important documents. The stamp ink that is on the market today is made from synthetic chemicals, some of which are poisonous and not eco-friendly. Gambier stamp ink is a product made from the local natural resources. It's a non-toxic nature, hypoallergenic, and more eco-friendly. Gambier stamp ink development will provide several benefits, namely increasing the added value of raw materials, increasing the income of gambier farmers, substituting imported products, and developing eco-friendly products.

The main ingredients of gambier are tannins (20%-50%) and catechins (7%-33%) [1]. Research about the use of gambier tannins has been carried out by earlier researchers, gambier tannins for textile dyes [2]–[5] and for leather tanner [6]. The use of gambier tannin as a raw material for ink products is due to tannin complex compounds that react with metal salts to form colors. Gambier tannins with saturated alcoholic $FeCl_3$ and saturated NaOH will produce a blue to black color. The complex compound of gambier tannin in alkaline conditions will produce a blood-red color. The reaction of tannins with $FeCl_3$ will form a complex compound as in Figure 1 [7], [8].

* To whom any correspondence should be addressed.

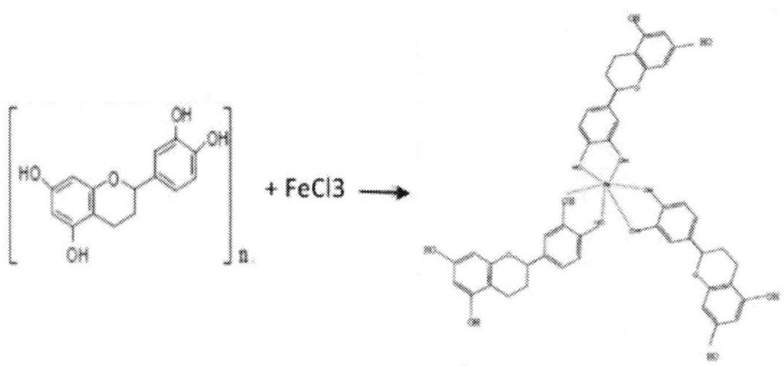

**Figure 1**. The reaction of tannin with $FeCl_3$ (Source : [9]).

The color obtained on stamp ink with saturated alcoholic complexing $FeCl_3$ and $FeSO_4$ was similar namely blue to black color. However, the ink obtained was still sticky with physical and chemical properties, namely pH (3.27), total solids (41.33%), tannins (2.58%) [1], [10]. Based on the physical properties and stamp ink deposits [11], the $Al_2(SO_4)_3$ complexing compounds are used because they have ionic bonds larger than the central atom Al ($^{+3}$). The complex bonds that are formed are expected to be more stable and the stamp ink will not coagulate. The purpose of this research was to see at the effect of $Al_2(SO_4)_3$ on the stamp ink products and to determine the optimum process by varying the concentration of $Al_2(SO_4)_3$.

## 2. Experimental

### *2.1Materials and tools*

The materials used were raw gambier (*Uncaria gambier* Roxb), ethanol for extracting fine gambier. $Al_2(SO_4)_3$ as complexing of tannin, glycerin, and distilled water. The gambier was in the dry form taken from gambier processing farmers in Harau Subdistrict, Lima Puluh Kota Regency, West Sumatra Province – Indonesia. The $Al_2(SO_4)_3$ used was in the form of white crystalline solid. It is soluble in water, non-combustible, and nontoxic. All chemicals used were technical grade which purchased from *brataco* chemical company. The equipment for analysis includes colorscan SS 6200 spectrophotometer, pH meter (HACH), and pycnometer. The research was carried out in the laboratory of the Institute for Research and Standardization of Industry - Padang, the laboratory of the Padang State University, the Laboratory of Institute of Textile Technology - Bandung.

### *2.2 Research implementation*

The process of making stamp ink was done through two stages. The first step was to make gambier extract by grinding the raw gambier and dissolve it in water with a ratio of gambier and water 1:4. The solution was heated at 60°C and then filtered to remove impurities in the raw gambier. The gambier solution is allowed to stand for 1 night and the residue was taken. The gambier residue was dried at room temperature, then ground and sieved with a 100 mesh sieve. The fine Gambier was dissolved in alcohol with ratio gambier and alcohol 1:4. It was stirred with 250 rpm and temperature 60°C for 5 hours, then precipitated for 1 night and then filtered. The filtrate was taken and the residue was removed.

In the second step, stamp ink formulations were made by varying the concentration of complexing $Al_2(SO_4)_3$, namely 15, 20, 25, 30, and 35% (v/v). $Al_2(SO_4)_3$ with the concentration according to the treatment was added to 35% (v/v) gambier extract solution and stirred at speed of 250 rpm for 2 hours. 20% glycerin (v/v) and water were then added to the ink solution while continuing to stir. The stamp ink was analyzed for its physical and chemical properties and tested for the quality of the stamp on white paper.

IOP Conf. Series: Materials Science and Engineering **845** (2020) 012012 doi:10.1088/1757-899X/845/1/012012

### *2.3 Analysis*
Analysis and observations made on the stamp ink were density, pH, direction and difference of color, and testing the stamp ink produced on the white paper to see the stamp quality.

#### *2.3.1 Analysis of pH*
A sample of 20 ml was measured by using a pH meter (*HACH*) at room temperature with constant agitation. pH shows the negative logarithm of the hydrogen ion concentration in the sample.

#### *2.3.2 Density*
Density is the ratio between the weight compared to the volume of a substance at a temperature of 25°C [12]. The density analysis was used to determine the specific gravity of a solution.

#### *2.3.3 Analysis of direction and difference of color*
Testing the color direction using the CIELAB 1976 method, namely color measurement based on three-dimensional color coordinate spaces in three axes, namely L* (brightness), a* (green - red direction), and b* (blue-yellow direction). The reading of the value L* (brightness) is 0 = black and 100 = white. Readings of a* + = red and - = green. Readings of b* + = yellow and - = blue [13], [14] . Testing the color direction of the gambier extract solution was done as a control of the stamp ink obtained by treating the percentage difference of $Al_2(SO_4)_3$.

## 3. Results and Discussion
### *3.1 Analysis results of gambier and gambier extract*
The analysis results of gambier and gambier extract are shown in Table 1. The tannin and catechin content of raw gambier used in this study was 33.25% and 51.75% respectively. Gambier with high tannin content is usually blackish brown, while gambier with high catechins content is usually yellowish brown. In the production of stamp ink, it is expected to use gambier with high tannin content. The levels of tannin in stamp ink from gambier at several concentrations of gambier extract ranged from 1.86% - 2.78% [10].

The gambier extract had tannin content 4.05% with a pH of 3.35. Tannins are compounds that play a role in textile coloring (Sofyan & Failisnur, 2016) and tanners (Ardinal et al., 2014). It is used in industry for drugs and ink as well (Muchtar, 2013). The tannin content of gambier extract is crucial in the complexes formed with $Al_2(SO_4)_3$ and produces ink dyes.

**Table 1**. Analysis results of gambier and gambier extract 35%.

| No. | Parameter | Sample | |
|---|---|---|---|
| | | Gambier (%) | Gambier extract (%) |
| 1. | Tannin | 33.25 | 4.05 |
| 2. | Catechin | 51.75 | - |
| 3. | pH | - | 3.35 |

### *3.2 Density*
The analysis results of the stamp ink density give fluctuating results (Figure 2). However, the difference in each treatment is not significant. The difference between the smallest and the largest density is 0.072. The density will affect the ink quality produced. The high density can cause the ink not to adhere properly on the paper. The density can be influenced by the particle size of gambier and supporting materials, as well as the conditions of the ink production process such as stirring speed and temperature.

The largest density was the treatment with adding 30% $Al_2(SO_4)_3$ which is 1.155 gr/cm3. This treatment was the best result with pH 3 having a more stable ink condition than other treatments. When compared to the stamp ink density values according to SNI 06-1567-1999 (minimum 1 gr/cm3), the density of the stamp ink produced has met this standard. The results of another study of ink from gambier were carried out by [15] who reported that the marker ink from gambier had a density 1.0254 g/cm³. The stamp ink in the market has a density of 1.0010 g/cm³.

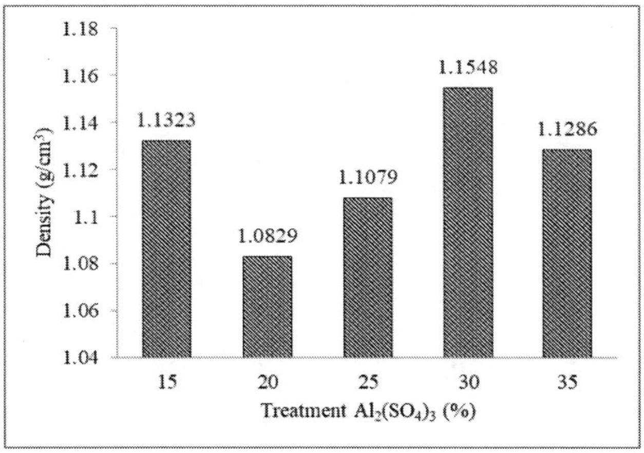


**Figure 2**. The density of stamp ink with varying concentration of $Al_2(SO_4)_3$.

*3.3 pH*

pH is the degree of acidity used to express the level of acidity or base owned by a substance, solution or object. pH value of the stamp ink produced is shown in Figure 3. The pH value of stamp ink for all treatments is acidic (3.739-4.069). The acidic pH conditions are caused by the gambier extract and the use of complexing $Al_2(SO_4)_3$ which is also acidic. Gambir contains katechu tannic acid compounds as its biggest component (20-55%) which are acidic [16]. The pH conditions will affect the color direction and color difference of stamp ink produced and can make the ink more stable and reduce clotting.

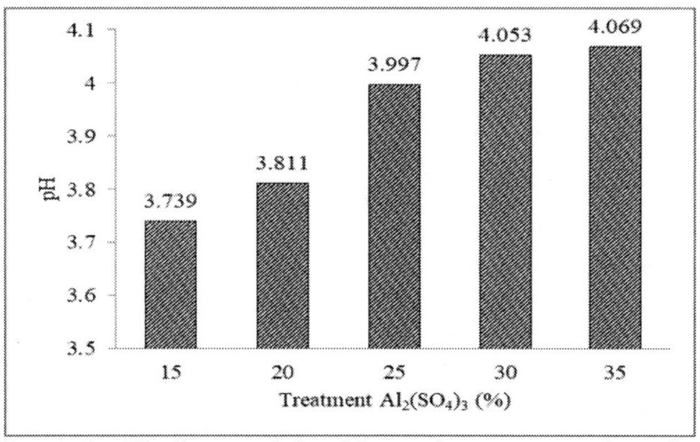


**Figure 3**. pH of stamp ink with varying concentration of $Al_2(SO_4)_3$.

One of the supporting material used in the stamp ink from gambier is glycerin. According to [10], glycerin has a neutral pH. The more glycerin added, the more easily oxidized the stamp ink formula so that the pH of the stamp ink is higher. The pH of stamp ink from gambier on the glycerin addition with several concentrations ranged from 3.39 to 3.98 [10].

*3.4 Analysis of color direction and color difference*

The analysis results of the color direction of the stamp ink show a positive value of L*, a*, and b* (Table 2). The L* value of gambier extract was 63.811 while the stamp ink ranged from 41.579-68.050. The value of a* is between 13.023-14.625. Gambier extract had b*value 34.583 and the stamp ink ranges from 21.132-40.234.

**Table 2**. Analysis results of L*, a*, b* the stamp ink with varying concentration of $Al_2(SO_4)_3$.

| Treatment | Wavelength (nm) | Absorbance | L* | a* | b* |
|---|---|---|---|---|---|
| Gambier extract | 400 | 1.422 | 63.811 | 12.104 | 34.583 |
| 15% $Al_2(SO_4)_3$ | 400 | 1.305 | 50.356 | 13.971 | 30.254 |
| 20% $Al_2(SO_4)_3$ | 420 | 1.157 | 62.354 | 13.237 | 39.987 |
| 25% $Al_2(SO_4)_3$ | 420 | 1.074 | 65.644 | 14.266 | 40.234 |
| 30% $Al_2(SO_4)_3$ | 400 | 1.323 | 41.579 | 13.023 | 21.132 |
| 35% $Al_2(SO_4)_3$ | 400 | 0.879 | 68.05 | 13.665 | 28.661 |

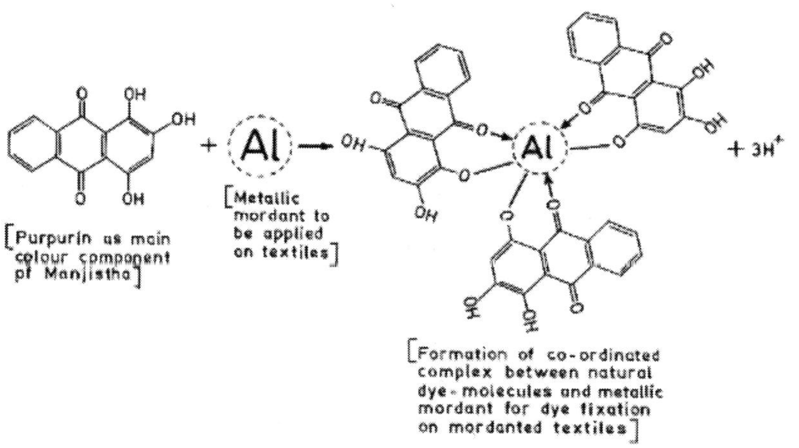


**Figure 4**. The reaction of tannin and $Al_2SO_4$ [17]

The results of color direction measurements of stamp ink and gambier extract had a reddish-yellow color. Gambier extract solution had a maximum wavelength or maximum light absorption at a wavelength of 400 nm with an absorbance value of 1.422 (Table 3). Stamp ink with treatment $Al_2(SO_4)_3$ concentration results in varying color intensity. The Color intensity can be seen from the value of absorbance at the same maximum wavelength. The stamp ink with the treatment of 15, 30, and 35% $Al_2(SO_4)_3$ have the same maximum wavelength in the visible light area which is 400 nm. However, the stamp ink with treatment 20 and 25% $Al_2(SO_4)_3$ have different maximum wavelengths that are 420 nm. This shows that in the treatment 20 and 25% $Al_2(SO_4)_3$, the batochromic effect occurs, which means a shift in wavelength from 400 nm to 420 nm.

The wavelength area 400 nm (gambier extract, 15, 30, and 35% $Al_2(SO_4)_3$, the color intensity decreases. The biggest color intensity reduction is at 35% $Al_2(SO_4)_3$ with absorbance value 0.879. The 15 and 30% $Al_2(SO_4)_3$ treatments are darker than gambier extract and other treatments, especially with treatments that have the same wavelength. This can be seen from the value of L* (brightness) which is lower than gambier extract. The measurement results of stamp ink color lead to reddish-yellow. this is in accordance by the results of the study [17] with the reaction shown in Figure 4.

**Table 3**. Color direction and color difference of the stamp ink with varying the concentration of $Al_2(SO_4)_3$.

| Treatment | Color difference | |
|---|---|---|
| 15% $Al_2(SO_4)_3$ | 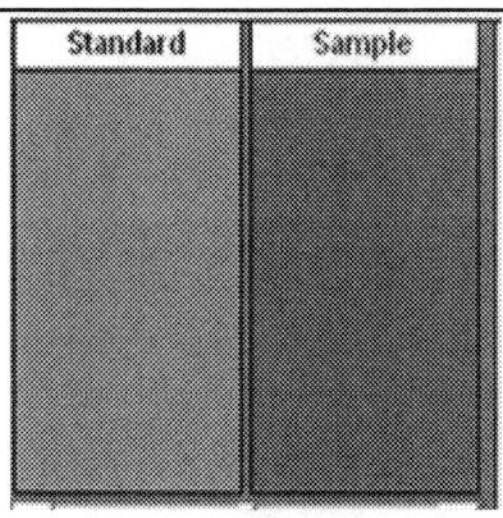  14.257 | 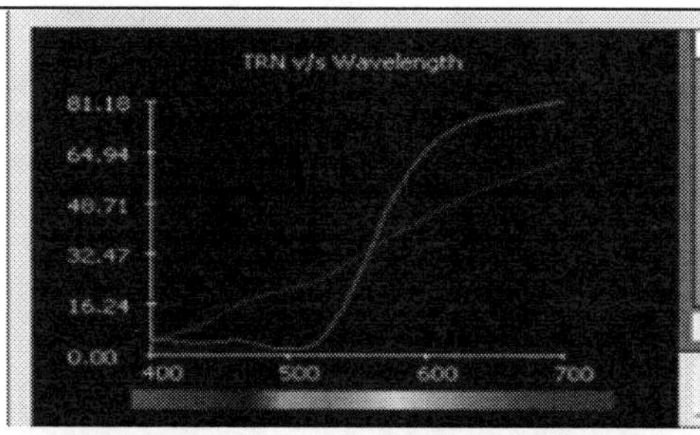  |
| 20% $Al_2(SO_4)_3$ | 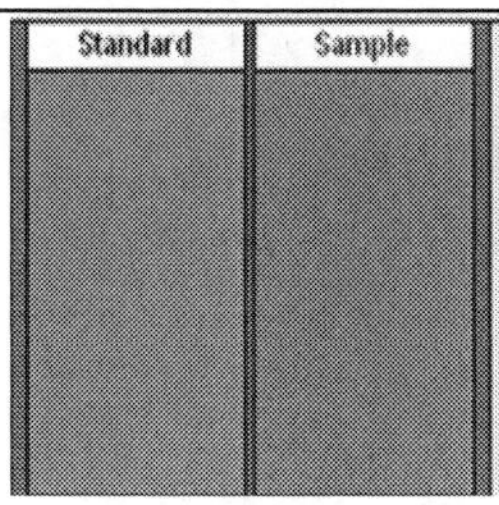  5.710 | 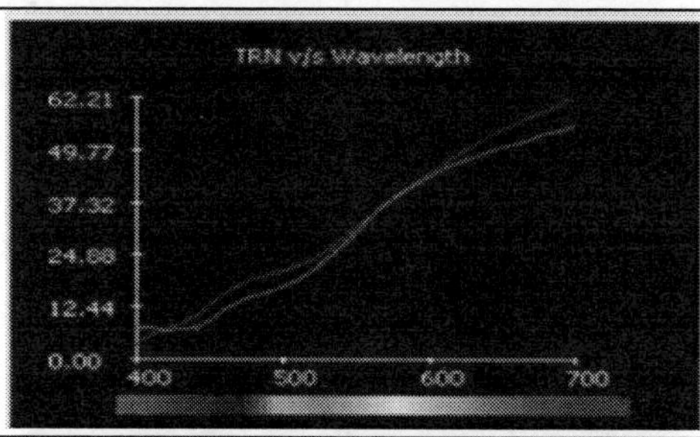  |
| 25% $Al_2(SO_4)_3$ | 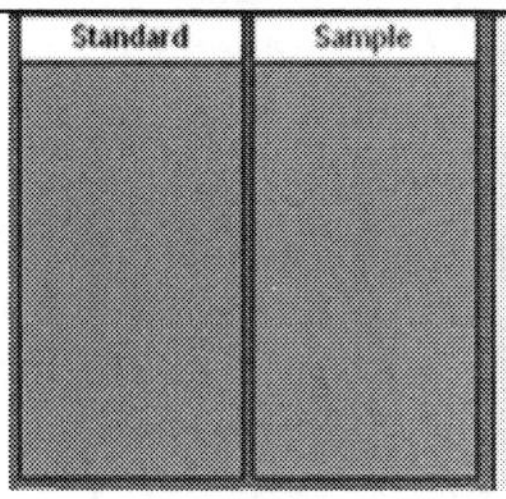  6.322 | 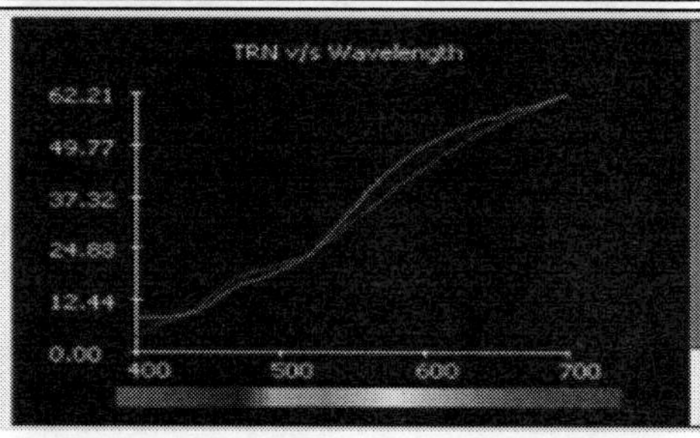  |
| 30% $Al_2(SO_4)_3$ | 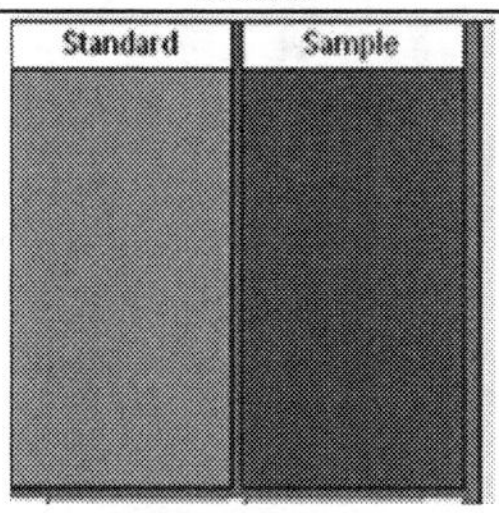  26.001 | 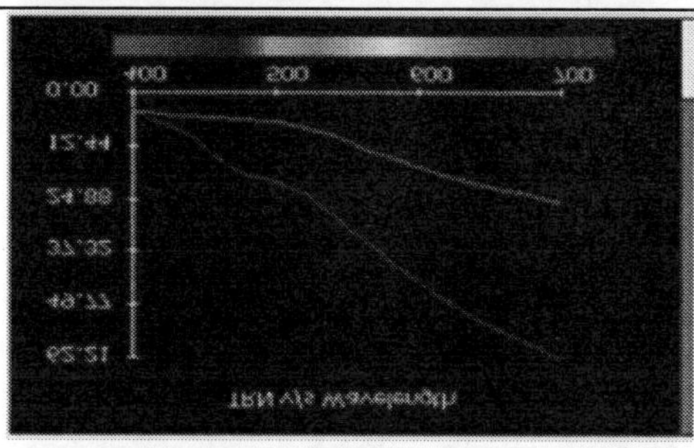 |
| 35% $Al_2(SO_4)_3$ | 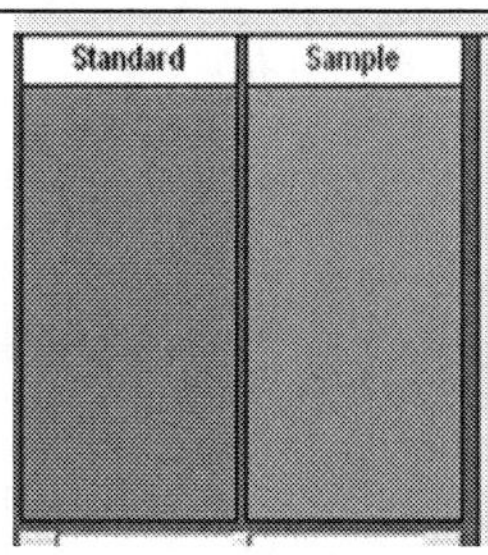  7.448 | 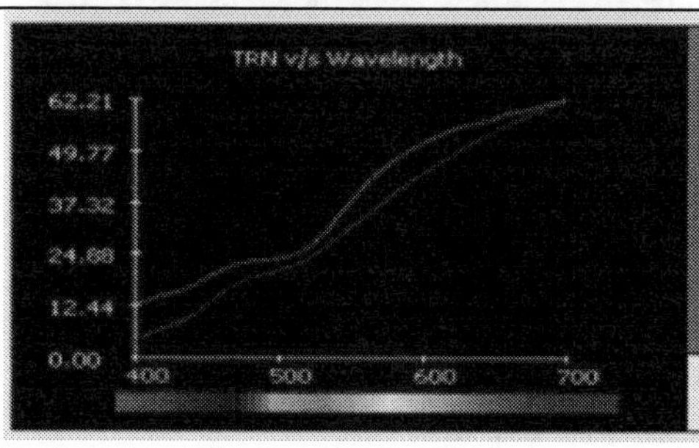  |

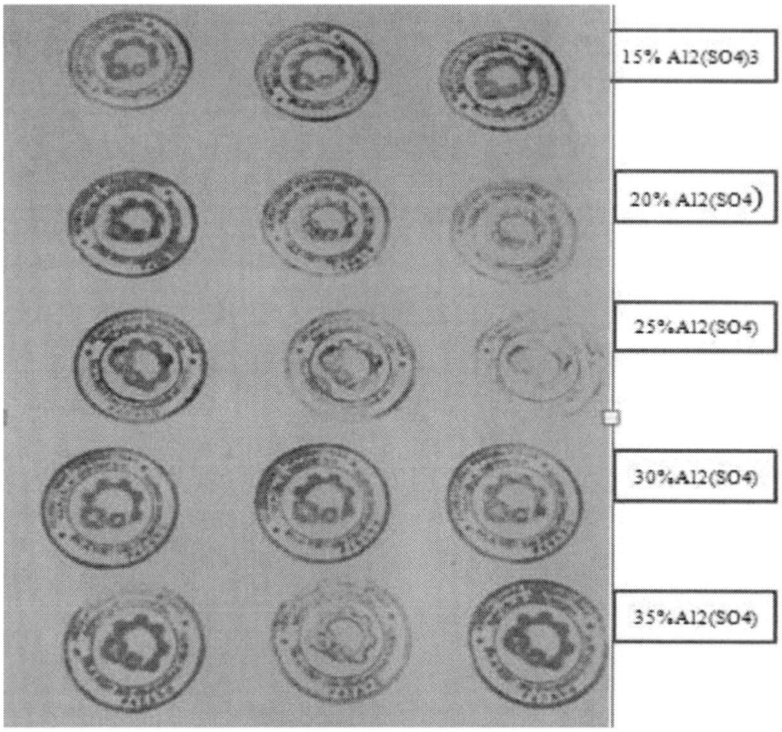


**Figure 5.** Test results of stamp ink on white paper

The reaction of tannin and $Al_2(SO_4)_3$ will form a complex compound as shown in Figure 4. The stamp ink color with $Al_2(SO_4)_3$ treatment is brownish Visually, this is in accordance with the results of the color direction testing. The color character can be seen from the value of the color difference and color direction. The color difference in the treatment $Al_2(SO_4)_3$ addition from the smallest value to the largest value is 20% $Al_2(SO_4)_3$ > 25% $Al_2(SO_4)_3$ > 35% $Al_2(SO_4)_3$ > 15% $Al_2(SO_4)_3$ > 30% $Al_2(SO_4)_3$. The colors of resulting stamp ink are shown in Table 3. The biggest color difference is 26.001 in the treatment of 30% $Al_2(SO_4)_3$. Variations in the treatment $Al_2(SO_4)_3$ concentration affected the color difference (light-dark).

Test results of stamp ink on white paper are shown in Figure 5. The results of the stamp on the treatment of 30% $Al_2(SO_4)_3$ appear to be more stable from the beginning to the third stamp compared to other treatments. The stamp is clearly visible, the ink is more compact, unified and does not expand. The stamp results of the other treatment are seen not compact, widen and not clear from the beginning of the stamp.

Stamp ink that is on the market today is made from synthetic chemicals and some of them are toxic so we need to find substitutes that are more eco-friendly. The use of gambier for stamp ink as a substitute for synthetic chemicals is an alternative with excellent prospects. Gambier is a natural organic material that is eco-friendly, renewable, and produces good stamp ink products technically. This plant has been cultivated and is available throughout the year so it is very feasible when used as raw material for the stamp ink industry. Its use as an industrial raw material will increase the added value of gambier in the market.

## 4. Conclusion

The results showed that the treatment variation of $Al_2(SO_4)_3$ concentration did not significantly influence the physical and chemical properties of the ink, but it did affect the color difference. Ink test results on white paper obtained the best results on the treatment of 30% $Al_2(SO_4)_3$ with the characteristic of pH 4.053; density 1.166 g/cm$^3$; color difference 26.001, and color direction with values L* 41.579, a* 13.023, and b* 21.132.

**Acknowledgment**
We would like to thank Hendri Yuhelminof and Awilda Mike Pertiwi for helping us to carry out the research activities, and Institution for Research and Standardization of Industry – Padang, Ministry of Industry for funding the research.

**References**

[1] Muchtar H. 2013 *J. Litbang Ind.* **3**(1) 59–65
[2] Failisnur F and Sofyan S 2016 *J. Litbang Ind* **6** (1) 25–37
[3] Failisnur F and Sofyan S 2014 *J. Litbang In.* **4** (1) 1–8
[4] Failisnur F, Sofyan S and Hermiant W 2017 *J. Litbang Ind.*, **7** (1) 19–28
[5] Failisnur and Yeni G "Sta2013 *Biopropal Ind.*, vol. 4, no. 1, pp. 7–16, 2013
[6] Ardinal, S. Salmariza, and A. Kasim 2014 *J. Litbang Ind* **4**(1) 59–66
[7] Silfia, Muchtar H and Failisnur 2015 *J. Litbang Indsutri* **5**(1) 53–60
[8] Sofyan S and Failisnur F 2016 *J. Litbang Ind* **6**(2) 89–98
[9] N. Chintya and B. Utami 2017 *JC-T (Journal Cis-Trans) J. Kim. dan Ter* **1** (1) 22–29
[10] Silfia S, Muchtar H and Failisnur F 2015 *J. Litbang Ind* **5** (1) 53-59
[11] Silfia S, Failisnur F and Sofyan S 2018 *J. Litbang Ind* **8**(1) 11–16
[12] A. L, Toruan, O. H. Kaseke, L. F. Kereh, and T. K. Sendov 2013 *J. Sipil Statik* **1**(3) 190–195
[13] Q. Zhao, H. Feng, and L. Wang 2014 *J. Clean. Prod* **80** 197–203
[14] Atika V, Farida F and Pujilestari T 2016 *Din. Kerajinan dan Batik* **33**(1) 25-32
[15] Anova I T and Muchtar H *J. Litbang Ind* **7**(2) 101-109
[16] Kasim A, Nurdin H and Mutiar S 2012 *J. Litbang Ind* **2**(2) 55-62
[17] A. K. Samanta and A. Konar *Nat. Dye* 29–56

# Where would our biodiesel stand in our technology?

**A Azizan**[1]
[1] Faculty of Chemical Engineering,
Universiti Teknologi MARA, 40450 Shah Alam, Selangor, MALAYSIA.

E-mail: amizon@uitm.edu.my

**Abstract.** Biodiesel market may be jeopardizing as compared to years before due to the change of fuel usage policies in few countries around the world. Despite that, naturally, there are still many diversified feedstocks available feasible for the production of biodiesel. Less environmental impacts by biodiesel can be approached by the use of biodiesel blends. Other possible feedstocks for the production of biodiesel with possible technological advancement are discussed aligned with the industrial revolutionary market scenario. For environmental sustainability, as Asian perspective, since palm oil as one of the biggest vegetable productions in the world, dominating the production of biodiesel and visioning the integration of IR 4.0 would be the sustaining future stand for global perspective.

## 1. Introduction

Where would diesel of biodiesel be in the next 10 years? In general, the diesel as petrol diesel, biodiesel and biodiesel blend have been of comparable use around the world since it was first introduced. In the early history of diesel, petrol diesel was extracted from petroleum source the crude oil. The higher carbon 14 to 20 ($C_{14}$-$C_{20}$) can be extracted at the temperature of 270°C in a fractionating column as seen in the figure 1a. The complex method of separating the lighter (smaller molecules) and heavier hydrocarbons (heavier molecules) conventionally is based on different boiling points or volatilities. The refinery gas, petrol, naphtha, kerosene, diesel, lubricating oil, fuel oil and residue like bitumen, are found to be useful in the production of primary and secondary petroleum products.

How is biodiesel produced? Conventionally, transesterification produces biodiesel from renewable oil like vegetable oil as the saturated long chains of fatty acids. The glycerin removal occurs from oil as separated from triglycerides. With mild temperature of 50 to 80°C, in the presence of alcohol, alkaline catalyst, 3 mono alkyl esters with glycerin are derived namely the so-called biodiesel [1] and [2].

For a technological commercial production, according to [3], biodiesel can be produced through these steps. In the flowsheet as shown by figure 1b, there are (1) the pretreatment of the vegetable oil, comprising of mixing and filtering and (2) the refining of plant oil, through degumming, neutralization and centrifugation. Impurities are removed. Then, (3) the transesterifications reaction to biodiesel and finally, (4) the ester washing and distillation of biodiesel towards as the purest biodiesel product.

Would biodiesel be still sustainable in the future and what is the future stand to the use of biodiesel or biodiesel blends? Due to the change of fuel usage policies around the world, biodiesel market maybe jeopardizing as compared to years before. With vegetable oils as the main raw materials, naturally, there are still many other diversified feedstocks available feasible for a sustainable production of biodiesel.

---

1 To whom any correspondence should be addressed.

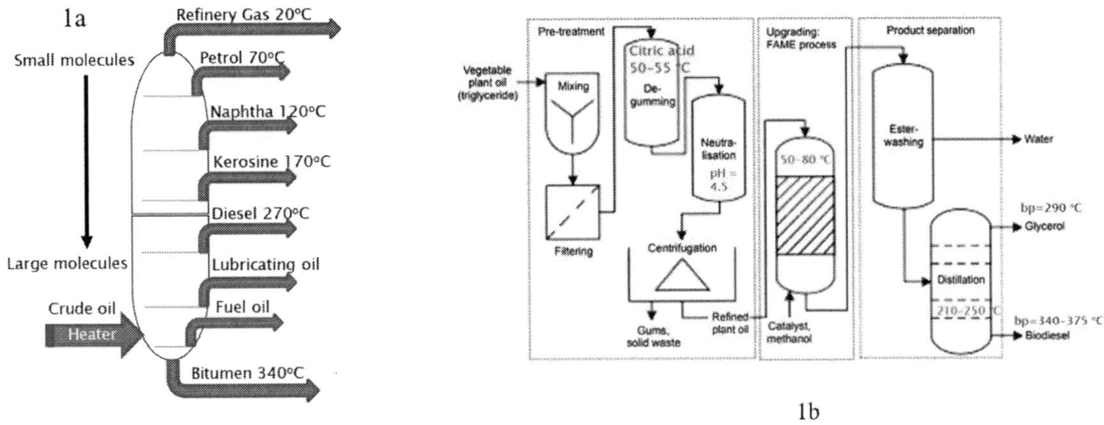


**Figure 1a and 1b.** General schematic diagram of a fractionating column of crude oil (1a: left) and general biodiesel production (1b: right) (adapted [3]).

## 2. Biodiesel materials and methods

### *2.1. Biodiesel from vegetable oil*

A portion of this biodiesel, can be blended with petrol diesel, to produce biodiesel blends, a clean burning alternative fuel, with the abbreviation of BX. For example, B10 means, 10% of biodiesel plus 90% of petrol diesel. It can be used in a diesel compression ignition engine with little or no modifications.

Historically, it started with in 1851, when Sir Samuel M Kier, discovered fractional distillation to produce carbon oil from crude oil. Two (2) years later, Sir Duffy and Sir Patrick, introduced transesterifications of triglycerides. From the year of 1893 to 1911, Sir Rudolph Diesel demonstrated his workable diesel compression ignition engine and made a declaration on the use of vegetable oil to the public. Unfortunately, in the 1920s, this diesel from vegetable oil was replaced by petrol diesel, by engine manufacturers. In 1980s, this so-called biodiesel, re-emerged as renewable fuel in Europe. At present, biodiesel is commercially produced for consumer use throughout the world.

*2.1.1. Biodiesel blends.* The types of biodiesel blends used available, ranges from B100 to B2. Most of the leading countries use B5 and B7. As for Malaysia, Envo Diesel/B5 was introduced in 2014 [4]. At present, B10 is planned to be introduced by the 1st of January 2019 [5]. Not recently, the use of B5 is reported to be used by the countries like Canada, Malaysia, Brazil, Netherlands, Poland China, Germany, Spain. Whilst, for B7, Argentina, Netherlands, Germany, France, Malaysia, are reported in usage.

*2.1.2. Biodiesel standards.* Table 1 shows the most commonly used biodiesel standards for respective biodiesel blends. As abbreviated: EN is from Europe and ASTM is from USA. EN14214, is applicable for all blends as shown. The biodiesel emission regulations range from Euro 1 (in 1991) to the most recently introduced which is Euro 6 in 2014. As for example, as shown in figure 2, Euro 5 releases of 180 mg/km diesel nitrogen oxide ($NO_x$) to that of 80 mg/km diesel $NO_x$ for Euro 6. Note that the regulations imposed on the emission of carbon monoxide (CO), $NO_x$ and particulate matter (PM) are becoming more stringent.

Ideally biodiesel, needs to have higher flash point, for safe use, and higher cetane number for higher combustion efficiency. The flash point is the temperature at which the fuel can ignite when being exposed to heat source. Cetane number is defined as the indicator of the combustion speed of diesel fuel and compression needed for ignition. As seen here, from Table 2, biodiesel from waste cooking oil successfully fulfils the specifications as stated by EN and ASTM standards.

**Table 1.** Biodiesel standard [6].

| **Biodiesel Blend** | **Europe** | **USA** |
|---|---|---|
| B2 to B5 | EN 14214 | ASTM D975 |
| B6 to B20 | EN 14214 | ASTM D7467 |
| B20 to B30 | EN 16709, EN 14214 | ASTM D 6751 |
| B30 to B100 | EN 14214 | ASTM D 6751 |

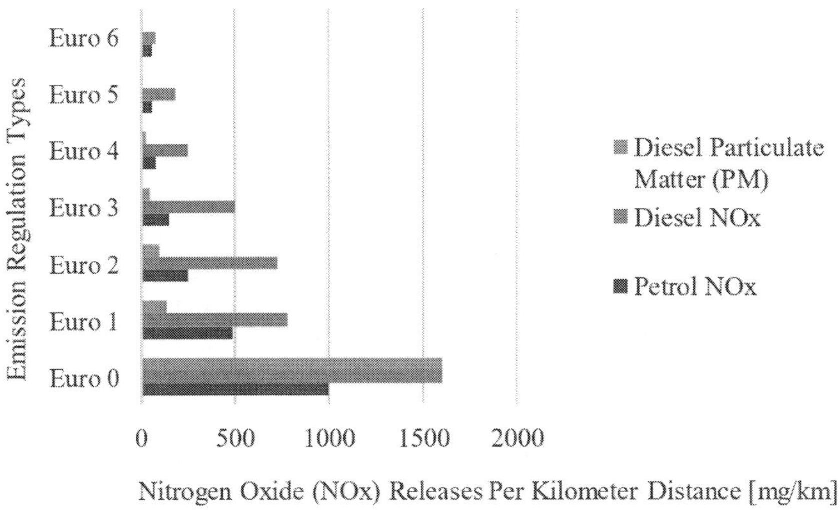


**Figure 2.** Reduction trend of petrol and diesel nitrogen oxides ($NO_x$) and diesel particulate matter (PM) as by different diesel emission regulation types from Euro 0 to Euro 6.

**Table 2.** Standards and specifications of biodiesel [2]

| **Properties** | **ASTM D6751** | **EN14214** | **Petro Diesel** | **Biodiesel (Waste Cooking Oil)** |
|---|---|---|---|---|
| **Flash Point, min (° C)** | 100-170 | ≥120 | 67-85 | 196 |
| **Pour point (° C)** | -3 to -12 | - | -19 to -13 | -11 |
| **Kinematic viscosity (40°C) ($mm^2/s$)** | 1.9-6.0 | - | 1.9-4.1 | 5.3 |
| **Density (15° C) ($kg/m^3$)** | 820-900 | 860-900 | 75-840 | 897 |
| **Cetane number, min** | 47 | 51 | 40-46 | 54 |
| **Ash content (%)** | 0.02 | - | 0.008-0.010 | 0.004 |
| **Carbon residue (%)** | - | - | 0.35-0.40 | 0.33 |
| **Sulphur content (%)** | 0.05% (m/m) | 10 mg/kg | 0.35-0.55 | 0.06 |
| **Water content (%)** | 0.03 (v/v) | 500 (mg/kg) | 0.02-0.05 | 0.04 |

*2.1.3. Biodiesel feedstock from plant oil and 1st generation: food crops.* These are some of the examples of oil feedstocks from plants, like soybean oil, corn oil, rapeseed oil, palm oil (PO), cottonseed oil and Jatropha. They are considered as the first-generation types of food crops in the production of biodiesel. There are also various feasible sources of feedstocks for the production of biodiesel.

### *2.2. Emission characteristics and sustainability compliance*
[7] reported on the emission characteristics that were not only on feedstock from plant oil, but also include the used cooking oil (UCO), animal fats (AF) and microalgae (MA). It is mentioned that, diesel has the highest CO and PM emission percentage (%). Whilst palm oil shows the lowest emission % for CO and NOx. With biodiesel from microalgae, animal fat, and Jatropa, the carbon dioxide ($CO_2$) emission % is comparably the same with that of diesel. Therefore, the need to further increase the reductions of CO, $CO_2$, $NO_x$ and PM by biodiesel is highly needed for sustainability compliance.

*2.2.1. Advantages of biodiesel: Sustainable energy source.* There are many published advantages about biodiesel, the sustainable energy source [2]. Those advantages are like, it has low greenhouse emission, low in ash content, lower sulfur content, and low in carbon residue. It is non-toxic, renewable, safe to handle, higher biodegradability, which degrades quickly in the environment. Most importantly, it is compatible to existing diesel engine without extensive engine modifications.

## 3. Diversifying dominating feedstock: Leading biodiesel producers worldwide 2017 (billion liters)
Referring to figure 3 [8], United States of America leads in the production in 2017 of approximately 6 billion liters (BL) of biodiesel, followed by Brazil of 4.3 billion liters and Germany of 3.5 billion liters, contributing to the leading biodiesel producers worldwide. Malaysia only produces about 0.4 billion liters of biodiesel in 2017, in comparison to other palm users of 1.3 BL of that for Thailand, 2.5 BL for Indonesia, 0.6 BL for Columbia, and 1.3 BL for Spain. According to [8], diverse types of plants as mentioned earlier like corn, soybean, rapeseed, tallow, palm, cottonseed and Jatropa leads the dominating feedstock diversifications. Besides that, there are other kinds of feedstocks which are also used for example used cooking oil (UCO) by India and Netherland, waste cooking oil (WCO) by China and used cooking oil methyl ester (UCOME) by Germany.

### *3.1. USA, EU and world diversifying dominating feedstock*
In United States, soybean contributes at 55% of the total feedstock type used in 2016. whilst, 47% of the European Union (EU) feedstock type used is rapeseed. Other feedstocks like poultry, tallow, white grease and yellow grease by both regions are also used other than that of corn, palm and sunflower [7].

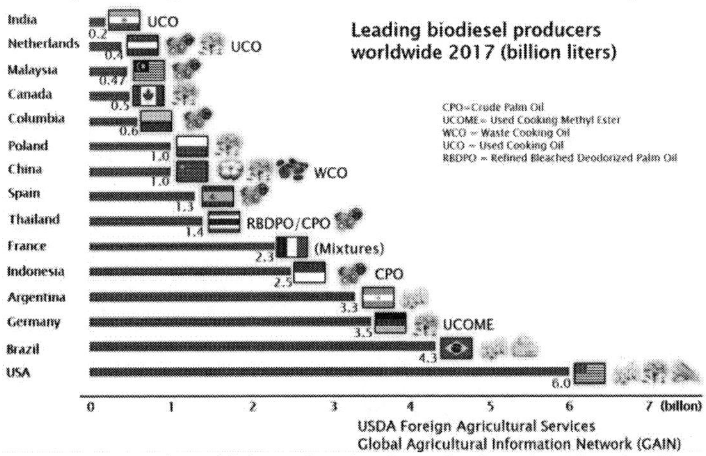


**Figure 3.** Diversifying dominating feedstocks for leading biodiesel producers worldwide in 2017 in billion liters [8]

*3.1.1. Germany feedstock.* According to German's office for agriculture and food, in Germany, in 2017, there is an increasing trend of the use of used cooking oil methyl ester, besides rapeseed, for biodiesel feedstock [9].

*3.1.2. Brazilian feedstock.* Whilst, in Brazil, soybean is still a dominating feedstock. As reported by GAIN 2017 for Brazil, however, other feedstocks like tallow and cottonseeds are also explored [10].

*3.1.3. Spain feedstock.* In Spain, palm oil dominates as the feedstock usage from the year of 2013 to the year of 2017. According to [11], in the year of 2015, there was an abrupt percentage increase of recycled oil. For the production of biodiesel. Additionally, sunflower, rapeseed and soybean are increasingly used in 2017 [11].

*3.1.4. China feedstock based on geographical regions.* Let us now look at China from Asia. It is shown that, a diverse feedstock use for biodiesel is based on geographical regions [12]. For example, cottonseed is planted at Northwest China, whilst, Jatropa and rapeseed are mainly at the South Central of China. Other than these plant sources, China also uses microalgae and waste cooking oil for the production of biodiesel which are concentrated at the East and North East China.

*3.1.5. Malaysia feedstock as biodiesel supply and demand.* There is an abundant supply of palm oil tree in Malaysia, therefore, the biodiesel feedstock mainly comes from palm. According to [13] of the year 2017, the production and consumption of biodiesel increased by 470 million liters and 360 million liters, respectively. However, the export of biodiesel decreased and there were no imports of biodiesel that year.

*3.2. Estimated kg of oil per unit hectare and price per tonne*

According to figure 4 [2], the estimated kg of oil per unit hectare is shown for oil tree, rapeseed, peanut, sunflower and soybean. Palm has the highest estimated kg of oil per hectare in comparison to other feedstocks, followed by coconut oil tree, rapeseed, peanut, sunflower and soybean. Additionally, the price of palm in USD per tonne is also reported to be the lowest followed by sunflower, soybean, rapeseed, coconut and peanut.

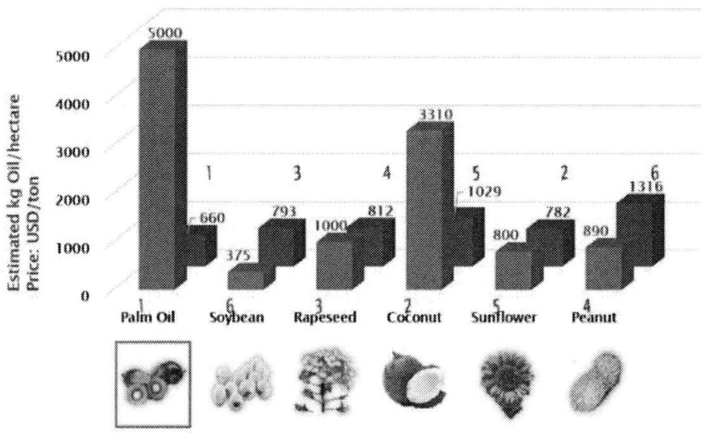


**Figure 4.** The estimated kilogram (kg) of oil per unit hectare and price per unit tonne [2]

### *3.3. Most probable future dominating feedstocks*

The most probable future dominating feedstocks can be camelina, castor algae in Brazil and Africa, Jatropha in Africa, animal fat and used cooking oil. Interestingly, apart from biodiesel, it is reported in 2017 that coffee wastes producing B20 blends will be run on double decker busses in London [14]. Therefore, the prospect of using diversified feedstocks for the production of biodiesel is still relevant. However, palm oil in Asia will still be dominating feedstocks due to the sustaining growth via palm plantations.

*3.3.1. Product/waste of each fresh fruit bunch.* Positively, palm product/waste can be alternatively used as alternative fuel in Malaysia. According to Index Mundi [15] as reported in the year of 2018, shows that Malaysia is still the second largest producer of palm oil after Indonesia. For each of fresh fruit bunch, 21% of palm oil can be extracted. Other than that, are from empty fruit bunch (EFB) (23%), palm oil methyl ester (POME) (28%), fibre (15%), palm kernel (7%), and shell (6%).

*3.3.2. Palm oil and biodiesel.* Historically, palm was originated from West Africa. Through this time until now, palm has become the biggest plantation estates replacing rubber trees. The largest palm plantations in Malaysia is Sime Darby Plantation. In 1980, the established Malaysian Palm Oil Board MPOB actively ventured biodiesel research. Test runs were conducted for 5 year from 1990 to 1995, on neat palm biodiesel and blends were conducted by MPOB Cycle and Carriage, on 30 Mercedez Benz buses [2].

## 4. Future technological expectation via life cycle and revolution

According to the world commission on environment and development, sustainable development is defined as when it meets the needs of the present without compromising the ability of future generations to meet their own needs'. $CO_2$ and biodiesel complete sustainable life cycle. The emitted $CO_2$ is absorbed by plant, and later to be used as biodiesel. With that, sustainability activities occur and the life cycle continues.

The Canadian lifecycle using rapeseed as feedstock for the production of biodiesel resulted in a great reduction of nearly 4.4 Million tonne of $CO_2$ equivalent, nearly 90% of the green house emission reduction in 2017 [16]. This highly supports sustainability policies relating to alternative fuel for future market.

### *4.1. Possible IR 4.0 integration: Home based standard compliance biodiesel production*

Imagine that one day an augmented reality concept on home based standard compliance aiding towards own home production of biodiesel. Sensors, automations and real time monitoring are embedded with augmented reality. This maybe expected to sustain alternative and sustainable supply of fuel for the future from the backyard. Further discussions on best strategies must be looked into.

## 5. Conclusion

In 1911 World's Fair in Paris, Dr. Rudolf Diesel once declared that 'The diesel engine can be fed with vegetable oils and would help considerably in the development of agriculture of the countries which uses it'. In a Biodiesel Sustainability webpage, the scientist to biodiesel declaration, which sounds like "Biodiesel from a variety of feedstocks can meet contemporary needs for environmental stewardship, economic prosperity, and quality of life without compromising the ability of future generations to meet these needs for themselves."

## References

[1] Demirbas A 2008 *Biodiesel: A realistic fuel alternative for diesel engines* (Springer London)
[2] Zahan K A and Kano M 2018 *Energies* **11** 1

[3] Neuling U and Kaltschmitt M 2017 *J. Oil Palm Res.* **29** 137
[4] THE STAR ONLINE, "Govt enlists oil firms to distribute Envo Diesel," *THE STAR ONLINE.*
[5] THE STAR ONLINE, "B10 biodiesel switch mandatory," *STAR ONLINE.*
[6] Jääskeläinen H "Biodiesel Standards and Properties," *Dieselnet.com.* [Online]. Available: https://www.dieselnet.com/tech/fuel_biodiesel_std.php.
[7] Kim D S, Hanifzadeh M, and Kumar A 2018 *Environ. Prog. Sustain. Energy* **37** 7
[8] USDA, "USDA Foreign Agricultural Services Global Agricultural Information Network (GAIN)," *GAIN.* [Online]. Available: https://gain.fas.usda.gov/Recent GAIN Publications/Forms/AllItems.aspx.
[9] GAF, "German Agriculture and Food," *GAF.* [Online]. Available: https://www.ble.de/EN/Home/home_node.html.
[10] GAIN, "Brazil: USDA Foreign Agricultural Services Global Agricultural Information Network (GAIN)," *GAIN BRAZIL.* [Online]. Available: https://gain.fas.usda.gov/Recent GAIN Publications/Biofuels Annual_Sao Paulo ATO_Brazil_9-15-2017.pdf.
[11] GAIN, "Madrid: USDA Foreign Agricultural Service: Global Agricultural Information Network (GAIN) 2017," *GAIN MADRID.* [Online]. Available: https://gain.fas.usda.gov/Recent GAIN Publications/Exporter Guide_Madrid_Spain_12-4-2017.pdf.
[12] Xu Y J, Li G X, and Sun Z Y 2016 *Renew. Sustain. Energy Rev.* **54** 318
[13] GAIN, "Kuala Lumpur: USDA Foreign Agricultural Services Global Agricultural Information Network (GAIN)," *GAIN KUALA LUMPUR.* [Online]. Available: https://gain.fas.usda.gov/Recent GAIN Publications/Biofuels Annual_Kuala Lumpur_Malaysia_10-24-2017.pdf.
[14] B. NEWS, "London buses to be powered by coffee," *BBC NEWS.* [Online]. Available: https://www.bbc.com/news/uk-england-london-42044852.
[15] IMM, "Index Mundi Malaysia," *IMM Malaysia*, 2018. [Online]. Available: https://www.indexmundi.com/energy/?product=biodiesel&graph=consumption.
[16] Canola, "Canola: The Perfect Candidate for Biodiesel and Renewable Diesel," *Canola Council of Canada.* [Online]. Available: https://www.canolacouncil.org/canola-biofuels/canola-biofuels/canola-the-perfect-candidate-for-biodiesel-and-renewable-diesel/.

# Precursor effect on the Fe-doped cryptomelane (Fe-OMS-2) prepared via one-step sol-gel route as heterogeneous Fenton catalyst for degradation of methylene blue

**A Awaluddin[1*], E Amiruddin, R Anggraini, N Indriyani and SS Siregar**
Department of Chemistry, Universitas Riau, Kampus Bina Widya Km 12.5 Pekanbaru, Riau 28293, Indonesia

*E-mail: amirawaluddin01@gmail.com

**Abstract.** The pristine cryptomelane (OMS-2) and Fe-OMS-2 (1%, 5%, and 10%) have been successfully synthesized by a one-step sol-gel method and characterized by XRD, SEM, BET surface area and AAS. The incorporation of Fe (III) within the structure of OMS-2 leads to an increase in surface area, pore-volume, and catalytic activity for the degradation of methylene blue (MB) and a decrease in the content of $K^+$ and average of pore diameter. The catalytic test indicated that the catalytic activity of Fe-OMS-2 1% is considerably higher that of pure OMS-2. The highest degradation of MB is 93.5% achieved using Fe-OMS-2 1% after 120 minutes of degradation time, the initial concentration of MB of 20 ppm, the catalyst concentration of 0.125 mg and the $H_2O_2$ volume of 15 mL. The enhanced catalytic activities of Fe-OMS-2 compared to pristine OMS-2 for catalytic degradation of MB are associated with an increase in surface area, pore-volume, and smaller particle size. The results from catalytic studies imply that the metal doping is an effective strategy to enhance the catalytic activities of OMS-2 for degradation of MB.

## 1. Introduction

Dye pollutions in water have become a major environmental issue worldwide due to the rapid development of textile industries and other related industries that use dyes as their coloring agents. The effluents from these industries contain dyestuffs that are very toxic, mutagenic, and carcinogenic to human and aquatic life. The dyes are unusually resistant to heat, light, and oxidizing agents and must be treated properly before being released into an environment. The chemical structures of most dyes contain aromatic compounds that are highly stable and are very difficult to biodegrade. The presence of a dye in water, even in a small amount, is very visible and undesirable for any use because it could interfere with the ecological balance. Several methods have successfully applied for dye removals such as biological, adsorption, coagulation, membrane filtration, oxidative degradation, and advanced oxidation processes [1,2,3]. Each method has its advantages and limitations for dye removal. Among these, biological treatment is considered to be the most economical. However, the method is time-consuming [4] and can not be applied if the dyes are toxic to the microbes used for dye removal.

The advanced oxidation processes (AOPs) have gained an increased interest for dye removal due to their efficient and environmentally friendly ways compared to other methods. Fenton processes-based advanced oxidation processes are one of the AOPs techniques, gaining more acceptance for treating

* To whom any correspondence should be addressed.

wastewater containing dye residues. Fenton processes are characterized by the formation of OH radicals as the main radical, following the reaction between Fenton catalyst ($Fe^{2+}$ ions) and hydrogen peroxides. The radical is the second most powerful oxidant after fluorine, non-selective, and is able to oxide most of the recalcitrant organic contaminants present in wastewater. Since the catalyst and the substrate being oxidized are the same phases, the separation between them is still challenging and requires further treatments; thus, additional costs are needed. To overcame this, heterogeneous Fenton processes using solid catalysts are introduced and gain extensive interests over the last decades. In this system, the hydrogen peroxide interacts with the surface catalyst, producing OH radicals. Thus, it is required to have a very active solid catalyst with high surface area and ease of electron transfer between the catalyst and hydrogen radical. Several solid catalysts such as clays, zeolite, and other metal oxides have been proposed as heterogeneous Fenton catalysts for the treatment of dye effluent.

Manganese oxide-based Fenton catalysts are known to have high catalytic activities for the degradation of organic contaminants [5,6]. Manganese oxides are ubiquitous in various environmental settings and play an important role in the mobility and the fate of organic contaminants. The oxides are strong oxidant, very stable, have low zero point charge ($pH_{zpc}$), high average oxidation state (AOS), and degrade most recalcitrant organic compounds as well as have high adsorption capacity for heavy metals. The presence of Mn (III) and Mn(II) in the frameworks of either tunnel or layer manganese oxides introduce more defects due to the more oxygen vacancies. The structure of manganese oxides mainly consists of layer and tunnel structures, with the basic structural unit of $MnO_6$ octahedra. The manganese oxides with tunnel structures are formed through sharing the edge of $MnO_6$ octahedra, resulting in single, double, and triple chains. These chains are linked together by sharing a corner, creating tunnel sizes of 1x 1 (pyrolusite), 2 x 2 (cryptomelane, OMS-2), and 3 x 3 (todorokite, OMS-1). The layer structures of manganese oxides are generated by stacked sheets of sharing edges of $MnO_6$ octahedra and are easily found in nature as birnessite minerals. Both cryptomelane and birnessite have been reported to possess high reactivity for catalytic degradation or oxidative degradation of organic molecules. Various factors govern the reactivity of manganese oxides such as morphology, the presence of defects by doping, surface areas, and oxidation potential.

Doping cryptomelane with transition metals (Cu, Co, Ni, Ce, Fe) have been successfully prepared via various synthetic methods such as hydrothermal, reflux, sol-gel, and solid-state. Doping in cryptomelane often introduces more defects and creates strains in the parent molecules and often transforms the cryptomelane into other phases such as birnessite, resulting in higher surface area, different surface morphology, and higher oxidation potential. This, in turn, may result in tremendous improvement of catalytic activities of doped OMS-2 for the degradation of dyes.

In the current study, OMS-2 and Fe-OMS-2 were prepared via the one-step sol-gel method and their catalytic performances were evaluated for the degradation of MB. Fe (III) ion is chosen as the dopant for the OMS-2 due to similar sizes, charges, and coordination tendencies with manganese. The one-step sol-gel synthesis for Fe-OMS-2 is adopted in this study because it is facile to prepare the oxides, cost-effective, requires simple equipment, and moderate synthesis condition.

## 2. Experimental

### *2.1. Synthesis*

The synthesis procedure of pure OMS-2 and iron doped-OMS-2 were the same as for Cu-doped cryptomelane in our previous work [7] except the $CuSO_4{\cdot}5H_2O$ was replaced by $Fe(NO_3)_3{\cdot}9H_2O$, with the molar ratios of 1%, 5% and 10% to $KMnO_4$. The obtained materials were named as Fe-OMS-2 1%; Fe-OMS-2 5% and Fe-OMS-2 10%. The properties of these materials were compared with pure OMS-2 that was synthesized using the same procedures except for the absence of dopant.

### *2.2. Characterization*

X-ray diffraction analysis of pure and Fe-dope OMS-2 was carried out on Mini Flex Rigaku Diffractometer using Cu Kα radiation (λ=1.5444nm). The diffractometer was operated with a scanning speed of 2° per min and a scan step of 0.1°. Morphologies of the samples were probed by scanning electron microscopy (SEM) with a JEOL JSM 6330F microscope. The specific surface area of the samples was measured by $N_2$ adsorption at 77 K using a BELSORP-max instrument and degassing at 423 K for 2 h. Elemental compositions for potassium, manganese, and iron were analyzed with atomic absorption spectroscopy Shimadzhu AA7000.

### *2.3. Catalytic activity test*

Methylene blue (MB) degradation was carried out in a glass Beaker. Briefly, 0.125 g of catalyst powder was loaded into glass Beaker, which contained methylene blue solution (initial MB solution 20 mg/L) and 60 mL deionized water. The suspension was stirred for 30 min and then added 30 wt% $H_2O_2$ (15 mL). The mixture was pipetted at a given time interval, then placed in an ice bath and centrifuge to remove the catalyst. The remain of MB solution concentration was determined using spectrophotometer Optima SP-300. The activity of the catalyst was calculated on the basis of the Equation (1).

$$\text{MB degradation} = \frac{C_0-C_t}{C_0} \text{x } 100\% \quad (1)$$

Where Co and Ct are initial and concentration of methylene blue at any reaction time (t), respectively.

## 3. Result and discussion

### *3.1 Characteristic of synthesized OMS-2 and Fe-OMS-2*

The pure OMS-2 and Fe-doped cryptomelane (Fe-OMS-2) are synthesized by the sol-gel method using $KMnO_4$ as oxidant and glucose as a reductant. The Fe-OMS-2 samples are prepared via a one-step sol-gel route by reacting all reactants simultaneously in one container, followed by washing, filtering, and finally calcinating at 450°C for 2 hours. XRD patterns of as-synthesized pure and doped manganese oxide with 1%, 5%, and 10% of Fe (III) are shown in Figure 1. The patterns are in good agreement with the tunnel structure of the octahedral molecular sieve cryptomelane-type manganese oxides (JCPDS (1977) 29-1020). No additional peaks associated with Fe species are detected following the incorporation of Fe into the pure cryptomelane. These results suggested that $Fe^{3+}$ ions are well distributed into the structure of the pure cryptomelane. A similar observation was reported by previous studies, where the XRD patterns of Fe-doped cryptomelane remained unchanged upon doping with Fe (III) [8,9,10,11,12,13]. The Fe-OMS-2 also displays less crystalline materials compared to pristine OMS-2, as indicated by wider peaks and low relative intensities. Thus, the incorporation of $Fe^{3+}$ ions in the structure of OMS-2 introduces more defects, which affect the physicochemical properties of Fe-OMS-2.

Several studies indicated that Fe(III) is replacing Mn(III) in the framework of OMS-2, while others revealed that Fe (III) is substituting $K^+$ in the tunnel of OMS-2. However, when OMS-2 is doped with other transition metals such as Cu ion [14], the OMS-2 is transformed into layer birnessite-type manganese oxide at high Cu concentration. Cu ions can be located in a different position, like in the surface, in the framework or in the tunnel, depending on the Cu concentration. Other transition metals such as Ce and Co result in the doped OMS-2 having the less crystalline phase compared to pure OMS-2 [10]. It is likely that the difference in doping strategy results in the formation of $Fe^{3+}$ species either in the surface, framework, or in the tunnel of the cryptomelane. In the current study, the Fe-OMS-2 are prepared by mixing $Fe^{3+}$ ions with $KMnO_4$ and glucose via mild sol-gel methods, whereas the previous report used the hydrothermal process using high temperature and pressure. The different precursor often leads to the preferential orientation of a certain plane, as reported by Awaluddin *et al.* [8]. The additional reflection plane (101) of the cryptomelane phase is shown when the citric acid is used as a reducing agent for the preparation of cryptomelane by the sol-gel method [8]. This plane (101) is undetected using other precursors such as glucose [7].

The SEM images of pure cryptomelane and Fe-OMS-2 at different Fe concentrations are shown in Figure 2. The pure OMS-2 manganese oxide consists of the aggregates of particles with different sizes. Upon doping with $Fe^{3+}$ ions, the aggregates became smaller in size and fragmented into small granular particles, especially in the high Fe doping concentration (Figure 2d). The previous studies reported a similar observation upon Fe doping in OMS-2 but using different precursors (citric precursor) [8] and Cu doping into OMS-2 with a similar precursor to the current study [7].

IOP Conf. Series: Materials Science and Engineering **845** (2020) 012014 doi:10.1088/1757-899X/845/1/012014

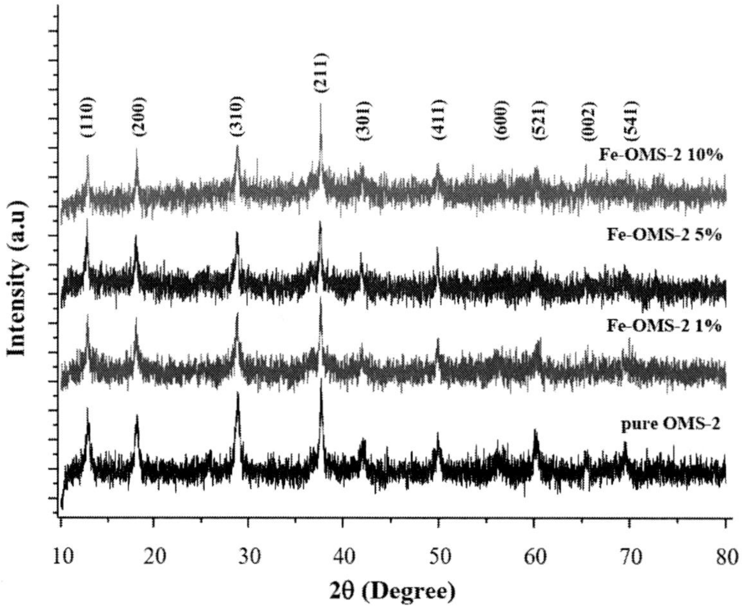


**Figure 1.** XRD patterns of pure OMS-2 and Fe-OMS-2 with a different dopant concentration

**Figure 2.** The SEM images of pure OMS-2 (a); Fe-OMS-2 1% (b); Fe-OMS-2 5% (c); Fe-OMS-2 10% (d) with 5000 x magnification.

Yin *et al.* reported the difference in morphology upon the Fe incorporation into OMS-2 [9]. The pure OMS-2 has a needle-like morphology and reduces in length and increases in width upon small Fe doping. Further increase in Fe concentration, Fe-OMS-2 becomes granular particles with aggregations. The same needle-like morphology for pristine OMS-2 was reported by Ma *et al.* [10], but the morphology altered into chunky morphology after the Fe addition. The difference in morphology between the current study and the previous reports is likely due to the difference in the preparation method. Ma *et al.* prepared OMS-2 and Fe-OMS-2 via the hydrothermal method [10], whereas Yin *et al.* synthesized OMS-2 and Fe-OMS-2 by reflux route [9]. These two previous studies, especially the hydrothermal route, require high temperature and pressure, thus affecting the rate of crystal growth.

Table 1 displays the selected data from BET and BJH measurements, which include BET surface area, total pore-volume, average pore diameter, pore-volume, and pore diameter of peak. The Fe-OMS-2 samples have a higher BET surface area than pure OMS-2. The previous studies also indicated similar trends of an increase in surface areas following the incorporation of dopant into cryptomelane. Similarly, the total pore volume of Fe-OMS-2 increases until 1% Fe doping into OMS-2, then decrease following the addition of 5 % Fe.

**Table 1.** Textural properties of pure OMS-2 and Fe-OMS-2

| Catalyst type | BET | | | BJH | |
|---|---|---|---|---|---|
| | Surface area ($m^2/g$) | Total pore Volume ($cm^3/g$) | Avarage pore Diameter (nm) | Pore Volume ($cm^3/g$) | Pore Diameter of *Peak* (nm) |
| Pure OMS-2 | 82.916 | 0.4911 | 23.689 | 0.4869 | 44.14 |
| Fe-OMS-2 1% | 135.4 | 0.6478 | 19.138 | 0.6363 | 59.00 |
| Fe-OMS-2 5% | 140.33 | 0.4689 | 13.365 | 0.4588 | 78.01 |

Table 2 displays the bulk analysis of pristine OMS-2 and Fe-OMS-2 obtained from atomic absorption spectroscopy (AAS). The content of $K^+$ ions drops significantly the following doping with Fe (III), while the Mn concentration, in general, remain unchanged. The Fe (III) concentration in the OMS-2 as expected increases following the incorporation of Fe (III). These results suggested that $Fe^{3+}$ ions are replacing $K^+$ in the tunnel.

**Table 2.** Bulk analysis of pure OMS-2 and Fe-OMS-2 using AAS

| Catalyst type | K | | Mn | | Fe | | K/Mn | Fe/Mn |
|---|---|---|---|---|---|---|---|---|
| | mg/L | (mmol) | mg/L | (mmol) | mg/L | (mmol) | | |
| Pure OMS-2 | 3.5284 | 0.0902 | 6.0975 | 0.1110 | 0.0000 | 0.0000 | 0.81 | 0.00 |
| Fe-OMS-2 1% | 1.6244 | 0.0415 | 5.5344 | 0.1007 | 0.6965 | 0.0125 | 0.41 | 0.12 |
| Fe-OMS-2 5% | 1.4386 | 0.0368 | 6.4849 | 0.1180 | 2.2918 | 0.0410 | 0.31 | 0.35 |
| Fe-OMS-2 10% | 1.1960 | 0.0306 | 6.639 | 0.1208 | 3.1718 | 0.0568 | 0.25 | 0.47 |

These results suggested that $Fe^{3+}$ ions substitute $K^+$ in the tunnel of OMS-2 instead of Mn in the framework of OMS-2. The small difference in Mn concentrations following the incorporation of $Fe^{3+}$ supports the claim that $Fe^{3+}$ ions are replacing $K^+$ in the tunnel of OMS-2. $K^+$ ions acting as balancing cations play a key role in stabilizing the structure of OMS-2. It is known that OMS-2 consists of mixed valent of Mn, such as Mn (IV), Mn (III), and Mn (II). The presence of Mn (III) or Mn (II) replacing Mn (IV) creates negative charges in the framework of OMS-2, that must be balanced by positive charges like $K^+$ or by removing oxygen atoms. The absence of oxygen (oxygen vacancy) in the OMS-2 structure introduces defects, which may cause a higher catalytic activity of Fe-OMS-2. The relative constant of Mn concentration of Fe-OMS-2 upon Fe doping could be due to the reduction of Mn (IV) into Mn (III) or Mn (II), also creating more oxygen vacancies (more defects). Several studies indicated that Fe (III) substituted Mn (III) in the framework of Fe-OMS, evidenced by an increase in AOS Mn upon Fe doping and higher Mn (IV) concentration from XPS studies. The different location of transition metals such as Fe and Cu due to different dopant concentration within the structure of OMS-2 has been reported by previous studies [14]. The Cu species are located on the surface of Cu-OMS-2 for low Cu concentration, while for medium and high Cu doping, Cu is located in the tunnel and in framework, respectively.

*3.2 Degradation performance of as-synthesized manganese oxide with different Fe concentration*

Figure 3 displays the catalytic activities of OMS and Fe-OMS-2 at different Fe concentration. In general, the catalytic activities of Fe-OMS-2 are higher than that of pristine OMS-2. For the 1% Fe-OMS-2, the rate of MB degradation reaches over 72% conversion of MB compared to only 39% conversion for pristine OMS-2 after 10 minutes of degradation time. For 5% and 10% of Fe-OMS-2, the rate of MB degradation (70% and 66%, respectively) is still higher than that of pristine OMS-2, even though lower than that of 1% Fe-OMS-2. The higher degradation rates of Fe-OMS-2 for MB degradation are correlated with the higher BET surface areas, smaller particle sizes, and lower crystallinity. For 1% Fe-OMS-2 with the highest degradation rate, the highest pore volume also play a significant role in the high catalytic activity. Data from bulk analysis also indicated that Mn contents decreased upon Fe incorporation in the 1% Fe-OMS-2, suggesting that some Fe (III) ions are likely to replace Mn (III) in the framework due to their similar size and charge. This, in turn, introduces more defects responsible for high catalytic activity. The effect of Fe doping on the less crystalline of OMS-2 is also responsible for the higher catalytic activity of Fe-OMS-2 compared to pure OMS-2.

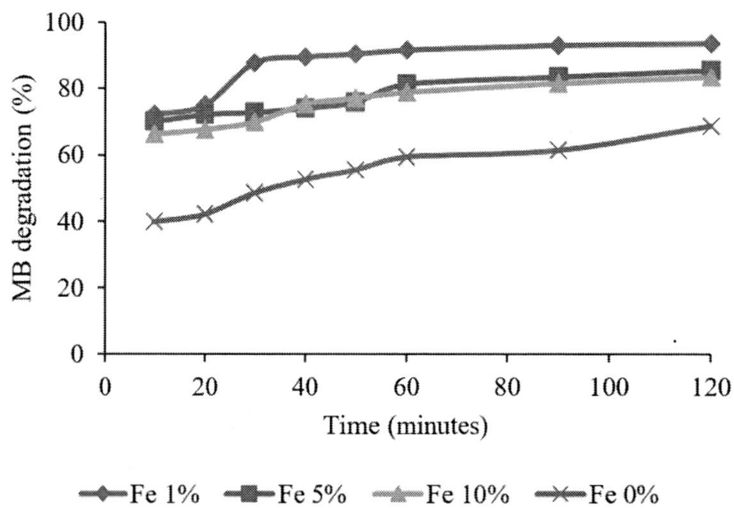


**Figure 3.** Degradation performance of as-synthesized manganese oxide with different Fe concentration, [MB] = 20 mg/L; catalyst = 0.125 g; $H_2O_2$ = 15 mL

**Conclusion**

The Fe-OMS-2 has been successfully synthesized via a one-step sol-gel route by a redox reaction between $KMnO_4$ and glucose with $Fe^{3+}$ ions as a dopant. The present of Fe dopant in the crystal structure of OMS-2 increases the surface area and pore volume and reduces the crystallinity and concentrations of $K^+$ and Mn. The catalytic activity test indicated that Fe-OMS-2 showed higher catalytic performance than OMS-2. The highest degradation of MB is 93.5% achieved using 1% Fe-OMS-2 catalyst over 120 minutes of degradation times under optimum condition.

**Acknowledgments**
The authors gratefully thank The Ministry of Research, Technology, and Higher Education of the Republic of Indonesia under PDUPT scheme research for financial support.

**References**

[1] Awaluddin A, Agustina M, Aulia R R and Muhdarina 2017 *AIP Conf. Proc.* **1823** 020108
[2] Siregar S S and Awaluddin A 2018 *IOP Conf. Ser.: Mater. Sci. Eng.* **345** 012005
[3] Awaluddin A, Zulfa R, Absus S, Nurhayati, Linggawati A and Siregar SS 2019 *IOP Conf. Ser.: Mater. Sci. Eng.* **509** 012011
[4] Permana I, Awaluddin A and Saryono 2019 *Biodiversitas* **20** 1693
[5] Zulfa R, Absus S, Awaluddin A, Anita S, Siregar S S and Zulfikri 2018 *IOP Conf. Ser. J. Phys.* **1097** 012051
[6] Absus S, Zulfa R, Awaluddin A, Anita S and Siregar S S 2018 *AIP Conf. Proc.* **2049** 020009
[7] Awaluddin A, Astuti L, Linggawati A, Siregar S S and Saputra L 2018 *AIP Conf. Proc.* **2026** 020075
[8] Awaluddin A, Anggraini R, Siregar S S, Muhdarina and Prasetya 2019 *MATEC Web Conf.* **276** 06005
[9] Yin H, Liu F, Feng X, Hu T, Zheng L, Qiu G, Koopal L K and Tan W 2013 *Geochim. Cosmochim. Acta* **117** 1–15
[10] Ma J, Wang C and He H 2017 *Applied Catal. B: Environ.* **201** 503
[11] Gao J, Jia C, Zhang L, Wang H, Yang Y, Hung S F, Hsu Y Y and Liu B *J. Catal.* 2016 **341** 82
[12] Stelmachowski P, Videla A H A, Jakubek T, Kotarba A and Specchia S 2018 *Electrocatalysis* **9** 762
[13] Cai J, Liu J, Willis W S and Suib S L 2001 *Chem. Mater.* **13** 2413
[14] Yang Y, Huang J, Zhang S, Wang S, Deng S, Wang B and Yu G 2014 *Appl. Catal. B: Environ.* **150–151** 167

# Cathodic protection on stuctures of carbon steel using sacrificial anode methode for corrosion control

**Komalasari***, **Evelyn***, **I D R Situmeang and D Heltina**
[1,2]Department of Chemical Engineering, Faculty of Engineering
Universitas Riau, Pekanbaru 28293, Riau, Indonesia

E-mail : komalasari@lecturer.unri.ac.id; evelyn@eng.unri.ac.id

**Abstract.** The pipes for land or sea are generally made of carbon steel and susceptible to corrosion. Cathodic protection can be used to control metal corrosion in water and soil. Cathodic protection is a metal surface protection system bypassing adequate direct current to the metal surface and converting sacrificial anode area on the metal surface into cathodic region. The aim of this research was to study the effect of cathodic protection in controlling the corrosion rate of carbon steel and determine the best immersing time for sacrificial anode in 25,000 ppm sodium chloride solution to protect the cathode. Zinc was used as the sacrificial anodes for the cathodic protection. Then, the effect of immersion time (10, 20, and 30 days) and distance of the anode from the cathode (1, 2, and 3 cm) were investigated. The best protection for corrosion was obtained at immersion time of 10 days and electrode distance of 1 cm, allowing corrosion rate of 0.694 mpy with an efficiency of 81.8%.

## 1. Introduction

In industrial engineering, especially in the field of engineering construction, corrosion is a major factor influencing the shelf life of pipes that transport important energy sources in each country. Judging from the cost of treatment in the chemical and petrochemical industries, the cost of overcoming the corrosion problem could reach 70 to 80 percent of all maintenance costs. Experts and practitioners tend to agree setting the cost of corrosion around 3.5 percent of Gross National Product. Some countries have calculated the cost of corrosion on a national scale in different ways, where the range is between 1.5 - 5.0 percent of the Gross National Product [1].

Various methods of preventing corrosion in the aquatic area had been developed. One of them is cathodic protection using sacrificial anode and impressed current method. Cathodic protection using sacrificial anode is a metal surface protection system by passing adequate direct current to the metal surface that have a lower potential and converting all anode areas on the metal surface to cathodic areas. This system is only effective for systems that are immersed in water or in the soil. This protection system has successfully controlled the corrosion process for marine vessels, seaside structures, pipeline and underground or marine tank installations and so on. The use of sacrificial anodes has advantages including simpler, more stable and lower maintenance costs [2].

Corrosion is a decrease in the quality of metals due to electrochemical reactions with their environment [3]. Corrosion can occur if it has the following conditions [2]:

* To whom any correspondence should be addressed.

a) Anode, where the oxidation reaction occurs where negative ions gather. Anodes are usually corroded by releasing electrons from neutral metal atoms to form ions.
b) The cathode, the site of the reduction reaction where positive ions gather at the cathode usually does not undergo corrosion, however it may suffer damage under certain conditions.
c) Electrolytes, which are solutions that have the property to conduct electricity. Electrolytes can be in the form of acid solutions, base solutions and salt solutions. Electrolyte solution has an important role in metal corrosion because this solution can make electrical contact between the anode and cathode.
d) Metal connections, between the anode and the cathode there must be a metal connection for electrical connections so that the current in the corrosion cell can flow.

The objectives of this research was to study the effect of cathodic protection in controlling the corrosion rate of carbon steel and determine the best immersing time for sacrificial anode to protect the cathode. The effect of immersion time and distance of the anode from the cathode were investigated.

## 2. Methodology

### *2.1 Preparation of spesimen and corrosion medium*

The materials used in this research was ASTM 36 carbon steel as a cathode with a size of 3 cm x 4 cm x 0.17 cm. Zinc was used as an anode or as a corrosion protection material with a size of 3 cm x 4 cm x 0.04 cm. Corrosion media used were sodium chloride solution with a concentration of 25,000 ppm.

### *2.2 Weight loss study*

The corrosion rate was determined from specimen weight loss of the carbon steel. Corrosion rate testing was carried out by anode and cathode pair dip test method connected by steel wires, referring to ASTM G 31-72 [4] and ASTM G1-90 [5] standards. The study was conducted on steel without protection and on steel with zinc protection (anode) mounted on the left and right sides of the cathode with the similar distance (1 cm). Experiments were also carried out for steel using 2 cathodes at a distance of 2 cm and 3 cm respectively. The length of immersed time of 10 days, 20 days and 30 days were also investigated.

### *2.3 Characterization of specimen*

The morphology and type of corrosion of the unprotected protectedcarbon steel ASTM36 were characterized by Scanning Electron Microscopy (SEM) and Energy Dispersive X-Ray (EDX) to identification of its.

## 3. Results and Discussion

### *3.1 The corrosion rate of carbon steel without protection vs. sacrificial anode protection*

Figure 1 shows a comparison of corrosion rate of protected versus unprotected carbon steel. The results of the immersion test on carbon steel without the protection of sacrificial anode tested on sodium chloride solution for 30 days showed the greatest weight loss and the corrosion rate was 9,252 mpy. This is due to the high dissolved salts causing an increase in the conductivity of the salt solution [6], the steel undergoes corrosion due to the presence of the Cl- ion where the Cl- ion will break the passive layer on carbon steel or prevent the formation of a passive layer on carbon steel [7]. With respect to protected carbon steel, less weight loss was observed than the unprotected carbon steel. It can be seen from Figure 1 that the rate of corrosion of steel protected by an anode of zinc sacrificed was slower compared to the rate of corrosion of unprotected steel.

Further corrosion of the carbon steel pipes in the salt solution with a protected structure (cathode) was observed at all distances of anodes (Figure 1). Corrosion rate with an immersion time of 10 days with the distance between the anode and cathode of 1 cm is 0.694 mpy. At the distance between the electrodes of 2 cm and 3 cm, an increase in corrosion in carbon steel occurs. Zinc sacrificial anode is not effective for protection with 2 cm and 3 cm distance between the electrodes because the energy produced by the ineffective sacrificial anode which affects the ion mobility becomes lower through the electrolyte and the influence of the electrode connecting steel wire resistors so that the electrical conductivity is lower [8]. This shows the effect of the distance between the electrodes, where the farther the electrode distance, the corrosion rate obtained will be greater, this is due to the slower movement of ions and the reduced number of ions moving to protect the cathode or carbon steel [9]. This is according to Juliana et al [10] due to the presence of resistance from the electrolyte and the connecting wire between the electrodes which results in the ineffectiveness of the energy produced by the electrodes. Besides, the effect of the potential is not so negative when compared to the metal that wants to be protected and the zinc anode is quickly corroded by dissolved sodium chloride ions which causes not effective electric current to flow into the electrolyte solution. This is in according to theory [11] where the anode corrosion potential must be more negative to the protected metal to encourage the protection current through the electrolyte.

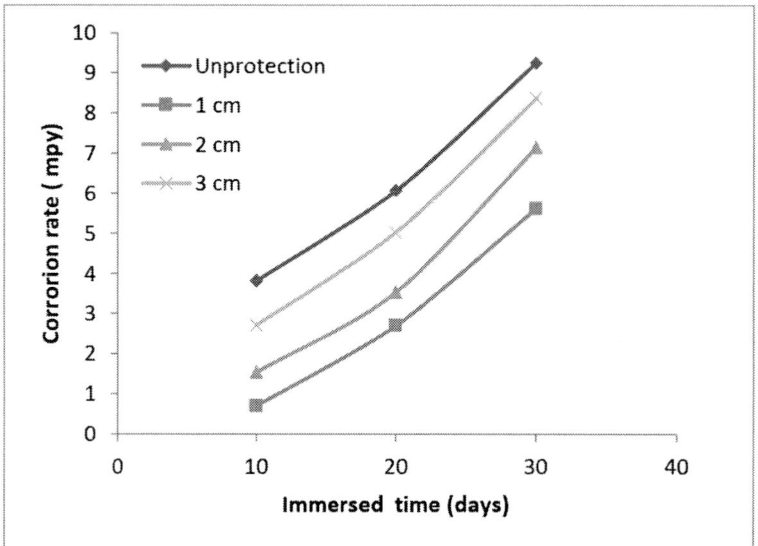


**Figure 1.** The rate of corrosion of steel to immersion time by comparison of the distance between the cathode and the sacrificial anode of zinc

The farther from the anode to the cathode will cause an increase in the corrosion rate of carbon steel, it can be seen that there is an increase in the corrosion rate of carbon steel by increasing the distance between the electrodes. The high content of Cl-ions at a concentration of 25,000 ppm with a corrosion rate of 8.337 mpy with the distance between the anode to the cathode 3 cm results in corrosion attacks so that resistance to the cathode increases and decreases the current from the anode. In that event, the surface area and the electrode distance are influential because they play a role in removing and delivering current to protect the surface of the cathode and affect the conductivity of the electrolyte solution. The sacrificial anode protects the cathode by giving current to the cathode, called the galvanic current. Anodes that have a more negative potential can protect more positive specimens by exchanging ions where the anode and cathode release their ions. Ion exchange that occurs in the specimen can form a layer that will interfere with the entry of oxygen to the surface of the cathode so that it can slow the corrosion process [10]. In galvanic current transfer, anodes which have more negative potential have better performance in protecting the cathode [8].

### *3.2 Efficiency of carbon steel corrosion rate with zinc as sacrificial anode*

Corrosion rate calculation is based on weight loss from carbon steel specimens before the dip test is carried out after the dip test is carried out in accordance with the immersion time to be performed. The corrosion rate efficiency is calculated by comparing the corrosion rate of unprotected carbon steel with the corrosion rate of carbon steel that is protected using a zinc as sacrificial anode.

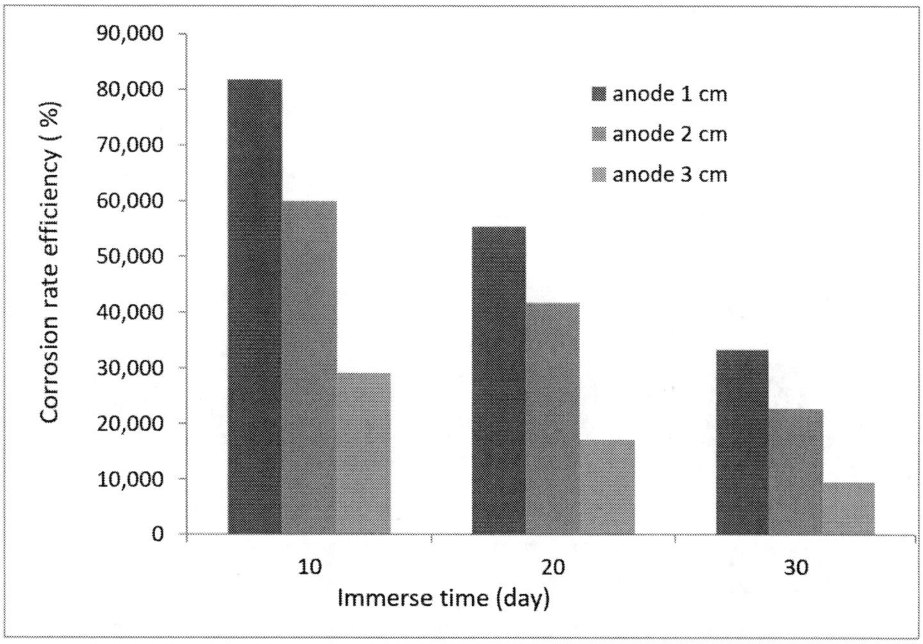


**Figure 2.** Efficiency of the corrosion rate using zinc as sacrificial anodes

Figure 2 shows the variation of the distance between the electrodes in sodium chloride solution 25,000 ppm to obtain the corrosion rate efficiency of carbon steel that is protected against unprotected carbon steel with the same immersion time for the same specimen. From the results of the study, it was found that the best protection on carbon steel protected by zinc sacrificial anode is at the time of immersion for 10 days with the anode to cathode distance of 1 cm with corrosion rate efficiency of 81.818%. From the decrease in corrosion rate, it can be seen that the decrease in corrosion rate is influenced by distance and time of immersion. According to Parthasarady [12] the amount of electric current flowing on the electrolyte is influenced by the potential difference given to the electrode, the resistance of the solution, the electrode distance and the electrode surface area.

A decrease in corrosion rate by using zinc sacrificial anode can occur because zinc metal has a more electronegative potential than steel metal [13]. Besides the influence of the distance between the electrodes also affects where the distance of the electrodes the lower the conductivity [8], this is due to the resistance of the electrolyte which results in ineffective energy produced by the electrodes [10]. In this study the longer the immersion time of the specimen, the higher the corrosion rate is due to the fact that the corrosion rate has not yet reached a certain point where it has not reached a critical point, so the corrosion rate tends to be constant [14].

The corrosion rate of carbon steel cannot be completely lost (zero corrosion rate) due to loss of protective current in the anode and cathode galvanic linkages. This can occur due to the following factors [10]:

a. There is a resistance in a steel wire that connects the anode and the cathode.
b. There is an electrolyte resistance between the anode and the cathode
c. The existence of resistance that occurs in corrosion products on metal surfaces
d. Effect of formation of hydrogen gas on the surface of the cathode, where the formation of hydrogen gas can increase resistance at the cathode which can reduce the current from the anode [4].

*3.3 SEM-EDX analysis*

SEM-EDX is a method of identifying morphology and functional groups in materials. Damage caused by corrosion can be in the form of metal oxides, morphological surface damage of metals, changes in mechanical properties and changes in chemical properties [15]. SEM-EDX analysis was conducted at the Integrated Laboratory of Diponegoro University, Semarang

In the SEM test the analyzed specimens are carbon steel specimens with immersion time of 30 days namely unprotected carbon steel specimens and carbon steel specimens protected with zinc sacrificial anodes with anode distance to cathode is 1 cm. SEM test aims to determine the type of corrosion and EDX to determine the chemical composition contained in the material.

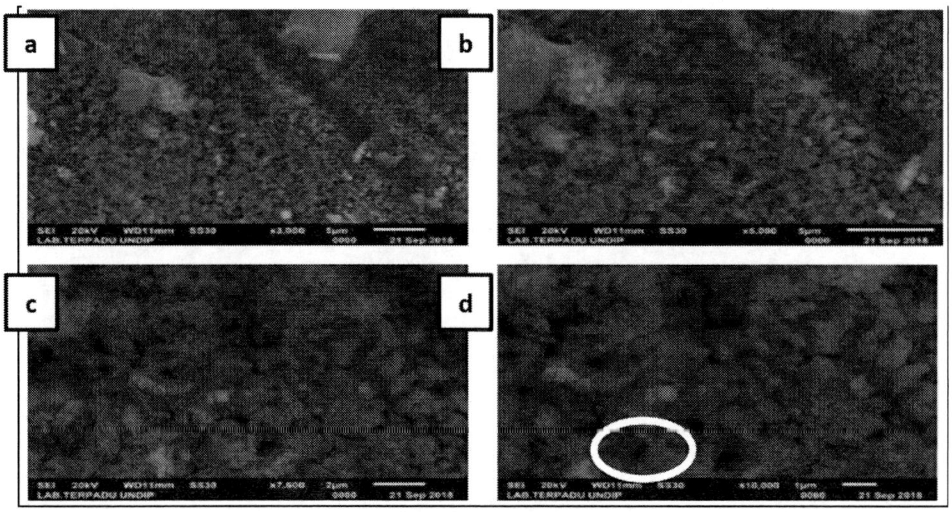


**Figure 3.** Unprotected carbon steel surface morphology (a) magnification 3.000x (b) magnification 5.000x (c) magnification 7.500x (d) magnification 10.000x

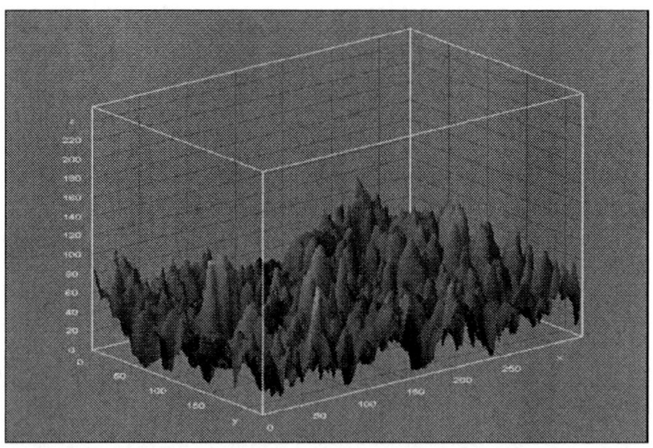

**Figure 4.** Unprotected carbon steel surface structures with 10,000x magnification using ImageJ

Figure 3 shows the results of SEM carbon steel which was exposed to corrosion due to immersion in a solution of NaCl for 30 days, where the solution is corrosive to the metal resulting in changes in the surface of the specimen due to corrosion. Figure 4 has a dark-colored part, this is often called pitting corrosion that occurs at grain boundaries that can be seen clearly in images with a magnification of 10,000x. In carbon steel, many dark-colored images are found which show that many are attacked by corrosion on the surface of carbon steel without protection, this is clarified by using the ImageJ program in figure 4 where the surface structure of the carbon steel specimen is uneven caused by corrosion of the well and in the presence of changes in the chemical composition of the material on the surface where there is an element of oxygen contained in the material. In this case, the breakdown of the passive layer by the Cl-ions occurs in certain parts of the metal surface (unevenly) which is indicated by the black part in figure 4. When oxygen reacts with the steel surface, a reaction between oxygen and iron will occur. will form iron (II) oxide (FeO) proven from the Energy Dispersive X-Ray (EDX) test results. The reaction that occurs is an oxidation reaction, which is the combining of oxygen with ferrous metals that form iron rust [16].

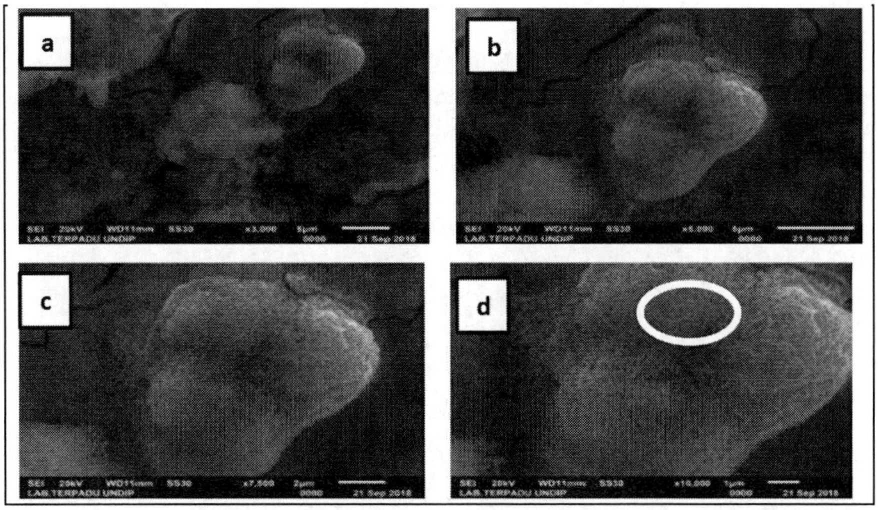


**Figure 5** Surface carbon morphology with zinc protection (a) magnification 3.000x (b) magnification 5.000x (c) magnification 7.500x (d) magnification 10.000x

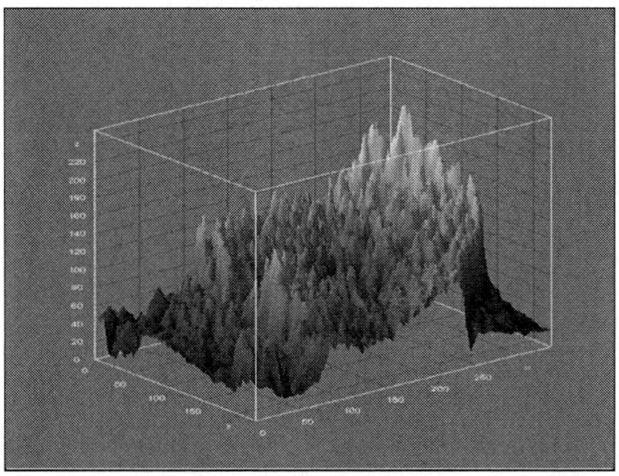

**Figure 6** Carbon Steel Surface Structure with Zinc Protection with magnification 10,000x using Image

Figure 5 shows the surface of the corroded specimen forming a needle that damages the specimen surface. The type of corrosion that is seen is pitting corrosion or corrosion where it arises and is concentrated in a particular area or locally caused by the deposition that acts as a center of corrosion [17]. Magnification to 10,000x, a number of needles were seen which indicated that the specimen was corroded but not as much as what had happened to unprotected specimens, seen from not as much black picture in figure 3. In figure 6 it can be seen that the specimen surface at 10,000x magnification using Image J the steel surface structure is uneven due to the formation of holes due to the corrosion process that has occurred. In pitting corrosion the corrosive agent (Sodium Chloride solution) will previously damage the protective layer of the carbon steel surface which causes the pores of the steel to open so that it enters the inside and the corrosive substance will damage the metal so that it forms a needle-like hole [18]. EDX analysis results show that the compounds that cause rust, namely iron (II) oxide or FeO are more abundant in carbon steel without protection that is equal to 93.38%, while carbon steel with zinc protection is 68.22%.

## 4. Conclusion

The highest efficiency of corrosion rate on carbon steel protected by zinc sacrificial anode was 81,8%. The longer immersion time produced more cathode weight loss. Immersion time of 10 days resulted in better efficiency. The performance of zinc sacrificial anode was effective to protect carbon steel from corrosion when the distance between the anode and the cathode electrodes was the smallest.

REFERENCE

[1] Sasono E J, Sulaiman, Darmanto S and Supriyo E 2014 *J. Universitas Diponegoro.* **1(9)** 28
[2] Afriani F, Komalasari and Zultiniar 2014 *J. Universitas Riau* **1(2)** 1.
[3] Abdoel G 2011 Rancangan dasar perhitungan proteksi katodik dengan menggunakan anoda korban pada struktur baja anjungan minyak di lingkungan air laut (*Majalah LPL* vol 45) (Jakarta: Lemigas) ed 1 pp 82.
[4] ASTM G31-72 2004 *Standard Practice for Laboratory Immersion Corrosion Testing of Materials* (United States: Reapproved).
[5] ASTM G1-90 1999 *Standard Practice for Preparing, Cleaning and Evaluating Corrosion Test Spesimens* (United States: Reapproved)
[6] Adiyanti Y R and Budi A 2011 *J. Teknik Material dan Metalurgi* **1(1)** 7.
[7] Novita S 2018 Analisis laju korosi dan kekerasan pada stainless steel 304 dan baja nikel laterit dengan variasi kadar Ni (0,3 dan 10% Ni) dalam medium korosif (Skripsi) (Bandar Lampung: Universitas Lampung)
[8] Ilham R, Komalasari and Rozanna S I 2015 *J. Universitas Riau.* **2(2)** 1
[9] Alphanoda A F 2016 *J. Teknologi Rekayasa.* **1(1)** 1.
[10] Juliana A, Tjitro S and Rizal V 1999 *J. Universitas Kristen Petra* **2(1)** 89
[11] Denny A J 1992. *Principles and Prevention of Corrosion* (New York: Macmillan Publishing Company,inc.)
[12] Pathasardhy N V 1998 *Practical Electroplating Handbook* (New Jersey: Prentice Hall Inc.)
[13] Utami I 2009 *J. Teknik Kimia* UPN Veteran **2(3)** 240
[14] Ali F, Desy S and Raka F N 2014 *J. Teknik Kimia* **20(1)** 35
[15] Muzkantri V R and Diah H K 2015 *J. Inovasi Fisika Indonesia* **4(3)** 61
[16] Bayuseno A P and Handoko E D 2012 Analisa Korosi Erosi Pada Baja Karbon Rendah dan Baja Karbon Sedang Akibat Aliran Air Laut (Tesis) (Semarang: Universitas Diponegoro)
[17] Ma J and Jiuba W 2010 *J. of Alloys and Compound* **496** 110
[18] Yanuar F and Dwi H S 2015 *J. Teknik Mesin Universitas Negeri Surabaya* **4(1)** 6

# The extraction of oil from cooling pond wastewater as a raw material for biodiesel

**T. E. Agustina[1*], I. M. Chandra[1], M. Nopriyansyah[1], S. Arita[1], F. Hadiah[1], T. I. Sari[1], T. Prakoso[2] and E. Heraldy[3]**

[1]Chemical Engineering Department, Faculty of Engineering, Universitas Sriwijaya
[2]Chemical Engineering Department, Faculty of Engineering, Bandung Institute of Technology
[3]Department of Chemistry, Faculty of Mathematics and Natural Sciences, University of Sebelas Maret

Email: tuty_agustina@unsri.ac.id

**Abstract**. The environmental pollution caused by the waste and the strict regulations of waste were currently a major concern for industries, such as the palm oil industry. Constraints experienced today are the difficulty of degradation process of the waste due to the high quantity and the content of contaminants in the waste. One of the efforts that can be prepared to reduce the negative impact is by using the waste as a raw material for biodiesel. Among the potential wastes to become a raw material for making the biodiesel is the wastewater from the cooling pond as a part of the wastewater treatment in palm oil industry. The objectives of this research are to study the process of oil separation from wastewater cooling pond by liquid-liquid extraction method to recover oil and use it as a raw material for making biodiesel. Extraction was carried out at room temperature. The solvent type, wastewater sample to solvent ratio, and extraction time were varied. Based on the results, the highest oil yield of 90% was reached by using n-hexane solvent, the wastewater sample to solvent ratio of 2: 3, and extraction time of 72 hours, while the acid number of 78.49 mg KOH/g oil was obtained. The separated oil can be used as a raw material for making biodiesel through the esterification and transesterification processes. The biodiesel produced has been met the standard of SNI 0471822006 for kinematic viscosity (0.83 cSt), acid number (0.54 mg KOH/g), methyl ester (99.40%), free glycerol (0.01%), and total glycerol (0.13%).

## 1. Introduction

The palm oil industry has significant benefits, yet it causes negative impacts, both to the quality and quantity of the natural resources and environment. The waste cause pollution and therefore should be treated before they are disposed to the environment. The waste in this regard refers to the by-products of the palm oil industry, usually in the liquid and solid forms. Generally, the solid wastes are directly released without being processed, and this is harmful to the environment. However, the study on utilization of waste from oil palm industry (empty fruit bunch) mixed with waste oil (used engine oil and used cooking oil) to produce solid fuel product have been reported [1]. Whereas the wastewater is often treated in moderation before being released to the water body.

* To whom any correspondence should be addressed.

The wastewater should be treated to meet the required standards of the Ministry of Environment before discharged to the environment. Table 1 shows the quality of wastewater based on the environmental parameter set by the Ministry of Environment of Republic of Indonesia [2].

**Table 1**. The general quality of wastewater released from palm oil mill

| No. | Environmental Parameter | Wastewater | |
|---|---|---|---|
| | | Estimation (mg/L) | Average (mg/L) |
| 1. | BOD | 8,200-35,000 | 21,280 |
| 2. | COD | 15,103-65,100 | 34,720 |
| 3. | TSS | 1,330-50,700 | 31,170 |
| 4. | pH | 3.3-4.6 | 4 |
| 5. | Total of nitrogen | 12-126 | 41 |
| 6. | Oil and fat | 190-14,720 | 3,075 |

The waste derived from the palm industry might be used after being processed. Importantly, these wastes, especially from the cooling ponds, have the potential to be used as raw materials for biodiesel. The waste comes from a fat pit pond flowed to the cooling pond meant to cool the hot wastewater to a temperature which activates the bacteria. Essentially, the cooling process is vital in preparing the life of bacteria needed in the waste processing [3], apart from precipitating the sludge. Also, the wastewater from the cooling pond flows to the anaerobic pond for the next biological processing [4].

The cooling pond wastewater contains oil and fat of around 6000 mg/L and should be extracted. This is due to the fact that it is classified as the low rank of oil and a high fatty acid content, apart from being cheaper and readily available [5]. Several methods have been used to separate the oil waste, such as the method of biological and chemical adsorption and extraction. Ahmad et al., (2008) conducted a study on the recovery process of oil and carotene from the POME waste by extracting the solvent to retake the oil left in the palm waste. The result showed that the oil derived from the solvent extraction of n-hexane was higher than petroleum ether, which was about 3280 mg/L using n-hexane solvent and 1710 mg/L with petroleum ether [6]. This finding showed that the more solvent used in the extraction process, the more oil would be obtained.

The purpose of this study, therefore, was to obtain the oil from the cooling pond which would be utilized as the primary material of biodiesel. The effect of solvent type, the ratio of raw material (wastewater sample) to the solvent, and the duration of the extraction of the oil from cooling pond wastewater were also studied.

## 2. Experimental

### *2.1 Sample Preparation*

The wastewater sample of the cooling pond used in this study was obtained from one of PTPN Plantation in South Sumatra. The samples and solvent (n-Hexane, ethyl acetate, and methanol) were measured and put in the beaker glass in the volume ratio of 1:1, 3:2, 2:3, and 2:1 with the mixed volume of sample and solvent of 300 ml. The mixture was stirred with a magnetic stirrer for 1 hour and put in the separating funnel for 24 hours, 48 hours, and 72 hours. The products formed layers which were separated by opening the faucet under the separating funnel for the impurities to be released. And only the mixture of oil and solvent remained, which is separated by a simple distillation method. The oil from distillation was measured and weighed, while the solvent separated was quantified to determine the amount recovered.

### *2.2 Extraction Process*

In this study, the extraction was conducted at the room temperature to prevent the excessive evaporation of the solvent and involved stirring using a magnetic stirrer. The stirring process was done to prevent the solvation of the waste preventing the solvent from penetrating the material and draw the oil since the waste used enormously. Additionally, the stirring was also carried out to fasten the balance of solvent and sample, broaden the contact wide, and equalize the extraction process.

This study was conducted to determine the effect of a solvent in the extraction process, the volume ratio of the cooling pond wastewater to the solvent, along with the duration of extraction, on the quantity and quality of the oil produced. Importantly, the quantity of the oil was determined by the yield, while the acid number indicated the quality. The solvent used was chosen based on its characteristics, such as cheap and readily available, low boiling point, and inertness. The solvent had low boiling point about 68.73°C (n-hexane), 77°C (ethyl acetate), and 64.7°C (methanol). The low boiling point aimed to ease the evaporation process without using high temperatures. The solvent did not have a low boiling point as this would sharply decrease the solvent during the evaporation process. The solvent should also be inert, which meant it should not react with the cooling pond wastewater. According to Krisyanti and Sukandar (2011), temperature was one of the factors affecting the extraction process. A higher temperature produced more oil and vice versa. However, the quality of the oil produced in this process was declining due to the heating process [7]. The acid number determined the quality of the oil. In general, the oil with a very high acid number was the one hydrolyzed and oxidized. The free fatty acid content was generated in the hydrolysis and oxidation process of the oil [8]. Moreover, the hydrolysis reaction occurred due to the excessive amount of water in the oil and the reaction cause a rancidity of the oil. The oil in the hydrolysis process was then changed into free fatty acid and glycerol.

## 3. Results and discussion

### *3.1. The effect of solvent type on the yield of oil and acid number*

The variation in solvent was meant to determine the effect of solvent on the quantity and quality of the oil obtained. Additionally, the variation was also meant to establish the best solvent used in the production stage of the biodiesel material. The yield of oil obtained by using different solvent during 24-72 hours of extraction is shown in Figure 1. Evidently, the solvent type used affected the yield of oil, where the highest yield of 71.67%, was achieved using n-hexane solvent in the sample to solvent volume ratio of 2:3, while the lowest yield of 47.00%, was obtained using methanol using the sample to solvent volume ratio of 2:1, during 24 hours of extraction as can be seen in Figure 1 (a). Since methanol is a polar solvent, it was not fit in recovering the oil from the cooling pond wastewater and is often used to extract plants [9]. This is because the polar solvent dissolved in water and was also hard to be recovered and re-used in the subsequent extraction process.

N-hexane produced a high yield of oil than methanol and ethyl acetate since they were non-polar solvents. h-hexane is a non-polar solvent that is most widely used for oil extraction because it has high susceptibility, ability to dissolve high oil, and low boiling point so that it is easy to recover [10]. It dissolved the oil, and it was easy to recover and use in the next extraction process. The use of n-hexane yielded a higher yield of oil compared to ethyl acetate because n-hexane had a smaller dielectric constant compared to ethyl acetate. The smaller the dielectric constant of a solvent led to the more non-polar of a solvent [11]. For this reason, n-hexane had a high rate of dissolved and dissolved the oil more than ethyl acetate.

The yield of oil achieved after 48 and 72 hours of extraction are shown in Figure 1 (b) and (c). The effect of the solvent type on the yield of oil showed the same tendency with a 24 hours of extraction, where the n-hexane solvent had higher yield of oil than methanol and ethyl acetate. Based on Figure 1 (b) and (c), the lowest yield of oil of 56.50% was found using methanol with the sample to solvent volume ratio of 2:1, while the highest yield of oil of 90.00%, was obtained using n-hexane with the sample to solvent volume ratio of 2:3.

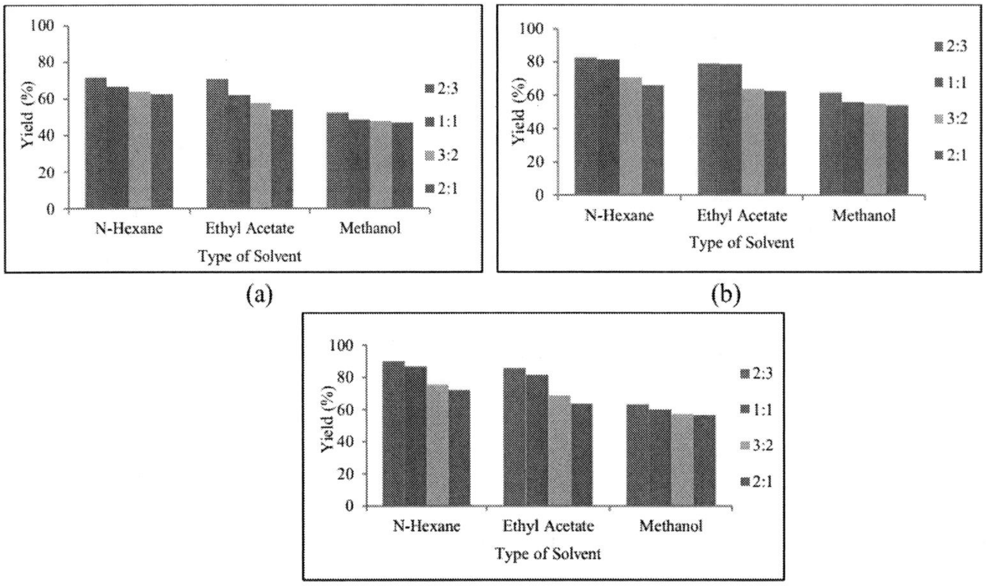


**Figure 1.** The effect of solvent on yield of oil in extraction process of (a) 24 hours; (b) 48 hours; and (c) 72 hours

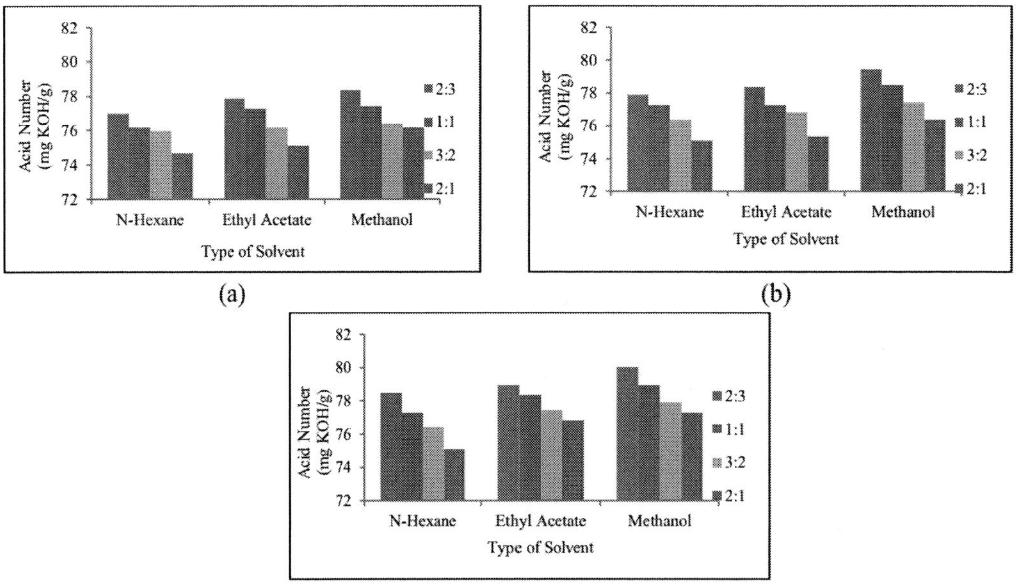


**Figure 2.** The Effect of solvent on the acid number obtained during (a) 24 hours; (b) 48 hours; and (c) 72 hours of extraction

Depending on the type of solvent, the acid number of the oil recovered during 24 hours of extraction is shown in Figure 2 (a). In the Figure, the highest acid number of 78.36 mg KOH/g, was formed using methanol with the sample to solvent volume ratio of 2:3, while the lowest acid number of 74.69% mg KOH/g was obtained using n-hexane with the sample to solvent volume ratio of 2:1. In this study, the high acid number during the extraction process was attributed to the fact that methanol was a polar solvent. The water caused triglycerides hydrolyzed into free fatty acid, leading to the increase in the acid number. The acid number of resulted oil from extraction using n-hexane and ethyl acetate lower than using methanol was due to the fact that both n-hexane and ethyl acetate were quite selective in dissolving a non-polar substance, such as triglycerides.

Figure 2 (b) and (c) showed that during extraction for 48 and 72 hours, respectively, the effect of solvent on the acid number showed the same result as the 24 hours of extraction, which meant the more polar the solvent used, the higher the rate of oil acid produced in the extraction process. After 72 hours of extraction, the lowest acid number of 75.10 mg KOH/g was obtained using n-hexane with sample to solvent volume ratio being 2:1. The highest acid number of 80.03 mg KOH/g was found by using methanol with sample to solvent volume ratio being 2:3.

### *3.2. The effect of wastewater sample to solvent volume ratio on the yield of oil and acid number*

The variation in sample to solvent volume ratio was meant to determine the effect of sample to solvent volume ratio on the quantity and quality of the oil produced in the extraction process. This variation was also used to establish the most optimum volume ratio of sample to solvent used during the production stage of the biodiesel. The ratio of waste volume and solvent in the process were 1:1, 2:1, 3:2, and 2:3 with a total volume of 300 ml.

The yield of oil in variation of sample to solvent volume ratio using n-hexane, ethyl acetate, and methanol, respectively, were shown in Figure 3 (a), (b), and (c). The result of the study showed the more volume of solvent used, the more oil obtained. Additionally, Figure 3 (a) showed a higher amount of solvent led to a more significant amount of oil being extracted.

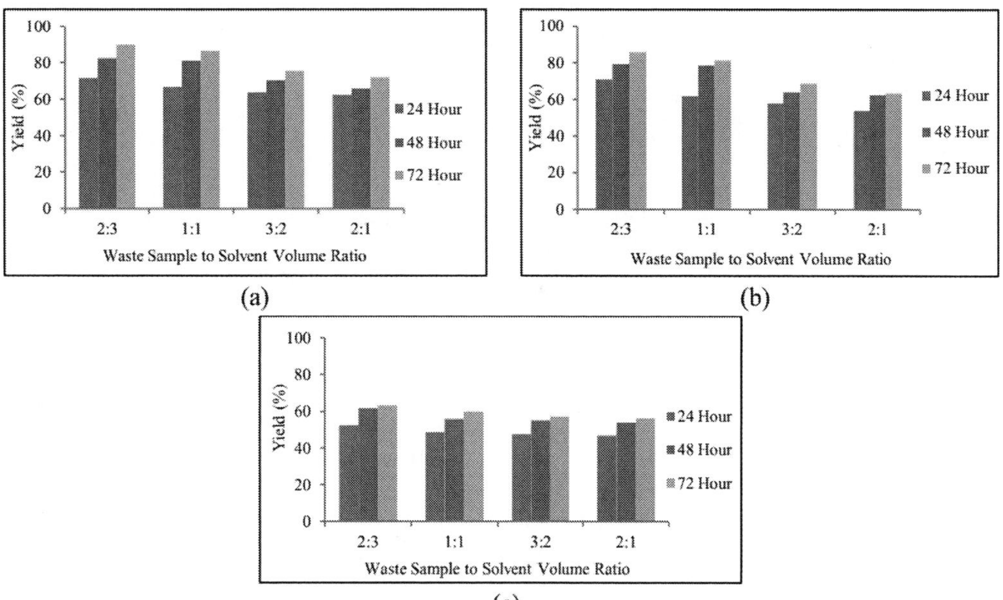


**Figure 3.** The Effect of sample to solvent volume ratio on the yield of oil produced using (a) n-hexane; (b) ethyl acetate, and (c) methanol solvent

The highest yield of oil of 90%, for the n-hexane solvent was obtained with the sample to solvent volume ratio of 2:3 during 72 hours of extraction as can be seen in Figure 3 (a). This high yield of oil resulted from the adequate amount solvent to penetrate the wastewater, and therefore the oil contained in the waste could be dissolved by the solvent optimally. The lowest yield of oil of 47%, was obtained using methanol solvent within the sample to solvent volume ratio of 2:1 during the 24 hours of extraction, as illustrated in Figure 3 (c). The low yield of oil was attributed to the inadequacy of the solvent to penetrate the waste and dissolve the oil. According to Ahmad et al., (2008), the yield of oil would increase with the upsurge in a solvent. However, after the solvent increased to a particular level, the increase of yield of oil would decline or even remain constant. If too many solvents are used, the yield of oil will decrease because the solvent not only dissolves the oil but also can dissolve other impurities. The use of too much solvent is also ineffective and inefficient because it can cause more and more other impurities to dissolve. The time used for the separation stage of the solvent and oil will also be longer, so there will be a decomposition of oil and other impurities obtained which can cause changes in the nature and composition of the oil.

The acid number gained using the n-hexane solvent was as shown in Figure 4 (a). A higher of volume of solvent compared to the volume of the waste would lead to a higher acid number. Moreover, the highest acid number of 78.49 mg KOH/g, was in the waste sample to solvent volume ratio of 2:3 during 72 hours of extraction. Contrastingly, the lowest acid number of 74.6 mg KOH/g, was produced in the sample to the solvent volume ratio of 2:1 during 24 hours of extraction. The highest acid number of 80.03 mg KOH/g, was obtained using methanol solvent in the waste to solvent ratio of 2:3 during 72 hours of extraction.

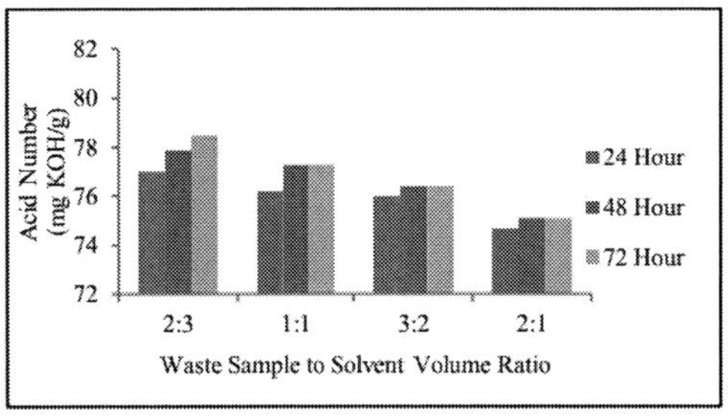


(a)

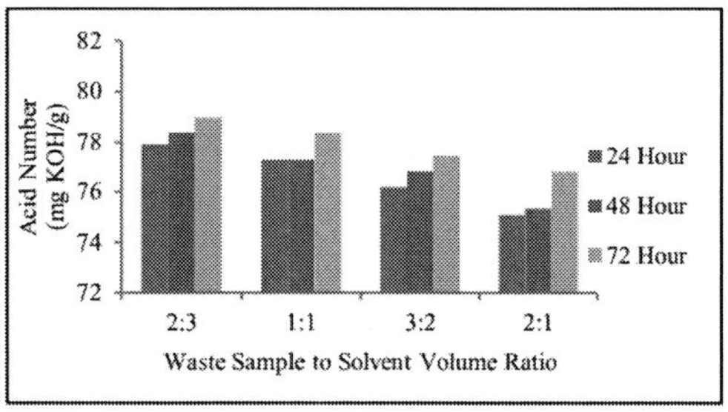


(b)

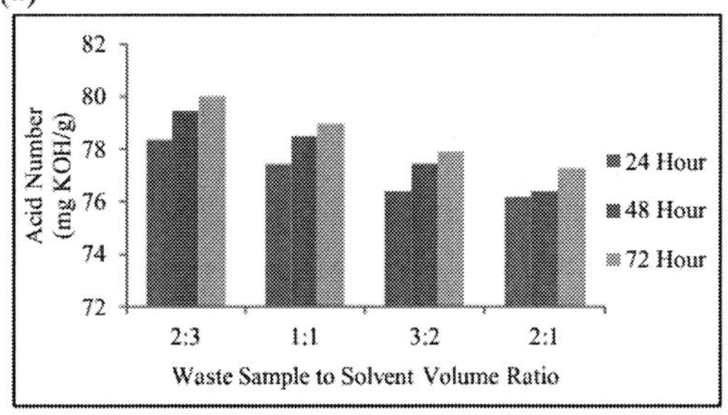


(c)

**Figure 4**. The Effect of Sample to Solvent Volume Ratio on the Acid Number Obtained using (a) n-hexane; (b) ethyl acetate; and (c) methanol Solvent

### *3.3. The effect of extraction time on the yield of oil and acid number*

The variation in extraction time was meant to determine the effect of extraction time between 24-72 hours on the quantity and quality of the oil obtained. Additionally, it was also purposed to establish the most optimum time, which was later used in the biodiesel production stage. The yield of oil produced in the variation of extraction time using the n-hexane solvent was as shown in Figure 5 (a).

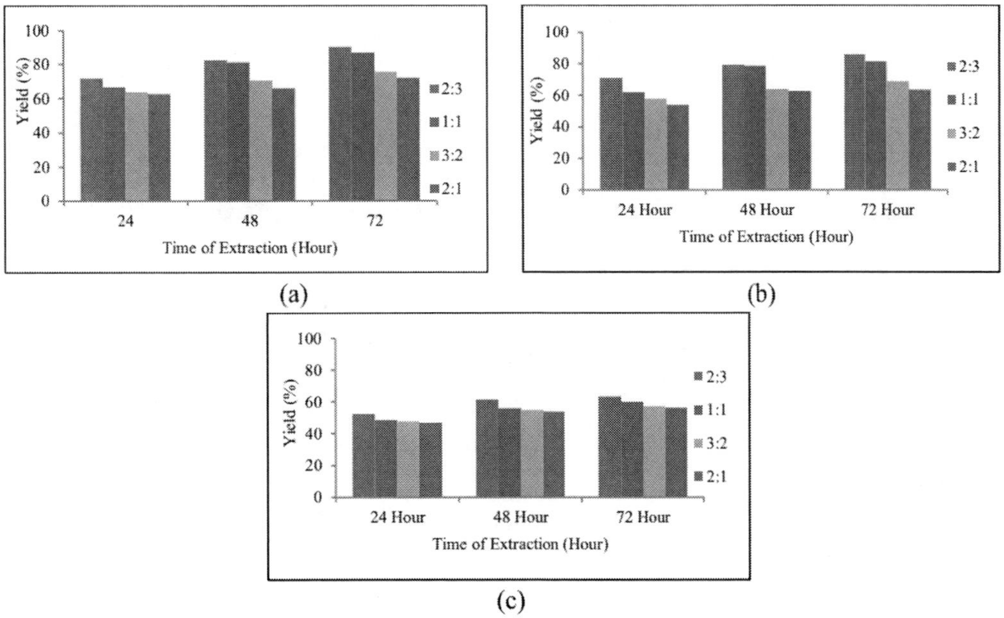


**Figure 5.** The Effect of Extraction Time on the Yield of Oil using (a) n-hexane; (b) ethyl acetate; and (c) methanol Solvent

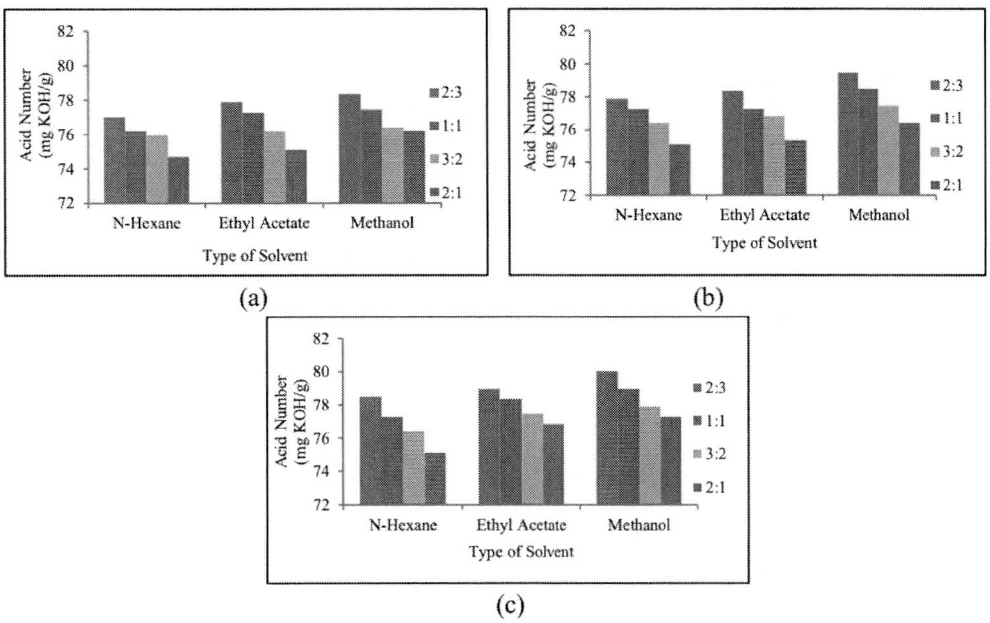


**Figure 6.** The Effect of Extraction Time on the Yield of Oil using (a) n-hexane; (b) ethyl acetate; and (c) methanol Solvent

The highest yield of oil of 90%, was found during 72 hours of extraction with the sample to solvent volume ratio of 2:1. Contrastingly, the lowest yield of oil of 62.50%, was obtained in the 24 hours of extraction with the sample to solvent volume ratio of 2:1, as can be seen in Figure 5 (a). Similar results were found when using ethyl acetate and methanol, the highest yield of oil was attained after 72 hours of extraction while the lowest yield of oil was reached after 24 hours of extraction, as demonstrated in

Figure 5 (b) and (c). Moreover, the longer the extraction time, the more the yield of oil produced. The long duration of extraction allowed the contact process between solvent and the wastewater sample. The dissolving rate of waste also increased with the increase in the extraction time, leading to solvent saturation. This caused the decrease of yield of oil that marked the optimum time of extraction since the amount of oil component was limited, and the solvent used had the ability to dissolve the waste.

The effect of extraction time on the acid number using n-hexane solvent was as shown in Figure 6 (a). It could be concluded that a longer extraction time would increase the acid number. Also, the lowest acid number of 74.69 mg KOH/g, was obtained during 24 hours of extraction with the sample to solvent volume ratio of 2:1. In contrast, the highest acid number of 78.49 mg KOH/g, was produced during 72 hours of extraction with the sample to solvent volume ratio of 2:3.

The effect of extraction time on the acid number produced using ethyl acetate and methanol were shown in Figure 6 (b) and (c), respectively. The finding was similar to the use of n-hexane solvent, showing the longer the extraction time would increase the acid number. Around of 5% increase in acid number was obtained during 72 hours of extraction

### *3.4. The study of cooling pond wastewater as the raw material of biodiesel*

The oil produced from the extraction of the cooling pond had the potential to be used as the raw material of biodiesel. The oil that will be processed into biodiesel is the oil that produces the highest yield of oil by utilizing n-hexane solvent with the sample to solvent volume ratio of 2:3 during 72 hours of extraction time. The oil obtained from the extraction of cooling pond wastewater still contained various fats, such as free fatty acid and triglycerides. The free fatty acid contained in the oil was one of the determining factors in making the biodiesel. The free fatty acid content in the cooling pond wastewater before esterification was 35.82%.

The oil still contained a higher acid number to be used as the material of biodiesel. It would produce soap through soaping reaction suppose it was used as the primary material during the transesterification process using a base catalyst. To decrease the free fatty acid content, esterification using acid catalyst HCI was needed. Later, the transesterification process was conducted using a base catalyst of NaOH to obtain the biodiesel product with characteristics which met the standard of SNI 04-7182-2006.

The esterification reaction was carried out at the temperature of 65°C using acid catalyst of HCI of about 1.25% (%v) for 1 hour. The acid number gained in this process was 2.09 mg KOH/g with the free fatty acid of 0.96 %. The oil from esterification process was taken to the transesterification process. According to Gafar et al., (2012), the raw material used in the esterification process should have a free fatty acid less of than 2% to prevent the soaping formation [12].

In this study, the transesterification process was carried out at the temperature of 65°C using NaOH catalyst about 1.5 % of the oil volume. The acid number produced was 0.54 mg KOH/g, viscosity of 3.83 cSt, density of 0.84 g/ml, methyl ester 99.40%, free glycerol of 0.01%, and total glycerol of 0.13% and therefore the biodiesel met the standard of SNI 04-7182-2006. Nevertheless, the density did not meet the standard but the result is very close. From the study, it could be concluded that cooling pond wastewater had the potential to be used as raw material for biodiesel production.

## 4. Conclusion

Based on the results, it was concluded that with the more polar solvent used, the yield of oil kept decreasing while the acid number increased. Moreover, the greater the ratio of cooling pond wastewater sample to the solvent, the yield of oil and acid number obtained will increase. Also, the yield of oil and acid number were increasing with the longer of extraction time. The oil obtained can be used as a raw material for making biodiesel through the esterification and transesterification processes. The biodiesel produced has been met the standard of SNI 0471822006 for density, kinematic viscosity, acid number, saponification number, methyl ester, free glycerol, total glycerol, and oxidation stability.

## ACKNOWLEDGEMENT

The authors are very thankful to the Ministry of Research, Technology and Higher Education of the Republic of Indonesia for financial support through the Consortium Research Grant of 2018 and to Waste Treatment Technology Laboratory Chemical Engineering Department of Universitas Sriwijaya.

## REFERENCES

[1] Lam S S, Tsang Y F, Yek P N Y, Liew R K, Osman M S, Peng W, Lee W H, and Park Y K 2019 *Process Safety and Environ. Protec.* **128** 30.
[2] Department of Agriculture 2006 *Guidelines for Palm Oil Industry Waste Management* (Jakarta: Subdit Management of the Environment Ditjen PPHP)
[3] Himmatuliza A 2016 *Pengolahan Limbah Cair Industri Minyak Kelapa Sawit Menjadi Biogas Menggunakan Digester Balok Sistem Fed Batch* (Palembang: Politeknik Negeri Sriwijaya)
[4] Rois M and Fresillia H 2017 Strategi Pengolahan Limbah Cair di PT. AMP Plantation Jorong Tapian Kandih Nagari Salareh Aia Kecamatan Palembayan Kabupaten Agam *Tunas Geografi* **6** 116.
[5] Leela D, Nur S M, Yandri E, and Ariati R 2018. Performance of Palm Oil Mill Effluent (POME) as Biodiesel Source Based on Different Ponds *E3S Web Conf* **67** 1
[6] Ahmad A L, Chan C Y, Shukor S R A, and Mashitah M D 2008 *Chem. Eng. J.* **141** 383.
[7] Krisyanti S and Sukandar 2011 Recovery Minyak dari Limbah Bahan Berbahaya dan Beracun (B3) Spent Bleaching Earth dengan Metode Ekstraksi Pelarut *Jurnal Teknik Lingkungan* **17** 35
[8] Permana D G 2009 *Recovery Minyak Sawit dari Limbah Bahan Pemucat dengan Metode Ekstraksi Menggunakan Pelarut Organik* (Bogor: Institut Pertanian Bogor)
[9] Diem Q D, Angkawijaya A E, Tran-Nguyen P L, Huynh L H, Soetaredjo F E, Ismadji S, and Ju Y H 2014 *J. of Food and Drug Analy.* **22** 296.
[10] Diphare M and Muzenda E 2014 *Int. J. of Res. in Chem.* **1** 75.
[11] Widiati S 2011 *Daya Hambat Ekstrak Ampas Teh Hitam (Camelia Sinensis L.) terhadap Pertumbuhan Staphylococcus Epidermidis* (Yogyakarta: Universitas Atma Jaya Yogyakarta)
[12] Gafar A, Nurdin M, and Natsir M. 2012. Sintesis dan Uji Kualitas Biodiesel dari Proses Transesterifikasi Limbah Cair Pabrik Minyak Kelapa Sawit *Jurnal Progres Kimia Sains* **2** 11

# The Effect of Plasticizers towards the Characteristics of Methylcellulose Film Packaging

**S F A Karim[1]*, N A N Hamzah[1], R A A Aziz[1] and U K Ibrahim[1]**

[1]Faculty of Chemical Engineering, Universiti Teknologi MARA, 40450 Shah Alam, Selangor, Malaysia

*E-mail: sitifatma9633@uitm.edu.my

**Abstract.** The usage of petroleum-based film packaging causes a serious environmental problem, which affects human health as well as marine life. The main objective of this study is to study the effect of different plasticizers on the methylcellulose (MC) based film food packaging. Sorbitol and red palm oil (RPO) were used as plasticizers. The films were prepared by using solution-casting method before being characterized. FTIR spectrum showed the presence of 3361 to 3305 $cm^{-1}$ peak as the presence of hydroxyl group in sorbitol and 2925 $cm^{-1}$ to 2853 $cm^{-1}$ peak corresponded to the C-H stretches of the alkanes of RPO. All films appeared in white colour for sorbitol while yellowish for films with RPO. Tensile strength and elastic modulus decreased over increasing amount of sorbitol or RPO due to the modified interactions between the plasticizers and the chemical structure of the film. Overall, RPO can be used as the substitute of sorbitol to incorporate with MC.

## 1. Introduction

Food packaging helps in extending the shelf life of the food and provides the physical barrier between the food and outside environment. The advantages of using bio-based food packaging are it is biodegradable, renewable and usually edible [1]. Food packaging can be in various forms such as films, plastics, trays and foamed products. The most commonly used bio-based polymers, as food packaging are polyhydroalkanoates (PHAs), polylactide (PLA), PLA blends, starch blends and cellulose derivatives. Cellulose can be gained from nature materials such as woods, hemp and plant-based materials as well as microorganisms and synthesis of tunicates [2].

Methylcellulose (MC) is one of the cellulose derivatives that are widely used as food packaging because it can be used as environmentally friendly products due to its large availability, inexpensive and easy processability [3]. Methylcellulose has various advantages such as biodegradable, biocompatible, non-toxicity and good film-forming polymer [4]. Methylcellulose are very effective oxygen, carbon dioxide and lipid barriers even though it has poor water vapour transport [5]. Methylcellulose also has the ability to form thermally induced coating, which can be used to cover fried foods **Error! Reference source not found.**. However, methylcellulose is very brittle. Adding plasticizers can improve the mechanical properties of the films by increasing extensibility, flexibility, elasticity and mechanical properties.

The most used plasticizers in films are glycerol followed by sorbitol and ethylene glycol [6]. Sorbitol is usually used in food packaging film because non-toxic and chemically stable. Sorbitol concentration in polymer based film affecting the hardness and elastic modulus (EM) of the film. Sorbitol is one of common plasticizers used in film food packaging application which is non-toxic and chemically stable [7]. Different concentration of sorbitol will affect the hardness and elastic modulus (EM) of the films [8]. The other plasticizer that is being investigated is red palm oil (RPO), which is a bio-based plasticizer [9]. The benefits of RPO are non-toxic, biodegradable, high resistance to heat, renewable and environmentally friendly. It also can improve the film appearance and expected to improve the elongation of the films because plasticizer able to improve the elongation of film [10]. It had been proven that RPO can replace the chemical-based plasticizers in the future [11]. Therefore, the objective of this paper was to synthesize and compare the characteristics, light barrier and tensile strength of MC based film with different type of plasticizer.

## 2. Methodology

### *2.1. Materials*

Methylcellulose (MC) type A4C (powder form, viscosity: 400 cP) was purchased from Take It Global Sdn. Bhd. Sorbitol with MW of 182.17 g/mol was purchased from Merck Sdn. Bhd (Millipore, Darmstadt, Germany) in white powder form. Pure red palm oil food grade (NaOH: 0.141, KOH: 0.1974, INS: 145) was purchased from IKO store, Malaysia.

### *2.2. Preparation of Methylcellulose based-film*

The films were prepared by using solution casting method. Three grams of MC was dissolved in 100 ml distilled water under continuous stirring. Different concentration of sorbitol, which were 1%, 2%, 3%, 4% and 5% (MCS1, MCS2, MCS3, MCS4 and MCS5 respectively) were incorporated in the homogenous MC solutions. The MC incorporated with red palm oil (RPO) were prepared by using same step with different concentrations of RPO, which were 1%, 2%, 3%, 4% and 5% (MCRPO1, MCRPO2, MCRPO3, MCRPO4 and MCRPO5 respectively). Pure MC film, which is the film control (MCSTD) and the MC solution films were prepared by pouring 30 ml of the solutions on the petri dishes. The films were then left in the oven at 80°C until completely dried. The dried films were peeled manually by using spatula and cut into desired size for further analysis **Error! Reference source not found.**.

### *2.3. Fourier Transformation Infrared (FTIR) Spectroscopy*

The film samples were cut into 3 x 3 cm sizes and IR spectra of the samples were taken in the range of 500 to 4000 $cm^{-1}$. The sample was placed on the metallic slit sample holder and analysed with over 10 cumulative scans at 2 $cm^{-1}$ resolution [12].

### *2.4. Tensile Strength*

The tensile strength, elongation at break (EAB) and Elastic Modulus were measured by using ASTM-D882 as the guideline. The machine was equipped with 2.5 kN static load cell and velocity of the crosshead used was 500 mm/min. The samples were cut into strips of 15 mm wide and 60 mm long and placed at the grip to be analysed. The test was carried out at room temperature with regular humidity conditions [13].

## 3. Results and Discussion

### *3.1. Thickness, Color Measurement and Transparency*

Figure 1 shows the average thickness of MC films at different concentration of sorbitol or RPO. The thickness was increased as sorbitol concentration increased except at 3% sorbitol having decreased condition. Similar situations occur for RPO except for 2% RPO where the thickness slightly decreased. The MC film at 5% sorbitol (MCS5) was the thickest where similar results found by [14]. The present of sorbitol and RPO had resulted in thicker films due to formation of hydrogen bonding between the plasticizer and MC film. The addition of sorbitol and RPO increased the solid content of the films. However, sorbitol having better thickness at same concentration of RPO because sorbitol had higher content of hydroxyl [13], [15]. Therefore, greater interaction occur between MC and sorbitol during gelatinization, resulting thicker films. The thickness of film was important for packaging industry because affecting the value of tensile strength and elastic modulus of the film [16].

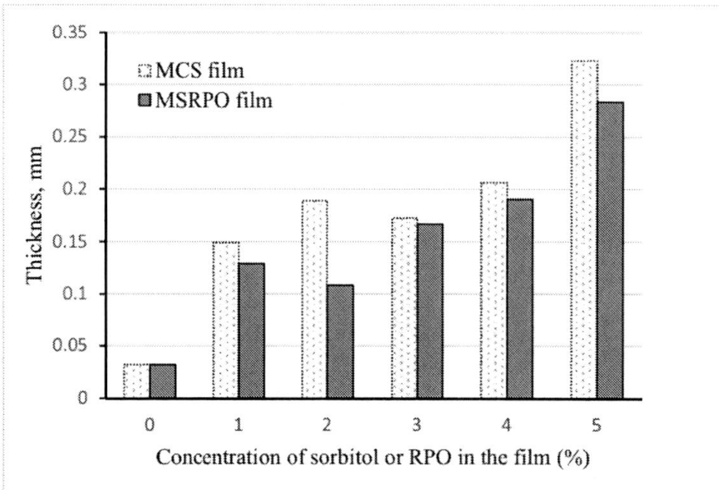


**Figure 1.** Thickness of MC films at different concentration of Sorbitol or RPO

*3.2. Fourier Transform Infrared Spectroscopy (FTIR)*

Figure 2 showed the FTIR result for MC films with sorbitol. From Figure 3, the peak at the 3558 $cm^{-1}$ shows the most relevant feature of MC structure, which this peak indicates O-H stretching band of hydroxyl group of MC. The broad absorption band of O-H stretching are getting narrowed and shifted from the peaks of 3361 to 3305 $cm^{-1}$ because the presence of hydroxyl group in sorbitol that lead towards formation of more hydrogen bonding [13], [15]. This indicate crosslink occur between sorbitol and MC. These changes can be seen in the MCS1 to MCS5 films [17] and also supported by increasing the thickness of the film when sorbitol was added into the film. Furthermore, with the addition of sorbitol, peak in the range of 2932 $cm^{-1}$ to 2923 $cm^{-1}$ that represent the C-H asymmetric and symmetric stretching of the sorbitol structure appear in MCS2 to MCS5 films. These similar spectra of sorbitol are shown in the study done by Lele Cao et al [17].

FTIR spectra of the MC/RPO films are shown in Figure 3. The peaks of the MC structure can be seen in all MC/RPO films. The spectrum that showed the present of RPO was the methyl and methylene groups at 2921 $cm^{-1}$ and 2853 $cm^{-1}$ as also being found by [18].The peaks in the range of 2925 $cm^{-1}$ to 2853 $cm^{-1}$ corresponded to the C-H stretches of the alkanes of RPO appear in all MC/RPO films and indicate successful crosslink between RPO and MC [19].

*3.3.* Mechanical *Properties*

Figure 4 showed the tensile strength of MC film at different concentration of sorbitol or RPO. MC control shows that the tensile strength of the film is 82.8 MPa. However, tensile strength value of MC film incorporated with 1% sorbitol (MCS1) is significantly decreased from 82.8 MPa to 16.8 MPa. The tensile strength value was then increased from 16.8 MPa to 20.4 MPa (MCS2). This might be due to the greater thickness of MCS2 (0.1890 mm) compared to MCS1 (0.1492 mm). Nevertheless, the tensile strength decreased from 20.4 (MCS2) to 3.23 MPa (MCS5) because the plasticizer amounts exceed the compatibility and sorbitol disturb the polypeptide chain [13].

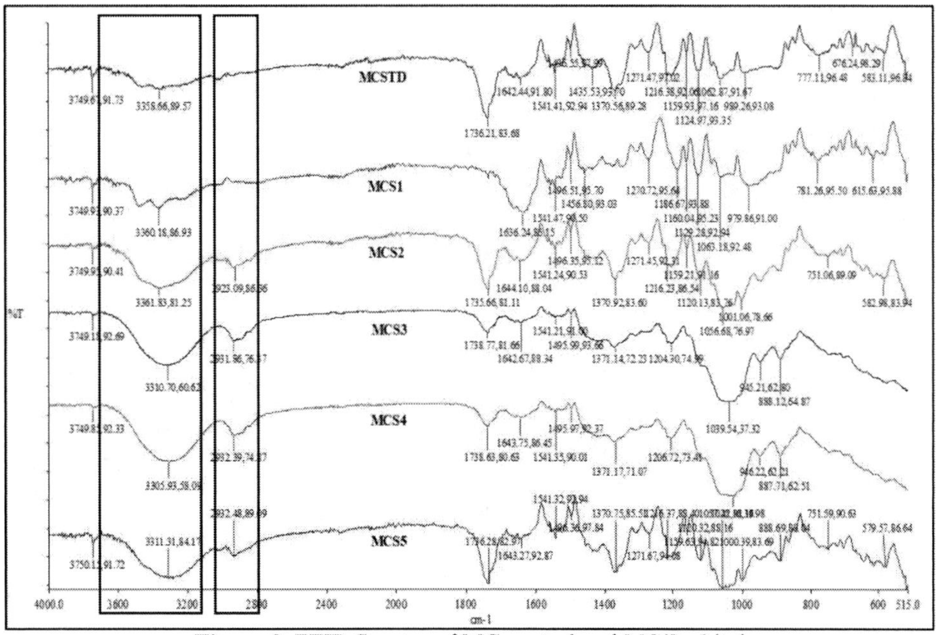


**Figure 2.** FTIR Spectra of MC control and MC/Sorbitol

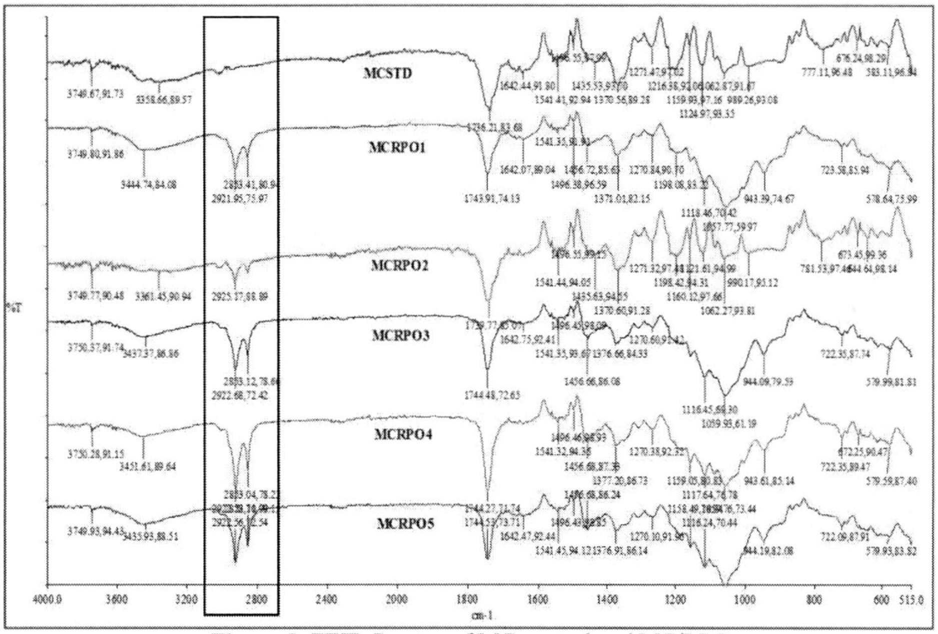


**Figure 3.** FTIR Spectra of MC control and MC/RPO

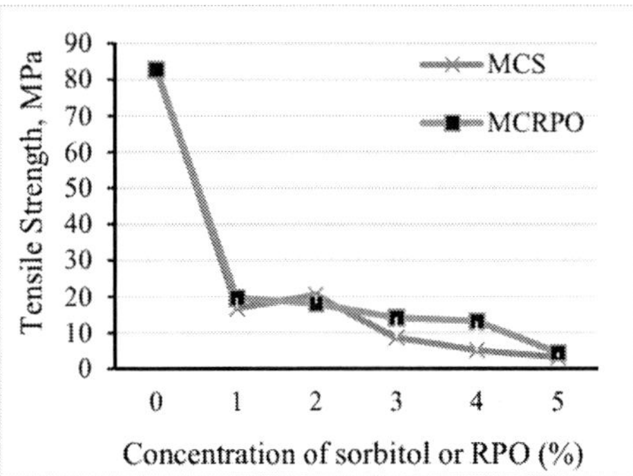


**Figure 4.** Tensile strength of MC films at different concentration of sorbitol or RPO

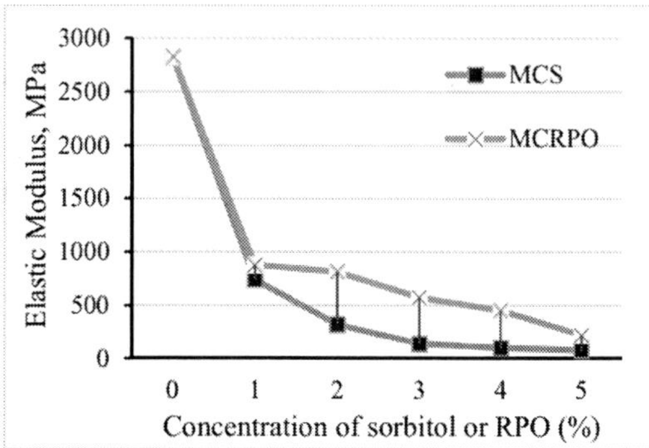


**Figure 5.** Elastic modulus of MC films at different concentration of sorbitol or RPO

For different concentration of RPO, tensile strength decreased from 19.7 (MCRPO1) to 4.55 MPa (MCRPO5) as the concentration of the RPO increased. The decreased value of tensile strength might occur due to the modified interactions between the plasticizers and the chemical structure of the film. These results are similar found by N. R. Saha et al **Error! Reference source not found.**. From the results in Figure 5, MC control film had the highest EM value of 2820 MPa while MC with addition of 5% sorbitol had the lowest EM value (80.50 MPa). The value of EM of MCRPO5 (225.00 MPa) was quite high compared to MCS3 (141.00 MPa), MCS4 (99.80 MPa) and MCS5 (80.50 MPa). This showed that the addition of the RPO to the film increased the elasticity of the MC film compared to incorporations of sorbitol. The smaller molecular weight of sorbitol compared to RPO had allowed increasing of molecular attraction and concurrently discourage the interaction of biopolymer chains. Therefore, plasticization better for sorbitol compared to RPO.

## 4. Conclusion

Sorbitol and RPO can be used as plasticizers for MC film. The present of solid content in plasticizer had improved the thickness of the film. The impact of redness of RPO had led the film to show yellowish colour but colourless film obtained for MC film with sorbitol. The thickness of the film and hydroxyl group of sorbitol had affected the transparency of MC film where the transparency level decreased over increasing value of sorbitol concentration. Higher concentration of RPO decreased the transparency level because lack of hydrogen bonding formed. The new peaks occurred at 3600 – 3020 $cm^{-1}$ were referred to formation of hydrogen bonding in the polymer chains of MC and sorbitol while addition of RPO showed new peak in the range of 2925 $cm^{-1}$ to 2853 $cm^{-1}$. This represent the C-H stretches of the alkanes of RPO. Elastic modulus of MC film with RPO gives better result compared to MC film with sorbitol. Therefore, this research need to be improved by adding compatibilizer to improve TS and EAB of film so that able to fulfil film packaging requirements.

**Acknowledgements**

Thank you for Faculty of Chemical Engineering for the equipment available and special appreciation to Kementerian Pendidikan Malaysia for the budget approval for RMI file number: 600-RMI/ST/FRGS 5/3/FST (37/2011).

## References

[1] Piñeros-Hernandez D, Medina-Jaramillo C, López-Córdoba A and Goyanes S 2017 *Food Hydrocoll.* **63** 488
[2] Nasatto P L, Pignon F, Silveira J L M, Duarte M E R, Noseda M D, and Rinaudo M 2015 *Polymers (Basel).* **7** 777
[3] Orasugh J T *et al.* 2018 *Carbohydr. Polym.* **188** 168

[4] Esmaeili A and Fazel E M 2016 *Flavour Fragr. J.*, **31** 341
[5] Suderman N, Isa M I N, and Sarbon N M 2018 *Food Biosci.* **24** 111
[6] Nguyen Vu H P and Lumdubwong N 2016 *Carbohydr. Polym.* **154** 112
[7] Ma X, Qiao C, Zhang J, and Xu J 2018 *Int. J. Biol. Macromol.*, **119** 1294
[8] Liu W *et al.* 2018 *Polymer (Guildf).* **148** 109
[9] Tee Y B, Talib R A, Abdan K, Chin N L, Basha R K, and Yunos K F M 2014 *Agric. Agric. Sci. Procedia* **2** 289
[10] Manshor N M, Rezali M I, Jai J, and Yahya A **2018** *IOP Conf. Ser. Mater. Sci. Eng.* **358** 1
[11] Halim A L A, Kamari A, and Phillip E 2018 *Int. J. Biol. Macromol.* **120** 1119
[12] Saha N R *et al.* 2016 *Carbohydr. Polym.* **136** 1218
[13] Bakry N F, Isa M I N, and Sarbon N M 2017 *Int. Food Res. J.* **24** 1753
[14] Gutiérrez T J and González G 2017 *Food Biophys.* **12** 11
[15] Noronha C M, De Carvalho S M, Lino R C, and Barreto P L M 2014 *Food Chem.* **159** 529
[16] Datta D and Halder G 2018 *Process Saf. Environ. Prot.* **114** 143
[17] Cao L, Liu W, and Wang L 2018 *J. Clean. Prod.* **175** 276
[18] Jia P, Xia H, Tang K, and Zhou Y 2018 *Polymers (Basel).* **10** 12
[19] Sharif Z I M *et al.* 2019 **507** 1

IOP Conf. Series: Materials Science and Engineering **845** (2020) 012018 doi:10.1088/1757-899X/845/1/012018

# Morphology and filtration performances of polyether sulfone membrane modified with carbon and $Fe_2O_3$

**R. Arisandi[1], A Fathurrahman[1], A. Fahrina[2,3], F Razi[1,2], Z Jalil[4] and N Arahman[1,5,6,7]*,**

[1] Department of Chemical Engineering, Universitas Syiah Kuala, Banda Aceh 23111, Indonesia;

[2] Magister Program of Chemical Engineering, Department of Chemical Engineering, Universitas Syiah Kuala, Banda Aceh 23111, Indonesia

[3] Doctor of Engineering, Graduate School of Engineering, Universitas Syiah Kuala, Banda Aceh 23111, Indonesia

[4] Material Physics Laboratory, Department of Physics, Faculty of Mathematics and Natural Science, SyiahKuala University, Banda Aceh, Indonesia

[5] Graduate School of Environmental Management, Universitas Syiah Kuala, Jl. Tgk Chik Pante Kulu No. 5 , Darussalam, Banda Aceh, 23111, Indonesia

[6] Research center for environmental and natural resources, Universitas Syiah Kuala, Jl. Hamzah Fansuri, No. 4, Darussalam, Banda Aceh, 23111Indonesia

[7] Atsiri Research Center, Universitas Syiah Kuala, Banda Aceh 23111, Indonesia;

E-mail: nasrular@unsyiah.ac.id

**Abstract.** Membrane separation technology has been widely applied in various industrial sectors. Some modification on membrane composition and structure is conducted to improve membrane filtration performance and mitigate fouling phenomena. Fouling on membrane surfaces lead to decrease permeabilty, increase energy consumtion, and shorten membrane life time. The aim of this study is to manufacture PES membrane which involved inorganic additives consits of Carbon and $Fe_2O_3$ particles to mitigate fouling tendency. All of membranes were prepared using phase inversion method with dissolving PES polymer into dimethyl formamide (DMF). The result membranes were characterized using Scanning Electron Microscopy (SEM) to investigate morphological structure and breaking strength to determine mechanical property. Membrane filtration performance were tested using soy protein solution in cross-flow system. The results showed that the addition of inorganic particles in PES membrane caused the increasing of pores which lead to increase pure water flux from 15.14 $L/m^2.h$ to 58.67 $L/m^2.h$. Fouling mitigation test towards soy protein also showed that carbon and $Fe_2O_3$ particles enhanced membrane flux recovery ratio from 64.5 to 76%. Furthermore, the inorganic particles presence in PES membrane also improved the tensile strength of membrane up to 128 Mpa.

## 1. Introduction

Polymer-based membranes are commonly used in the separation industry because it requires lower energy for operation and maintenance. The membrane can be formed from a polymer, inorganic materials, or the combination of organic and inorganic chemicals. One of the polymers used as a membrane material is polyethersulfone (PES). PES contains ether and sulfone chains which alternate between benzene rings to provide a unique structure with high mechanical, and resistance to extreme liquids [1-3]. Conversely, the hydrophobic nature of PES is the drawback of its application in ultrafiltration for purification of water or concentrated of protein. Flux declined is easily occur on membrane filtration due to fouling phenomena.

Many researchers modify the membrane structure to maintain the stability of filtration. Several inorganic substances have been used, such as $Al_2O_3$, $SiO_2$, ZnO, $TiO_2$. Uzal et al [4] produced membranes via non-solvent induced phase separation technique by blending Al2O3. The membranes with increased mechanical strength, higher porosity, large flux rates, and lower water contact angle were successfully obtained. Other authors used $Fe_2O_3$ as an additive to produce polyvinyl chloride membrane. It is confirmed that the addition of $Fe_2O_3$ nanoparticles greatly enhances the overall pore structure of the membrane. The hydrophilicity properties of the membrane were also increased. The addition of $Fe_2O_3$ into polymer solution may increase the mechanical strength, water permeability, and antifouling performance [5].

In this research, nanocarbon obtained from palm kernel shells and $Fe_2O_3$ synthesised from local minerals are tried to modify the PES membrane. The addition of carbon in the membrane solution is expected to increase the pore size of the membrane, and the addition of $Fe_2O_3$ is expected through physical mixture will result in increased membrane hydrophilic properties. The aim of this study is to analyze the effect of those additive on the morphology and filtration performance of blended PES membrane. The performance of membrane filtration for soy protein concentrate is also discussed.

## 2. Materials and Methods

### *2.1 Materials*

PES with an average molecular weight of 65,000 Da (BASF, Ludwigshafen, Germany) was used as the main polymer to prepare flat sheet membrane. Dimethyl formamide purchased from Merck KgaA, Germany was supply as solvent to dissolve the polymer. $Fe_2O_3$ and nanocarbon was kindly provided by Physic Science Laboratory and Energy laboratory of Syiah Kuala University, respectively.

### *2.2 Membrane Preparation*

For original membrane, the dope solution is prepared by mixing 16wt% of PES in DMF. In other dope solution, nanocarbon and $Fe_2O_3$ of 0.05wt% was added separately to prepare the modified PES membrane. The detail composition of membrane solution are summarized in Table 1. All dope sample is stirred for about 24 hours to obtain the homogeneous solution. After releasing air bubble, the solution is poured onto a glass plate using an applicator knife with a thickness of 300 μm. The casted solution and the glass plate then dipped in a coagulation bath containing pure water for solidification of membrane.

**Table 1.** Dope composition

| Composition (wt%) | | | | Sample |
|---|---|---|---|---|
| PES | DMF | $Fe_2O_3$ | Carbon | |
| 16 | 84 | - | - | M1 |
| 16 | 83.95 | 0.05 | - | M2 |
| 16 | 83.95 | - | 0.05 | M3 |

### *2.3 Membrane characterization*

The cross-section structure of the membrane is visualized using Scanning Electron Microscopy (SEM, JSM-7500F, JEOL Ltd., Tokyo, Japan). A standard sample preparation procedure was provided consist of drying in a freeze dryer, immersing in liquid nitrogen for membrane fracture, and coating with osmium for sample imaging.

In order to analyze mechanical property of original and modified PES membrane, a tensile strength measurements was carried out by using a Universal Testing Machine (Autograph AGS-J, Shimadzu Co. Japan). A membrane samples were cut into pieces of 5 cm lenght and at least 2 tests for each type of the membranes were made and the average was reported in this paper for the tensile strength.

*2.4 Filtration performance*

Filtration performance of the fabricated membrane was analyzed in term of water flux and flux recovery ratio (FRR). Pure water flux is calculated from the volume of water that passes through the effective surface area of the membrane during filtration period. Filtration experiment was carried out at constant pressure (1 bar) on the cross-flow module and the permeate is taken every 10 minutes.

Moreover, a 5000 ppm soy protein solution was also filtrated using the same module to investigate the FRR profile of the membrane. Flux of membrane during filtration of water was calculated using Equation (1).

$$J = \frac{V}{A \times t} \quad (1)$$

Where:

J = Flux (L/m$^2$.hour)

V = permeate volume (L)

A = membrane surface area (m$^2$)

t = Time (hour)

## 3. Result and Discussion

*3.1 Membrane Structure*

Scanning Electron Microscopy (SEM) is an instrument used in observing the morphology of membranes. Membrane morphology is one of the crucial parameter for manufacturer in determination of the specific application in the separation industry. In this study, the imaging of SEM was taken only on the cross-section part of the membrane. Fig. 1 shows the cross-sectional structure of the resulted membranes with magnification of 2000X.

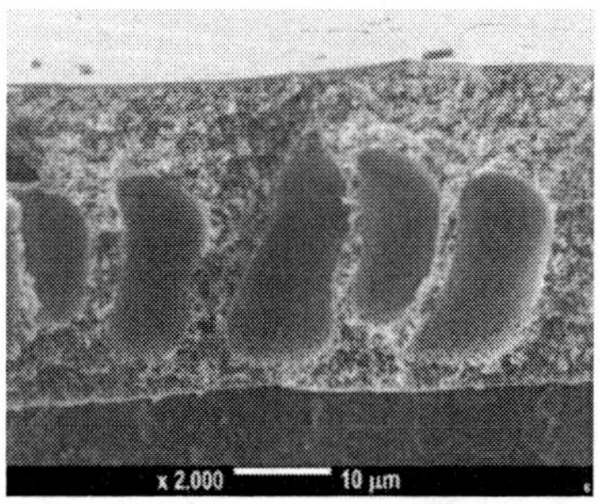


(M1) PES 16%

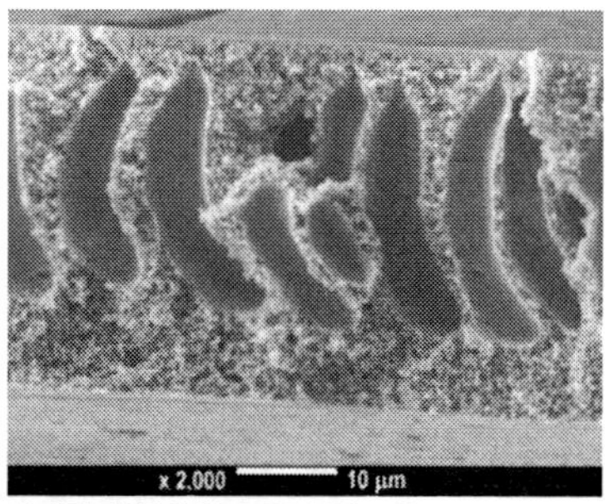


(M2) PES/0.05% $Fe_2O_3$

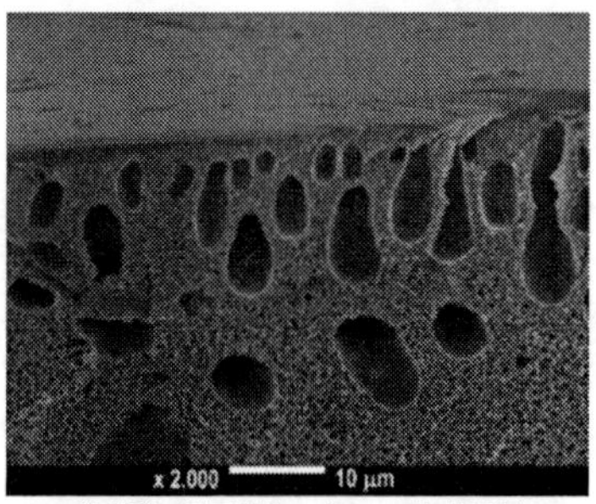


(M3) PES/carbon 0.05%

**Figure 1**. SEM image of membrane morphological structures

The modified membrane with addition of $Fe_2O_3$ (M2) posseses longer fingerlike pores than the pristine membrane (M1), while the presence of carbon particles in membrane solution lead the increasing number of finger pores [6].

### *3.2 Mechanical Properties of Membranes*

Tensile strength of membrane is an important parameter to determine the mechanical property of membrane. Membrane with a denser structure commonly has better breaking strength than the porous membrane [6]. Tensile strength of prepared membranes are presented in Fig. 2.

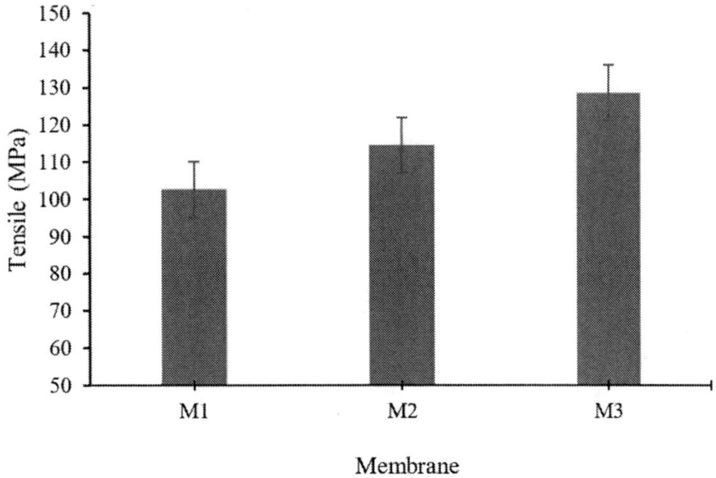


**Figure 2**. Tensile strength of prepared membranes

Fig. 2 showed the increasing of PES membrane breaking strength after modification with $Fe_2O_3$ and carbon particles. The PES membrane with the addition of 0.05% carbon (M3) showed the strongest tensile strength with 128.56 MPa compared to PES/$Fe_2O_3$ (M2) and PES (M1) membrane with 114.5 Mpa and 102.626 MPa, respectively. It is due to the adhesion forces between carbon particles and PES polymer that lead the increasing of membrane tensile strength [7]. Moreover, the presence of $Fe_2O_3$ particles scattered in membrane solution can limit the movement of polymer chains leading to the improvement of membrane mechanical strength. The same result was reported by Ng, et al [8].

### *3.3 Filtration Performance*

Fig. 3 showed the pure water flux of all membranes conducting in cross-flow filtration system. The pure water flux of PES membrane increased from 15.14 L/m$^2$.h to 51.98 and 58.67 L/m$^2$.h after addition of $Fe_2O_3$ and carbon additives, respectively. The presence of the hydroxyl and oxygen group in the particles increase the hydrophilicity of PES membrane solution and has strong interaction with water in coagulation bath [9]. As a consequence, the exchange rate between solvent and non-solvent during precipitation increased, leading the formation more porous structures (as shown in Fig 1). It is well-known that more porous structure in membrane produced higher permeation.

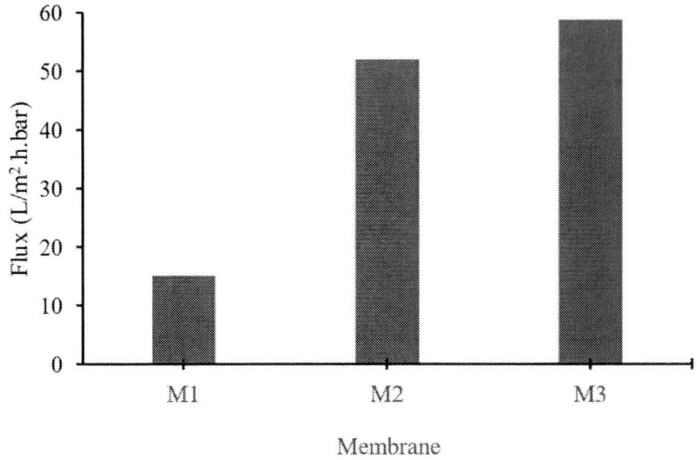


**Figure 3.** Pure water flux of resulted membranes

The filtration of soy protein was also carried out using a cross-flow filtration module for 60 minutes recirculation process. A 5000 ppm soy protein solution was pumped to the module at filtration pressure of 1,0 bar. The concentration of soy protein in permeate and retentate tank was analyzed by using spectrophotometer UV-Vis. It is found that, the rejection of soy protein filtrated with M2, and M3 membrane was 30.40 and 40.50 %, lower than the original PES membrane (M1) which is rejected about 50%. However, the filtration duration of M2, and M3 was better than the M1 membrane. The presence of inorganic additive in membrane surface brought about the reducing the interaction between protein and membrane surface leading to easier passes of water through the membrane [11-12].

## 4. Conclusions

Modification of PES membrane has been successfully fabricated by blending of nanocarbon and $Fe_2O_3$ into the polymer solution. PES/carbon membrane exhibited more porous and highest pure water flux of 58 L/m$^2$.h. The addition of $Fe_2O_3$ particles improved membrane mechanical properties. Filtration of soy protein was better employed by modified membranes.

## Acknowledgement

We acknowledge the Indonesian Ministry of Research, Technology, and Higher Education for funding support of this research. (Basic research grant: 215/SP2H/LT/DRPM/2019).

## References

[1] Fitradi, Ryan Bagus. (2015). Preparasi dan modifikasi membran untuk pengolahan air. *Teknik Kimia Institut Teknologi Bandung*,01-15
[2] Muchtar S, Wahab M Y, Fang L F, Jeon S, Rajabzadeh S, Takagi R, Mulyati S, Arahman N, Riza M, Matsuyama H. 2019 *J. Appl. Polym. Sci.* **136** 47312.
[3] Nawi N I M, Bilad M R, Zolkhiflee N, Nordin N A H, Lau W J, Narkkun T, Faungnawakij K, Arahman N, Mahlia T M I 2019 *Polymers* **11** (865) 1.
[4] Nigmet U, Nuray A, Seda Saki Y. Emre B and Yongsheng C 2017 *Sep. Purif. Technol* **187** 118
[5] Elif D, Bopeng Z, Marc P, Shuman X and Yongsheng C *J. Memb. Sci.* **529** 170
[6] Nita K and dan Septiana T 2012 *Molekul* **7** 43
[7] Nasrul A, Mukramah, Syawaliyah, Teuku M and Muhammad R B 2018 *Int. J. Geomate* **15** 51
[8] Law Y Ng, Abdul W M, Choe P L and Nidal H *Desalination* **308** 15

[9] Sabri N S M., Hasbullah H., Said N, Ibrahim N, Kasmani R M., Ali R R and Rahman S A 2017 *J. Appl. Memb. Sci. Technol.*. **20** 29
[10] Alla A, Mohamed M, Kerry N McPhedran and Mohamed G El-Din 2015 *J. Memb. Sci.* **490** 227
[11] Suffriandy S, Fachrul R and Nasrul A 2017 *J. Rekayasa dan Lingkungan*. **12** 46
[12] Arahman N, Satria S, Razi F, Bilad M R 2018 *Water* **10** (1207) 1

# Torrefaction of Empty Fruit Bunches: Evaluation of Fuel Characteristics Using Response Surface Methodology

**Z Helwani[1*], Zulfansyah[1], W Fatra[2], A Q Fernando[1], G M Idroes[3], Muslem[4], R Idroes[4]**
[1]Departement of Chemical Engineering, Universitas Riau, Pekanbaru 28293, Indonesia
[2]Departement of Mechanical Engineering, Universitas Riau, Pekanbaru 28293, Indonesia
[3]Department of Chemical Engineering, Faculty of Engineering, UniversitasSyiah Kuala, Kopelma Darussalam, Banda Aceh 23111, Indonesia
[4]Department of Chemistry, Faculty of Mathematics and Natural Sciences, UniversitasSyiah Kuala, Banda Aceh 23111, Indonesia

*E-mail: zuchra.helwani@lecturer.unri.ac.id

**Abstract.** Empty Fruit Bunches (EFB) can be used as an alternative energy source by torrefaction process. Torrefaction is a treatment process of biomass into solid fuel within temperature range of 200-300 °C in an inert condition. This research aims to study the effect of process variable interaction process and to study the characteristic of solid fuel resulted. Torrefaction of EFB was performed by using horizontal fixed bed reactor at three different temperatures, times and particle sizes, that are 225, 250 and 275 °C, for 15, 30 and 45 minutes, and particle size 2, 4 and 6 cm, respectively. Response surface methodology (RSM) was used for to optimization optimize of torrefaction conditions so that biofuel of with high energy density, maximum energy properties, and minimum weight loss could be manufactured. The analysis showed that increase in heating values was affected by treatment severity (cumulated effect of temperature, time and particle size). Our results clearly demonstrated an increased degradation of the material due to the combined effects of temperature and treatment time. While the reaction time had less impact on the energy density of torrefied biomass, the effect of reaction temperature was considerably stronger under the torrefaction conditions used in this study. The most significant factor affecting all responses was temperature of torrefaction,then followed by time and particle size. The optimized torrefaction conditions were verified empirically and the applicability of the model was confirmed. The torrefied biomass occurred more suitable than raw biomass in terms of calorific value, physical and chemical properties.

## 1.Introduction

As an agrarian country, Indonesia is abundant in plant biomass resources from plantation and agricultural wastes. This biomass has many benefits because it contains beneficial components such as cellulose, hemicellulose, lignin, monosaccharide, pectin, amino acids, aliphatic acids, ether, fats, oils,

* To whom any correspondence should be addressed.

waxes, resin, pigments and proteins[1–3]. The biomass can be utilized by extraction of beneficial components, conversion to other forms, or use as raw material for fuel production.Pectin can be isolated from coffee biomass [4] for membrane matrix material [5,6]. Biomass can be converted to char chemical feedstocks for soil repaired purpose [7]. Biomass is also a promising material for fuel production [8]. Especially for chemical feedstock and fuel production purpose,these utilizations are achieved by using conversionmethods of polymer materialsuch as thermal conversion(pyrolysis, hydrothermal, gasification) [9,10], and also chemical, biochemical and electrochemical conversion [11–13].

As one of the largest producer and exporterof palm oil, Indonesia has abundant availability of oil palm biomassas a promising source of lignocellulosic biomass. This biomass is a primary source of renewable carbon that can beutilized as a feedstock for biofuels production in order to achieveenergy independence. Additionalquantities of lignocellulosic biomass can be yielded from energycrops. These crops are fast-growing plants which are harvested specifically for energy production. Innature, lignocellulosic biomass has high moisture content, low bulkenergy density.It is also difficult to transport, to handle, to store and to feedinto existing combustion systems without pretreatments. Whenlignocellulosic biomass is employed for bioenergy, pretreatmentsof lignocellulosic biomass are essential procedures for achieving high efficiencies of fuel production or consumption. For example, moisture content from the lignocellulosicbiomass was driven off to enhance the combustion efficiency andprevent microbial degradation during storage [14].

While, in liquid biofuel production, acid pretreatments have beenadopted widely to facilitate the conversion of hemicelluloses andcellulose into soluble sugars for lignocellulosic ethanol production[14,15]. In addition to the methods mentioned above, torrefaction oflignocellulosicbiomass is another notable pretreatment method that enablesfor energy densification of biomass and biomass homogenization [16–18]. This pretreatment has been recommended as an efficient way to enhance solid biofuel properties bywater removal, to reduce the hygroscopic range, and to increasegrindability. Torrefaction is a thermal pretreatment process bysubjecting wood to temperature levels between 200 and 300°Cin the absence of oxygen. During this process, the cell walls aredegraded and the products produced are between wood and charcoal. The residual product is in solid form, which isreferred to as torrefied biomass.

Torrefaction product quality is influenced by several factor like temperature, reaction time, biomass type, and particle size [18]. In the previous research, it showed that temperature and time give a significant effect on mass and energy yield-reducing and enhance the energy density [19,20]. This is caused by the increase inthe degradation rate of hemicellulose when temperature and time raised. Mass of lignocellulosic biomass mostly decreases in the range of 30-45 minute torrefaction [20].Particle size is also one of the factors that influence the product. Different particle size in term length, give different result for mass and energy yield [21].

One of lignocellulosic biomass from oil palm biomass is Empty Fruit Bunches (EFB). This biomass is mill waste that remains from oil pressing of fresh fruit bunches. Thus, the purpose of this study was to build a model of torrefaction procedure applicable in the production of solid biofuel fromlignocellulosic biomass of EFB. Changes in the energy properties of thelignocellulosic biomass when subjected to torrefaction in the temperature range of 225–275 °C with different treatment time wereinvestigated. Weight loss is an important parameter for optimizingthe design and operation of a biomass torrefaction plant. So thatunderstanding of the process variables for optimization is required.Torrefaction condition was optimized by using the technique of response surface methodology (RSM) that recently report enables to find the optimum torrefaction conditionsfor a high-density solid fuel.

## 2.Experimental

*2.1 Material Preparation,* The oil palm biomass used was EFB obtained from CPO-mill nearest in Riau province. EFB was cut with dimensions of length 2,4,and 6 cm. EFB was dried by sun for a few days and remove the residual from EFB. The moisture content was measured in an oven at 105°C for 24 hours.

*2.2 Torrefaction Process,* 50 g dried wood fine was placed in a furnaceand torrefied under the conditions as follows: temperature 225 °C,250 °C and 275 °C for 15 min, 30 min, and 45 min. Torrefaction wasperformed in a horizontal fixed bed reactor of 6 cm diameter and 60 cm length, (Fig. 1) controlled by a continuous nitrogenflowrate at 100 ml/minutes. The carrier gas was continuously blown into the reactor to keep the system stay in inert condition and to remove the volatiles produced during the torrefaction.

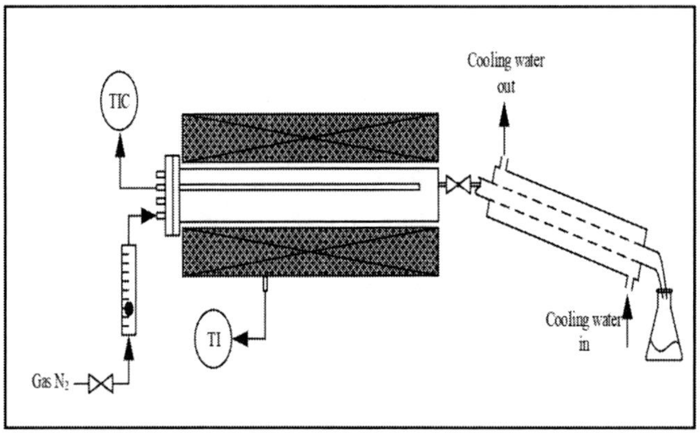


**Figure 1.** Torrefaction Reactor

The reactor is also equipped with condenser to condensevolatiles and bubble shop flow meter to control the flow rate of nitrogen. For each treatment, the mass yield and energy yield ofthe torrefied wood were calculated using the following expression (Eq (1) and (2)):

$$MY = \frac{Mass\ after\ torrefaction}{Mass\ before\ torrefaction} x\ 100\% \quad (1)$$

$$EY = \%MYx\frac{HHV\ product}{HHV\ biomass} \quad (2)$$

*2.3 Determination of higher heating value (HHV),* The calorific value was determined according to BSI standardEN 14918 by using bomb calorimeter, in which 0.50 g of oven-driedbiomass was completely combusted under an oxugen pressurized (3000 kPa).*Proximate Analysis*: moisture content was measured by drying in an oven at 105°C according to ASTM D-3173. For ash content, a prescribed amount (1 g) of the sample was calcined in an electric furnace at 750°C for 2 hours, according to ASTM D-3174. The volatile matter was also measured using furnace which 1 g of sample was calcined at 950°C for 7 minutes according to ASTM D-3175.

*2.4 Response Surface Analysis,* Second-degree polynomials were calculated using Design Expert Version 7.0 software (Stat-Ease., USA) to estimate the response of the dependent variables. Temperature ($X_1$, 225-275°C); time ($X_2$, 15-45 minutes); and particle size ($X_3$, 2-6 cm) were chosen as the independent variables. HHV, mass yield, and energy yield of lignocellulosic biomass of EFB were used asthe dependent output variables.Central Composite Design (CCD) was used as a design experiment.

## 3. Results and Discussion

The results of mass measurement, calorific value, and proximate analysis for the sample and torrefied samples are listed in Table 1.In this study, EFB color became darker with increasing of torrefactiontemperature and time. A similar tendency was observed from previous research [19].

*3.1 Calorific Value (HHV)*

The HHV and weight loss depended on the torrefaction conditions listed in Table 1. Calorific value as energy density is one of the important parameter of the product of torrefaction. It gives the amount of energy released when unit mass of torrefied product is burnt [18]. Compared to fossil fuel, biomass contains less carbon and has lower calorific values on similar weight basis.HHV significantly increased with increasing of torrefaction temperature and time. The highest HHV obtained at 292 °C. This tendency is consistent with the previous study[14,18,20,22,23]. HHV increased mainly due to the increase of carbon content and decreasing ofoxygen and hydrogen contents. The oxygen and hydrogen content decrease bydecomposition of hemicellulose and organic compounds such aswater, acetic acid, and phenol during torrefaction [20,24]. In term of increasing in length, the HHV decreased while particle size increased. The tendency was similar to the previous study,and it's mainly due to the surface area of biomass decrease along with increasing particle size. The smaller the surface area, the more difficult the volatiles formed during torrefaction to get out from inner core of biomass [21].HHV represents the maximum amount ofenergy embedded per gram of material. It is important to mentionthat HHV increases with the extent of treatment.The energy value ranged from 18.08 to 28.94 MJ/kg depending onthe torrefaction conditions. This implies that the HHV in the torrefied biomass increased by 6.5-30% compared to theuntreated material.

**Table 1.** Results of mass yield, energy yield, and proximate analysis of the product

| | Raw EFB | Torrefied EFB |
|---|---|---|
| HHV (kJ/kg) | 16034 | 17091-20697 |
| *Mass yield* (%) | - | 52,63-84,52 |
| *Energy yield* (%) | - | 64,12-99,05 |
| Moisture (%) | 11,50 | 5,00-8,50 |
| Ash (%) | 1,45 | 1,75-3,41 |
| Volatile(%) | 69,45 | 47,73-63,20 |
| Fixed carbon(%) | 17,60 | 26,26-44,31 |

HHV was more affected by the temperature thanthe reaction time. The HHV significantly increased with a torrefaction temperature exceeding 250 °C, regardless of the reactiontime. HHV slowly increased as the temperature raised until 250 °C,then drastically increased as the temperature further increasedfrom 250 °C to 292 °C. It is clear that the characteristics of the thermaldegradation of the biomass are different. Furthermore, from the viewpoint of practical applications oftorrefaction, if the light torrefaction (225 °C) is applied for the pretreatment of biomass, its main function will be to remove moistureand light volatiles from the biomass (Pierre et al., 2011). Therefore, it is inferred that the heating value of the torrefied biomass isimproved to a small extent. If biomass is pretreated using the severetorrefaction conditions (above 300 °C), the role of lignin (acid-insoluble fraction) in energy contribution might be more importantthan holocellulose resulting from a large portion of holocellulose being consumed in the severe torrefaction. High lignincontent in torrefied biomass is associated with high energy density due to the ether and carbon linkages in lignin, whichhas higher energy than the C O or C H bonds (Lehtikangas, 2001). Therefore, high lignin content in the torrefied biomass results in a higher HHV.

*3.2 Mass and Energy yield*

Mass yield gives a measure of the solid yield of the torrefaction. It defines what fraction of the original mass of biomass would remain in the torrefied product. While energy yield gives the fraction of energy in biomass retained after torrefaction [18]. Solid fraction remained after torrefaction.Most of it consists of cellulose and lignin that was not degraded. based on composition, carbon and oxygen are the main constituent of the solid product, followed by a small amount of nitrogen and hydrogen [24]. The measure of mass and energy yield based on equation 1 and 2.

Mass and energy yield decreased when torrefaction temperature and time increased. Its mainly due to more and more degraded hemicellulose in range of torrefaction temperature and time followed by a litle bit of cellulose and lignin decreased gradually [19,20]. Energy yield obtained in this study also decreased when temperature and time increasing. This is due to the increase in the calorific value that was not able to offset the decrease in mass yield when temperature and time raised [14,19,20]. It can be also caused by the fibrous structure of EFB compared to the oil palm waste like shell, tree, and frond thus resulting in decreasing solid fraction significantly.

The particle size effect is less significant than the torrefaction temperature and time. Increase inparticle size in length will increase mass and energy yield. It's mainly due to the smaller surface area of biomass, the higher the heat resistance,the more difficult the volatiles formed to get out through the biomass [21].

*3.3 Proximate Analysis*

Moisture content is an important property of biomass fuels because a high moisture level in a fuel leads to a high energy loss in the course of burning [25]. As a feedstock, it is desirable to lower the moisture content of biomass through drying process before thermochemical conversion. In addition to increasing energy in solid fuel, biomass which has a lower water content can be stored for a long time and minimize transportation costs. In this study, moisture content decreased when torrefaction of temperature and time increased. Its mainly due to dehydration during torrefaction [18].

Ash is the inorganic solid residue left after biomass is completely burned. Its primary ingredients are silica, aluminium, iron, and calcium with small amounts of magnesium, titanium, sodium, and potassium [18,25]. Effect of increasing of torrefaction temperature, time, and particle size gives the positive effect for increasing ash content. Maybe, the ash content increases because temperature does not reach the ashing temperature on biomass in range 550°C and for coal is 780°C. in addition, the mass reduction occuredduring the process is not accompanied by degradation of the components forming ash.

Volatile matter is closely related tofixed carbon in biomass. Devolatilization was affected by torrefaction temperature and time. when temperature and time increased, volatile matter decreasedbutfixed carbon increased. This is due to the torrefaction temperature which is also in the temperature range of hemicellulose degradation [25]. Hydrogen and oxygen as partsof major components in biomass arepartsof the most widely lost component during torrefaction. The reduced fraction of hydrogen and oxygen resulted in a decrease volatile matter and increase in the ratio of carbon to hydrogen and oxygen so that calorific value of the product also increased[18,25].

*3.4 Response Surface Analysis for HHV, mass and energy yield*

The primary goal of torrefaction is to refine raw biomass into upgrade solid fuel, improved handling qualities and enhanced combustible properties comparable to fossil coal, reading to reduce costs and to raise economic gains[26]. Regarding fuel efficiency, it's necessary to increase calorific value, mass yield, and energy yield.Those responses were fitted to the response surface model provided by the mathematic models in Table 2, in order to analyze the effect of the torrefaction factor on responses. $Y_1$is the predicted calorific value (kJ/kg), while $Y_2$ and $Y_3$were predicted for mass yield and energy yield. Where $X_1$, $X_2$, and $X_3$were temperature, time, and particle size.

**Table 2** Summary of the results of various research responses to the heating value ($Y_1$); Mass yield ($Y_2$); Energy yield ($Y_3$)

| Run | *Natural Variables* | | | *Coded Variables* | | | *Responses* | | |
|---|---|---|---|---|---|---|---|---|---|
| | $\xi_{1\ (°C)}$ | $\xi_2$ (menit) | $\xi_{3\ (cm)}$ | $X_1$ | $X_2$ | $X_3$ | $Y_{1\ (kJ/kg)}$ | $Y_{2\ (\%)}$ | $Y_{3\ (\%)}$ |
| 1 | 225 | 15 | 2 | -1 | -1 | -1 | 18124,40 | 82,99 | 93,81 |
| 2 | 275 | 15 | 2 | 1 | -1 | -1 | 19038,35 | 70,02 | 87,86 |
| 3 | 225 | 45 | 2 | -1 | 1 | -1 | 19029,91 | 72,00 | 85,46 |
| 4 | 275 | 45 | 2 | 1 | 1 | -1 | 20068,94 | 66,82 | 87,62 |
| 5 | 225 | 15 | 6 | -1 | -1 | 1 | 17954,39 | 83,86 | 93,90 |
| 6 | 275 | 15 | 6 | 1 | -1 | 1 | 19659,74 | 70,38 | 86,29 |
| 7 | 225 | 45 | 6 | -1 | 1 | 1 | 18811,49 | 72,92 | 85,55 |
| 8 | 275 | 45 | 6 | 1 | 1 | 1 | 20078,40 | 57,65 | 72,20 |
| 9 | 208 | 30 | 4 | -1.682 | 0 | 0 | 18322,67 | 82,40 | 95,96 |
| 10 | 292 | 30 | 4 | 1.682 | 0 | 0 | 20641,13 | 66,31 | 85,57 |
| 11 | 250 | 5 | 4 | 0 | -1.682 | 0 | 18527,37 | 82,39 | 95,21 |
| 12 | 250 | 55 | 4 | 0 | 1.682 | 0 | 19543,13 | 53,19 | 64,83 |
| 13 | 250 | 30 | 0,7 | 0 | 0 | -1.682 | 18896,72 | 82,17 | 96,83 |
| 14 | 250 | 30 | 7,3 | 0 | 0 | 1.682 | 18056,56 | 75,92 | 85,52 |
| 15 | 250 | 30 | 4 | 0 | 0 | 0 | 19506,95 | 64,10 | 78,03 |
| 16 | 250 | 30 | 4 | 0 | 0 | 0 | 19186,38 | 66,60 | 79,69 |
| 17 | 250 | 30 | 4 | 0 | 0 | 0 | 18953,61 | 65,62 | 77,57 |
| 18 | 250 | 30 | 4 | 0 | 0 | 0 | 19229,87 | 66,52 | 79,78 |
| 19 | 250 | 30 | 4 | 0 | 0 | 0 | 19330,98 | 68,18 | 82,20 |
| 20 | 250 | 30 | 4 | 0 | 0 | 0 | 19340,74 | 66,14 | 79,79 |

To match the model to experimental data obtained, the regression analysis carried out on the second-degree polynomial using Analysis of Variance (ANOVA), which can be seen in Table 3. Themathematical model as shown in Table 2 correlates well that can be seen from R-squared that is high. The probability value (p-value) used for each model is <0,05 which states the model is significant at 95%. Based on results obtained, all variables; torrefaction temperature, time, and particle size havea significant influence over 95%.

The responses for calorific value, mass yield, and energy yield from torrefaction were plotted as three-dimensional surface plots of two factor; temperature vs time. Particle size was plotted on a fixed axis due to level of significantly smaller than the torrefaction temperature and time. Fig 2 shows the effect of torrefaction operating conditions of responses were analyzed.

**Table 3**. Analysis of Variance (ANOVA) quadratic model for calorific value, mass yield, and energy yield.

| Source | Sum of Squares | Degrees of Freedom | Mean Square | F-Value | p-Value (Prob>F) |
|---|---|---|---|---|---|
| Calorific Value | | | | | |
| Model | 1,744E+07 | 9 | 1,938E+06 | 39,67 | < 0,0001 |
| A-Temp | 1,014E+07 | 1 | 1,014E+07 | 207,51 | < 0,0001 |
| B-Time | 3,545E+06 | 1 | 3,545E+06 | 72,57 | < 0,0001 |
| C-Size | 2,007E+05 | 1 | 2,007E+05 | 4,11 | 0,0579 |
| AB | 24548,62 | 1 | 2,455E+04 | 0,50 | 0,0285 |

| Source | Sum of Squares | Degrees of Freedom | Mean Square | F-Value | p-Value (Prob>F) |
|---|---|---|---|---|---|
| AC | 2,597E+05 | 1 | 2,597E+05 | 5,32 | 0,4841 |
| BC | 1,090E+05 | 1 | 1,090E+05 | 2,23 | 0,1460 |
| $A^2$ | 7,276E+05 | 1 | 7,276E+05 | 14,89 | 0,0006 |
| $B^2$ | 1,402E+05 | 1 | 1,402E+05 | 2,87 | 0,1010 |
| $C^2$ | 2,059E+06 | 1 | 2,059E+06 | 42,14 | < 0,0001 |
| Residual | 1,417E+06 | 29 | 48851,66 | | |
| Lack of Fit | 9,126E+05 | 19 | 48033,93 | 1,03 | 0,5007 |
| Total | 1,891E+07 | 38 | | | |
| | | | | | R-Square: 0,925 |
| **Masss Yield** | | | | | |
| Model | 2574,87 | 9 | 286,10 | 27,09 | < 0,0001 |
| A-Temp | 729,90 | 1 | 729,90 | 69,10 | < 0,0001 |
| B-Time | 1107,65 | 1 | 1107,65 | 104,87 | < 0,0001 |
| C-Size | 44,91 | 1 | 44,91 | 4,25 | 0,0483 |
| AB | 9,00 | 1 | 9,00 | 0,85 | 0,3636 |
| AC | 28,14 | 1 | 28,14 | 2,66 | 0,1134 |
| BC | 22,47 | 1 | 22,47 | 2,13 | 0,1555 |
| $A^2$ | 213,17 | 1 | 213,17 | 20,18 | 0,0001 |
| $B^2$ | 0,06 | 1 | 0,06 | 5,29,E-03 | 0,9425 |
| $C^2$ | 466,75 | 1 | 466,75 | 42,14 | < 0,0001 |
| Residual | 306,31 | 29 | 10,56 | | |
| Lack of Fit | 263,93 | 19 | 13,89 | 3,28 | 0,293 |
| Total | 2884,71 | 38 | | | |
| | | | | | R-Square: 0,894 |
| **Energy Yield** | | | | | |
| Model | 2333,75 | 9 | 259,31 | 17,18 | < 0,0001 |
| A-Temp | 261,05 | 1 | 261,05 | 17,29 | 0,0003 |
| B-Time | 987,82 | 1 | 987,82 | 65,44 | < 0,0001 |
| C-Size | 188,14 | 1 | 188,14 | 12,46 | 0,0014 |
| AB | 1,42 | 1 | 1,42 | 0,094 | 0,7611 |
| AC | 73,75 | 1 | 73,75 | 4,89 | 0,0351 |
| BC | 47,92 | 1 | 47,92 | 3,17 | 0,0853 |
| $A^2$ | 400,09 | 1 | 400,09 | 26,5 | < 0,0001 |
| $B^2$ | 0,16 | 1 | 0,16 | 0,011 | 0,9182 |
| $C^2$ | 431,63 | 1 | 431,63 | 28,59 | < 0,0001 |
| Residual | 437,78 | 29 | 15,1 | | |
| Lack of Fit | 354,36 | 19 | 18,65 | 2,24 | 0,0965 |
| Total | 2771,72 | 38 | | | |
| | | | | | R-Square: 0,842 |

**Table 4**. The mathematical models derived from the experimental results for calorific value, mass yield, and energy yield.

| Responses | Model Calculated ($Yi$) |
|---|---|
| Calorific Value | $Y_1$=19256,71 + 609,21$x_1$ + 360,27$x_2$ – 285,71$x_3$ + 127,41$x_1x_2$ – 39,17$x_1x_3$ – 85,24$x_2x_3$ + 158,88$x_1^2$ – 69,75$x_2^2$ – 267,24$x_3^2$ |
| Mass Yield | $Y_2$=66,27 – 6,37$x_1$ – 5,17$x_2$ – 1,28$x_3$ + 0,75$x_1x_2$ – 1,33$x_1x_3$ – 1,19$x_2x_3$ + 2,72$x_1^2$ + 0,044$x_2^2$ + 4,02$x_3^2$ |
| Energy Yield | $Y_3$=79,55 – 6,01$x_1$ – 3,09$x_2$ – 2,62$x_3$ + 0,3$x_1x_2$ – 2,15$x_1x_3$ – 1,73$x_2x_3$ + 3,73$x_1^2$ – 0,075$x_2^2$ + 3,87$x_3^2$ |

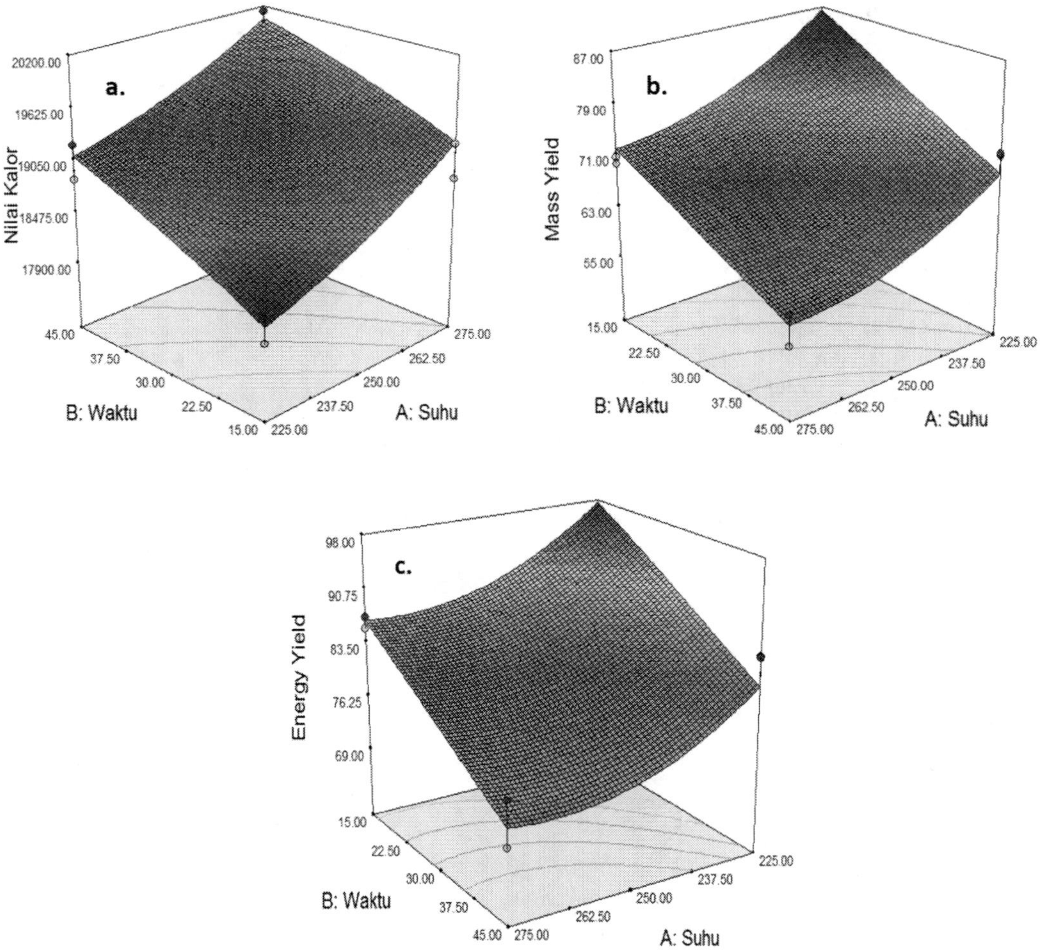


**Figure 2.** Response surface and countour plot of torrefaction temperature vs. time at 4 cm for; (a) calorific value (b) mass yield and (c) energy yield.

## 4. Conclusion

Biomass especially EFB can be upgraded and used as a fuel by torrefaction process. The torrefied biomass is more suitable than raw biomass in terms of calorific value, energy yield, and proximate analysis. The reaction temperature had a strong impact on the responses analyzed followed with torrefaction time and particle size.

**Acknowledgements**
The support from the University of Riau under Hibah Guru Besar Riau University Grant is gratefully acknowledged.

## REFERENCES

[1] Earlia N, Rahmad, Amin M, Prakoeswa C.R.S, Khairan 2019 *Sains Malaysiana* **48** 1019.
[2] DeMartini J D, Pattathil S, Miller J S, Li H, Hahn M G and Wyman C E 2013 *Energy*

*Environ. Sci.***6** 898
[3] Earlia N, Suhendra R, Amin M, Prakoeswa C R S and Idroes R 2019 *Sci. World J.***2019** 1
[4] Hasanah U, Setyowati M, Edwarsyah, Efendi R, Safitri E, Idroes R, Heng L Y and Sani N D 2019 *IOP Conf. Ser. Mater. Sci. Eng.***523** 012014
[5] Hasanah U, Setyowati M, Efendi R, Muslem M, Sani M, Diyana N, Safitri E, Yook Heng L and Idroes R 2019 *Biosensors* **9** 60
[6] Hasanah U, Sani N D M, Heng L Y, Idroes R and Safitri E 2019 *Biosensors* **9** 135
[7] Werner C, Schmidt H-P, Gerten D, Lucht W and Kammann C 2018 *Environ. Res. Lett.***13** 044036.
[8] Yaman S 2004 *Energy Convers. Manag.***45** 651.
[9] Muslem M, Kuncaka A, Himah T N and Roto R 2019 *Indones. J. Chem.***19** 835.
[10] Akhtar A, Krepl V and Ivanova T 2018 *Energy & Fuels* **32** 7294.
[11] Liu G, Larson E D, Williams R H, Kreutz T G and Guo X 2011 *Energy & Fuels* **25** 415
[12] Linares-Pasten J, Andersson M and Karlsson E 2014 *Curr. Biotechnol.***3** 26–44
[13] Munnings C, Kulkarni A, Giddey S and Badwal S P S 2014 Biomass to power conversion in a direct carbon fuel cell *Int. J. Hydrogen Energy* **39** 12377.
[14] Chin K L, H'ng P S, Go W Z, Wong W Z, Lim T W, Maminski M, Paridah M T and Luqman A C 2013 *Ind. Crops Prod.***49** 768.
[15] Mosier N, Wyman C, Dale B, Elander R, Lee Y Y, Holtzapple M and Ladisch M 2005 **96** 673.
[16] Liu C, Lv P, Yuan Z, Yan F and Luo W 2010 *Renew. Energy***35** 1531.
[17] Liu X, He H, Wang Y, Zhu S and Piao X 2008 *Fuel* **87** 216.
[18] Basu P 2013 *Biomass Gasification, Pyrolysis and Torrefaction* (New York: Elsevier Inc)
[19] Uemura Y, Omar W N, Tsutsui T and Yusup S B 2011 *Fue l***90** 2585.
[20] Sabil K M, Aziz M A, Lal B and Uemura Y 2013 *Biomass and Bioenergy* **56** 351.
[21] Basu P, Rao S and Dhungana A 2013 *Can. J. Chem. Eng.* **91** 466.
[22] Asadullah M, Adi A M, Suhada N, Malek N H, Saringat M I and Azdarpour A 2014 *Energy Convers. Manag.* **88** 1086.
[23] Kong S-H, Loh S-K, Bachmann R T, Rahim S A and Salimon J 2014 *Renew. Sustain. Energy Rev.* **39** 729.
[24] Ciolkosz D and Wallace R 2011 *Biofuels, Bioprod. Biorefining* **5** 317.
[25] Chen W-H, Peng J and Bi X T 2015 *Renew. Sustain. Energy Rev.* **44** 847.
[26] Chen W-H, Lu K-M, Liu S H, Tsai C-M, Lee W-J and Lin T-C 2013 *Bioresour. Technol.* **146** 152.

# Utilization of Avocado Seed Waste as Raw Material for Producing Biodiesel with CaO Catalyst from Eggshell

**D Valensya[1], I Rozalia[1], Zuhra[1] and Y Syamsuddin[1*]**
[1]Chemical Engineering Department, Engineering Faculty, Syiah Kuala University, Banda Aceh 23111, Indonesia

[*]E-mail: yanna_syamsuddin@unsyiah.ac.id

**Abstract.** Avocado is one of the plants that thrives in the tropical area such as Indonesia. Avocados produce waste in the form of avocado seeds because it is less used. The oil from avocado seeds contains triglyceride as raw materials in biodiesel synthesis. Biodiesel is an alternative diesel oil from the transesterification of vegetable oils or animal fats with the addition of catalyst. The objective of this study was to utilize avocado seed waste as raw material in the production of biodiesel and eggshell waste as source of CaO catalyst. The various effect of reaction conditions, such as methanol-to-oil molar ratio (3:1–12:1), reaction time (30–120 min), and amount of catalyst loading (1–7% wt) were investigated. Catalyst reusability was also studied. Result showed that the highest biodiesel oil was obtained from reaction with the condition of 6:1 methanol-to-oil ratio, 60 min reaction time, 5% wt amount of catalyst loading, and the catalyst can be reused up to 3 reaction cycles with the yield above 75%. The physico-chemical properties of biodiesel produced met the SNI standar of biodiesel quality.

## 1. Introduction

Currently, the need to develop renewable alternative fuels has been increasing due to the increasing in energy demand as the population and industry increased [1]. Regarding alternative fuels, biodiesel is a promising type of fuel because of its enviromental friendliness such as low toxicity, low $CO_2$, sulfur and benzene-free emission product, and also has high cetane number [2]. Biodiesel is a mixture of methyl esters resulting from the transesterification reaction of vegetable oil or animal fats with alcohol with the aid of an acid or base catalyst [3].

Transesterification reaction can be conducted homogenously or heterogenously. A heterogenous catalytic process is more adventageous than a homogenous one. A heterogenous catalyst easier to separate and reuse, the homogenous catalyst presents problems in separation and purification of the product, the use of homogenous catalyst is not environmentally benign because the washing step for catalyst removal produces a large amount of waste water [4]. CaO, an active and base catalyst, is widely used for transesterification of vegetables oil. Eggshells, which contain $CaCO_3$ for about 94%,canbeused as raw material to obtain CaO through the calcination process [5].

The purpose of this study is to obtain the optimum condition for biodiesel production by transesterification process of avocado seed oil using methanol as solvent and CaO from eggshells as

[*] To whom any correspondence should be addressed.

catalyst, to study there usability of the catalyst, and to characterize the physicochemical properties of the biodiesel and compare to the SNI standard.

## 2. Materials and method

### *2.1. Materials*

Avocado seeds and egg shells were obtained from food sellers around Banda Aceh. Chemicals used to produce biodiesel were methanol that purchased from PT. Jeungki Mali Medical, n-hexane as solvent for the extraction process and FFA analysis process also purchased from PT. Jeungki Mali Medical. Other materials used were deionized water for biodiesel washing, phenolpthalein indicator and NaOH 0.1 N for FFA analysis process.

### *2.2. Oil extraction from avocado seeds*

Avocado seed was cut manually using knife and then crushed in blender. The sample was dried in oven at 110°C for 2 h. A 50 g sample was extracted using n-hexane 250 ml at 65°C for 4 h [5]. The avocado seed oil produced was characterized for free fatty acid (FFA), viscosity, and density.

### *2.3. Catalyst preparation*

Eggshell were dried in oven at 110°C for 2 h, crushed in blender and sieved at 60 mesh. Egg shell powder was calcined in furnac eat 900°C for 2 h [6]. The catalyst was characterized using XRD and SEM method.

### *2.4. Transesterification process*

Various parameters were investigated to study their effect on the biodiesel yield. The parameters studied were methanol-to-oil ratio (1:3–1:12), reaction time (30–120 min), and amount of catalyst loading (1–7%wt). The operating temperature was kept constant at 65°C for all the parameters studied.

Reactions were conducted in a three-neck rounded flask. The reactor was equipped with a magnetic stirrer and thermometer. Specific amount of oil, methanol, and catalyst were placed in reactor, and the reactions were run at the various conditions stated above. After the reaction, product was centrifuged to separate the catalyst, then the biodiesel and gliserol were separated using separating funnel, left for 12 hours and biodiesel produced was washed with aquadest to remove methanol from the product [7].

## 3. Result and discussion

### *3.1 Avocado seed oil analysis*

The avocado seed oil that obtained from extraction was analyzed for FFA, density, and viscosity. FFA is the most important thing in producing biodiesel. For transesterification reaction, FFA should be below 1% [8]. The result from characterization shown in Table 1.

**Table 1.** Characterization of avocado seed oil

| Analysis | Result |
|---|---|
| FFA (%) | 0.42 |
| Viscosity (poise) | 0.47 |
| Density (g/mL) | 0.86 |

*3.2 Catalyst characterization*

CaO catalyst from eggshell was analyzed using XRD to identify the components in catalyst after calcination process. Figure 1 shows the diffractogram of the anaysis.

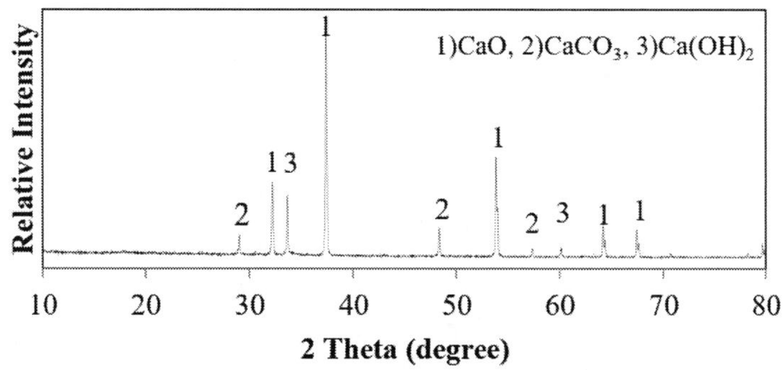


**Figure 1.** Catalyst difractogram after calcinationat 900°C for 2 h

As shown in Fig.1, XRD spectrum showed CaO component (JCPDS No.82-1690), $CaCO_3$ (JCPDS No.47-1743), and $Ca(OH)_2$ (JCPDS No.01-073-5492) [9]. CaO is the active phase for transesterification process available in high intensity than $CaCO_3$and $Ca(OH)_2$.

*3.3 Transesterification of Avocado Seed Oil*

*3.3.1. Effect of methanol-to-oil molar ratio on biodiesel yield*

The stoichiometric ratio for transesterification using methanol requires 3 moles of methanol for 1 mole of triglyceride. Reactions were performed at 65°C for 60 min with 5 wt % catalyst by varying the molar ratio of methanol-to-oil from 3 to 12. The result is shown in Figure 2.

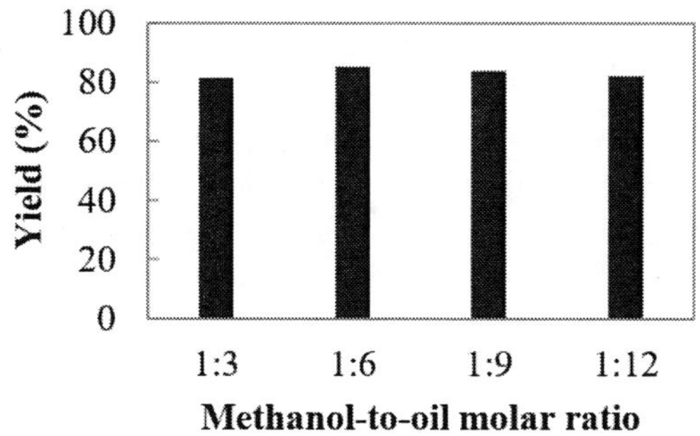


**Figure 2.** Effect of methanol-to-oil molar ratioon biodiesel yield at 60 min reaction time and 5 wt % catalyst loading

The result, as illustrated in Fig. 2, showed that the best methanol-to-oil molar ratio was 6:1, resulted in 85.3% yield. At 9:1 and 12:1 methanol-to-oil molar ratio, biodiesel yield decreased because of the high amount of methanol used in the reaction, which increased solubility and made the glycerol barely separated [10]. At the methanol-to-oil molar rasio of 3:1, the yield also decreased because the amount of methanol used in the reaction was less than the 6:1 molar ratio reaction and produced excess triglyceride[11].

*3.3.2. Effect of various reaction time on biodiesel yield*

Biodiesel yield from the transesterification reaction was also influenced by reaction time. The reaction was conducted at 65°C, 6:1 methanol-to-oil molar ratio, and 5 wt % catalyst loading amount.The result is shown in Figure 3.

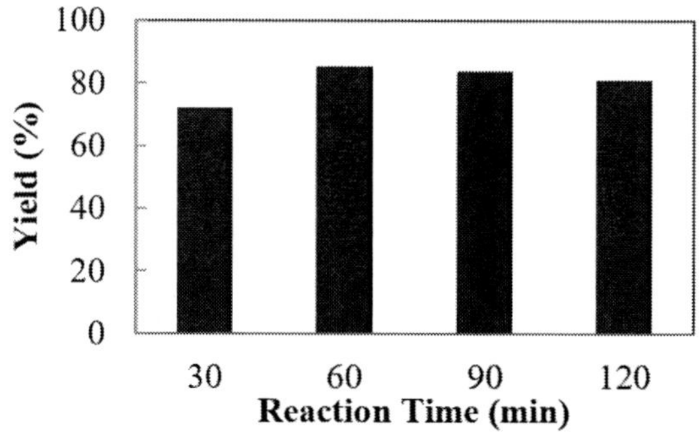


**Figure 3.** Effect of various reaction time on biodiesel yield at the condition of 6:1 methanol-to-oil molar ratio and 5 wt % catalyst

Figure 3 shows that the best reaction time for transesterification was 60 minutes with biodiesel yield of 85.3%. Biodiesel yield was decreased at reaction time 90 and 120 minutes because the transesterification process was heated too long that affected the quality of the oil.When the reaction run for 30 minutes, the biodiesel yield was also decreased because the mixing process of methanol, oil, and catalyst has not been completed and the reaction run slowly [12].

*3.3.3 Effect of various catalyst loading amount on biodiesel yield*

Various catalyst loading amounts ranged from 1 to 7 wt % onthe basis of oil weight was investigated. The reaction was conducted at 65°C, methanol-to-oil molar ratio of 6:1, and the reaction time of 60 minutes. The result is presented in Figure 4.

Figure 4 shows that the best catalyst loading amount was 5 wt %, resulted in 85.3% yield. Biodiesel yield decreased at 1 and 3 wt% catalyst because of insufficient active phase available for the reaction to occur. Reaction with 7 wt% catalyst loading also yielded lower biodiesel, due to high amount of catalyst produced emulsion and gel that increased viscosity[13].

Based on the above result, the best biodiesel yield was 85.3% resulted from the reaction with the condition of 6:1 methanol-to-oil molar ratio, 60 min reaction time, and 5 wt% catalyst loading.

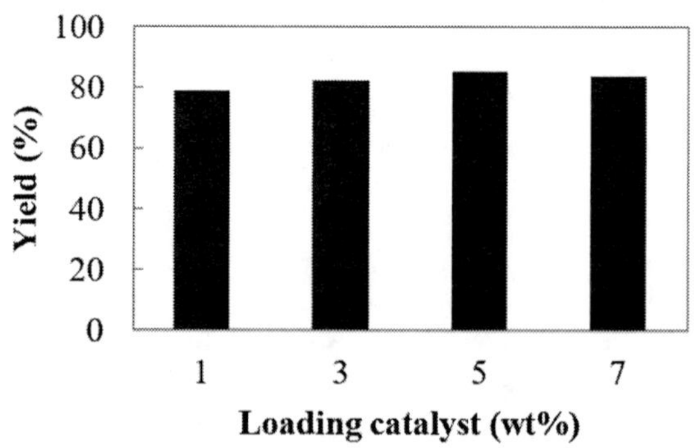


**Figure 4.** Effect of various catalyst loading amount on biodiesel yield at the condition of 6:1 methanol-to-oil molar ratio and 60 min reaction time

*3.4 Characterization of biodiesel*

*3.4.1 Physical properties of biodiesel*

The analysis of the physical properties of biodiesel according to SNI 7182-2012 includes viscosity and density. Based on the analysis of the best product, viscosity value was 4.23 cSt and the density was 0.86 g/mL. This value shows that biodiesel is not only made from good raw materials, but also with good processing.The characterization result is shown in Table 2.

**Table 2.** Characterization of physical properties of biodiesel

| **Analysis** | **Analysis Method** | **Result** | **SNI** |
|---|---|---|---|
| Viscosity (cSt) | ASTM D-445 | 4.23 | 2.3–6.0 |
| Density (g/mL) | ASTM D 1298 | 0.86 | 0.85–0.89 |

*3.4.2 Chemical properties of biodiesel*

Methyl ester that produced from the best operating condition was analyzed using GC-MS to determine the components of the fatty acid in biodiesel. The result from characterization shown in Figure 5 and Table 3.

Figure 5 shows that there are five peaks of the avocado seed oil methyl ester compund. Then each peak on the GC chromatogram was identify by a mass spectrophotometer, where each compound has a specific mass fragmentation pattern. Based on Table 3, the highest esters of avocado seed biodiesel is methyl oleat that indicated by peak number 4 withresult of 57.39%.

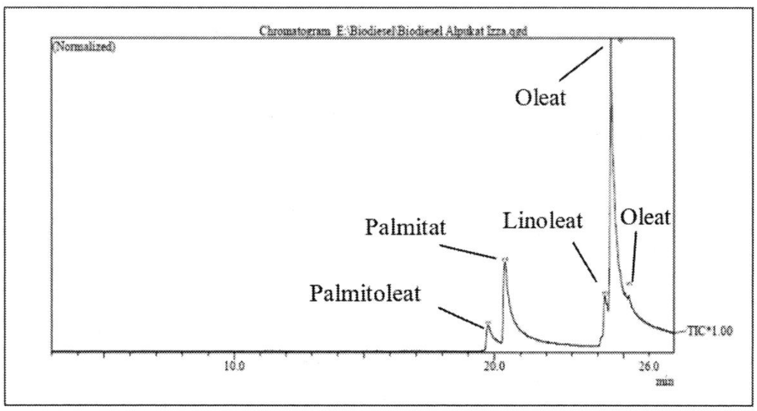


**Figure 5.** Chromatogram of avocado seed biodiesel from GC-MS analysis

*3.5. Reusability of catalyst*

Heterogeneous catalyst can be reused in the reaction after being separated from the product. The result of reusability test of the catalyst is shown in Figure 6.

Based on the figure, biodiesel yield slightly decreased from 85.3% from reaction with fresh catalyst to 81.77% and 77% from reaction with the second and third cycle, respectively. This was caused by the decreasing of the active phase of CaO on the catalyst surface, which can be proven by the XRD analysis in Figure 7.

The figure shows that the large CaO content in fresh catalyst decrease significantly after the catalyst was used for three times in the transesterification process. A small amount of CaO reacts with the atmospheric $CO_2$ and produce $CaCO_3$ which is a passive component in transesterification reaction[14].

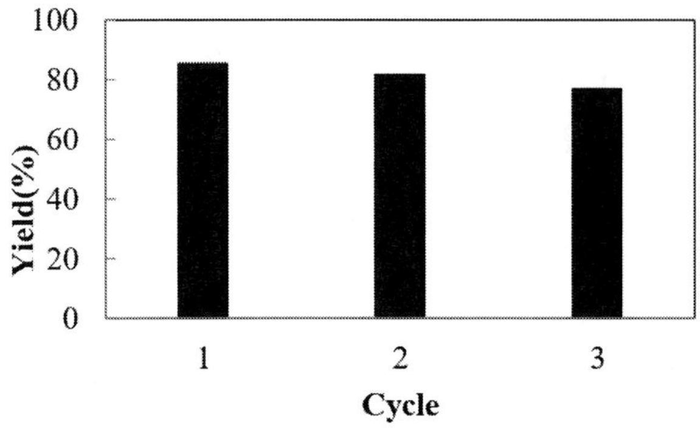


**Figure 6**. Transesterification using reusable catalyst

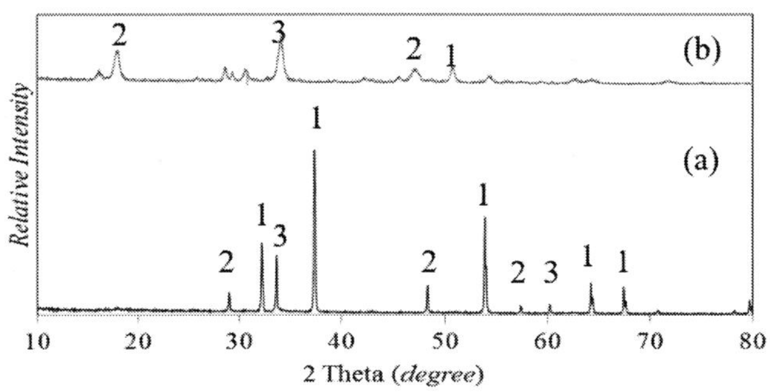


**Figure 7**. Diffractogram of CaO catalyst; (a)fresh catalyst, (b)reused catalyst; 1)CaO, 2)$CaCO_3$, 3)$Ca(OH)_2$

In addition, biodiesel yield decreased by the deactivation of the catalyst due to blockage of catalyst pores by triglyceride from the oil during the reaction [15], there by reducing the amount of active pores on the catalyst surface as seen from the SEM analysis that shown in Figure 8.

Figure 8 (a) shows the surface of the catalyst before reaction, the pores of the catalyst still visible. Meanwhile, in Figure 8 (b) the catalyst pores have been covered by oil during the reaction process.

(a) (b)

**Figure 8**. Result of SEM analysis of CaO from eggshell; (a) fresh CaO, (b) reusable CaO

IOP Conf. Series: Materials Science and Engineering **845** (2020) 012020 doi:10.1088/1757-899X/845/1/012020

## 4. Conclusion

Avocado seed oil can be used as raw material for biodiesel synthesis over CaO from eggshell as catalyst. The best operating condition for the transesterification reaction was 6:1 methanol-to-oil molar ratio, 60 min reaction time and 5 wt% catalyst loading amount. The catalyst can be reused up to 3 reaction cycles with the biodiesel yield more than 75%. Characterization of the product showed that the highest ester compounds of biodiesel from avocado seed oil was methyl oleat and the physical properties of biodiesel produced met SNI 7182-2012.

## References:


[1] Miskah S, Moeksin R and Aprianti T 2018 *IOP Publishing.* **298** 1

[2] Devita L 2015 *Agresia Ekstensia.***9**23–26

[3] Adhani L, Isalmi A, Siti N and Christie A O 2016 *Jurnal Kimia Valensi.* **2** 71

[4] Helwani Z, Othman M R, Kim J and Fernando W J N 2015 *A Rev. Appl. Catal., Ayst: A General.* **4** 13

[5] Miskah S, Andika A and Gunadi 2016 *Jurnal Teknik Kimia.* **22** 54

[6] Sundari D, Zuchra H and Drastinawati 2015 *Jom Fteknik.* **2** 11

[7] Mahfud, Muharto and Pramudita R A 2012 *Jurnal Teknik Kimia.5 219*

[8] Armalita R D, Bahri S and Yusminar 2015 *Jom Fteknik.* **2** 20

[9] Wendi, Cuaca V and Taslim 2015 *Jurnal Teknik Kimia USU.* **4** 35

[10] Imaduddin M, Yoeswono, Wijaya K and Tahir I 2008 *Bul. of Chem.Reac. Eng. & Cat..* **3** 14

[11] Karnanim 2010 *Sintesis Biodiesel dari Bahan Baku Minyak Sawit Menggunakan Reaktor Jet Bubble Column, UI, Depok*

[12] Dewi T P P, Usman T and Harlia 2015 *JKK.* **4** 31

[13] Syamsuddin Y, Murat M N and Hameed B H 2015 *J. Energ Convers Manage* **106** 1356

[14] Syamsuddin Y and Hameed B H 2016 *J Taiwan Inst Chem E* **58** 181

[15] Ngamcharussrivichai C, Totarat P and Bunyakiat K 2018 *Appl Catal A-Gen A: General.* **341** 14

IOP Conf. Series: Materials Science and Engineering **845** (2020) 012021 doi:10.1088/1757-899X/845/1/012021

# Utilization Palm Shell Bioelastomeric as Composite Foams for the Removal $Hg^{2+}$ in Liquid Waste

**N N Munawarah[1], A Gani[2*], Mahidin[2]**
[1]Graduate School of Chemical Engineering, Universitas Syiah Kuala, Banda Aceh, 23111, Indonesia
[2]Department of Chemical Engineering, Syiah Kuala University, Banda Aceh, Indonesia, 23111

[*]E-mail: asri_gani@unsyiah.ac.id

**Abstract.** Efforts to deal with waste using the adsorption method showed effective results. However, it was discussed of the sorbent after the remediation process is cumbersome. An alternative to overcome this problem can be done by fixing the adsorbent into a solid form consisting of a porous polymer without changing its chemical affinity with metal ions. In this study, the adsorbent used was bioelastomeric containing activated charcoal from palm shell waste and elastomeric, which would then be contacted with $Hg^{2+}$ metal ions. The process variable in this study was the composition of heavy metal $Hg^{2+}$ 8, 12, 16, 20, and 24 mg/L with the contact time until it reached equilibrium. Based on the calculation, the palm shell bioelastomeric used three adsorption isotherm models, Freundlich isotherm ($R^2$= 0.991), Langmuir isotherm ($R^2$= 0.967) and Dubinin-Radushkevich isotherm ($R^2$= 0,871). The results showed that the efficiency of adsorption of heavy metal $Hg^{2+}$ with bioelastomeric palm shell was almost close to 100%. In conclusion, palm shell bioelastomeric can be a potential adsorbent with a high heavy metal removal ability.

## 1. Introduction

Individual activities, as well as the rapid development of industry, cause many environmental problems, one of which is water pollution. Water pollution can occur because of waste from various industrial process activities. In Aceh, there are many industries that operate in the food processing industry and the mining industry. One of them is the gold mining industry in Krueng Sabee, Aceh Jaya, which becomes a highly discussed topic because it uses the amalgamation method. Gold mining activities that contaminate the environment have resulted in adverse impacts on the community and the surrounding environment marked by contamination in soil, aquatic ecosystems, and so forth [1].

Contamination caused by liquid waste can be overcome by various methods, one of which is the adsorption method. An adsorption process is an event that occurs on the surface of the stationary phase, where material from the free period is adsorbed. This process is generally used to reduce heavy metals, ions, and compounds that are harmful to humans and the environment [2]. Research into the use of palm shell's activated charcoal as an adsorbent in the adsorption process showed efficient results for the removal of heavy metals from liquid waste. However, after the procurement process that enhances complicated processes, activated charcoal requires methods such as centrifugation and load screening [3, 4].

* To whom any correspondence should be addressed.

An alternative to overcome this problem was by configuring activated palm oil into a solid form (in the form of porous polymer composites) without changing its chemical affinity using metal ions. It became an attractive solution for increasing the use of more specific activated charcoal applications. Based on research that had been done on the utilization of spent coffee ground as elastomeric composite foams, it was found that spent coffee ground successfully removed heavy metal ions from water. Besides, research on the utilization of elastomeric nanocomposite foams in water contaminated with $Pb^{2+}$ or $Hg^{2+}$ succeeded in removing $Pb^{2+}$ ions exceeding 98% and $Hg^{2+}$ ions approaching nearly 100%. Utilization of biomass waste into bio-based polymer products derived from agricultural raw materials and renewable biomass became more important because this environmentally friendly and sustainable product reduced greenhouse gas emissions [3, 4, 5].

Indonesia is one of the biggest oil palm producers in the world. The area of oil palm plantations in Indonesian is 14,677,560 ha in 2019. Sumatra has 8,762,135 ha of oil palm plantations in 2019, which is the largest oil palm producing area in Indonesia. Aceh is one of the provinces that has extensive oil palm plantations in Sumatra. The area of oil palm plantations in Aceh is 566,378 ha in 2019. Besides that, palm shell waste from palm oil production, which is gradually increasing so far, has been used mainly as fuel. Therefore, palm shell can be used as an adsorbent in the form of polymer composites to add value to it, as it is easily handled and environmentally friendly [6, 7]. It can be used as a source of sorbent due to its organic components, such as cellulose, hemicellulose, and lignin [8].

## 2. Materials and Methods

*2.1 Materials,* The palm shell waste was obtained from the palm oil industry in Aceh, Indonesia, and used as the material for making activated charcoal, Acetoxy-Polydimethylsiloxane (PDMS), Poly (ethylene oxide) -b-PDMS, Hexane and Sanding Sugar used as the material for bioelastomeric manufacturing. The other supporting materials were $HgCl_2$ as an artificial contaminant sample and distilled water were also used in this work.

*2.2 Preparation of Activated Carbon Sorbent from Palm Shells,* The palm shell was dried under sunlight and then carbonized at 350°C for 120 minutes using a muffle furnace until the carbon was formed. The resulting carbon was crushed using ball mill then sifted using a +100/-200 mesh sieve. Then the physics activation process will be carried out by heating using a muffle furnace at a temperature of 800°C for 120 minutes. After the activation process is complete, the carbon is allowed to cool, stored in a desiccator.

*2.3 Bioelastomeric Fabrication Process,* The foam was made using a sugar leaching technique [9,4], 16 g elastomer (acetoxy-PDMS) and 0.16 g poly surfactant additive (ethylene oxide) -b-PDMS (wt. Ratio 10.0: 0.1) mixed in 12 ml hexane. Furthermore, 0.4 gram of activated charcoal is slowly added to the polymer solution under constant stirring for 1 hour. Then, 12 g of sugar is added to the solution slowly. The mixture is then poured in a Petri dish and placed in a fume hood for 24 hours at room temperature to allow the evaporation of solvents and the formation of polymers. Composites that have been formed are then placed in hot water under an ultrasonic for 2 hours to dissolve sugar, and bioelastomer foam will form.

*2.4 Adsorption Test on Artificial $HgCl_2$ Wastewater,* Artificial $HgCl_2$ wastewater solution was prepared with varying initial concentrations of 8, 12, 16, 20 and 24 mg/L. A total of 250 ml sample was contacted with 0.8 gram sorbent with the contact time until it reached equilibrium, the process was carried out with a stirring speed of 250 rpm. The concentration of $HgCl_2$ in feed solution and after adsorption was measured Hg using Atomic absorption spectrometer (AAS).

IOP Conf. Series: Materials Science and Engineering **845** (2020) 012021 doi:10.1088/1757-899X/845/1/012021

## 3. Results and Discussion

*3.1 Effect of contact time and concentration variation on the adsorption capacity of $Hg^{2+}$* ions

Contact time and concentration variation were factors affecting the adsorption process. Determination of the optimum contact time for adsorption aimed to determine the time needed by the bioelastomeric palm shell sorbents to adsorp the maximum $Hg^{2+}$ ion. The relationship between contact time and concentration variation on the adsorption capacity of $Hg^{2+}$ ions can be seen in Figure 1.

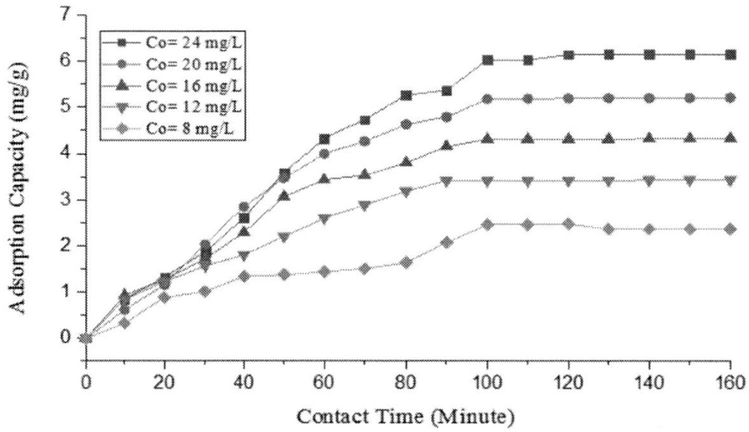


**Figure 1.** Effect of contact time and concentration variation on the adsorption capacity of $Hg^{2+}$ ions

Based on Figure 1, it can be seen that the adsorption capacity increased along with the increasing adsorbate concentration and the longer contact time. At a high concentration of 24 mg/L, an absorption capacity of 6,171 mg/g was obtained. Meanwhile, at a lower concentration of 8 mg/L, an absorption capacity of 2,371 mg/g was obtained. The higher the concentration of adsorbate, the greater the amount of substance that is collected on the surface of the adsorbent. It happened since there were more dissolved substances dispersed in the solvent, so that the likelihood of contact between the adsorbate and the adsorbent was greater [10]. The maximum adsorption capacity for each variation an average concentration were in 100 minutes, then in the 110 minutes to the 160 minutes the value of the adsorption capacity starts to reach equilibrium ie it has no significant changes. At the contact time of 10 to 100 minutes, the pores on the active surface were large so that the adsorption ability was also high, but starting from the 100th minute, the adsorbent started to saturate. This was indicated by the constant value of absorption capacity and the equilibrium between the concentration of metal ions in activated carbon and metal ions in solution. Thus, the adsorption capacity at contact time above 100 minutes became constant [11].

*3.2* Effect of contact time and concentration variation on the adsorption efficiency of $Hg^{2+}$ ions

Adsorption efficiency was the percentage of adsorbate absorption on the surface of the adsorbent. The relationship between contact time and concentration variation on the adsorption efficiency of $Hg^{2+}$ ions can be seen in Figure 2.

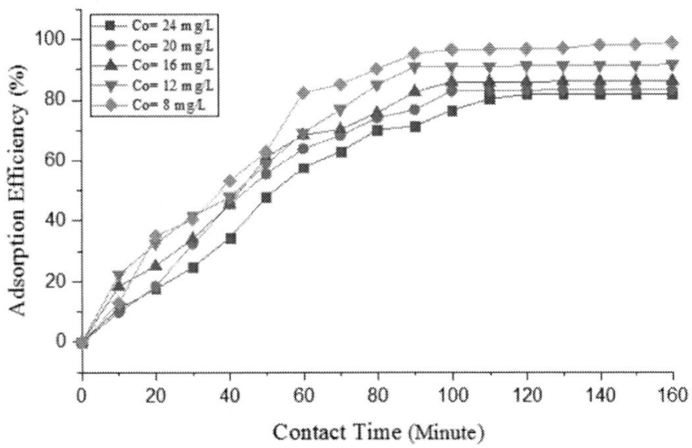


**Figure 2.** Effect of contact time and concentration variation on the adsorption Efficiency of $Hg^{2+}$ ions

Based on Figure 2, adsorption efficiency was increasing gradually from the contact time of 10 minutes to 100 minutes. The longer the contact time, the greater the ability to bind to the adsorbent and adsorbate. Therefore, more adsorbates were bound to the surface of the adsorbent [12]. The efficiency value from time to time was increasing gradually, but after some time, i.e. above 100 minutes, it showed a constant value that indicated the achievement of equilibrium. At a concentration of 24 mg/L with a contact time of 160 minutes, an absorption efficiency of 81.94% was obtained. Meanwhile, at a concentration of 8 mg/L with the same contact time, an adsorption efficiency value of 94.61% was obtained. This showed that adsorbate concentration that was too high caused a decrease in absorption efficiency. The decrease occurred due to the higher concentration, the amount of metal ions in solution was not proportional to the amount of available adsorbent. Thus, the surface of the adsorbent reached saturation point and absorption efficiency did not increase significantly [10, 11].

*3.3 Adsorption isotherm*

Adsorption isotherm was used to describe the relationship between the adsorbent and the adsorbed substance in equilibrium [4, 13, 14]. This research tested three adsorption isotherm models, namely Langmuir isotherm, Freundlich isotherm and Dubinin-Radushkevich isotherm. The Langmuir $Hg^{2+}$ adsorption isotherm was obtained by making a comparison curve between the equilibrium of the liquid phase (Ce) against the equilibrium of the solid phase (Ce/qe). The comparison graph of Langmuir isotherm can be seen in Figure 3.

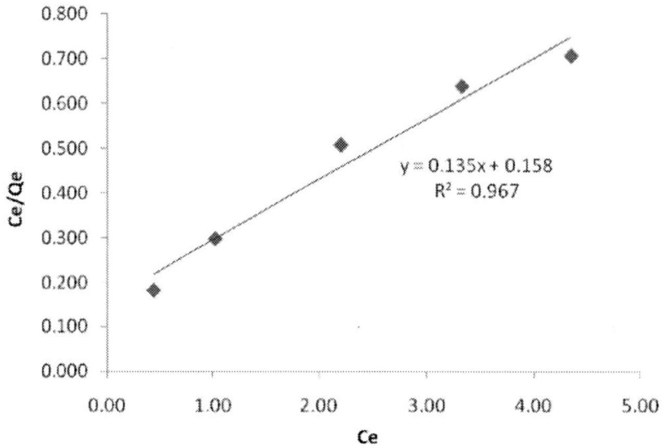


Figure 3. Langmuir isotherm model on the adsorption of $Hg^{2+}$ ions

Freundlich adsorption isotherm was obtained by determining the relationship between log Ce and log qe. The Freundlich isotherm graph can be seen in Figure 4.

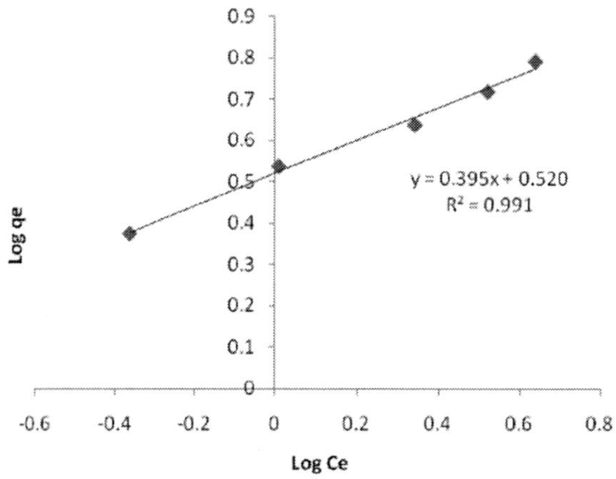


**Figure 4**. *Freundlich isotherm model on the absorption of* $Hg^{2+}$ *ions*

Dubinin-Radushkevich adsorption isotherm was obtained by determining the relationship between $\varepsilon^2$ and ln qe. The Dubinin-Radushkevich isotherm graph can be seen in Figure 5.

Based on Figure 3, Figure 4, and Figure 5, it can be seen that the adsorption of $Hg^{2+}$ ions using palm shell bioelastomeric foam followed the Freundlich isotherm model. This can be seen from the value of the determinant coefficient ($R^2$) of the Freundlich isotherm model, which was equal to 0.991. It was higher than the determinant of the Langmuir and Dubinin *–Radushkevich* isotherm model, which was value to 0.967 and 0,871. The Freundlich isotherm assumed that the surface of the adsorbent was heterogeneous; the adsorption formed many layers. This allowed the adsorbate to move freely until the adsorption process physically occurred in those layers [14].

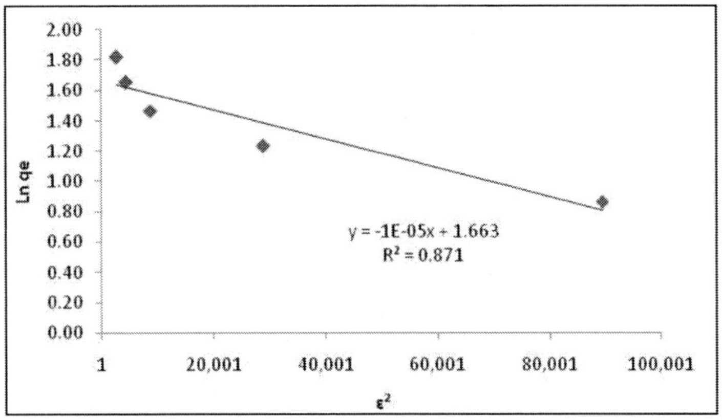


**Figure 5**. Dubinin-Radushkevich isotherm model on the absorption of $Hg^{2+}$ ions

Dubinin-Radushkevich isotherm model described the adsorption relationship using the Gaussian energy distribution on the heterogeneous surface of the adsorbent [14]. From Figure 5, it can be determined that the free energy was E = 0.7 kJ/mol, which showed the physisorption process. Based on the three isotherm models analyzed, the value of the adsorption capacity and its constant in each isotherm model can be seen in Table 1

**Table 1** The results of the calculation of adsorption isotherm

| **Langmuir Isotherm** | | | **Freundlich Isotherm** | | | **Dubinin-Radushkevich isotherm** | | |
|---|---|---|---|---|---|---|---|---|
| $q_m$ | $K_L$ | $R^2$ | n | $K_F$ | $R^2$ | $q_m$ | $K_{ad}$ | $R^2$ |
| **7,4074** | 0,854 | 0,967 | 2,5316 | 3,3113 | 0,991 | 1,508 | $1 \times 10^{-5}$ | 0,871 |

## 4. Conclusion

In conclusion, the present study demonstrates that the utilization of bioelastomeric palm shell waste as an adsorbent for the adsorption of heavy metal $Hg^{2+}$ produces good adsorption efficiency. The adsorption efficiency of $Hg^{2+}$ increases with increasing contact periods and the lower concentration of the solution. At a concentration of 24 mg/L with a contact time of 160 minutes an absorption capacity of 6.171 mg/g was obtained and an absorption efficiency of 81.94% while at a concentration of 8 mg/L with the same contact time an absorption capacity of 2.337 mg /g was obtained and an efficiency value adsorption of 94.61% was almost close to 100%. The adsorption process is more likely to follow the Freundlich isotherm model with a determinant value of $R^2$ of 0.991.

## References

[1] Sofia and Husodo H A 2016 *J. of People and Env.* **23** 310
[2] Rodriguez-Narvaez O M, Peralta-Hernandez J M, Goonetilleke A and Bandala E R 2017 *A Rev. Chem. Eng. Journal* **323** 361
[3] Chavan A A, Pinto J, Liakos I, Bayer S I, Simone L, Athanassiou A and Fragouli D 2015 *ACS Appl Mater & Interfaces* **27** 14778
[4] Chavan A A, Pinto J, Liakos I, Bayer S I, Simone L, Athanassiou A and Fragouli D 2016 *ACS Sustainable Chem. & Eng.* **10** 5495
[5] Wang R, Zhang J, Kang H and Zhang L 2016 *Composites Sci. and Tech.* **133** 136

[6] Ditjenbun 2019 *Luas Areal Kelapa Sawit Menurut Provinsi di Indonesia, 2015-2019,* http://ditjenbun.pertanian.go.id/
[7] Saber S, Ezzuldin M, Rahim A B S, Yussof W H, Olalere A O and Habeeb A O 2019 *Chemical Data Collection* **21** 100
[8] Iberahim N, Sethupathi S, Goh L C, Bashir K J M and Ahmad W 2019 *J. of Env. Mangt.* **248** 109
[9] Choi J S, Kwon H T, Hwon I, Dong-II M, and Baek J D 2011 *ACS Appl Mater & Interfaces* **3** 4552
[10] Handayani K, and Yusnimar 2013 *Jurnal Teknobiologi* **4** 117
[11] Zikra N R Y, Chairul C and Yenti S R 2016 JOM FT UNRI. **3** 1
[12] Chen S, Qin C, Wang T, Chen F, Li X, Hou H and Zhou M 2019 *J. of Mol. Liq.* **285** 67
[13] Dawlet A, Talip D, Mi Y H and Malikezhati 2013 *Proc. Env. Sci.* **18** 800
[14] Nashine A L and Tembhurkar A R 2016 *J. of Env. Chem. Eng.* **4** 3267

IOP Conf. Series: Materials Science and Engineering **845** (2020) 012022 doi:10.1088/1757-899X/845/1/012022

# Production Anti Diabetes Flour from *Tanjung* Fruit (*Mimusops elengi L*)

**R Maula[1], Nurhaliza[1], Nurlaila[2], Mukhlishien[1], Sofyana[1] dan Y Syamsuddin[1*]**
[1]Chemical Engineering Department, Faculty of Engineering, Syiah Kuala University, Banda Aceh
[2]Agricultural Technology Department, Faculty of Agriculture, Syiah Kuala University, Banda Aceh

[*]E-mail: yanna_syamsuddin@unsyiah.ac.id

**Abstract.** Flavonoids are phenol compounds found in various types of vegetables, cereals and fruits. Phenol compounds have activity as antidiabetic substances. *Tanjung* fruit waste, contains flavonoid compounds. The purpose of this study was to extract flavonoid compounds from the fruits and to investigate the effect of temperature and drying time on the quality of the flour produced. The flour products obtained were analysed for the levels of fat, ash, water, protein and carbohydrate content, and the results were compared with standard values. The results of the analysis showed that the best flour was obtained from a process with drying time and temperature conditions of 40 minutes and 100°C, respectively. In this condition, 65% flour is produced from raw materials that met the SNI-01-3751-2006 standard of wheat flour, has a long durability from mold and microbial growth, bright colour and soft texture. Flavonoids obtained from the extraction were 105.01 mg QE / mg extract. Giving flour to white mice with the dosage of more than 1 g/kg body weight showed significant reduction of its blood glucose level.

## 1. Introduction

Indonesia is a country rich in biodiversity. Indonesia has 30,000 species of plants that live in tropical forests and some of them are medicinal plants with 9600 species. Of the many medicinal plants, only 200 species are newly processed and used for natural herbal medicines. Judging from the quality and side effects, there are still many Indonesians taking herbal medicines because they are considered to be safer, especially after popping up medicinal products that declare back to nature to avoid side effects [1]. *Tanjung* fruits (*Mimusops elengi L*) are a familiar plant because there are many in surrounding environment. *Tanjung* plants are thought to originate in Southeast Asia. Then it spread to Indochina, Sri Rare, Myanmar, Malaysia and most of Indonesia [2].

Flavonoid contained in *Tanjung* fruits makes this fruit has a potential to be processed into food (flour) for diabetics. Flavonoid compounds can bind blood sugar, as blood sugar inhibition. Antioxidants from this fruit has function as anti-diabetes, anti-stroke, anti-cholesterol and prevent blood pressure. Flavonoids are natural phenol compounds that are found in plants. Flavonoid

[1] To whom any correspondence should be addressed.

compounds and their derivatives which can inhibit blood sugar, are found in many leaves, wood, bark, flowers, pollen, seeds, fruit flesh and roots of plants/herbal plants [3]. Due to human needs, research on usability of flavonoids for food and medicine are increasingly developed. Research on the processing of *Tanjung* fruits into flour has been done previously by [4] who correlated total phenolics and total flavonoids in the fruits. The general objective of this research is to process *Tanjung* fruit into flour as food for diabetics. The specific purpose of this study is to look at the effect of temperature and drying time on the quality of flour and flavonoid extracts produced. The quality of flour is determined through analysis of the products and compared with food standards according to SNI.

## 2. Materials and Method

The tools used in this study were a set of extraction equipment, rotary evaporator, 80 mesh sieve, 250 ml measuring cup, oven dryer, blender, knife, plastic and bucket, porcelain and furnace cups, fat flask, filter paper, desiccator, balance sheet high analytic, a set of test tubes, spatulas, 100, 250 and 500 mL measuring flasks, 1000 ml measuring cups, 10-100 μl micro pipettes, cuvettes and UV-Vis spectrophotometer. The ingredients used were *Tanjung* fruit, ethanol 96%, hexane solvent, aquadest, 5% $NaNO_2$, NaOH 4%, quersetin (Sigma Aldrich) and $AlCl_3$ Reagent 10%.

*Tanjung* fruits (*Mimusops elengi L*) were obtained from the area around Syiah Kuala University, Darussalam, Banda Aceh. The fruits were washed thoroughly and peeled. A 500 grams of fruit flesh was dried in an oven dryer at temperature of 40, 60, 80 and 100°C and the drying time of 20, 40, 60, 80 and 100 minutes. Samples were then blended and sieved with a particle size of 80 mesh. The flour products were analyzed for water content, ash content, and fat content. The best product was then analyzed for its protein and carbohydrate levels. The extraction process was then carried out using 96% ethanol solvent. A total of 150 grams of *Tanjung* fruit flour was extracted using 500 ml of 96% ethanol solvent. The extraction process was carried out at a temperature and time of 60°C and 6 hours, respectively. The filtrate obtained was concentrated using a rotary evaporator and then tested for flavonoid levels using a UV-Vis spectrophotometer.

## 3. Results and Discussion

### *3.1 Effect of Temperature and Drying Time on Physical Changes of Tanjung Fruit Flour*

The effect of drying temperature from 40 to 100 °C on the physical appearance of the flour with 80 mesh sieve size and information on its physical appearance can be seen in Figure 1.

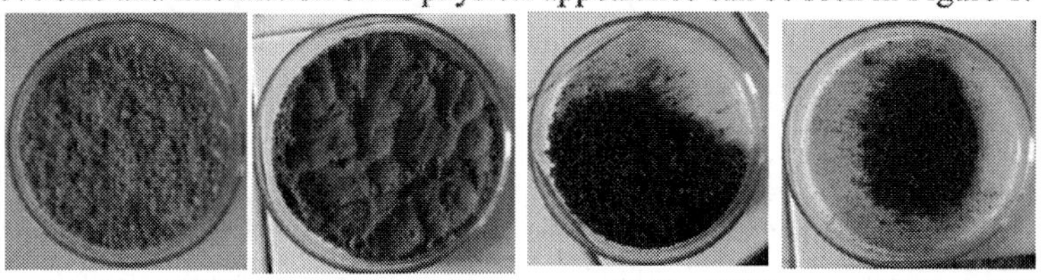

(a) T = 100°C (b) T = 80°C (c) T = 60°C (d) T = 60°C

**Figure 1.** Physical changes (color and texture) of *Tanjung* fruit flour at various drying temperatures and 60 min drying time.

Drying temperature greatly affects the texture of the acquisition of *Tanjung* fruit flour. At low temperature (40°C and 60°C), the texture of the resulting flour was not soft, easy to stick to the storage and the color of the flour is very dark. This was probably caused by the high water content in the flour. Otherwise, at high drying temperatures (80°C and 100°C), the *Tanjung* fruit flour obtained was brighter and the texture of the resulting flour was softer and more durable from the growth of fungi and other microbes. Microbes damage foods and unsafe for consumption. The activity of microbes in low water content food was hampered so as longer the storage time [5]. The effect of drying time from 20 to 100 minutes on the physical changes of the flour can be seen in Figure 2.

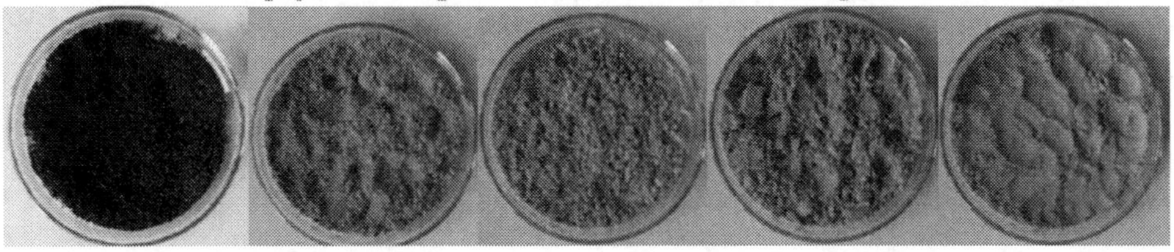

(a) t = 20 min (b) t = 40 min (c) t = 60 min (d) t = 80 min (e) t = 100 min

**Figure 2.** Physical changes (color and texture) of *Tanjung* fruit flour at various drying time at a temperature of 100°C

Drying time affects the acquisition of the flour. Short drying time resulted in moist flour and dark brown color. In addition, the durability of flour was also low. Brown color resulted from the reaction of oxygen with phenol compounds contained in flour lowering the quality of the flour. The color was also influenced by enzymes contained in the flour [5]. The longer the drying time, the brighter the color of the flour.

*3.2 Effect of Temperature and Drying Time on Water Content*

Effect of temperature and drying time on water content of the flour is presented in Figure 3.

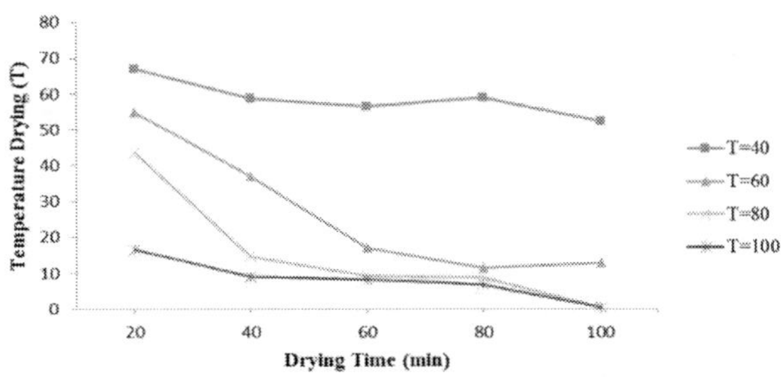


**Figure 3.** Effect of Temperature and Drying Time on water content

Figure 3 shows that the highest water content was 67.15%, resulted from the drying process at 40°C for 20 min. While the lowest water content of 0.48% was obtained from drying process at 100°C for 100 min. The longer the drying time, the lower the water content in the flour. Likewise, the higher the drying temperature, the lower the moisture contained in the material. [5] said that the higher the drying temperature, the faster the rate of water transfer to the environment, thereby reducing the water content in food. This causes the higher temperature and drying time to accelerate the rate of water transfer or the process of evaporation of water in a food material so that the water content of the material reduces. At temperatures below 45°C, microbes and fungi that damage food is still active [6].

From Figure 3, it can be concluded that the water content at the drying temperature (T) of 100°C and the drying time (t) of 40 minutes is the same as the water content at T = 80°C and t = 60 minutes. Based on the SNI for wheat flour (SNI-01-3751-2006) as food ingredients, the water content of flour is a maximum of 14.5%. From the results of the study, it was found that the drying temperature of 80 and 100°C and the drying time of more than 60 minutes produce flour with a moisture content lower than 14.5%, which fulfilled the SNI standards [7].

*3.3 Effect of Temperature and Drying Time on ash content*

Effect of Temperature and Drying Time on ash content of the flour can be seen in Figure 4.

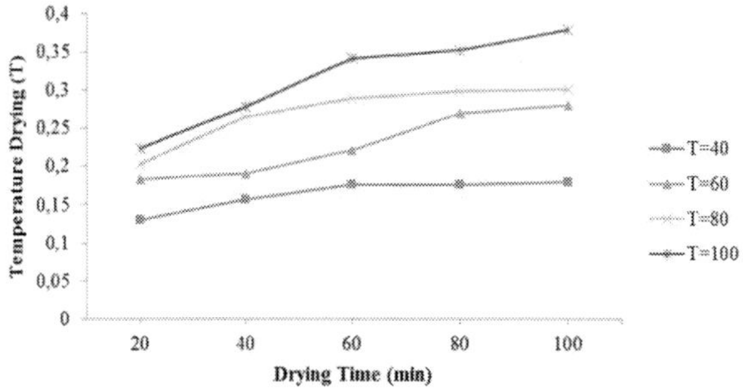


**Figure 4.** Effect of Temperature and Drying Time on ash content

The above figure shows that the higher the temperature and the drying time, the ash content in the material also increases significantly. This is caused by an increase in mineral levels in a food [5]. The higher the mineral content, the higher the ash content in the food. The mineral content in food such as phosphorus, magnesium, potassium, calcium and iron. From the ash content can be determined the nutritional value of a food. The higher the mineral content, the ash content contained in a material is also high, otherwise if the mineral content is low, the ash content contained in the food is also low [6]. From the research results, the highest ash content was obtained at a treatment temperature of 100°C with a drying time of 100 minutes of 0.37%, while the lowest ash content was obtained at 40°C with 20 minutes drying time of 0.12%. The higher the drying temperature, the higher the ash content obtained. This is due to at high drying temperature, water content is low. Therefore, food becomes concentrated so that the minerals left behind in the food material increase. With the high drying temperature, the ash content is not damaged, it is just the reduced water content in the foodstuff [8]. Based on the SNI standard for wheat flour (SNI-01-3751-2006) as a food ingredient, the ash content of flour is set to a maximum of 0.6%. From the research results obtained, *Tanjung* fruit flour with ash content less than 0.6% for all variations of operating conditions, meet the SNI-01-3751-2006 standard.

*3.4 Effects of Temperature and Drying Time on fat content*

The Effect of Temperature and Drying Time on fat content is shown in Figure 5.

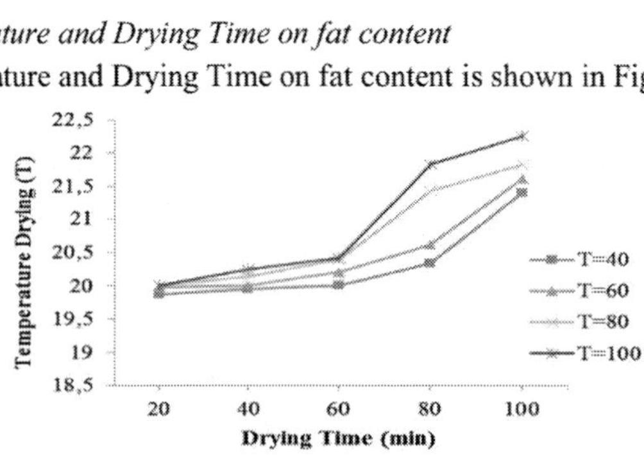


**Figure 5.** Effect of Temperature and Drying Time on fat content.

Figure 5 shows that fat content increases with increasing drying time until it reaches 100°C. This is because the higher the temperature and drying time, the fat content will increase and inversely proportional to the water content in the *Tanjung* fruit flour which decreases as the temperature and drying time increase. This result is in lined with Nuraeni research [6], which stated that when the temperature and duration of drying time are high, the water content decrease, the fat content in the material increase. In this study, the highest fat content of 22.25% was obtained from the treatment at 100°C drying temperature with 100 minutes drying time, while the lowest fat content of 19.86% was obtained at a treatment temperature of 40°C drying with 20 minutes drying time. Fat occurs due to the esterification reaction between glycerol and fatty acids. The use of high temperature resulted in the breaking of double bond in fat and break down into fatty acids and glycerol. Too high fat in the ingredients causes rancidity in the storage process [5].

*3.5 Analysis of Protein and Carbohydrate*

The analysis was carried out on samples obtained from the best operating conditions, that was drying time and temperature of 40 minutes and 100 °C, respectively. The results show that protein and carbohydrate in *Tanjung* fruit flour were 1.69% and 10.24%, within SNI limit for wheat flour of max. of 7.0% and 27.3%, respectively. Table 1 shows the quality of *Tanjung* fruit flour resulted from the process at 100°C temperature and 40 minutes drying time.

**Table 1.** Quality of *Tanjung* fruit flour at the operating condition of 100°C temperature and 40 minutes drying time

| No | Parameter | *Tanjung* fruit flour (Research result) | wheat flour (SNI-01-3751-1994) | Unit | Method Test |
|---|---|---|---|---|---|
| 1 | Water content | 9 | Maks 14,5 | % | - |
| 2 | Ash content | 0,27 | Maks 0,7 | % | - |
| 3 | Fat content | 20,24 | - | % | Pumpkin fat |
| 4 | Protein content | 1,69 | Maks 7,0 | % | Kjeldahl |
| 5 | Carbohydrate content | 10,24 | Maks 27,3 | % | Luff Schoorl |

*3.6 Analysis of Flavonoid Concentration*

Flavonoid are phenol compounds that is a lot on the part of stems, roots, wood, skin, flowers, pollen, fruits, seeds and leaves of plants or herbal plants [9]. [10], reported that flavonoid compounds act as anti-diabetes compounds and flavonoid levels can reduce blood sugar. Flavonoid concentrations are expressed in milligrams of Quersetin/gram extract (mg QE/mg extract). The analysis result of flavonoid levels obtained from *Tanjung* fruit flour extracted under various conditions of temperature and drying time is presented in Figure 6.

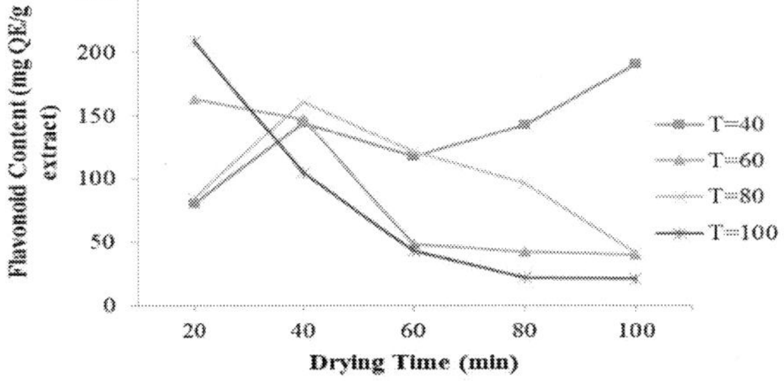


**Figure 6.** Effect of Temperature and Drying Time on Flavonoid Content

From Figure 6, it can be seen that at various temperature and drying time the flavonoid levels obtained also vary. Flavonoid levels obtained at drying temperature of more than 60°C decrease with increasing drying time. Whereas at a drying temperature of 40°C the flavonoid levels increase with increasing drying time. High temperatures damage the flavonoids in food. Drying time that is too long also damage the flavonoids in food because exposure to high temperatures during the drying process can interfere the air circulation in food. Also stated that one of the phenolic compounds, flavonoids, are damaged by heat. Phenolic compounds are very sensitive to high temperature and solar radiation [11].

From the variations of drying operating condition, it can be concluded that the optimum levels of flavonoids are taken at a drying temperature of 100°C and a drying time of 40 minutes. This conclusion is also reviewed from various analysis of flour products such as the acquisition of water content, ash content and fat content. Seen on the temperature and drying time variables (100°C and 40 minutes), water and ash content have fulfilled the SNI-01-3751-2006. In this variation flour has a long-lasting resistance to fungal and microbial growth such as mold and has a very bright and soft colour.

*3.7 Analysis of glucose blood of mice*

**Table 2.** Result analysis of glucose blood of mice at various conditions

| Mencit | Glucose Blood Content | | | | | | |
|---|---|---|---|---|---|---|---|
| | Before Injecting Glucose | After Injecting Glucose | Dosage of Tanjung flour | After 30 min | After 60 min | After 90 min | After 120 min |
| Mice 1 | 96 | 264 | 0,5 g | 206 | 196 | 176 | 166 |
| Mice 2 | 87 | 304 | 1 g | 276 | 201 | 198 | 176 |
| Mice 3 | 99 | 256 | 1,5 g | 152 | 146 | 126 | 96 |
| Mice 4 | 91 | 271 | 2 g | 196 | 186 | 176 | 156 |
| Mice 5 | 89 | 300 | 2,5 g | 222 | 198 | 186 | 147 |

Table 4 shows that injection of glucose with a certain dosage to mice showed positive results, glucose injection with dosage above 1 g/kg body weight showed a significant decrease in blood sugar.

## 4. Conclusion

The best condition in processing *Tanjung* Fruit flour with good quality in terms of water, ash, fat, protein and carbohydrate content was temperature 100°C and drying time 40 minutes. The results met the SNI-01-3751-2006 standards for wheat flour. It was resistance to mold growth, has bright colour and soft structure. The maximum flavonoid content at those conditions was 105.01 mg QE/mg extract. Giving flour to white mice with the dosage of more than 1 g/kg body weight showed significant reduction of its blood glucose level.

**Acknowledgments**

The authors acknowledge the research fund provided by The Ministry of Research, Technology and Higher Education, Indonesia through Student Creativity Program.

## References

[1] Ekor M 2014 *The growing use of herbal medicines: issues relating to adverse reactions and challenges in monitoring safety* **7** 1.
[2] Lim T K *Mimusops elengi Fruits* 2013 **6** 119
[3] Abdelmoaty M A, Ibrahim M A, Ahmed N S and Abdelaziz M A 2020 *Confirmatory Studies on The Antioxidant And Antidiabetic Effect of Quercetin In Rats* **2** 188
[4] Perwiratami C, Suzery M dan Cahyono B 2014 Korelasi Fenolat Total dan Flavonoid Total Dengan Antioksidan dari Beberapa Sediaan Ekstrak Buah Tanjung (*Mimusops Elengi*), *Jurnal Chemistry Progress* **7**
[5] Mulyati S, Pramesthy F A, Meutia F, Rinaldi A, Siregar S M and Muchtar S 2019. **523** 1
[6] Nuraeni L 2017 *Pengaruh Suhu dan Lama Pengeringan terhadap Karakteristik Tepung Terubuk* universitas Pasundan, Bandung.
[7] Badan Standarisasi Nasional 2006 Standar Nasional Indonesia (SNI). *SNI-01-3751-2006 Tepung Terigu Sebagai Bahan Makanan.* Dewan Standarisasi Indonesia. Jakarta.
[8] Yuniarti D W, Sulistiyati T D dan Suprayitno E 2013 *Pengaruh Suhu Pengeringan Vakum terhadap Kualitas Serbuk Albumin Ikan Gabus (Ophiocephalus striatus)* **1** 1
[9] Rahayu S and Tania N 2019 *Flavonoid of some antioxidant plants in Taman Wisata Alam Pangandaran* **1402** 1
[10] Gustina N N R A 2012 *Aktivitas Ekstrak Fraksi Pelarut dan Senyawa Flavonoid Daun Sukun (Artocarpus Altilis) Terhadap Enzim A-Glukosidase Sebagai Antidiabetes*, Institute Pertanian Bogor, Bogor
[11] Sulasmi E S, Wuriana Z F, Sari R S and Rohmawati U 2019 *Flavonoid characterization in Fern From Baluran National Park* **276** 1

# Canning of Traditional Acehnese Food Made by Dried Little Tuna (*Euthynnus affinis*) Using Two Sterilization Methods

**S H Anwar[1*], A A R M Poena[1], M Muzaifa[1] and H Hasan[2]**
[1] Department of Agricultural Product Technology / Faculty of Agriculture, University of Syiah Kuala, Indonesia, Banda Aceh, 23111
[2] Department of Electrical Engineering / Faculty of Engineering, University of Syiah Kuala, Indonesia, Banda Aceh 23111

[*]E-mail: sri.haryani@unsyiah.ac.id

**Abstract**. Dried fish of *Euthynnus affinis* which is also known as "little tuna" or "tongkol" is called *ikan kayu* in Bahasa Indonesia or *keumamah* in Acehnese. *Keumamah* is similar with Japanese *katsuobushi* which is prepared from dried and smoked bonito. Traditional dishes which made from *keumamah* are quite many and the one that is famous among Acehnese is called *keumamah tumis* (cooked *keumamah* or CK) which is cooked with special Acehnese herbs and spices. This study aimed to determine the effect of sterilization methods on the quality of canned CK. Two variables examined were the treatment of fish as raw material, i.e., B1 = boiled and sun-dried little tuna, B2 = boiled and fried little tuna and the sterilization method: M1 = water bath sterilization for 120 minutes and M2 = sterilization using pressure canner at 115 °C for 60 minutes. Heat penetration study and calculation of sterilization time (Fo value) were conducted using thermocouple data logger for canned CK sterilized using pressure canner. Proximate, total plate count (TPC) analysis and pH were done for fresh fish. Canned CK were analyzed for proximate, peroxide value, and microbiological analysis (TPC). The results showed that the canned CK had an average water content of 49.85%, ash, protein and fat content were 2.18%, 17.74%, and 21.0%, respectively. The pH was 4.85, Peroxide Value (PV) of 5.87%, and the products' TPC were < 1 x $10^1$ colony/g. Heat penetration study indicated that the Fo for canned CK sterilized at 115 °C for 60 minutes was 5.76 min and at 100°C for 120 minutes was 0.54 min.

## 1. Introduction

The Indonesia is one of the largest archipelagic countries in Asia where fish is the greatest commodity to be explored. The Province of Aceh is located in the westernmost part of Indonesia with a coastline of up to 1,660 km and sea area of 295,370 $km^2$. With this condition, the fishery processing industries have the brightest potential to grow in Aceh Province.

In the world of Indonesian fisheries, the term "cob or tongkol in Bahasa Indonesia" is for small tuna, namely the eastern little tuna (*Euthynnus affinis*), the frigate and bullet tunas (*Auxis sp.*), and long-tailed tuna (*Thunnus tonggol*) [1]. The most common types of fish found in Aceh Province are tuna (*Thunnus sp*), skipjack (*Katsuwonus pelamis*) and little tuna (*Euthynnus affinis*). In certain season, the amount of little tuna caught by fisherman abundant and exceeding the need of the community, therefore it is necessary to maintain the quality of the fresh fish. In order to reduce the risk

[*] To whom any correspondence should be addressed

of spoilage, proper handling and processing are needed to preserve and increase the value added of the products obtained.

The emergence of pollution cases is due to imperfect processing, where it is not hygienic, does not comply with commercial standards and does not follow the procedure of "good manufacturing practice". All of these will trigger the emergence cases of dangerous food poisoning. However, if the sterilization process is successful, the product value added will increase significantly compared to the sale or export of fresh fish. Processed fish can also reduce the amount of fish that is wasted and deteriorated when the amount of fish caught by fishermen is an abundance.

One of fish preservation that is commonly done in Aceh Province since long time ago is making *keumamah*. *Keumamah* or *ikan kayu* in Bahasa Indonesia is made from boiled and sun-dried "tongkol" or little tuna. *Keumamah* is similar with Japanese *katsuobushi* which is prepared from dried and smoked bonito. Traditional dishes which made from *keumamah* are quite many and the one that is famous among Acehnese is called *keumamah tumis* which is cooked by stir-frying *keumamah* with special herbs and spices. For the rest of this article *keumamah tumis* will be written as cooked *keumamah* and abbreviated as CK. To maintain the quality of CK, proper sterilization process is needed so that it can last for a longer time without reducing taste, color, aroma and the nutritional value.

Sterilization is a heat process carried out at high temperatures, usually >100°C with the aim to eliminate all types of microorganisms including molds, yeast, bacteria and its spores that are formed. Unproper sterilization process can change the color, taste, aroma and nutrients of food. In order to obtain a good quality of CK the selection of the appropriate sterilization method is critical.

The sterilization method using a pressure canner (retort) is the most common sterilization process used for canning or bottling food. The sterilization process is designed to kill microbes, especially *Clostridium botulinum* which is heat resistance. Sterilization is carried out at a sterilization temperature at 121°C at a certain time until the heat given is sufficient to destroy pathogenic bacteria but will minimize the reduction of the nutritional and organoleptic qualities [2].

Sterilization using water bath or heating with boiling water is one method of sterilization that uses a temperature of 100°C at a certain time to kill microbes and avoid contamination of a product. The sterilization process using boiling water is usually applied to products that contain acids [3].

Canned foods must be commercially sterile, which means if there are microorganisms survive, these microbes will not be able to grow and destroyed canned foods in storage conditions at room temperature. Canning CK is a way of preserving and storing processed little tuna in canned containers that can be hermetically and sterilized so that the product is not easily to deteriorate physically, chemically, or biologically. In addition, canning also aims to extend the shelf life, to increase the sale value of little tuna compared to fresh little tuna before it is processed.

Hot filling is a common method used to fill the product into a can or a jar before sterilization process. The product is previously heated until it reaches ± 82°C. The results of a study showed that the products packaged in bottles and filled with hot filling method thus sterilized using boiling water can last for five weeks if the product stored at 5-10°C [4].

Hasnidar (2004), conducted a study using two methods of sterilization in the production of canned CK. The 1st method was heating CK at 80°C then filled into a bottle and sterilized at 118°C in an autoclave for 30 minutes. In the 2nd method, the CK was filled into a bottle (without preheating), then boiled in water at 100°C and sterilized at 118°C for 30 minutes. The results showed that only the first method was successful to preserve the canned CK for two weeks [5].

Sterilization using the boiling water method can only be used for food products that contain acids, while for low acids food the sterilization process must be done in a pressure canner (retort). The main problem that often occurs in the process of food canning is safety and quality of food. The nutritional value of foods with high protein content such as fish (pH > 4.5) can be reduced after canning [6]. Changes in nutritional value after canning can be caused by improper temperature and time combination during sterilization. Sterilization using high temperature (± 121 ° C) for a long time in the canning process can cause protein denaturation [7]. In addition, canned fish can be contaminated as well with microbial heat-producing toxins. Other microbiological hazards are contamination by histamine (scromboid toxins) and algal toxins which are also heat-resistant [8].

This research aimed to find the best sterilization method to preserve CK so that it can be stored for a longer period of time at ambient temperature. The best method indicated minimal changes in nutritional value and fulfill the microbial standard for canned food.

## 2. Materials and methods

This study examined two variables i.e., 1) the treatment of fresh little tuna as raw material, which consisted of two factors: B1 = boiled and sun-dried little tuna, B2 = boiled and fried little tuna and 2) the sterilization method, which also consisted of two factors: M1 = water bath sterilization for 120 minutes and M2 = sterilization using pressure canner (retort) at 115 °C for 60 minutes.

### *2.1. Preparation of boiled and sundried little tuna*

The fresh little tuna was first washed thoroughly with running tab water then removing the head, tail, thorns and bones so that only the meat was obtained. The fish was cut into four parts, then boiled for 20 minutes, drained and dried for 2 to 3 days under the sun. The dried little tuna was immersed 12 hours before the surface can be cleaned and cut into small pieces (± 2 cm x 2 cm x 3 cm) then ready to be cooked.

### *2.2. Preparation of boiled and fried little tuna*

Fresh fish was purchased from the Lampulo Baru Market in Banda Aceh. Weeded fish are washed thoroughly with running water and fish meat was cut into four parts. Five (5g) of salt and 300 ml of lime juice then added into the fish meat, turned over every 5 minutes so that the salt and lime juice were absorbed. Water was finally added until the meat was completely submerged and then boiled for ±10 minutes and drain. After cooled and dried, the fish parts were fried into 2000 ml of cooking oil at 170-200°C for ± 10 minutes. Finally, the dried and fried fish was cut into small pieces and ready for cooking.

### *2.3. Process for making cooked keumamah*

The cooking process of CK was done first by heating 300 ml cooking oil then stir-fried the spices. The blended spices were consisted of onion, garlic, red chili, cayenne, *asam sunti* (salted, fermented and dried bilimbi - *Averrhoa bilimbi*), ginger, and turmeric which had been blended previously. After blended spices were half cooked then curry leaves and chopped green chili were added into the mixture. Finally, small pieces of boiled and sundried little tuna or boiled and fried little tuna was added, mixed and cooked thoroughly.

### *2.4. Sterilization, cooling and ripening processes*

Before the sterilization process began, the jar (9.2 cm x 6.5 cm) and the lid were heated in boiling water for 15 minutes. It was drained and while the jar was still hot, the warmed CK (± 170 g) was filled into the jar and closed immediately with the hot lid. Sterilization were conducting using two methods: first, in a water bath at 100°C for 120 minutes and second, in a retort (pressure canner) at 115°C for 60 minutes. Upon completion of sterilization, all jars were drained and stored at room temperature.

### *2.5. Thermal penetration study for the Fo calculation*

For this purpose, the thermocouples gland with build-it logger (TSP Micro, Ellab, Denmark) were placed at two positions. The retort temperature measurement was done by placing a thermocouple right above the water level at the base of pressure canner (retort). For the core product temperature measurement, a thermocouple was placed inside a can before the can was seamed. This thermocouple was positioning in the middle of the can pointing upward to the geometric center and was held by stuffing the product (warmed CK). The recorded temperatures were converted into data (Lethality rate,

L) using a docking system (Pro Multi Reader, Ellab, Denmark) and an application software (ValSuite Pro, Ellab). The experiments were repeated several times and the data obtained were compared and evaluated carefully.

### *2.6. Analysis of fresh little tuna and canned CK*
Analysis of fresh little tuna included proximate analysis: water content (method of SNI 01-2891-1992, point 5.1), ash content (SNI 01-2891-1992, point 6.1), protein content (SNI 01-2891-1992, point 7.1), fat content (AOAC method 938.06-33.6.04.2005), total plate count (TPC) analysis and pH. For canned CK similar analysis were performed which were proximate analysis, peroxide value, and microbiological analysis (total plate count or TPC).

### *2.7. Statistical Analysis*
All the experiments in this study were performed in triplicates. The significant differences in water, ash, protein, and fat content as well as the results of Peroxide Value (PV) were analyzed using two-way analysis of variance (ANOVA). Significantly different groups were compared at a 95% confidence level ($P < 0.05$).

## 3. Results and discussion

### *3.1. The quality of fresh little tuna*
Proximate analysis and total plate count (TPC) were conducted for the fresh little tuna before it was processed to be CK. The results can be seen in Table 1.

Table 1. The analysis results of fresh little tuna

| Parameters | Percentage (%) ± SD |
|---|---|
| Water / moisture | 75.25 ± 0.18 |
| Ash | 2.0 ± 0.07 |
| Fat | 0.54 ± 0.01 |
| Protein | 24.20 ± 0.14 |
| pH | 5.64 ± 0.10 |
| Total Plate Count (TPC) | $3.3 \times 10^5$ cfu/g |

A study reported that the water content of fresh tuna was 80.40%, fat content of 0.28%, protein content of 20.9% and ash content of 1.20% [9]. If this is compared with the results of this study as in Table 2, the water content is slightly lower, the ash content is slightly higher, the fat and protein content are still in the range. Usually, the pH level of fresh fish ranges from 6.4 to 6.6 while in this study the pH of fresh fish is 5.64. This might be due to glycolysis occur in fish [10]. Based on SNI 7388:2009 the limit of microbial contamination of fresh fish is $5.0 \times 10^5$ cfu/g. Therefore, the fresh fish as raw material for the CK was still meet the SNI standard [11].

### *3.2. The quality of canned CK*

*3.2.1. Water content.* The most important property in a food is water content. Water content contained in food will affect the appearance, texture, taste and will determine the durability of the food itself. The results of the analysis showed that the canned CK water content ranged from 44.53 to 53.50% with an average of 49.85%. The results indicated that the sterilization method (M) had a significant effect ($P \leq 0.05$) on the canned CK water content (Figure 1). It can be seen that the highest water content was obtained when the canned CK sterilized using waterbath for 120 minutes (M1) and decreased when sterilization carried out using a pressure canner at 115°C for 60 minutes (M2).

All food ingredients contain different amounts of water. Water is the most important component in a food. The analysis result showed that the water content of fresh fish was 75.25% which far higher than the canned products' water content. These can be cause by the treatment received by fresh fish, since the fish were dried into *keumamah*, boiled, fried and because of the heat received by the products during cooking, hot-filling process as well as when sterilizing for a long time. Water content in food will shrink after experience the cooking process using high temperatures. The higher the sterilization temperature used, the more the water content decreased.

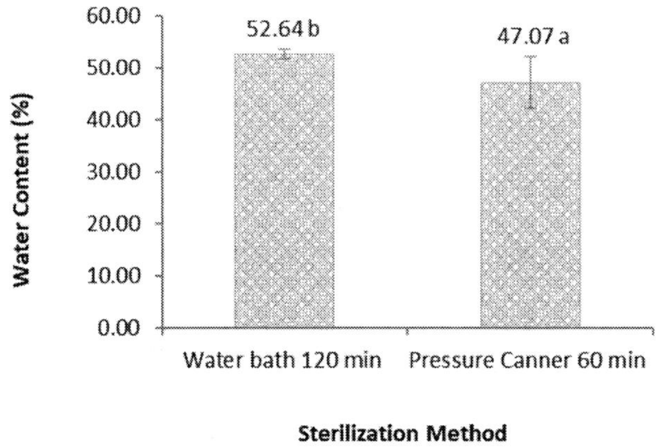


**Figure 1.** Influence of sterilization methods on the water content of canned CK

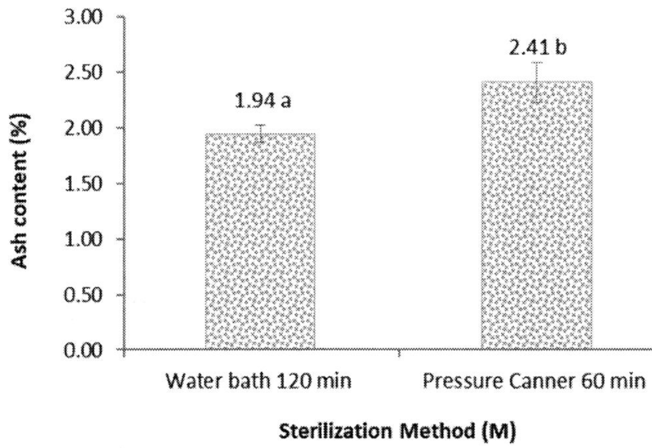


**Figure 2.** Effect of method and time of sterilization (M) on the ash content of canned CK

*3.2.2. Ash content.* Ash content is closely related to mineral content in food. The higher the ash content contained in food ingredients the higher the mineral content in these materials [12]. In principle, the analysis of minerals in food can be divided into two, namely the analysis of total minerals or total ash and analysis of each type of mineral. In this study the ash content analysis was carried out in the form of total ash analysis.

The results of the analysis showed that the sterilization method (M) had a significant effect ($P \leq 0.05$) on the ash content of canned CK, as can be seen in Figure 2. The graph indicated that the lowest ash content was for the product that sterilized using water bath at 100°C for 120 minutes (M1) and increased when sterilized using pressure canner at 115°C for 60 minutes (M2).

The ash content indicates that there is inorganic mineral contain in the foodstuff. The slight increment of ash content from fresh little tuna to canned CK can be caused by the addition of spices when cooking. The ash is obtained from a material that is first dried and then bonded and the residue is mineral. About 96% of the food is made up of organic matter and water, the rest are minerals. The mineral component can be determined by the determination of the residual combustion of the mineral salt or known as aromatic.

*3.2.3. Protein content.* Protein is one of the food substances in the form of amino acids which have functions as builders and regulators for the body. The results of data analysis showed that the protein content of canned CK ranged from 16.97-18.72% with an average of 17.74%. The effect of raw materials and methods as well as time of sterilization on canned CK can be seen in Figure 3.

The use of heat in food processing such as boiling, steaming and frying can affect the nutritional value of food, especially protein content. The protein in the food consumed is absorbed by the intestine in the form of amino acids. Protein is a source of amino acids consisting of carbon, hydrogen, oxygen and nitrogen. Protein plays an important role which functions as a builder and regulator of the human body [12].

In the cooking process, the treatment of fish as raw materials which was boiled and fried then sterilized using a pressure canner or water bath causing a decrease in protein content significantly from 24.20% (fresh fish) to averagely 17.74% (canned CK). As can be seen in Figure 5, the reduction was more when the canned CK made from fish that was boiled and fried compared to the boiled and sun-dried fish. Protein reduction occurs during food processing at high temperatures. During the sterilization process, around 40% of the total amino acids in fish meat are degraded depending on how high the temperature and how long the sterilization process are conducted.

Figure 5 also demonstrated that there was no significant reduction of canned CK protein content when sterilized using water bath (100°C) for 120 min. These indicated that there was a reduction in protein content when a higher sterilization temperature used (115°C) even though the processing time was shorter (only 60 min). However, the overall reductions were statistically not significant.

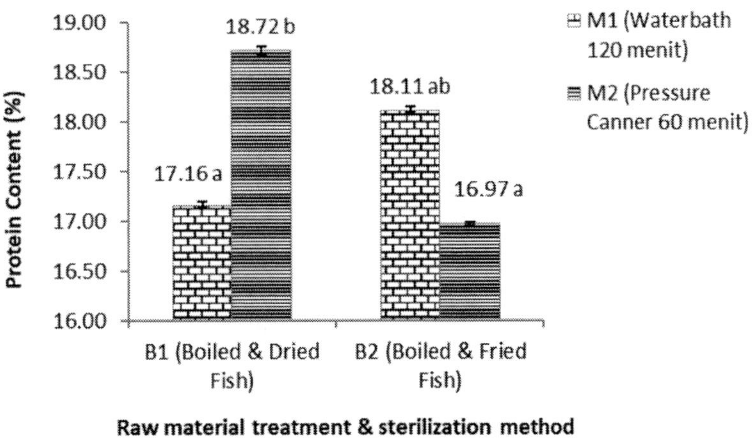


**Figure 3.** Effect of raw material treatment and method of sterilization on the protein content of canned CK

*3.2.4. The fat content.* Fresh little tuna contains omega-3 fatty acid of 1.5g /100g and omega-6 fatty acids of 1.8g /100g [13]. Determination of fat content in a food is important, especially in fish. Fish fat contains omega-3 polyunsaturated fatty acids, especially EPA and DHA which are good for the human body. In food, fat acts as a carrier for vitamins, flavor and as important as the calories source [12].

The results of the fat content on canned CK ranged from 4.49 to 6.27% with an average of 5.19%. The results of analysis of variance showed that the raw material treatment (B) and the sterilization method had no significant effect (P> 0.05) on the fat content.

*3.2.5. Peroxide value.* Peroxide value is an index of the amount of fat or oil that has undergone oxidation. Oils that contain unsaturated fatty acids can be oxidized to produce a peroxide compound. A high peroxide value indicates fat or oil has oxidized. Peroxide compounds can accelerate the process of rancidity and food deterioration.

The results of the analysis showed that the peroxide value of canned CK ranged from 5.83 to 5.90 mEq/kg with an average value of 5.87 mEq/kg. The effect of raw material treatment on the fat content can be seen in Figure 4. As can be seen in the graph, the highest PV was found in canned CK prepared from boiled and dried fish (B1) and decreases with when boiled and fried fish was used (B2).

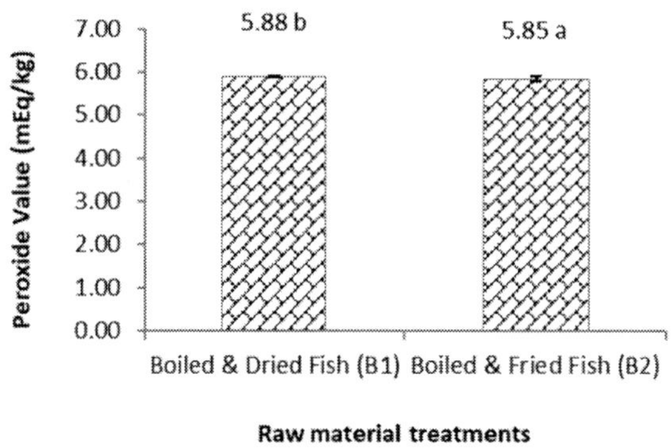


**Figure 4.** Effect of raw material treatment on the peroxide value (PV) of canned CK

Rancidity in oil is caused by the oxidation of fatty acids to form free radicals thus produce peroxide which is an intermediate product that will decompose to form carbonyl compounds (aldehydes and ketones) [12]. An increased of peroxide value indicates that the oil has undergone oxidized. Deterioration of oil is caused by repeating used of oil for heating and or frying. Based on Indonesian National Standard-SNI (2013) the maximum limit for cooking oil peroxide value is 10 meq/kg [14].

The PV of canned CK in this study was 5.85-5.88 meq/kg therefore it still meets the SNI-3741-2013. The low PV indicated that the heating process has less effect on the rancidity of the canned CK. This was because during fish frying, the oil used was the fresh oil and it was still in its best quality and has never been rancid. Rancidity in food products can cause products to decline in quality and selling power. Rancidity occurs due to the interaction of fat and air (oxygen) which causes the formation of peroxides that are unstable and easily decomposed into derivatives such as aldehydes, ketones and other compounds with low molecular weight.

*3.2.6. The microbiological analysis.* TPC test for canned CK was done to find out the number of aerob mesophillic microbes per gram sample if they are still existed in the products after sterilization and to find out the quality of the products during storage. The tests were carried out to determine whether or not the canned CK was safe to be consumed for a long time.

The occurrence of microbial contamination in a canned food due to imperfect processes during product sterilization in a pressure canner or water bath. A recommended temperature to kill microbes according UN-FAO is 121°C [15]. A recommended sterilization time is depending on the amount and types of initial microorganisms that are possibly contaminated the food.

The sterilization condition used in this study was 115 ° C for 60 minutes. The vegetative cells can be killed, but for bacterial spores which highly heat resistance may not be able to be killed completely. The results of the TPC on the canned CK in week 0 showed that there was no microbial contamination. Based on the Indonesian National Standard (SNI-7388-2009) for food processing product that use sterilization such as canned fish, the microbial contamination limit is $<1x10^1$ colony/g. In this study, no single colony grew on solid media in petri dishes. Therefore, it can be said that the sterilization processes have produced canned CK which have met the SNI-7388-2009 at week 0, based on the TPC analysis.

*3.2.7. Calculation of the $F_o$ value.* The standard parameter used to indicate the success of a thermal treatment is called the $F_0$ value (the thermal process severity). The $F_0$ value of a thermal process in this research was determined by means of physical method which used the temperature changes during thermal process at the slowest heating point (SHP) of the product. The results are related to the thermal destruction rate at a reference temperature [15]. The $F_0$ value was identical with time in minute if the SHP of a product reached 121.1 °C.

Heat penetration study resulted that the Fo for canned CK sterilized at 115 °C for 60 minutes was 5.76 min. This Fo value has meet the safety standard for the survival of *C. botulinum*, pathogenic bacteria that are commonly caused serious illness if present in food. According to UN-FAO, "the minimum thermal process required to provide safety from the survival of *C. botulinum* is equivalent, in sterilizing effect, to 2.8 min at the slowest heating point (the SHP) of the container. This process is commonly referred to as a "botulinum cook" [15].

The Fo for canned CK sterilized using water bath at 100 °C for 120 minutes, however, only 0.54 min. The differences between Fo for the two sterilization methods caused by the heat generated by each method. Pressure canner (retort) applied pressure during its process thus the temperature can increase far above 100°C in a short time. This equipment allows sterilization temperature to be set at 121°C. Water bath sterilization on the other hand, generates maximum heat only as a boiling water temperature (±100°C) at all time. Therefore, the heat that penetrates into the product particularly at the slowest heating point (SHP) will reach 100°C as the maximum temperature after a long sterilization time.

Theoretically, the Fo obtained by water bath sterilization was not sufficient yet for a "botulinum cook" to kill *C. botulinum* completely which required minimum Fo of 2.8 min. However, the result of total plate count (TPC) analysis indicated that no microbial treats found in canned CK. Samples taken from the two products showed no single colony grew on solid media in petri dishes. The TPC analysis were done in the same week with the production of canned CK. This result might be caused by the lower pH (< 5) of the CK due to the addition of specific herbs and spices which can be functioned as natural preservatives for the product. However, a further study is essential to examined the product quality during storage for at least six months to one year after production. Therefore, a storage test at room temperature is necessary to be done to prove if there is a potential threat, microbiologically and to make sure that the sterilization process is already sufficient and successful or further experiment with longer sterilization times are needed.

## 4. Conclusion

The sterilization method used in preserving the canned CK reduced the water content in a significant amount when compared to the water content of fresh fish. Water content in food will shrink after experience the cooking process using high temperatures. The treatment of fish as raw materials which was boiled and fried then sterilized using a pressure canner or water bath causing a decrease in protein content significantly from 24.20% (fresh fish) to averagely 17.74% (canned CK).

The PV of canned CK in this study was 5.85-5.88 meq/kg therefore it has meet the Indonesian National Standard-SNI. The low PV indicated that the heating process has less effect on the rancidity of the canned CK. The microbial contamination limit according to the SNI-7388-2009 for food processing such as canned fish is $<1x10^1$ colony/g. In this study, no single colony grew on solid media in petri dishes for canned CK right after sterilization and before the storage test. The sterilization process, either using water bath at 100°C for 120 min or pressure canner at 115°C for 60 min have met the SNI-7388-2009 based on TPC analysis at week 0. The fish treatment which is recommended to be used is the boiled dan fried little tuna based on the results of peroxide value (PV) analysis.

## 5. Acknowledgement

All authors thank to The Ministry of Research, Technology, and Higher Education as well as University of Syiah Kuala for funding this research through the scheme of "Penelitian Unggulan Universitas Syiah Kuala - PUU" (contract No. 581/UN11/SPK/PNBP/2019).

## References

[1] Proctor C H, Merta I G S, Sondita M F S, Wahju R I, Davis T L O, Gunn J S and Andamari R 2001 *A Review of Indonesia's Indian Ocean Tuna Fisheries, Part 1. ACIAR Project FIS/2001/079: Country Status Report.* (CSIRO Marine Research Australia)

[2] Fellows P 2009 *Food Processing Technology: Principle and Practice* (UK: Woodhead Publishing)

[3] Csuros M and Csuros C 1999 *Microbiological Examination of Water and Wastewater* (New York: CRC Press)

[4] Pratama Y and Abduh S B M 2016 Perlakuan Panas Mendidih Pada Pembuatan Milk-Tea Dalam Kemasan (Kajian Pada Industri Skala Kecil) [Hot Boiling Treatment on Packaged Milk-Tea (Study on Small Scale Industry) *J. Pangan Dan Gizi* **07**

[5] Hasnidar 2004 *Kajian Pembuatan Ikan Kayu (Keumamah) Tumis Instan Dengan Dua Metode Sterilisasi* [Study of Making Keumamah Instant with Two Sterilization Methods] (Banda Aceh: Universitas Syiah Kuala)

[6] Lewis M J 2006 Thermal Processing *Food Processing Handbook* ed J G Brennan (Wiley-Blackwell) pp 33

[7] Stephen N M, Jeya Shakila R, Jeyasekaran G and Sukumar D 2010 Effect of different types of heat processing on chemical changes in tuna *J. Food Sci. Technol.* **47** 174.

[8] Dewanti-Hariyadi R 2011 Pengendalian dan Pengujian Pangan Steril Komersial [Controlling and Testing Commercial Sterile Food]

[9] Anjarsari B 2010 *Pangan Hewani Fisiologi Pasca Mortem dan Teknologi* [Post Mortem Physiology and Technology of Food from Animal Sources] (Yogyakarta: Graha Ilmu)

[10] Buckle K A, Edwards R A, Fleet G A and Wooton M 1987 *Ilmu Pangan. Penerjemah: Hari Purnomo dan Adiono* [Food Science] (Jakarta: Universitas Indonesia Press)

[11] Badan Standardisasi Nasional Indonesia 2009 SNI-7388-2009-Batas-maksimum-cemaran-mikroba-dalam-pangan [The Maximum Limit of Microbial Contamination in Food]

[12] Winarno F G 2008 *Kimia Pangan dan Gizi* [Food Chemistry and Nutrition] (Bogor: M-Brio Press)

[13] Khomsan A 2004 *Peranan pangan dan gizi untuk kualitas hidup* [The Role of Food and Nutrition for the Quality of Life] (Jakarta: Gramedia Widiasarana Indonesia)

[14] Badan Standardisasi Nasional Indonesia 2013 SNI-3741-2013-Minyak-goreng [Frying Oils]

[15] Warne D 1988 Manual on fish canning - 1. Principles of canning *FAO Fish. Tech. Pap. - 285 Food Agric. Organ. U. N.*

# The Effects of Magnesium Hydroxide ($Mg(OH)_2$) on the Characteristics and Performance of Polyethersulfone Membranes (PES) for Water Treatment

**U Fathanah[1,2], I Machdar[1], M. Riza[1*], N Arahman[1] , M R Lubis[1], F Nurfajrina[1], A Yolanda[1] , M Y Wahab[1]**

[1] Department of Chemical Engineering, Faculty of Engineering, Universitas Syiah Kuala, Darussalam, Banda Aceh 23111, Indonesia

[2] Doctoral Program School of Engineering, Universitas Syiah Kuala, Darussalam, Banda Aceh 23111, Indonesia

[*]E-mail: medyan_riza@unsyiah.ac.id

**Abstract.** Surface water is commonly used as a source of clean water, but increasingly, surface water is contaminated with various foulants. Humic acids (HAs) are a natural organic matter predominant in ground and surface waters. HAs can reduce the quality of water. This study aims to investigate the results adding magnesium hydroxide, or $Mg(OH)_2$, to modify polyethersulfone (PES) membranes by determining the characteristics and performance of the membranes. The mixed-matrix membranes were prepared using PES and various amounts of modified $Mg(OH)_2$ (0%, 1%, and 1.5% wt) as nanoparticle additives. A scanning electron microscope (SEM), Fourier-transform infrared spectroscopy (FTIR), and water contact-angle analysis were used to characterize the membranes, followed by a membrane performance test (flux, permeability, and selectivity of the membrane) using a dead end ultrafiltration module. It can be concluded that $Mg(OH)_2$ can improve the hydrophilicity of PES membrane with decreasing the water contact angle from 84.2° (pure PES membrane) to 68° (modified membrane), where ultrafiltration membranes with the highest permeability coefficient are obtained at 1.5% wt of $Mg(OH)_2$ with a value of 41.94 $L/m^2.h.bar$ and a rejection percentage of 67.72%.

## 1. Introduction

The usage of surface water as a source of clean water is increasing howevermost of a majority of surface water has been contaminated with soluble foulants such as wastewater. It is necessary to remove major organic contaminants in surface water such as humic acids [1]. Humic acids containing carboxyl and phenolic hydroxyl, which can cause the water to become darker, hampering the availability of light in the water [2]. Humic acids are also a substrate in microorganism growth and so make the water more polluted [3].

Membrane technology can be used as a water management alternative in developing countries like Indonesia. There are many advantages of using membrane technology, such as high water quality, consistent water quality, and lower chemical usage [4]. Using membrane technology also allows for selective separation and aseptic operations without generating any byproducts [5].

The membrane used in this study is a polymer membrane of polyethersulfone. A phase inversion technique is used in the preparation process based on the principle of membrane-solution dope phase-change from liquid to solid. PES is a polymer used in membrane manufacturing, known for its chemical stability, chemical resistance, and good mechanical properties. However, PES polymer is hydrophobic, so the membrane flux will tend to decline during the operation due to the occurence of fouling on the membrane[6]. Foulant build-up on the membrane surface can cause the pore blockage of the membrane, thereby reducing the filtration performance. Foulant can be either organic depositions (e.g., macromolecules or biological substances), precipitated inorganic depositions (e.g., metal hydroxide or calcium salt), or particulates. Foulant accumulates on the surface of the membrane, resulting in reduced effectiveness and membrane flux [7].

To improve the membrane performance, the modification is necessary, for instance by adding additive material to the dope during fabrication of the PES membrane. Previous studies have investigated the development of polymer membranes by modifying PES with tetronic 1307 [8], using PES with chitosan to remove nitrate in the water [9], and adding graphene oxide/chitosan to the PES for processing chromium and improving the membrane's antifouling properties [10].

This study will discuss the modification of PES membranes by adding an inorganic compound, magnesium hydroxide or $Mg(OH)_2$, to improve the membranes' performance. $Mg(OH)_2$ is added as an agent to increase the hydrophilic nature of the polymer PES. Hydrophilic $Mg(OH)_2$ is used as an inorganic filler to prepare PES/$Mg(OH)_2$ membranes using a phase inversion method to prevent a decrease in the flux caused by membrane fouling [11].

$Mg(OH)_2$ is one of the inorganic compounds that have hydrophilic OH groups, making it attractive as an additive in the membrane [12]. In this study, $Mg(OH)_2$ is added to the PES membrane to improve the membrane's hydrophilicity for water treatment processes. The effects of adding $Mg(OH)_2$ will be investigated by examining the characteristics and performance of the resulting membrane.

## 2. Materials and Methods

*2.1 Materials,* Some of the chemicals used to manufacture the polymer membrane are polyethersulfone (PES, Ultrason E6020 BASF), N-methyl-2-pyrrolidone (NMP) solvent from Merck, and additive $Mg(OH)_2$ purchased from Xinglu Chemical Co. Other materials used are deionized water as a non-solvent, a sample of humic-acid solution to test the membrane's rejection ability, and a set of dead-end filtration equipment.

*2.2 Membrane Preparation,* All of the membranes were prepared by a phase-inversion method. A dope solution was prepared by dissolving the PES in the NMP solvent with a concentration of 18% wt for all types of membranes. For membrane modification, $Mg(OH)_2$ was added at 1% and 1.5% wt (see Table 1). The dope solution was stirred for 24 hours until homogeneous. Then, the dope solution was sonicated for 30 minutes to prevent the agglomeration of the $Mg(OH)_2$ and to ensure extraction for air bubbles. The homogeneous PES solution was placed onto a glass plate in a casting machine to produce a wet membrane with a thickness of 250 μm. This membrane was immersed in a coagulant bath, where the polymer precipitated and the flat membrane formed. The composition of the membrane dope solution with $Mg(OH)_2$ is presented in Table 1.

Table 1. Dope solution composition and coded name of membranes.

| No. | PES (%) | $Mg(OH)_2$ (%) | NMP (%) | Membrane Name |
|---|---|---|---|---|
| 1 | 18 | 0 | 82 | A-0 |
| 2 | 18 | 1 | 81 | A-1 |
| 3 | 18 | 1.5 | 80.5 | A-2 |

*2.3 Membrane Characterizations,* Membrane characterization was done by observing its morphology (membrane cross-section) using SEM and the chemical composition of the membrane surface using FTIR equipment, and measuring the hydrophilicity of the membrane using water contact angle analysis (water contact angle meter, Drop Master 300, Kyowa Interface Science Co., Japan.

*2.4 Membrane Performance (Pure Water Permeation and membrane rejection coefficient),* The membrane performance in terms of pure water permeability and humic acid selectivity was assessed using a filtration test with the dead-end module. Experiments were carried out using membrane pressure 1; 1.5; 2 and 2.5 bar as the driving force. The membrane was first compacted for 30 minutes to minimize compaction effects. After the flux reached a steady state, the flux was calculated with the following equation [13]:

$$J = \frac{1}{A}\frac{dV}{dt} \qquad (1)$$

Where J is pure water flux (L / $m^2$ h); dV is permeated volume (L); dt is the time (hours), and *A* is the membrane area ($m^2$). While the equation used to calculate the membrane permeability coefficient ($L_p$) [13] is:

$$L_p = \frac{J}{\Delta_p} \qquad (2)$$

Wherein, J is pure water flux (L / $m^2$.h); Δp is trans membrane pressure (bar)
Rejection coefficient (R) is the concentration fraction of dissolved substances that cannot pass through the membrane. The equation of rejection coefficient (R) [13]is:

$$R = \left(1 - \frac{Cp}{Cf}\right) x\ 100\% \qquad (3)$$

Where,
R = coefficient of rejection (%)
Cp = concentration of solute in the permeate (mg/L)
Cf = concentration of solute in the feed (mg/L)

The relative selectivity of membranes was determined with a humic acid solution containing 50 mg/L humic acid. The filtration condition was exactly the same as that of the permeability test. The concentration of humic acid in the permeate was measured with UV-Vis spectroscopy at a wavelength of 254 nm.

## 3. Result and discussion

### *3.1 FTIR analysis*

Analysis to determine the difference in the chemical composition of the modified membrane was performed using a Fourier Transform Infra-Red (FTIR) Spectroscopy. FTIR spectroscopy is an instrument that is used to determine the surface functional groups contained in a sample based on molecular interactions in the form of absorbance or transmittance from infra-red rays given to the sample.

The FTIR spectra of PES/$Mg(OH)_2$ 1% w/w membrane, PES/$Mg(OH)_2$ 1.5% membrane, and the pure PES membrane are presented in Fig.1. It can be seen that all peak positions of the spectra are precisely the same, including the 1100 $cm^{-1}$ for sulfone ($SO_2$) bending vibration, 1230 $cm^{-1}$ for ether the aromatic (C-O-C), 1465 $cm^{-1}$ for C-H, and the 1510 $cm^{-1}$ for aromatic ring [14]. Figure 1 also shows a peak at a wavenumber of 3697 $cm^{-1}$ on the membrane (A-1) to membrane (A-2), wherein the peak indicates O-H bonds of the $Mg(OH)_2$ stretching vibration. This peak does not appear on A-0 membrane (pure PES membrane). The O-H group indicates that the $Mg(OH)_2$ compound added interacts with the PES membrane. O-H can make PES membrane more hydrophilic [15].

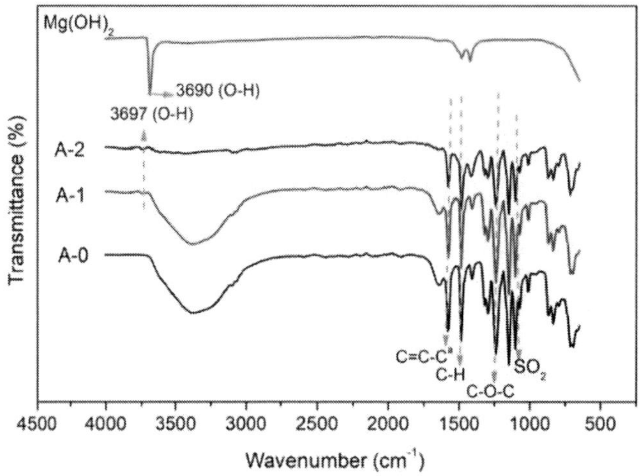


**Figure 1.** FTIR spectra of pure PES membrane and modified membranes

*3.2 Membrane Morphology*

Membrane performance is not only influenced by the pore size but also the number of pores on the surface [16] as seen in the structure of the membrane morphology. The cross-sectional morphology was characterized bySEM, as shown in Figures 2.

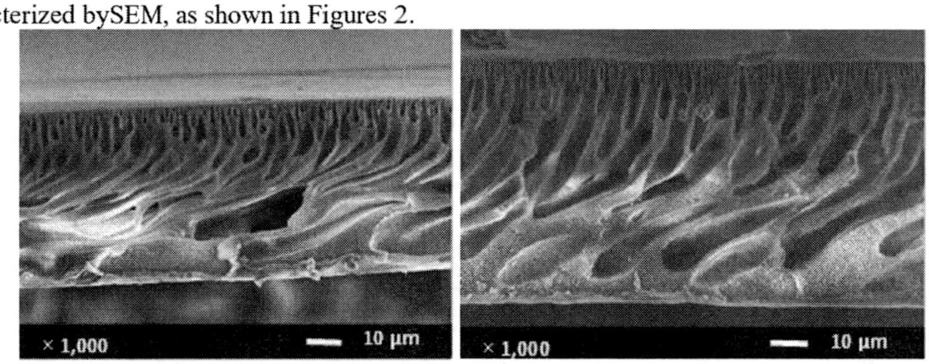


(a) (b)

**Figure 2.** The cross-section morphologies of (a) PES and (b) 1.5% w/w $Mg(OH)_2$/PES membranes

As shown in Figure 2, the PES membranes possess an asymmetric structure and a micro-porous finger-like sub-layer. The membrane samples are distinguished by the various amounts of $Mg(OH)_2$ dispersed in the PES solution. In Fig.2 (a) there is no additive. In Figure 2 (b) a membrane with additives added, it can be seen that the macrovoid size of this membrane is greater than that of the pure PES membrane.

Based on the images, it can be concluded that the $Mg(OH)_2$ addition changed the morphology of the membranes by increasing the sizes of the pores. The changes in the pore structure of the membrane are related to the phase inversion process between polymer matrix, solvent, and non-solvent during membrane fabrication. [17]. Previous researcher [18] states that the formation of $Mg(OH)_2$ produces macro pores and more pores than PES membrane because $Mg(OH)_2$ is hydrophilic, provides more space for water penetration and further accelerates the rate of water penetration.

*3.3 Water Contact Angle*

Hydrophilic property of the membranes is an important characteristic that greatly affects the permeability and other membrane performances. It can be investigated by measuring the water contact angle (WCA). A higher WCA indicates that the membrane is more hydrophobic, while lower values can be an indication of more hydrophilic membranes.

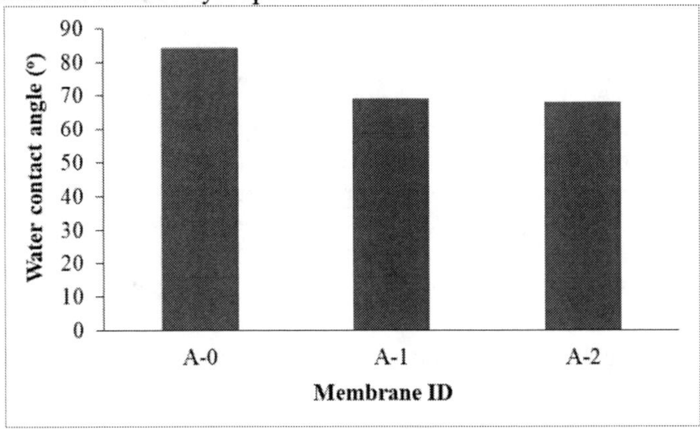


**Figure 3**. Water contact angle on various types of membrane

As shown in Figure 3, the pure PES membrane has the highest contact angle, 84.2° indicating the most hydrophobic membranes of all evaluated. As the $Mg(OH)_2$ content increases, the contact angle on membrane surface increases, when the $Mg(OH)_2$ concentration increases 1% and 1.5% w/w, the contact angle on the membranes surface becomes 69° and 68°, respectively. The decrease of water contact angle indicated the increasing of hydrophilic property of the membrane [19]. Peningkatan sifat hidrofilisitas ini disebabkan oleh kehadiran gugus hidroksil –OH dari aditif MgOH pada matrix membran. Hal ini dapat dikonfirmasi melalui analisa FTIR (Gambar 1).

### *3.4. Pure Water Flux and Permeability Analysis*

The pure water flux (PWF) of the fabricated bare and blended PES membranes is shown in Fig.4. Filtration test equipment used is called a membrane module. The membrane module used is the dead-end ultrafiltration module. Meanwhile, the distilled water is used as the feed in the flux performance test.

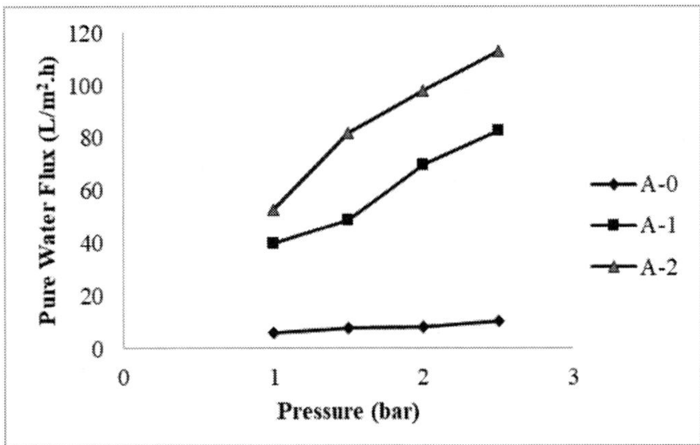


**Figure 4.** The pure water flux (J) on various types of PES membrane

The pure water flux properties of the prepared membranes were studied by collecting the pure water through a certain time at various pressures in the range of 1 to 2.5 bar. In Fig. 5 it can be seen that the magnitude of the pressure is proportional to the flux generated in the same type of membrane. In A-0 membrane, the flux value is at the operating pressure 1; 1.5; 2 and 2.5 bar by 5.88; 7.59; 8.29 and 10.27 $L/m^2{\cdot}h$, respectively. Increasing the flux value increases the pressure exerted, because of the formation of the membrane deformation as a result of the high pressure on the feed stream and make the membrane pores become wider [20].

In addition, in Fig. 4 also shows the effect of $Mg(OH)_2$ in various concentrations to the flux value produced. The higher concentration of $Mg(OH)_2$ is added, the higher the value of the flux. The flux of membrane A-1 and A-2 at a pressure of 1 bar respectively 40.01 and 52.89 L/m$^2$·h. Increasing the flux value because of the addition of $Mg(OH)_2$ made a hydrophilic membrane surface easier to absorb water molecules, and increase in the amount of permeate [21].

The relationship between permeability and selectivity is a well-known trade-off in membranes. The coefficient of permeability (Lp) declares the membrane's ability to pass the pure water at the membrane operating pressure [22]. Fig.5 compares membrane permeability between pure PES, PES/1wt% $Mg(OH)_2$, and PES/ 1.5% $Mg(OH)_2$ membranes. The addition of $Mg(OH)_2$ particles increased the pure water permeability of the PES/$Mg(OH)_2$ membranes. As a control, the pure PES membrane permeability was only 4.12 L/m$^2$·h·bar. However, this value increased to 30.62 and 41, 94 L/m$^2$·h·bar for A-1 and A-2, respectively. Based on the overall pure water permeability obtained, the A-1 and A-2 membranes are included in the category of ultra-filtration membranes with permeability coefficient in the range of 10–50 L / m$^2$·h·bar [13].

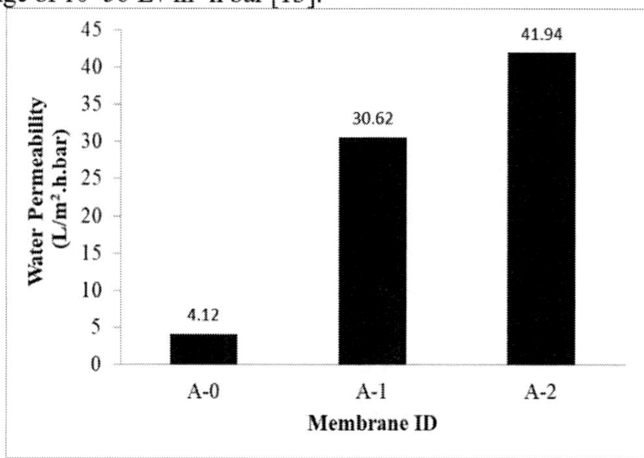


**Figure 5.** The water permeability coefficient (Lp) of pure water on various types of membranes

Water permeability of the membrane is influenced by the value of water contact angle. The increase in water permeability of the membrane A-1 and A-2 could be explained by the results of water contact angle measurement in Fig.3. There is a tendency for a decrease in the value of water contact angle on the modified membrane with the addition of $Mg(OH)_2$ additive shown in the Figure 3. This shows that the membrane modification with $Mg(OH)_2$ can improve the hydrophilic properties of PES membranes. It affects on the value of the resulted membrane permeability and is a reason for the increase of the membrane permeability.

*3.5 Humic Acid Rejection*

Membrane selectivity performance is tested through membrane filtration with a humic acid solution. Humic acid is one of the natural organic matters containing water. Filtration membranes with good performance are a membrane with a high water permeability and high solute rejection [17]. The results of flux measurements and rejection analysis of humic acid can be seen in Figure 6.

Based on Figure 6 it can be seen that the highest percentage of humic acid separation with PES membrane is on the membrane A-0 with a value of humic acid rejection of 77.6%. The presence of membrane modification with the addition of $Mg(OH)_2$ additive makes the rejection coefficients decreased in A-1 and A-2 membrane as 68.63% and 67.72%, respectively.

Figure 6 also shows that the flux value of humic acid feed solution increased as the increase of $Mg(OH)_2$. The highest flux of humic acid solution is on the A-2, namely 36.23 L/m$^2$·h. This is influenced by the $Mg(OH)_2$ addition, thus increasing membrane pore make more humic acid passes through the membrane pores. In general, there is a reverse comparison between selectivity and flux membranes.The increase of the membrane pore size would lead to a decline in the selectivity and an improvement in flux [16]. It can be seen in studies conducted, wherein the A-1 and A-2 produce the rejection value declining at each additive addition. The increase of the additive causes an increase in the hydrophilic property of the polymer membrane and allows the high swelling on the polymer chain to withhold part of the small pores in the membrane (when the large pores keep functioned). In summary, these results show that the higher the value of flux, the lower the selectivity is produced. It results in the lower amount of water to permeate and ultimately the far water flux [16].

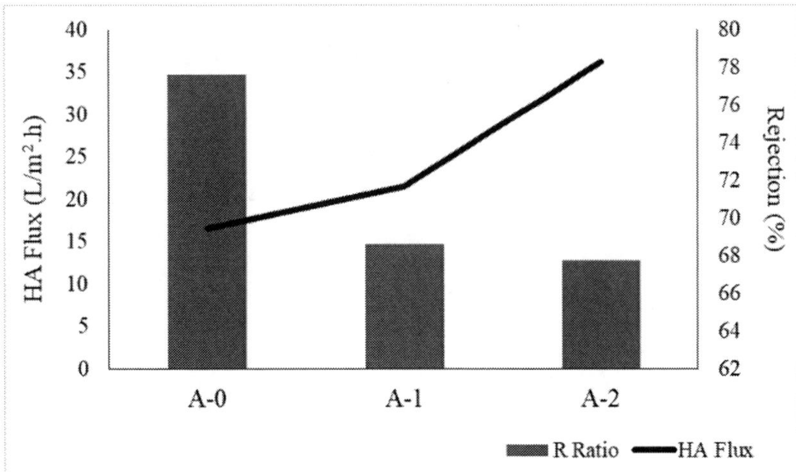


**Figure 6**. Flux and rejection coefficient of humic acid solution on various types of membranes

## 4. Conclusion

Modifying the PES membrane with additives $Mg(OH)_2$ has been carried out. The results showed that the presence of hydroxyl group (-OH) of $Mg(OH)_2$ which can be seen from the results of the FTIR analysis. Increased hydrophilic properties can be confirmed with a reduction in the water contact angle of the membrane modified with $Mg(OH)_2$. Characterization with SEM showed that the membrane is modified with a $Mg(OH)_2$ solution, asymmetric shape, with the top has a thinner dense layer, while the bottom (support) in the form of finger-like macrovoid. It has more pores and larger size than the original PES membrane without modification. The modified membrane performance gives a permeability coefficient of 41.94 $L/m^2{\cdot}h{\cdot}bar$ and a rejection coefficient of 67.72%.

## References

[1] Guo J, Khan S, Cho S H and Kim J 2019 *J. Ind. Eng. Chem.* **4555** 1
[2] Son M, Kim H, Jung J, Jo S and Choi H 2017 *Chemosphere.* **179** 194
[3] Li Y, Qu G, Zhang L, Wang T, Sun Q, Liang D and Hu S 2017 *Sep. Purif. Technol.* **180** 36
[4] Mirwan A, Indriyani V and Novianty Y 2017 *Konversi.* **6** 11
[5] Taleghani H G, Ghoreyshi A A and Najafpour G D 2018 *Biochem. Eng. J.* **132** 52
[6] Arahman N 2014 Pengaruh jenis non-pelarut dan penambahan polimer hidrofilik terhadap struktur morphologi membran polietersulfon *Prosiding Seminar Nasional Aplikasi Sains & Teknologi (SNAST) (Yogyakarta, 15 November 2014)* 249
[7] Fauzzia M, Rahmawati I and Widiasa I N 2013 *J. Teknol. Kim. dan Ind.* **2** 155
[8] Arahman N 2014 *Int. J. Appl. Eng. Res.* ***9*** *10453-62*
[9] Ghaemi N, Daraei P and Akhlaghi F S 2018 *Carbohydr. Polym.***191** 142
[10] Bagheripour E, Moghadassi A R, Hosseini S M, Van der Bruggen B and Parvizian F, 2018 *J. Ind. Eng. Chem.***62** 311
[11] Dong C, He G, Li H, Zhao R, Han Y and Deng Y 2012 *J. Memb. Sci.* **387** 40
[12] Jochym P T, Oleś A M, Parlinski K, Łazewski J, Piekarz P and Sternik M 2010 *IOP Conf. Ser.:J. Phys. Condens. Matter.* **22** 1
[13] Mulder M 1991 *Basic Principles of Membrane Technology* 1st ed. vol 53 (USA: Kluwer

Academic Publishers)
[14] Coates J 2000 *Encyclopedia of Analytical Chemistry* (Chichester: John Wiley & Sons Ltd) 10815
[15] GuoW, Chen J, Sun S and Zhou Q 2018 *J. Mol. Struct.* **1171** 600
[16] Yuan S, Li J, Zhu J, Volodine A, Li J, Zhang G, Puyvelde P V and Bruggen B V 2018 *J. Memb. Sci.* **563** 655
[17] Aprilia S and Arahman N 2016 *Mns. dan Lingkung.* **23** 149
[18] Han S, Mao L, Wu T and Wang H 2016 *J. Memb. Sci.* **516** 47
[19] Farahani M H D A, Rabiee H and Vatanpour V 2019 *J. Water Process Eng.* **27** 47
[20] Kusumawati N and Tania S 2012 *Molekul.* **7** 43
[21] Zhao Q, Xie R, Luo F, Faraj Y, Liu Z, Ju X J, Wang W and Chu L Y 2018 *J. Memb. Sci.* **549** 151
[22] Mulyati S, Razi F and Zuhra 2017 *Biopropal Ind.* **8** 55

IOP Conf. Series: Materials Science and Engineering **845** (2020) 012025 doi:10.1088/1757-899X/845/1/012025

# Utilization of Hydroxyapatite from Tuna Fish Bone Waste as Adsorbent for Cadmium from Aqueous Solutions

**R A Herawati[1,2] , M Faisal[1*], Mahidin[1]**
[1]Chemical Engineering Department Faculty of Engineering, Syiah Kuala University, Darussalam, Banda Aceh 23111, Indonesia
[2] Balai Riset dan Standarisasi Industri Banda Aceh, Banda Aceh 23232, Indonesia

*E-mail: mfaisal@unsyiah.ac.id

**Abstract.** This study aims to determine the performance of hydroxyapatite adsorbents from tuna bone waste against cadmium metal ion removal. Hydroxyapatite adsorbents have been activated physically (calcined at 500°C and 700°C) and chemically (using $H_3PO_4$). The adsorbent was then characterized using Scanning Electron Microscopy (SEM) and Fourier Transform Infrared Spectroscopy (FTIR). A batch experiment was conducted to study the effect of initial concentration and contact time on cadmium adsorption. The results of SEM and FTIR analysis showed that the best adsorbent was obtained from calcination results at 700°C. The observation shows that the contact time and initial concentration affect the efficiency of adsorption. The highest removal efficiency of cadmium was 86.28% with an adsorption capacity of 44.62 mg/g obtained at the contact time of 150 minutes and an initial concentration of cadmium 100 mg/L. The results of the isotherm study show that the adsorption equilibrium in the cadmium solution follows the Freundlich isotherm model with values $R^2 = 0.923$, N = 9.4339, and $K_F = 7.92$. The adsorption kinetics of cadmium adsorption followed the pseudo-second-order equation with values of $K_2$ and $q_e$ at 0.021534 g.mg⁻1.min⁻1 and 47.61 mg/g, respectively. The results obtained indicate that hydroxyapatite from tuna bone waste is a promising alternative for cadmium adsorption.

## 1. Introduction

Environmental damages and decrease of quality of public health are negative impacts of heavy metal liquid waste pollution [1]. This pollution can contaminate the soil, groundwater, rivers, ponds, lakes and oceans. Sources of heavy metal waste mostly come from industry, domestic waste, agriculture, and laboratories. The biggest source of heavy metal waste is found in industrial activity discharges [2]. Environmental damages and decrease of quality of public health are negative impacts of heavy metal liquid waste pollution [1].

One of the heavy metal contents in industrial waste is cadmium ($Cd^{2+}$) [2]. Many industries contribute to producing $Cd^{2+}$ pollution, such as textiles, phosphate fertilizer, plastics, battery, cement, and petrochemicals [3]. $Cd^{2+}$ is a toxic waste and difficult to decompose naturally, so it is dangerous for human health and the environment [4].

* To whom any correspondence should be addressed.

According to the World Health Organization (WHO), $Cd^{2+}$ very easily contaminates the environment. The maximum acceptable limit of $Cd^{2+}$ content in water is 0.005 mg/L [5]. The handling of heavy metal pollution in liquid waste must be specifically managed by reducing or eliminating the hazardous properties [6]. There are several methods for removing $Cd^{2+}$, such as flotation [7], liquid-liquid extraction [8], co-precipitation [9], membrane process [10], adsorption [11], electro-coagulation [12], and solid phase extraction [13].

In this paper, we use the adsorption method for removing heavy metal ions because it is highly effective in $Cd^{2+}$ adsorption, cheap, and easily applied [14]. The common adsorbents used in heavy metal removals include bentonite, biomass cells, activated carbon, clay, zeolite, and rock [15].

Hydroxyapatite ($Ca_5(PO_4)_3COH$) is a promising candidate as an agent of heavy metal adsorption which can be easily extracted from fish bones [16]. The hydroxyl group contained in hydroxyapatite can adsorb heavy metal ions [17]. Content of inorganic substances in fish bones reaches 60% to 70% of its bulk, with high portions of hydroxyapatite and calcium phosphate [18].

The purpose of this study was to investigate characteristics, adsorbent performances, adsorption isotherms, and adsorption kinetics of hydroxyapatite produced from tuna skeletons. In this study, hydroxyapatite was produced by physical and chemical activation, namely calcination in tuna bones and addition of $H_3PO_4$, respectively [19]. These activations aimed to form a micropore structure with a large effective surface so that it could be used to adsorb $Cd^{2+}$ in liquid waste [20].

## 2. Materials and Methods

*2.1 Fabricating adsorbent,* The raw material for producing hydroxyapatite is tuna fish bone. Tuna bone is firstly washed, boiled, and dried at 105°C for 5 hours. Calcination was carried out in furnace (F6000 - Thermolyne) at 500°C and 700°C for 30 minutes. The bone mass during calcination is 500 grams. After calcination, the bone was cooled and then mashed using mortar and sieved by a mesh size of 100. The nano-size was made using a ball mill (pulverisette 6 - Fristsch) for 10 hours. The activation process is carried out after carbonization and the mashing process, and the product then inserted into a 6 M $H_3PO_4$ solution for chemical activation with a ratio of 1:10 and heated at 90°C for 1 hour. Then, 1 M NaOH is added to raise pH to 10 and allowed to stand for 24 hours. At the final step, activated hydroxyapatite is washed with distilled water to neutralize pH and dried in an oven at 105°C until dry.

*2.2 Characterization,* An iodine adsorption test was carried out to determine the absorption capacity of hydroxyapatite. This test using FTIR (Prestige 21 - Shimadzu) was carried out to find the dominant functional groups of the adsorbent.

*2.3 Adsorption test,* Adsorption testing is conducted by batch method using a 250 mL beaker glass to contact the adsorbent with an artificial solution of $Cd(NO_3)_2$. Concentration of artificial solutions are 25, 50, 75, 100, 125, and 150 mg/L. Contact time is varied for 30, 60, 90, 120, and 150 minutes. Stirring speed is 100 rpm. The adsorption solution was tested using atomic adsorption spectrophotometer (AAS - 6300) to determine adsorption efficiency, adsorption capacity, adsorption isotherm, and adsorption kinetics. The equations used to calculate adsorption efficiency, adsorption capacity, adsorption isotherms, and adsorption kinetics are as follows.

*2.3.1 Adsorption efficiency,*In order to determine adsorption efficiency, the initial concentration of adsorbate ($C_0$) before adsorption and the final concentration of adsorbate ($C_e$) after adsorption are calculated. Calculation of adsorption efficiency uses the following equation [15]:

$$\%E = \frac{(C_0 - C_e)}{C_0} \, x \, 100\% \quad (1)$$

*2.3.2 Adsorption capacity,* Calculation of adsorption capacity uses the equation below:

$$q_e = \frac{(C_0 - C_e)v}{m} \quad (2)$$

where $q_e$ is amount of the adsorbate, v is volume of the adsorbate, $C_0$ is concentration before adsorption, $C_e$ is concentration after adsorption, and m is mass of the adsorbent [21].

*2.3.3 Langmuir's isotherm,* For Langmuir's adsorption isotherm, adsorption-adsorbate equilibrium occurs on the surface of the adsorbent monolayer. Langmuir's isotherm model follows the equation:

$$\frac{1}{q_e} = \frac{1}{Q} + \frac{1}{K_L Q C_e} \quad (3)$$

where $C_e$ is equilibrium concentration of the adsorbate (mg/L), $q_e$ is the adsorbate equilibrium concentration (mg/g), Q is maximum capacity of the monolayer (mg/g), and K is Langmuir's isotherm constant. The magnitude of $q_{max}$ and $K_L$ are determined from the slope and intercept of the Langmuir curve versus $\frac{1}{C_e}$ [1].

*2.3.4 Freundlich's isotherm,* This is a characteristic of adsorption equilibrium on heterogeneous surfaces. The Freundlich isotherm model follows the equation:

$$\log q_e = \log K_F + \frac{1}{n} \log C_e \quad (4)$$

where $K_F$ is Freundlich's isotherm constant (mg/g), $n$ is adsorption intensity, $C_e$ is adsorbate equilibrium concentration (mg/L), and $q_e$ is the amount of adsorbed adsorbate (mg/g) [3].

*2.3.5 Temkin's isotherm,* Temkin's isotherm describes the types or factors that influence interactions between adsorbent and adsorbate. Temkin's isotherm model follows the equation:

$$qe = B \ln A_T + B \ln Ce \quad (5)$$

where $B = (RT/b)$ is the constant associated with the heat of adsorption (J/mol), $A_T$ is the equilibrium constant of Temkin's isotherm bond (1.075 L/g), $b$ is Temkin's isotherm constant, R is the gas constant (8.314 J/mol.K), and T is temperature [3].

*2.3.6 First-order kinetics model,* This kinetics model means that the reaction rate is directly proportional to the first order of the reactant concentration expressed by the following equation:

$$\log(q_e - q_t) = \log q_e - \frac{k_1}{2.303} t \quad (6)$$

where $qe$ is concentration of adsorbate at equilibrium (mg/g), $qt$ is concentration of adsorbate at time $t$ (mg/g), $t$ is contact time (minute), and $k_1$ is the corresponding pseudo constant ($min^{-1}$) [1].

*2.3.7 Second-order kinetics model,* This kinetics model has a reaction rate that is directly proportional to the second order of the reactant concentration, which can be expressed as:

$$\frac{t}{qt} = \frac{t}{qe} + \frac{1}{k_2 qe^2} \quad (7)$$

where $q_e$ is concentration of adsorbate after equilibrium (mg/g), $q_t$ is concentration of adsorbate at time $t$ (mg/g), and $k_2$ is the corresponding pseudo constant (g.mg$^{-1}$.min$^{-1}$) [1].

## 3. Results and Discussions

### *3.1 Functional group analysis*

FTIR analysis on hydroxyapatite adsorbents aims to identify functional groups and components contained in the adsorbent. FTIR spectra of the samples with calcination temperature of 700°C and addition of chemical activation is depicted in Figure

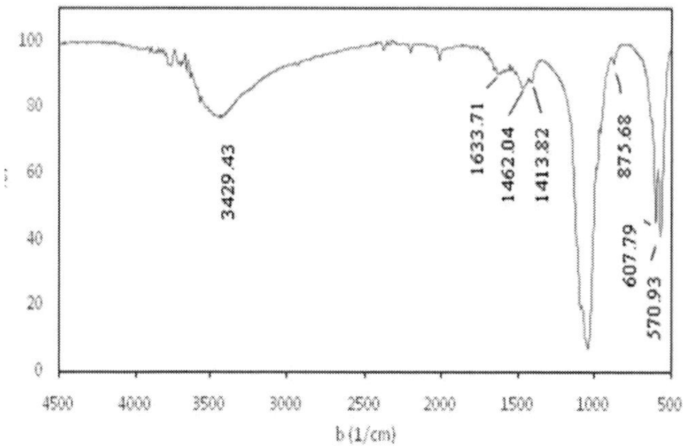


**Figure 1.** FTIR spectra for analyzing functional groups of nano hydroxyapatite by calcination temperature of 700°C and chemical activation by $H_3PO_4$.

The wavelengths of FTIR analysis are in the range of 500 cm$^{-1}$ to 4000 cm$^{-1}$. The results identified the presence of functional groups in the form of hydroxyl (OH), phosphate ($PO_4$) and carbonate ($CO_3$), where their corresponding wavelengths are: 3429.43 cm$^{-1}$ and 570.93 cm$^{-1}$; 601.79 cm$^{-1}$, 1041.56 cm$^{-1}$ and 1413.82 cm$^{-1}$; and 1462.04 cm$^{-1}$ and 1633.71 cm$^{-1}$, respectively. Also present is a C = H bond at a wavelength of 875.68 cm$^{-1}$ [24]. These findings prove that tuna fish bone produces hydroxyapatite which can adsorb heavy metal ions [15].

Previous works also reported the same results, but with slightly different wavelengths for the three functional groups (OH, $PO_4$, and $CO_3$), which occur at 3570 cm$^{-1}$ and 628 cm$^{-1}$ for hydroxyl [23], 1415 cm$^{-1}$, 1090 cm$^{-1}$ and 600 cm$^{-1}$ for phosphate [15], and 1466 cm$^{-1}$ and 1411 cm$^{-1}$ for carbonate [23].

### *3.2 Iodine adsorption analysis*

Adsorption activity of hydroxyapatite to iodine solution had a range of 318.99 to 331.06 mg/g. The concentration of iodine adsorption can be seen in Figure 2. Hydroxyapatite with chemical activation resulted in higher iodine adsorption compared to the others. Higher calcination temperature and addition of $H_3PO_4$ activator establish a pore with a larger surface area so that the iodine adsorption is higher [20]. The highest adsorption value is 331.06 mg/g at calcination temperature of 700°C. By considering the values of iodine adsorption in this section, the adsorption isotherm and kinetics can be evaluated.

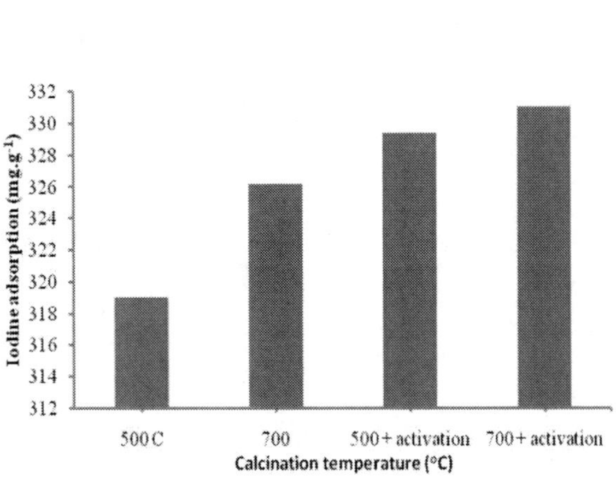


**Figure 2.** Results of iodine adsorption analysis

*3.3 Adsorption effectivity*

Adsorption efficiency of $Cd^{2+}$ over contact time is shown in Figure 3. With longer adsorption time, the effectiveness of adsorption will increase as shown by sample with initial concentration of 100 mg/L. For variables with initial concentrations of more than 100 mg/L, adsorption effectiveness will tend to be stable after 90 minutes. The best adsorption efficiency is 86.28%, obtained by the variable with initial concentration of 25 mg/L and contact time of 150 minutes. However, if the initial concentration of 25 mg/L and the contact time is increased, the adsorption efficiency can still increase until it reaches the equilibrium time. High calcination temperature and chemical activation cause a large adsorbent pore surface area so that the adsorption of the adsorbate is higher. The longer adsorption times also provide more adsorbate, which is adsorbed by the adsorbent until stable at equilibrium time [1].

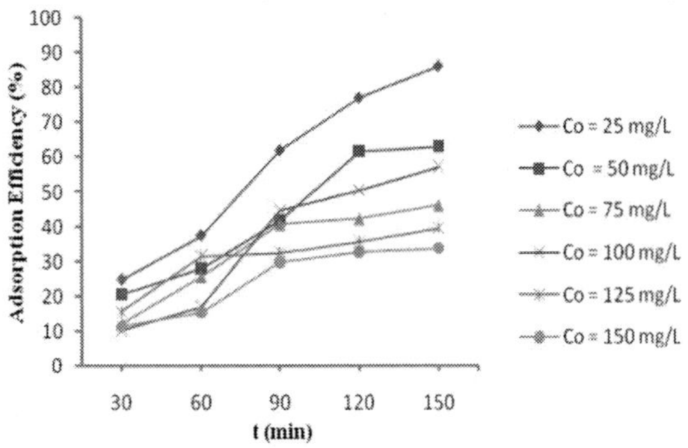


**Figure 3.** Effect of contact time on adsorption efficiency.

*3.4 Adsorption capacity*

As depicted in Table 1, the highest adsorption is achieved by the sample with initial concentration of 100 mg/L at contact time of 150 minutes, i.e. 44.62 mg/L. From this observation, when contact time between adsorbent and adadsorptionsorbate increases, the adsorption capacity will increase up to the equilibrium [24]. When contact time is longer, the more adsorbent pore is filled by the adsorbate until saturation time, when there is no longer free space inside the adsorbent [21]. Based on Table 1, the highest adsorption capacity is obtained at initial concentration of 100 mg/L. For variations of lower than 100 mg/L, the sample still experienced an increase in adsorption capacity up to 150 minutes.

Meanwhile, for samples with initial concentrations of more than 100 mg/L, adsorption capacity tended to be stable at 120 minutes and 150 minutes. At this point, the samples are close to saturation and no longer able to adsorb $Cd^{2+}$ [21].

**Table 1.** Adsorption capacity the hydroxyapatite to $Cd^{2+}$

| Time (minutes) | Capacity (mg/L) | | | | | |
|---|---|---|---|---|---|---|
| | $C_o$ = 25 mg/L | $C_o$ = 50 mg/L | $C_o$ = 75 mg/L | $C_o$ = 100 mg/L | $C_o$ = 125 mg/L | $C_o$ = 150 mg/L |
| 30 | 6.35 | 9.7 | 8.76 | 10.18 | 18.83 | 16.76 |
| 60 | 9.08 | 12.43 | 17.62 | 16.05 | 35.8 | 21.36 |
| 90 | 14.2 | 17.59 | 26.37 | 39.45 | 34.87 | 38.69 |
| 120 | 16.6 | 24.51 | 25.84 | 42.02 | 36.07 | 39.86 |
| 150 | 17.4 | 23.42 | 26.43 | 44.62 | 37.39 | 38.65 |

*3.5 Isotherm model of the adsorption*

Figure 4 depicts three equilibrium models which describe adsorption process of the adsorbent. Freundlich's isotherm model was found to be more suitable for this adsorption, indicated by the highest $R^2$ value (0.923) in linear regression approximation. It also shows that a complex layer occurs when cadmium ions are adsorbed by the hydroxyapatite.

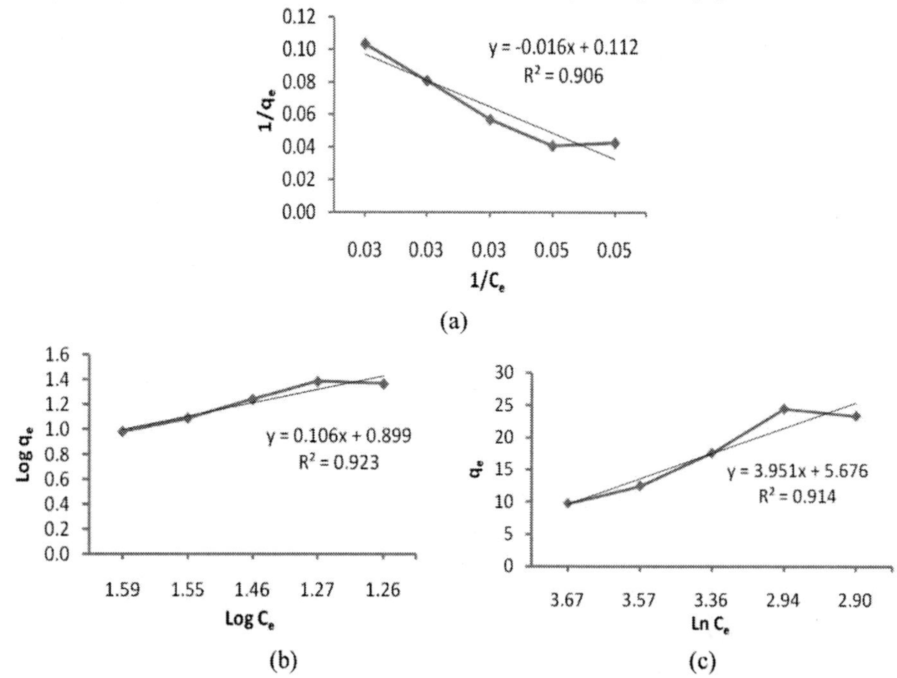


**Figure 4.** Isotherm model with chemical activation, calcination temperature T = 700°C, and concentration $C_o$ = 50 mg/L. (a) Langmuir (b) Freundlich, and (c) Temkin.

This indicates that adsorption is more physically dominant and van der Waals force triggers adsorbate ($Cd^{2+}$) to stick with surface of the adsorbent. Table 2 provides detailed parameters for Langmuir, Freundlich, and Temkin models from each model which has the highest $R^2$ value [1].

**Table 2.** Isotherm model on adsorption of $Cd^{2+}$

| Isotherm type | Parameters |
|---|---|
| Langmuir | $Q_m$ (mg/g) = - 62.50 |
| | $K_L$ (mg/g) = - 7.0 |
| | $R^2 = 0.906$ |
| Freundlich | N = 9.4339 |
| | $K_F = 7.92$ |
| | $R^2 = 0.923$ |
| Temkin | B = 3.951 |
| | $A_r = 4.186$ |
| | $R^2 = 0.914$ |

*3.6. Adsorption kinetics*

Figure 5 shows adsorption kinetics for first- and second-order models of adsorption of $Cd^{2+}$ metal using hydroxyapatite. The calculation results can be seen in Table 3, where $q_e$ value of the second order of the pseudo model is closer to the experimental results with higher $R^2$ than to the first order. This proves that pseudo-second-order model more accurately describes the kinetics of $Cd^{2+}$ using hydroxyapatite. A high $R^2$ value can determine diffusion rate of adsorption [25]. This result is in line with [21], which used nano hydroxyapatite for adsorption of metal $Pb^{2+}$, $Cd^{2+}$, and $Cu^{2+}$. The study also stated that experimental value of $q_e$ obeys pseudo-second-order model.

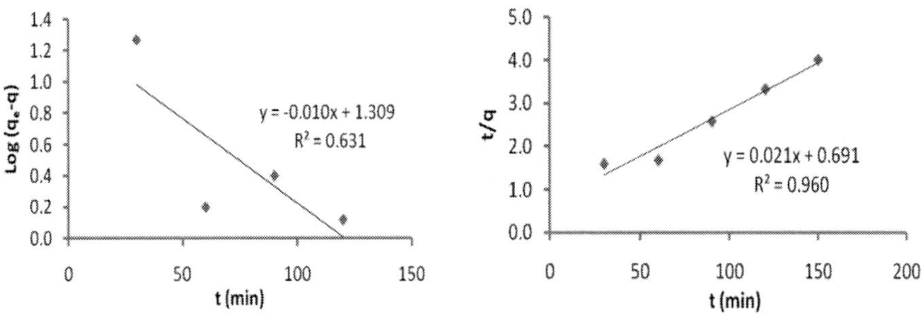


(a) (b)

**Figure 5.** Kinetics of adsorption on metal $Cd^{2+}$ with chemical activation, calcination temperature T = 700°C, and concentration $C_o$ = 125 mg/L. (a) pseudo-first-order and (b) pseudo-second-order model.

**Table 3.** Adsorption kinetics of the hydroxyapatite to $Cd^{2+}$

| $q_e$(Experiment) (mg/g) | First order of the pseudo model | | | Second order of the pseudo model | | |
|---|---|---|---|---|---|---|
| | $K_1$ ($min^{-1}$) | $q_e$ (Calculation) (mg/g) | $R^2$ | $K_2$ ($mg.mg^{-1}\ min^{-1}$) | $q_e$ (Calculation) (mg/g) | $R^2$ |
| 37.39 | 0.02303 | 20.37042 | 0.631 | 0.021534 | 47.61905 | 0.960 |

In Table 3, it is known that K is the adsorption rate that occurs between each $Cd^{2+}$ ion in the solution to the surface of the adsorbent. The pseudo-second-order model assumes adsorption capacity is proportional to the surface of hydroxyapatite and depends on its ability to adsorb $Cd^{2+}$ [25]. The obtained value is in agreement with the second order [3].

## 4. Conclusion

Hydroxyapatite from tuna fish bone with physical (calcination at 700°C) and chemical activation (using $H_3PO_4$) is the best adsorbent for $Cd^{2+}$ adsorption in this study. Adsorption by batch method shows that the adsorption of $Cd^{2+}$ using hydroxyapatite is very dependent on the contact time and initial concentration of adsorbate. The optimal adsorption value of $Cd^{2+}$ by hydroxyapatite adsorbents was 86.28% and the optimal adsorption capacity was 44.62 mg/g. The adsorption isotherm obeys Freundlich's isotherm, and the reaction kinetics satisfies the second order of the pseudo model. Based on this analysis, tuna skeletons can be used as a raw material for fabricating hydroxyapatite for $Cd^{2+}$ removal.

**Acknowledgement**

The authors would like to acknowledge the Ministry of Research, Technology and Higher Education of Indonesia for the research grant provided (No. 215/SP2H/LT/DPRM/2019).

**References**

[1] Iqbal M, Iqbal N, Ijaz Bhatti A, Ahmad N and Zahid M 2016, *Ecol. Eng.* **88** 265
[2] Sardella F, Gimenez M, Navas C, Morandi C, Deiana C and Sapag K 2015 *J. Environ. Chem. Eng.* **3** 253
[3] Eshraghi F, Motavassel M, Roozbehani B and Fard N J H 2016 *Environ. Friendly Process* **4** 4
[4] Bediako J K, Wei W and Yun Y -S 2016 *J. Ind. Eng. Chem.* **43** 61
[5] Fernández S, Cotos-yáñez T, Roca-pardiñas J and Ordóñez C 2016 *Geoderma* **31** 120
[6] Lu P, Feng Q, Meng Q and Yuan T 2012 *Sep. Purif. Technol.* **98** 216
[7] Guajardo A E, Liamas J C, Maqueira L, Andrade C S, Alves K B and Melo C P 2015 *Chem. Eng.* **281** 826
[8] Rahman M, Z. Begum Z A and Hasegawa H 2013 *Microchem. J.* **110** 485
[9] Barakat M A and Kumar R 2015 *Indus. Eng. Chem.* **23** 93
[10] Wang Z, Ma Y, Hao X, Huang W, Guan G and Abudula A 2014) *Electrochim. Acta.* **130** 40.
[11] Ali I 2014 *Sep. Purif. Rev.* **43** 175
[12] Ali I, Asim M and Khan T A 2013 *Int. J. Environ. Sci. Technol.* **10** 377
[13] Wu Y, Zhang J, Liu J, Chen L, Deng Z, Han M, Yu A, Zhang H and X. Wei X, 2012 A*ppl. Surf. Sci.* **258** 6772

[14] Baláž M., Bujňáková Z, Baláž P, Zorkovská A, Danková Z and Briančin J 2015 *J. Colloid Interface Sci.* **454** 121
[15] Sheha R R, Moussa, S I, Attia, M A, Sadeek S A and Someda H H 2016 *J. Environ. Chem. Eng.* **4** 4808
[16] Santhosh S and Prabu S B 2013 *Mater. Lett.* **97** 12
[17] Herpandi N H, Rosma A and Wan Nadiah W A 2011 *Compr. Rev. Food Sci. Food Saf.* **10** 195
[18] Kamari A, Yusof N, Ishak C F, Phillip E and Yuliani G 2014 The Effects Of Micro and Nano Hydroxyapatite Application in Metal Contaminated Soil On Metal Accumulation In Ipomeoa Aquatica and Soil Metal Bioavailability *Proc. International Conference On Research, Implementation And Education Of Mathematics And Sciences 2014 (Yogyakarta, Indonesia, 18-20 May 2014),* 18
[19] Mustafa N, Ibrahim M H I, Asmawi R and Amin A M 2015 *Appl. Mech. Mater.* **774** 287.
[20] Brundavanam R K, Poinern G E J and Fawcett D 2013 *Am. J. Mater. Sci.* **3** 84
[21] Chen S B, Ma Y B, Chen L and Xian K 2010) Geochem. J. **44** 233
[22] Lakherwal D 2014 *Int. J. Environ. Res. Dev.* **4** 41
[23] Mu Y, Saffarzadeh A and Shimaoka T 2016) *Procedia Environ. Sci.* **31** 345
[24] Zayed E M, Sokker H H, Albishri H M and Farag A M 2013 *Ecol. Eng.* **61** 390
[25] Verma R and Dwivedi P 2013 *Recent Res. Sci. Technol.* **5** 98

# Evaluation on the UV-Shielding Performance of Self-polymerized Dopamine-blended Membrane

**S Mawarni[1], F Meirisa[1], S Muchtar[2], M Y Wahab[2], S Mulyati[3*],Suraiya[3], C M Rosnelly[3], Sofyana[3], Y Syamsuddin[3], N Arahman[3]**

[1]Graduate School of Chemical Engineering, Syiah Kuala University, Banda Aceh, Indonesia 23111
[2]Doctoral School of Engineering Science, Syiah Kuala University, Banda Aceh, Indonesia 23111
[3]Department of Chemical Engineering, Syiah Kuala University, Banda Aceh, Indonesia, 23111

[*]E-mail: sri.mulyati@unsyiah.ac.id

**Abstract.** Herein we reported the UV resistance evaluation of PVDF membrane which modified through a new blending combined polymerization method. The membrane was prepared by means of non-solvent induced phase separation process in which dopamine was blended altogether with polyvinylidene fluoride polymer in a dimethylacetamide solvent. The blended membrane was further modified by immersion into tris-HCl system to trigger the self-polymerization of dopamine into polydopamine. The UV stability performance was studied by exposing the membrane to the UV light for 5 days. The morphological change of the membrane was observed by using scanning electron microscopy analysis. The results revealed that, with only blending the dopamine into the membrane system, the membrane was severely damaged after UV exposure. However, by simply immersing the blended membrane into buffer Tris-HCl solution, the membrane showed minimal morphological damage after similar UV exposure treatment. This membrane can be a promising choice for the photocatalytic membrane reactor application.

## 1. Introduction

Recently, it has been reported that polydopamine (PDA) has the ability to scavenge UV-induced free radicals and protects materials against photodegradation [1-3]. The ability of PDA in shielding materials from UV radiation is due to its catechol chain components consisting of very long OH and NH groups [4]. These groups are electron donors and are able to deactivate UV-induced free radicals through electron donor reactions [3]. Modification of PDA for this purpose has been applied in several fields of research such as fabrication of fiber [5], clay materials [6], membrane [1, 2], etc. In membrane technology itself, the application of this research is very useful for photocatalytic membrane reactor application because direct and continuous exposure of the membrane to UV light will morphologically damage the membrane and affect on mechanical properties and overall performance [7].

[*] To whom any correspondence should be addressed.

The ability of PDA as a UV-protector for photocatalytic membrane has been previously reported by Feng et. al [1] and Wu et. al [2]. Feng and team [1] used a coating modification method in which the PDA layer formed directly on the membrane surface. Their modified PSf/PDA/$TiO_2$ membrane revealed to have an excellence UV resistance and self-cleaning property. However, the presence of the PDA layer blocked the membrane surface pores thereby declining the permeation performance. On the other hand, Wu et al [2] continued the modification with a similar goal but through blending technique. They prepared PDA and $TiO_2$ into nanohybrid particles then blended in casting solutions. This process produced a membrane with desirable pore properties resulting in high flux production. However, because $TiO_2$ as a photocatalyst particle was blended in the membrane, the photocatalysis reaction does not take place effectively hence the self-cleaning performance was slightly low.

In this study, we report a new method of dopamine-based modification to produce membranes with good UV-shielding performance without abandoning their permeation and antifouling properties. The membrane is prepared by triggering the polymerization of polydopamine from a dopamine-blended polyvinylidene fluoride (PVDF) membrane. We have previously reported in detail the permease and antifouling performances of this membrane [8], hence, for this reason in this paper we will continue the study by only reporting its UV resistance evaluation. This membrane has the potential to be used for the membrane photocatalytic reactor as there is a thin PDA layer growing on the membrane surface from the polymerization of dopamine which has been blended in the membrane system. The presence of PDA layer can be used as a medium for attaching photocatalysts and also as a free radical scavenger. The evaluation of membrane UV resistance performance is evaluated by observing the morphological changes after 5 days of exposure to UV light.

## 2. Experimental

*2.1. Materials,* Polymer PVDF (Solef 6020, Solvay Polymer, USA), and Dimethylacetamide (DMAc, Wako Pure Chemicals, Japan) were the basic materials used for membrane preparation. As an additive, dopamine hydrochloride (DA) which purchased from Sigma Aldrich (Germany) was used. Tris-HCl (Wako Pure Chemical Industries, Japan) with a pH of 8.8 was employed as a base buffer to trigger dopamine polymerization. DI water was used as a non-solvent during membrane preparation and for other solvent necessities.

*2.2. Preparation and Modification* ,The membrane preparation procedure was carried out exactly as we reported in the previous study [8]. Membrane casting solution consisting of PVDF, DA, and DMAc with a respective mass of 15, 2, and 85 grams were dissolved altogether until homogeneous and then prepared using the non-solvent induced phase separation method. The blended dopamine membrane was immersed in a Tris-HCl solution with a pH of 8.8 for 24 hours.

*2.3 UV Radiation experiment,* This experiment was carried out by adapting that reported in our another work [7]. UV lamp (SUV-16, AS ONE, Japan) with a power of 22 W and a wavelength of 254nm was used as a source of radiation. The experiment was carried out in a closed condition with the sample-to-lamp distance of 5cm. The exposure and observation were conducted for 5 days.

*2.4 Membrane Characterization,* Field-Emission Scanning Electron Microscopy (FE-SEM, JSF-7500F, Jeol Co. Ltd., Japan) was used to observe changes in membrane morphology before and after UV exposure. Before analysis, the membrane was freeze-dried overnight then coated with an osmium coating. the data recorded was only surface morphology at magnification 1000x.

## 3. Results and Discussion

To see whether the PVDF membrane modified by PDA through polymerization techniques after blending succeeded in protecting the membrane from UV exposure, an evaluation was carried out by observing the changes in morphological structure. Morphological changes after the membrane was exposed to UV light were analyzed using SEM instrument. The results of the analysis are shown in Fig. 1. From Fig. 1 it can be seen that the dopamine-blended membrane (P-D0) has a porous surface which supported by the porosity evaluation results reported previously [8]. After being exposed for 5 days to UV light, the P-D0 UV-5 membrane has a much more porous surface. The surface pore is seen to have a sponge-like shape which is due to the damaging effect of photodegradation [9]. The visible dust-like piles on the membrane surface are presumed to be degraded membrane layers. This kind of morphological condition will negatively impact on the membrane separation performance. As for the P-D24 UV-5 membrane, the damage experienced after UV exposure is almost identical to that of the P-D0 UV-5 membrane, which is in form of pore enlargement on the membrane, however, the damage is not as severe. This is because initially (before UV exposure) the P-D24 membrane has a less porous surface due to the presence of a PDA layer formed through tris-triggered polymerization [7].

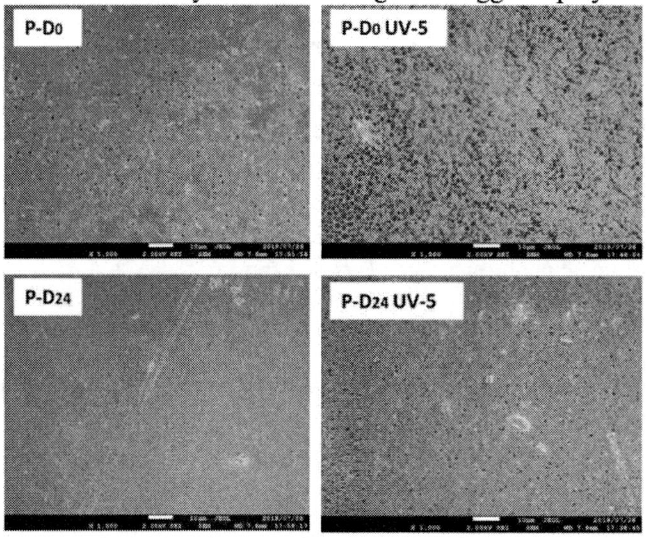


**Figure 1.** The SEM photographs of P-D0 and P-D24 membranes Before and after 5 days exposure to UV

From the result, it can be concluded that the existence of the PDA layer is succeeded in protecting the membrane from photodegradation. Though, when compared with the results of experiments reported in our previous work [7] using the common coating method, the results are not as outstanding. This is because, with the usual coating process in dopamine + Tris-HCl solution, the PDA layer formed is thicker than this method. However, because the PDA layer is not too thick this method can produce membranes with good permease performance and decent selectivity as well.

The UV resistance ability of the polymerized dopamine-blended membrane (P-D24) comes from the presence of catechol groups from PDA which function as a radical scavenger. After modification, the presence of these groups consisting of NH and OH can be observed from the results of the IR analysis shown in Fig. 2 [8]. These groups minimize the occurrence of photodegradation by binding to radical compounds with H atoms through the electron donor reaction [3, 7].

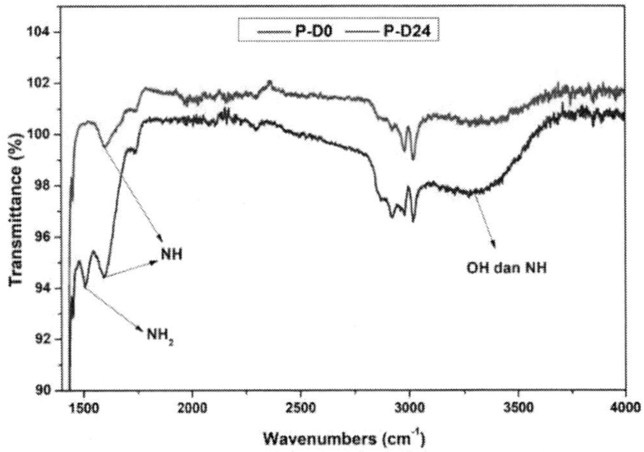


**Figure 2.** Infrared spectra of the PVDF membrane which dopamine-blended and combined with polymerization as reported in Muchtar et al [8].

## 4. Conclusions

Herein we reported the UV resistance evaluation of PVDF membrane which modified through a new blending combined polymerization method. The UV stability performance was studied from the membrane morphological change after exposure to UV light for 5 days. The results confirmed that, with only blending the dopamine into the membrane system, the membrane was severely damaged after UV exposure. However, by simply immersing the blended membrane into buffer Tris-HCl solution, the membrane showed minimal morphological damage after similar UV exposure treatment. This membrane can be a promising choice for the photocatalytic membrane reactor application.

**Acknowledgment**

The funding support from the Ministry of Research, Technology and Higher Education of the Republic of Indonesia under 'Penelitian Tesis Magister Grant Founding Year of 2019" is greatly acknowledged by authors.

## References

[1] Feng K, Hou L, Tang B and Wu P 2015 *J. Membr. Sci.***490** 120
[2] Wu H, Liu Y, Mao L, Jiang C, Ang J a.nd Lu X 2017 *J. Membr. Sci.* **532** 20
[3] Yen G C and Hsieh C L 1997 *Biosci* ***Biotech*** *Bioch.* **61** 1646
[4] Liebscher J, Mrówczyński R, Scheidt H A, Filip C, Hădade N D, Turcu R, Bende A and Beck S 2013 *Langmuir* **29** 10539
[5] Liu Z, Hu J, Sun Q, Chen L, Feng X and Zhao Y 2017 *Macromol. Res.* **25** 431
[6] Phua S L, Yang L, Toh C L, Guoqiang D, Lau S K, Dasari A and Lu X 2013 *ACS Appl Mater Interfaces* **5** 1302
[7] Muchtar S, Wahab M Y, Fang L-F, Jeon S, Rajabzadeh S, Takagi R, Mulyati S, Arahman N, Riza M and Matsuyama H 2019 *Reactor* **136** 47312
[8] Muchtar S, Wahab M Y, Mulyati S, Arahman N and Riza M 2019 *J.Water Process Eng.* **28** 293
[9] Rånby B 1989 J **Anal** Appl Pyrol. **15** 237

# Improving Physical Properties of Polyvinyl Alcohol Film through The Addition of Nanocellulose Prepared from Palm Oil Solid Waste

**D Syafrina[1], S Aprilia[1*], B Arifin[2], N Razali[1], H P S A Khalil[2] and A Amin[3]**
[1] Department of Chemical Engineering, Universitas Syiah Kuala, Banda Aceh, Indonesia 23111
[2] School of Industrial Technology, Universiti Sains Malaysia, Penang 11800, Malaysia
[3] Department of Mechanical Engineering, Engineering Faculty, University of Abulyatama, Aceh Besar, Indonesia

E-mail: sriaprilia@unsyiah.ac.id

**Abstract.** Cellulose from the palm oil solid waste (oil palm shells and stems) be isolated and used as a filler material in a biodegradable plastic, in order to improve its mechanical properties. Polyvinyl Alcohol (PVA) is a biodegradable polymer and compatible for nanocellulose fillers, where its mechanical properties can reach the same level as the conventional plastic. In this study, nanocellulose has been successfully prepared from the oil palm shells and stems with acid hydrolysis method and applied as a filler for PVA/ nanocellulose film composite. The characterizations include Fourier Transform Infrared (FTIR), Differential Scanning Calorimetry (DSC) and mechanical properties (tensile and elongation). The FTIR and DSC analysis confirm the improvement of the mechanical properties after the addition of nanocellulose, where the tensile strength reaches 14.64 kgf/mm$^2$ with the optimal thermal ratio at 203.5°C. Nevertheless, the addition of 20% nanocellulose reduces the value of tensile strength and elongation.

## 1. Introduction

Indonesia is the largest palm oil producing country in the world. However, a large quantity of palm oil production also results in an equally proportionally enormous solid waste, found in the form of oil palm empty fruit bunches, shells, stems, fibers, and etc. [1]. The solid waste contributes to the release of organic pollutant to the environment. In order to overcome the problem, solid waste can be taken as biomass resources. The solid waste, including the palm shells and stems, can be processed into biomass material due to their high fiber contents. Cellulose, hemicellulose, and lignin are the three major components of the fiber. The cellulose content in palm shells can reach up to 26.27%, followed by the hemicellulose and the lignin that respectively give 12.61% and 42.96% [2]. Palm stems are the likewise, having a high content of cellulose, hemicellulose, and lignin. On the other hand, natural fibers are the main composites, consisted of cellulose fibrils, which can be found in the lignin matrix [3]. Agricultural wastes such as shells, stems, coconut husk, coconut shell, and oil palm empty fruit

* To whom any correspondence should be addressed.

bunches, can be taken as natural fiber resources, which make them to be very abundant and diverse. Nowadays, the natural fiber has been widely developed for its use as fillers and reinforcement in composite films, owing to its renewable and environmentally friendly properties.

This study evaluates the effect of the addition of fibers in polyvinyl alcohol (PVA) composites, where the fibers are from palm oil nanocellulose. PVA is a polymer with a molecular weight of 26,300-30,000, a melting point of 180-190°C, degree hydrolysis of 86.5-89%, and biodegradability [4]. Due to the stated properties it has, PVA has been widely utilized as a promising alternative for packaging. It is also due to the fact that PVA is easy to shape, resistant to oils and fats, and high in tensile strength and flexibility. However, PVA has a weak property against high humidity. The ambient water can be absorbed, causing the PVA film to experience a decrease in tensile strength and tear strength and an increase in elongation. According to Roohani et al. [5], PVA exhibits high compatibility with nanocellulose fillers, where the composites are found to be environmentally friendly. Thus, the addition of cellulose nanofiber in PVA-based films is expected to improve the mechanical properties of the final product [6].

Previous studies have reported the use of acid hydrolysis in the preparation of nanocellulose. Xiong, et al. [7] has successfully obtained cellulose nanofibers (with the size of 10-65 nm) through an acid hydrolysis method employing strong acid, namely sulfuric acid Moreover, Tang and Liu [8] reported that the addition of high cellulose nanofibers (> 40% by weight) in a PVA-based film, can increase its mechanical strength by up to 50%. This study aims to determine the effect of the nanocellulose addition to the mechanical and thermal properties of PVA-based film composites.

## 2. Methodology

### *2.1. Preparation of cellulose isolation*

Palm oil solid waste (oil palm shells and stems) was shredded to a size of 1-5 mm. Then the sample was washed and dried in an oven (24 h, 60°C). Next, 10 g sample was put into a solution of 300 mL NaOH 0.5 M and simultaneously evaporated at 80 °C. After that, the sample was washed 3 times, filtered, and dried in an oven at 60 °C for 5 h. The product of this procedure is cellulose.

### *2.2. Preparation of nanocellulose*

6 grams cellulose, obtained from the oil palm shells and stems, was mixed in 180 mL formic acid 50%. The mixture was hydrolyzed at 90 °C and 400 rpm for 4 h. After that, it was washed with distilled water 3 times and centrifuged for 20 minutes. The nanocellulose pellet was collected from the centrifuge tube and dried in an oven at 70 °C.

### *2.3. Preparation of PVA composite films*

PVA crystals (1.5 g) were dissolved in 15 mL distilled water, and the mixture was heated at 80 °C for 2 h and left to cool down. Nanocellulose was added into the PVA mixture with the w/w (of PVA) ratio variations of 1%, 5%, 10%, and 20%. The mixture was stirred for 2 h [9]. Then, it was poured onto a mold (Petri dish) for 48 h at the ambient temperature until the plastic (film) can be removed.

### *2.4. Characterizations*

Fourier Transform Infrared (FTIR) analysis was carried out with electromagnetic radiation within the wavelength region of 0.75–1,000 μm. For palm oil solid waste (oil palm shells and stems), the functional group analysis was carried out with Shimadzu Prestige FT-IR 6400 at 4000–400 $cm^{-1}$. Analysis of the tensile strength and elongation of the samples was carried out with HT-8503 Computer Type Universal Testing Machines. Prior to the tensile strength analysis, the samples were cut according to ASTM-D638 standards. The analysis conducted by pinning both ends of the sample to a tensile testing machine, then the tool pulled the sample to break, and the data of tensile strength and elongation were recorded after the breaking. The thermal analysis was performed by Differential

Scanning Calorimetry (DSC), using a Shimadzu DSC-60 type Detector tool. The sample analysis was carried out at the range of 20–200 °C with a heating rate of 10 °C/min.

**Table 1.** Composition of composite film with fixed variable of 1.5 g PVA

| Sample code | Nanocellulose (%) |
|---|---|
| PVA | 0 |
| Shell1 | 1 |
| Shell2 | 5 |
| Shell3 | 10 |
| Shell4 | 20 |
| Stem1 | 1 |
| Stem2 | 5 |
| Stem3 | 10 |
| Stem4 | 20 |

## 3. Results and discussion

### *3.1. Morphology of the PVA/ nanocellulose composite*

FTIR spectrum of PVA film (without filler), as shown in Figure 1, indicates the presence of hydroxyl groups at 3528 cm$^{-1}$. It also suggests the hydrophilic properties of the PVA, which can be associated with its affinity with water. The absorbance peak is observed at 1741 cm$^{-1}$, associated with the presence of ketone functional group (C=O). Meanwhile, the presence of C=C aromatic rings can be observed at 1568 cm$^{-1}$. The results of the FTIR analysis can be seen in Figure 1, where the wavelengths of each functional group are presented in Table 2 [10].

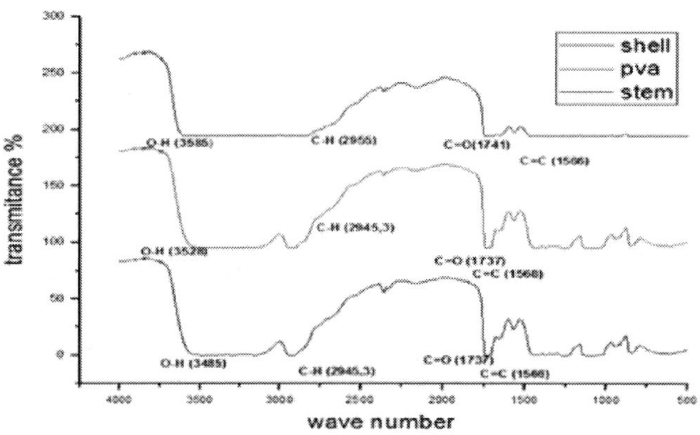


**Figure 1.** Nanocomposite FTIR analysis results from the palm stem and shell

The FTIR analysis is also carried out for PVA/ nanocellulose composites filled with the nanocellulose from oil palm shells and stems. The FTIR spectra of the PVA/ nanocellulose composites exhibit a similar pattern with the PVA film spectrum, indicating the absence of chemical reaction during the composite preparation. However, shifts of wavelength are observed in the composites spectra, which can be ascribed to the presence of hydrogen bonds and ester bonds. Those interactions are responsible for the improvement of the adhesion properties between the fillers and the PVA polymer matrix.

**Table 2**. Interpretation of IR spectra in pure polyvinyl alcohol nanocomposites

| Wavelenght ($cm^{-1}$) | Bond | Functional Group |
|---|---|---|
| 3600 – 2000 | O – H | Alcohol, hydrogen bond |
| 3000-2800 | C-H | Alkana, aromatic rings |
| 1760-1690 | C = O | Keton, Ester |
| 1600 – 1500 | C = C | Aromatic rings |

3.2 *Thermal properties of the PVA/ nanocellulose composite*

DSC analysis of PVA/ nanocellulose composite shows the transition sequence of the thermal properties of the polymer, namely melt onset, Tg and Tm. As can be seen in Table 3, melt onset temperatures of PVA films were obtained between the range of 67.81 to 200.3 °C, the Tg temperatures were between the range of 52.58 to 157.8 °C and Tm temperatures were between the range of 187.7 to 203.5 °C. It can be observed that the higher transition glass temperatures (Tg) were given by the PVA/ nanocellulose composites. It can be ascribed to the presence of strong bonds between the matrix (PVA polymer) and the filler (nanocellulose from oil palm shells and stems). Higher Tg contributes to a longer reaction required to change the materials from glassy to rubbery, thus indicating an increase of stability.

**Table 3**. Results of DSC analysis of polyvinyl alcohol nanocomposites

| Sample | *Melt onset (T °C)* | *Tg (T°C)* | *Melt peak* ™ |
|---|---|---|---|
| PVA | 84.35 | 52.58 | 203.48 |
| shell 1% | 80.11 | 54.50 | 201.45 |
| shell 5% | 67.81 | 63.07 | 119.12 |
| shell 10 % | 82.52 | 58.31 | 201.76 |
| shell 20% | 88.76 | 64.83 | 201.49 |
| stem 1% | 71.69 | 54.16 | 201.44 |
| stem 5% | 77.90 | 53.63 | 187.73 |
| stem 10% | 83.13 | 56.99 | 203.50 |
| stem 20% | 200.39 | 157.81 | 201.50 |

The stated results can also explain the thermal characteristics of the composite. Below the Tg, the polymer is found to be fragile and nearly rigid. It is due to the fact that the polymeric chains are locked in a circular and irregular position. Whereas, when the temperature is above the Tg, the rigid and glassy properties are removed, transforming the polymer to be more elastic. As an addition, Tg depends on the preparation process, structures, bonds, and molecular weight of the materials. Meanwhile, for Tm, samples received constant heating, thus causing a stronger vibration. Eventually, it allows the sample to decompose, where the bonds between the chains are broken, thus reducing the heat resistance of the sample and reaching the melting point. It is important to study the thermal properties of a sample in order to obtain the information pertaining to the preparation process and the maximum storage temperature (below the melting temperature of the sample).

3.3 *Mechanical properties of the PVA/ nanocellulose composite*

As shown by Figure 2, the tensile strength of the composites increases along with the increase of the nanocellulose filler content, either from the oil palm shells or stems. For the addition of nanocellulose from oil palm shells of 1%, 5%, and 10%, the tensile strength values are 7.5, 11.29, and 13.5 $kgf/mm^2$. Meanwhile for nanocellulose from oil palm stems, with the same % weight ratio, the tensile strength values are 8.21, 14.64, and 13.93 kgf/mm$^2$. The increase of tensile strength is owing to the closer distance between particles after the addition of nano fillers (< 10%) [11]. Nevertheless, the addition of 20% nanocellulose, from either oil palm shells or stems, results in a decrease of tensile strength. It can be ascribed to the reduction of the nanocellulose particle homogeneity in the mixture, where in fact, the mixing process was carried out with the same conditions and treatments [12].

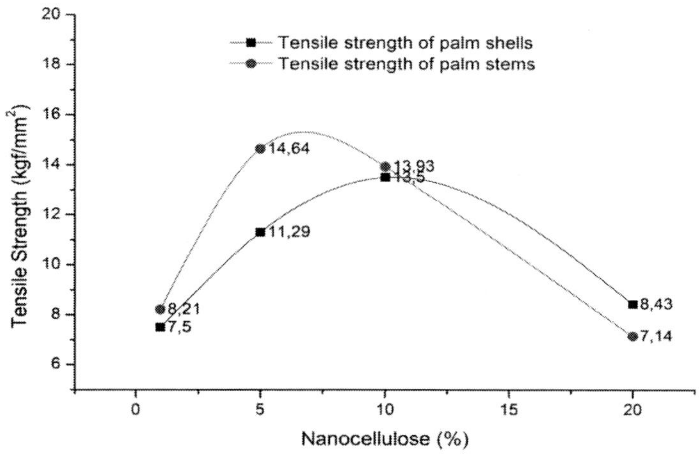


**Figure 2.** Effect of filler contents on the tensile strength of the PVA/ nanocellulose composites

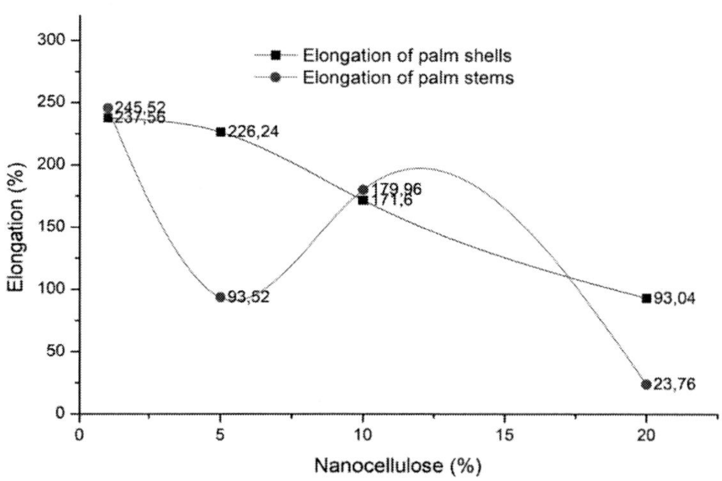


**Figure 3.** Effect of filler contents on the elongation of PVA/ nanocellulose composites

Elongation is defined as an extension of a film, measured from the initial time until the film breaks [13]. The elongation analysis results can be seen in Figure 3, where the addition of the filler affects the elongation. For nanocellulose from the oil palm shells, the addition of 1%, 5%, and 20% gives 237.56%, 226.24%, and 171.6% elongation, respectively. Meanwhile, with the same variations, the addition of nanocellulose from the oil palm stems gives 245.52%, 93.52%, and 179.96% elongation, respectively. The clearly observable decrease of elongation with the addition of 20% nanocellulose can be ascribed to the distribution of nanocellulose particles that occupy the pores of the composites, allowing a tighter filler network arrangement and a closer distance between fillers. The obtained values increase the strength of the composites but reduce the elongation.

## 4. Conclusion

Nanocellulose has been successfully prepared from palm oil solid waste (oil palm shells and stems) and applied as a filler for PVA/ nanocellulose film composites. DSC analysis suggests better thermal properties by the increase of Tg in PVA film after the addition of nanocellulose, either from oil palm shells or stems. It is ascribed to the presence of strong bonds of hydrogen and ester between the matrix and the filler, confirmed by the FTIR spectra profile. These bonds are also responsible for the improvement in mechanical properties, indicated by the increase of tensile strength value. The addition of nanocellulose filler improves the mechanical properties (up to 245.52 kgf/mm$^2$), with the optimal thermal ratio at 203.5°C. However, the homogeneity of the nanocellulose distribution is reduced by the addition of 20% nanocellulose, resulting in a drawback in the tensile strength. The addition of 20% nanocellulose also results in the reduction of elongation due to the formation of a tighter filler network arrangement and a closer distance between fillers. This study, in conclusion, provide evidence that the nanocellulose that is prepared from palm oil solid waste can be effectively used as a filler to improve the mechanical properties of PVA film.

## 5. Acknowledgement

The authors are grateful for technical support from the Department of Chemical Engineering, Faculty of Engineering, University of Syiah Kuala and Directorate of Research and Community Service and Directorate General of Research and Development, Ministry of Research, Technology and Higher Education who have funded this research in accordance with the Research Contract No. 099/SP2H/LT/DPRM/2019.

## 6. References

[1] Warsito J, Mulyani S and Mustapa K 2016 *J. Akademika Kimia* **5** 8
[2] Meisrilestari Y, Khomaini, Rahmat, Wijayanti and Hesti 2013 *J. Konversi* **2** 46
[3] John M J and Thomas S 2008 *Carbohydr. Polym.* **71** 343
[4] Saxena A, Elder T J, Kevin J and Ragaukas A J 2010 *Nano-micro Lett.* **2** 235
[5] Roohani M, Habibi Y, Belgacem N M, Ebrahim G, Karimi A N and Dufresne A 2008 *Eur. Poly. J.* **44** 2489
[6] Cherlina D I, Gea S and Nainggolan H 2017 *J. Kimia Mulawarman* **14** 121
[7] Xiong R, Xinxing Z, Dong T, Zehang Z and Canhui L 2012 *Cellulose* **19** 1189
[8] Tang C and Liu H 2008 *Compos. Part A Appl. Sci. Manuf.* **39** 1638
[9] Fortunati E, Puglia D, Monti M, Santulli C, Maniruzzaman M and Kenny J M 2012 *J. Appl. Polym. Sci.* **10** 1
[10] Coates J 2000 *Encyclopedia of Analytical Chemistry* **12** 10815
[11] Camargo P H C, Satyanarayana K G and Wypych F 2009 *Mater. Res.* **12** 1
[12] Iriani E S, Savitri E and Wahyuningsih 2015 *J. Penelitian Pascapanen Pertanian* **12** 11
[13] Hasanah Y R and Haryanto 2017 *Techno.* **18** 96

# Reduction of Zn(II) ions in Acid Mine Drainage by Adsorption Using Adsorbent Prepared from Bone Coal

**N R Arifah, A Muslim*, D S Syahiddin, W Rinaldi, H Meilina, A Salamun and Mariana**
Chemical Engineering Department, Faculty of Engineering, Universitas Syiah Kuala, Banda Aceh, Indonesia

E-mail: abrar.muslim@che.unsyiah.ac.id

**Abstract.** This study proposed a method to reduce Zn(II) ions concentration in the acid mine drainage using a novel activated bone coal. Bone coal was activated using NaOH solution with varied concentration. The effect of independent variables such as adsorption time, activator NaOH concentration and initial concentration of Zn(II) ions on adsorption capacity of activated bone coal (A-BC) was investigated. Metal content analysis using AAS (Atomic Absorption Spectroscopy), functional group analysis using Fourier Transformation Infrared Spectroscopy (FTIR) and morphological analysis using Scanning Electron Microscopy (SEM) were conducted. The Zn(II) ions adsorption onto the A-BC had excellent fitting with correlation coefficient ($R^2$ = 0.86) to the pseudo second-order adsorption kinetic. It followed Langmuir isotherm with $R^2$ being 0.89. Optimum condition of A-BC and Zn(II) ions solution system was obtained at 27°C consisted of 1 g of adsorbent with 0.4 M NaOH activator in 100 mL Zn(II) ions solution at pH 4.5 stirred at 150-rpm, and Zn(II) ions adsorption capacity of activated bone coal was 11.223 mg/g.

## 1. Introduction

The use of mining materials in Indonesia has increased in recent decades, indicating the development of a better mining industry that provides benefits to the country economy growth. Mining industry also necessarily results in an increase in the content of heavy metals in the environment including top soil, sediments, aquifers and groundwater, and rivers [1-7]. One of the problems that occur in the mining industry is acid mine drainage (AMD). It was produced by interacting between rainwater or groundwater interacting with rocks containing certain sulfides present in mining materials. The AMD acidity is high containing sulfate and heavy metal ions including Zn(II) [8-9]. It is hazardous heavy metal because it is not biodegradable material, and it can accumulate in human organs causing serious adverse effects on human health [10-11]. It should be reduced before discharging into the environment.

To reduce heavy metals concentration in acid mine drainage, several methods have been applied such as hydrolysis, chemical precipitation, solvent extraction, oxidation [12], neutralization [13], membrane5separation [14] and adsorption [15]. Adsorption has become the more effective and popular method for the heavy metals reduction in the AMD and wastewaters [16]. Adsorption method has advantages which are simpler operation, eco-friendly materials and chemicals to use [17-18].

* To whom any correspondence should be addressed.

Many low-cost adsorbents have been developed from natural materials for heavy metal adsorption, especially for Zn(II) ions, such as dairy manure compost [19], coal-mine drainage sludge [20], natural bentonite [21], zeolite and vermiculite [22]. Silica-based adsorbents were proposed for Zn(II) ions adsorption [23-24]. However, bone coal has not be used for the adsorption of Zn(II) ions. Meawhile, it consist almost entirely of silica, [25] and this impure coal has been noticed as waster in coal mining industries. Thus, it is necessary to use it as raw material of adsorbent for the adsorption of Zn(II) ions.

The main objective of present study was to prepare adsorbent from the A-BC (Activated Bone Coal) for Zn(II) ions adsorption. The chemical functional groups and structure of the A-BC were analyzed and characterized using Fourier transform infrared spectroscopy (FTIR) and scanning electron microscopy (SEM). The effect of independent variables such as adsorption time, NaOH concentration and initial concentration of Zn(II) ions on adsorption capacity of activated bone coal was investigated. Isotherm and kinetic adsorption studies were conducted, and the maximum Zn(II) ions adsorption capacity in optimum condition of adsorption was highlighted.

## 2. Method

### *2.1. The A-BC preparation*

The adsorbent used in this study was bone coal from PT Mifa Bersaudara, Meulaboh, Aceh Barat. Bone coal that was still in the form of chunks was crushed to be the size of a gravel, and milled using a ball mill to get the smaller size. Then, the bone coal was sieved into 150-200 mesh. Then, it was chemically activated using NaOH solution to enlarge the pore, to provide a hydroxyl group (OH) on the adsorbent surface as the active site. The method for activating adsorbent in the previous studies was taken into account [26-28]. Bone coal of 25 grams was activated separately using 100 mL of NaOH with various concentration at 0.1 M in a beaker glass. The solution is stirred for 5 hours. After that, it was washed with distilled water until the used washing water pH becoming neutral. The activated bone coal (A-BC) was filtered and dried using an oven at 110 °C until the A-BC being constant. The procedure was repeated for 0.25 and 0.4 M of NaOH to produce three type of the A-BC (A-BC0.1; A-BC0.25; A-BC0.4). The chemical functional groups and structure of the A-BC were analyzed and characterized using the FTIR and SEM.

### *2.2. The experiments of Zn(II) ions adsorption on the A-BC*

To make 1000 mL of Zn(II) ions solution stock, 4,399 mg $ZnSO_4.7H_2O$ (99% pure, from Aldrich) was diluted into 1000 mL of distilled water in an Erlenmeyer and stirred at 150 rpm. The Zn(II) ions concentration in solution stock was determined using an atomic absorption spectrophotometer (AAS, AA-6300 Shimadzu, made in Japan). The procedure of adsorption experiments in the previous studies was taken into account [28]-[29]. The initial Zn(II) ions concentration for each run of batch mode experiment was obtained by dilution to make 6, 10, 15, 30, 90, and 180 mg/L of solution, and the real Zn(II) ions concentration was determined by the AAS. The Zn(II) ions adsorption capacity was investigated over the AAS-based initial Zn(II) ions concentration (6.02–180.03 mg/L), activator NaOH concentration (0.1–0.4 M) at constant pH of 4.5, temperature of 27 °C and 1 gram of the A-BC. The adsorption kinetic and isotherm of Zn(II) ions was determined based on the optimum condition.

## 3. Results and Discussion

### *3.1 Functional groups of the activated bone coal*

Figure 1 shows the FT-IR transmission spectra of the A-BC0.1, A-BC0.25 and A-BC0.4. There are many bands shown in the spectra, but only 3 functional groups can be identified. The 1st band is assigned to O-H stretching at a wavenumber of 3502-3782 $cm^{-1}$ [27-29] with 2 peaks at 3745 and 3612 $cm^{-1}$. The 2nd band has 2 peak at 1697 and 1541 $cm^{-1}$ which efers to the C=C stretching at wavenumber of 1529-1715 $cm^{-1}$ [30], and the last one is ascribed to C–C stretching 400-700 $cm^{-1}$ with a peak at 694 $cm^{-1}$ [31].

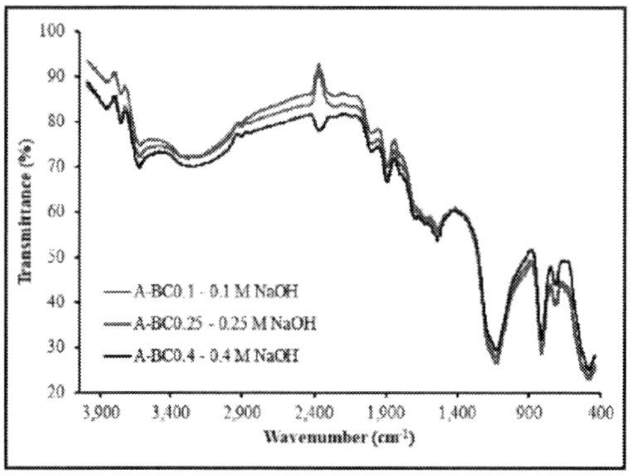


**Figure 1.** The FT-IR spectra of the activated bone coal, the A-BC0.1; A-BC0.25; A-BC0.4

As can be observed in figure 1, increasing NaOH concentration for activation from 0.1 M to 0.25 and 0.4 M decreased the transmission spectra from 80 % to 77 and 68 %, respectively. It is shown by the A-BC0.4 peak being at bottom of the A-BC0.1 and A-BC0.25. It means that hydroxyl groups were formed more on the A-BC0.4surface compared to the A-BC0.1 and A-BC0.25 surface. Therefore, there are more active sites of O-H stretching being formed on the AC-CBB0.4 leading to higher adsorption capacity expected.

*3.2 Surface Morphology of Adsorbent*

Surface morphology of the A-BC0.1, A-BC0.25 and A-BC0.4 are shown in Figures 2(a), (b) and (c), respectively. As can be seen in Figure 2, the A-BC0.1 wall was thicker compared to the A-BC0.25 and the A-BC0.4. A-BC0.1 and A-BC0.25 had less pores compared to the A-BC0.4. More pores should be on the on the A-BC0.4 because NaOH acted as dehydrating agent [27-29]. A-BC0.4 showed the best surface morphology, and Zn(II) ions adsorption capacity of the A-BC0.4 should be highest compared to the A-BC0.1 and A-BC0.25.

(a) (b) (c)

**Figure 2.** The SEM micrographs of (a) the A-BC0.1, (b) A-BC0.25 (c) the A-BC0.4

*3.3 Effect of contact time on adsorption capacity*

As can be seen in figure 3, Zn(II) ion adsorption capacity of the A-BC0.4 over contact time increased exponentially. The adsorption capacity with 29.72 and 92.45 mg/L of Zn(II) initial concentration had the same trend whereas it was almost stable after 20-min contact time. It was approximately 2.82, 2.83 and 2.86 mg/g at 60, 70 and 80 min of contact time for 29.72 mg/L of Zn(II) initial concentration, and it was 6.10, 6.30 and 6.43 mg/g, respectively for 92.45 mg/L of Zn(II) initial concentration. In contrast, it increased moderately from 0.27 to 4.15 mg/g at 10 to 40 min of contact time, respectively for 180.03 mg/L of Zn(II) initial concentration. It jumped to 11.19 mg/g at 50 min of contact time, and it did not change too much for the rest of contact time. It was reasonable because the adsorption capacity of adsorbent reaches an equilibrium condition faster for the lower initial concentration of adsorbate [26-28]. Overall, Zn(II) ion adsorption capacity of the A-BC0.4 increased exponentially growth over time, and the contact time of 60 min was chosen as an equilibrium contact time for isotherm adsorption experiments.

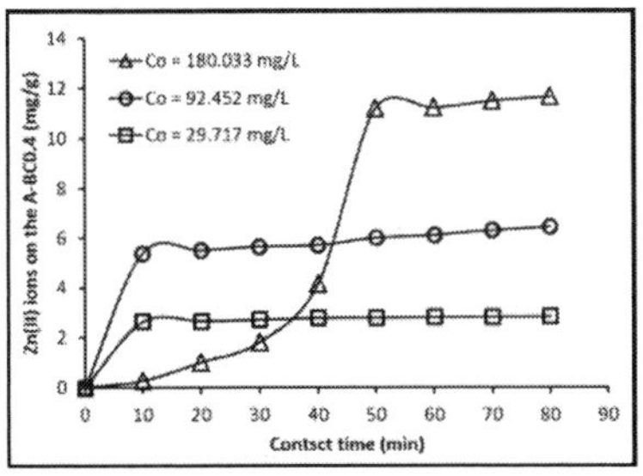


**Figure 3.** Zn(II) ions adsorption capacity of the A-BC0.4. Experimental conditions: 100 mL Zn(II) ions solution at 180.03 mg/L, pH 4.5, 27 °C, 1 g A-BC0.4, 150-rpm stirring speed, 1 atm and 0–60 min contact time.

### *3.4 Effect of activator NaOH concentration on adsorption capacity*

Zn(II) ions concentration on different activated bone coal (the A-BC0.1, A-BC0.25 and A-BC0.4 is shown in figure 4. As revealed in figure 4, the effect of activator NaOH concentration on adsorption capacity was unclear until 40 min of contract time. As can be seen clearly in figure 4, the A-BC0.4 had much higher Zn(II) ions adsorption capacity which was 11.22 mg/g at 60 min of equilibrium contact time compared to the A-BC0.1 and the A-BC0.25. It was 7.39 and 7.16 mg/g for the A-BC0.1 and A-BC0.25, respectively.

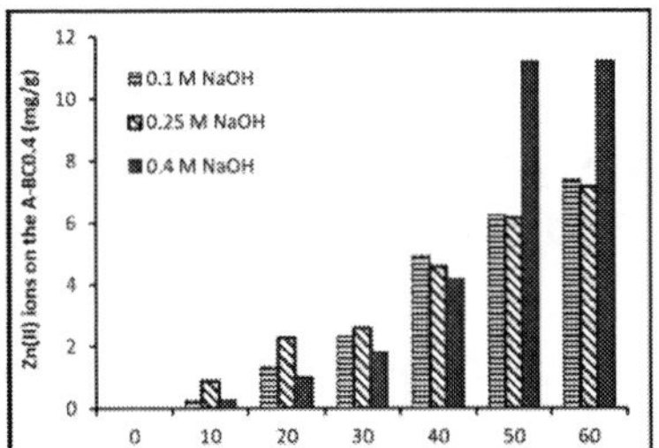


**Figure 4.** Zn(II) ions concentration on different activated bone coal (the A-BC0.1, A-BC0.25 and A-BC0.4). Experimental conditions: 100 mL Zn(II) ions solution at 180.03 mg/L, pH 4.5, 27 °C, 1 g A-BC0.4, 150-rpm stirring speed, 1 atm and 0–60 min contact time.

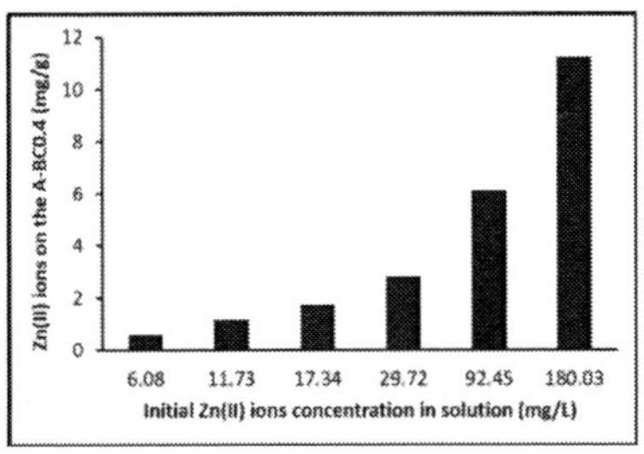


**Figure 5.** Zn(II) ions adsorption capacity of the A-BC0.4 over initial Zn(II) ions concentration. Experimental conditions: 100 mL Zn(II) ions solution at 6.08–180.03 mg/L, pH 4.5, 27 °C, 1 g A-BC0.4, 150-rpm stirring speed, 1 atm and 60 min contact time.

### *3.5 Effect of initial Zn(II) ion concentration on adsorption capacity*

In general, adsorption capacity of adsorbent increases with the increase of initial heavy metal ions concentration ion solution [20,23,26-29]. This trend also occurred in the present study, as can be seen in figure 5. It increased gradually from 0.61 to 2.82 mg/g for the increase in the initial Zn(II) ions concentration from 6.08 to 29.72 mg/L, respectively. Then, it inclined shaprly to 6.10 and 11.22 mg/g for the initial Zn(II) ions concentration in solution being 92.45 and 180.03 mg/L. It should be stable when the initial Zn(II) ions concentration in solution is increased because the active sites on the A-BC0 might be occupied by the Zn(II) ions, highlighted in the previous studies [26-29].

*3.6 Zn(II) ions adsorption kinetic and isotherm of the A-BC0.4*

The common linearized pseudo first-order kinetic model (PFOKM) and pseudo second-order kinetic model (PSOKM) [29,32] were used in the present study. The PFOKM and PSOKM is expressed as Eqs. (1) and (2), respectively:

$$\log(q_e - q_t) = \log q_e - \left(\frac{k_L t}{2.303}\right) \tag{1}$$

$$\frac{t}{q_t} = \frac{1}{k_H q_e^{\,2}} + \frac{t}{q_e} \tag{2}$$

where $q_t$ (mg/g) denotes as Zn(II) ions adsorption capacity of activated bone coal at time of $t$ (min), $q_e$ (mg/g) is the $q_t$ value at equilibrium time ($t = e$); $k_L$ (/min) is the PFOKM rate constant; and $k_H$ (g/mg.min) is the PSOKM rate constant. All the kinetic constants value were obtanied using the slope and intercept of the related equations. The common Langmuir and Freundlich models were used in present study, it can be expressed linearly in Eqs. (3) and (4), respectively[29,32]:

$$\frac{C_e}{q_e} = \frac{1}{q_m K_L} + \frac{1}{q_m} C_e \tag{3}$$

$$\log q_e = \frac{1}{n} \log C_e + \log K_F \tag{4}$$

where $C_e$ (mg/L) is the concentrations of Zn(II) ions in solution at $t = e$; and $q_m$ (mg/g) represents adsorption capacity based on Langmuir; $K_F$ (L/mg) and $K_L$ (L/mg) represent the Freundlich and Langmuir constants, respectively; and $1/n$ denotes the adsorption intensity of Freundlich. The adsorption isotherms constants are worked out using intercepts and the slope of Eqs. (3) and (4).

The adsorption kinetic trend lines are shown in figure 6. As the result, the best fitting was obtained for the PSOKM with correlation coefficient, $R^2$ being 0.86. Meanwhile, the adsorption isotherm plots are shown in figure 7. Langmuir fitting is the best fit for the adsorption isotherm of Zn(II) ions onto the activated bone coal with the $R^2$ value being 0.89, the $K_L$ and $q_m$ value being 10.38 mg/g and 0.29 L/mg, respectively. Langmuir isotherm of the monolayer of adsorbate ions being on the surface adsorbent was reasonable because the monolayer of adsorbate ions being adsorbed on the surface activated bone coal is suitable for the PSOKM adsorption kinetic [26,33].

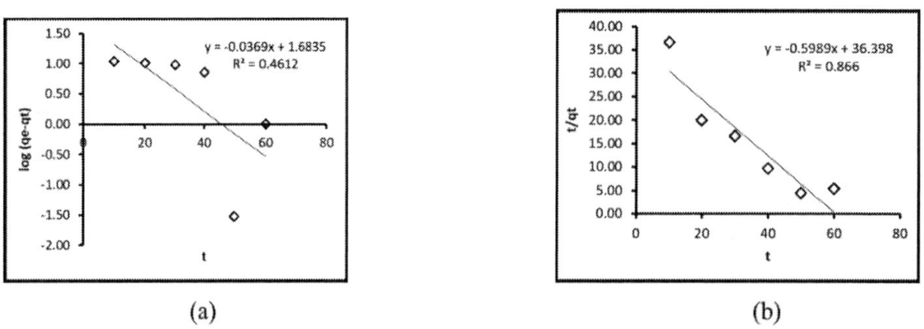


**Figure 6.** Plots of adsorption kinetic in the linearized form (a) PFOKM and (b) PSOKM. Experimental conditions: 100 mL Zn(II) ions solution at 180.03 mg/L, pH 4.5, 27 °C, 1 g A-BC0.4, 150-rpm stirring speed, 1 atm and 0–60 min contact time.

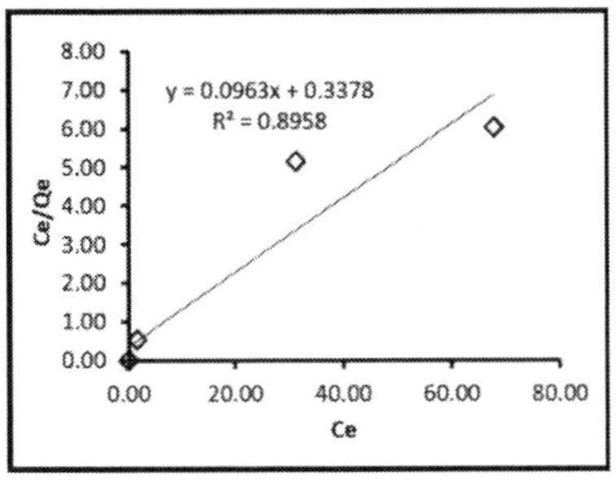


(a)

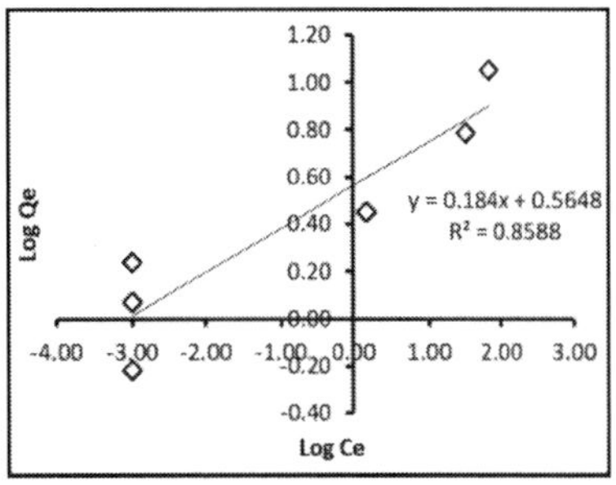


(b)

**Figure 7.** Plots of adsorption isotherm in the linearized form (a) Langmuir and (b) Freundlich. Experimental conditions: 100 mL Zn(II) ions solution at 6.08–180.03 mg/L, pH 4.5, 27 °C, 1 g A-BC0.4, 150-rpm stirring speed, 1 atm and 60 min contact time.

## 4. Conclusions

This study proposed novel adsorbent prepared from bone coal. Bone coal was activated using NaOJ. The active site and surface morphology of activated bone coal (A-BC) were analysed using Fourier Transform Infrared Spectroscopy and Scanning Electron Microscopy. Zn(II) ions adsorption capacity of the AC-CBB were investigated over independent variable such as NaOH activator concentration, over contact time and initial Zn(II) ions concentration. The Zn(II) ions adsorption onto the A-BC fitted well to the pseudo second-order adsorption kinetic with correlation coefficient being 0.86. Langmuir isotherm fitting with correlation coefficient being 0.89 is the better one compared to the Freundlich fitting. The A-BC and Zn(II) ions solution system with 1 g of activated carbon with 0.4 M NaOH activator in 100 mL Zn(II) ions solution at pH 4.5 stirred at 150-rpm was observed as the optimum condition 27°C.

## 5. Acknowledgment

The authors are grateful to Chemical Engineering Department and Mechanical Engineering Department, Faculty of Engineering, Universitas Syiah Kuala for technical support.

## 6. References

[1] Viers J, Jose A G, Cyril Z, Rémi F, Jérémy M, Teresa V, Maria-Lusia de L T, Christine D and Oleg S P 2018 *Chemosphere* **193** 1071
[2] Fu F and Wang Q 2011 *J. Environ. Manage.* **92** 407
[3] Modoi O C, Carmen R, Zoltán T and Alexandru O 2014 *Environ. Eng. Manag. J.* **13** 2325
[4] Moodley I, Sheridan C M, Kappelmayer U and Akcil A 2018 *Miner. Eng.* **126** 207
[5] Trumm D 2010 *New Zeal. J. Geol. Geop.* **53** 195
[6] Simate G S and Ndlovu S 2014 *J. Env. Chem. Eng.* **2** 1785
[7] Zhang T, Zhihong T, Guining L, Xingchun D, Xiaoyun Y, Chuling G and Zhi D 2017 *J. Environ. Manage.* **188** 1
[8] Liao J, Ru X, Xie B, Zhang W, Wu H, Wu C and Wei C 2017 *Ecotox. Environ. Safe.* **141** 75
[9] McCauley C A, O'Sullivan A D, Milke M W, Weber P A and Trumm D A 2009 *Water Res.* **43** 961
[10] Cui M, Jang M, Cho S -H, Khim J and Cannon F S 2012 *J. Hazard. Mater.* **215** 122
[11] Rakotonimaro T V, Neculita C M, Bussière B, Benzaazoua M and Zagury G J 2017 *Environ. Sci. Pollut. Res.* **24** 73
[12] Eslami H, Ehrampoush M H, Esmaeili A, Ebrahimi A A, Salmani M H, Ghaneian M T and Falahzadeh H 2018 *Chemosphere* **207** 303
[13] Tolonen E -T, Sarpola A, Hu T, Rämö J and Lassi U 2014 *Chemosphere* **117** 419

[14] Kalhor M M, Rafati A A, Rafati L and Rafati A A 2018 *J. Mol. Liq.* **266** 453
[15] Mohan D and Chander S 2006 *J. Hazard. Mater.* **137** 1545
[16] Motsi T, Rowson N and Simmons M 2009 *Int. J. Miner. Process.* **92** 42
[17] Eren E, Afsin B and Onal Y 2009 *J. Hazard. Mater.* **161** 677
[18] Khosravi R, Eslami H, Zarei A, Heidari M, Norouzian-Baghani Abbas, Safavi N, Mokammel A, Fazlzadeh M and Adhami S 2018 *Desalin. Water Treat.* **116** 119
[19] Zhang M 2011 *Chem. Eng. J.* **172** 361
[20] Cui M, Jang M, Cho S -H, Khim J and Cannon F S 2012 *J. Hazard. Mater.* **215** 122
[21] Sen TK and Gomez D 2011 *Desalination* **267** 286
[22] Malamis S and Katsou E 2013 *J. Hazard. Mater.* **252** 428
[23] Fuangfa U, Benjawan W, Nuchnicha P, Mahitti P and Juwadee S 2007 *J. Hazard. Mater.* **142** 455
[24] Mona K, Ahmad K, Hanafy H and Zakia O 2014 *Energy Procedia* **50** 113
[25] Anon 1990 *Surface Mining*, 2nd Edition, Edited by B A Kennedy (Colorado: Society for Mining, Metallurgy and Exploration)
[26] Muslim A 2017 *JESTEC* **12** 280
[27] Muslim A, Aprilia S, Suha T A and Fitri Z 2017 *J. Korean Chem. Soc.* **61** 89
[28] Syahiddin D S and Muslim A 2018 *J. Korean Chem. Soc* **62** 79
[29] Muslim A, Ellysa and Syahiddin DS 2017 *J. Eng. Technol. Sci.* **49** 472
[30] Hesas R H, Daud W M A W, Sahu J N and Arami-Niya, A 2013 *J. Anal. Appl. Pyrol.* **100** 1
[31] Chakravarty P, Sarma N S and Sarma H P 2010 *Desalination* **256** 16
[32] Muslim A 2017 *Iran J. Chem. Chem. Eng.* **36** 115
[33] Kiran B and Thanasekaran K 2011 *Int. Biodeterior. Biodegradation* **65** 840

# Effect of operating conditions on the efficiency of adsorption of Manganese (Mn) metals using Nano Zeolite adsorbents

**Y Syamsuddin[1], Zulfikar[2], F Meirisa[1], S Muchtar[1], A Ambarita[2], S Mulyati[1,3*], Suraiya[1], C M Rosnelly[1] and N Arahman[1,3]**
[1] Department of Chemical Engineering, Universitas Syiah Kuala, Banda Aceh, Indonesia, 23111
[2] Graduate School of Chemical Engineering, Universitas Syiah Kuala, Banda Aceh, Indonesia 23111
[3] Doctoral School of Engineering Science, Universitas Syiah Kuala, Banda Aceh, Indonesia 23111

E-mail: sri.mulyati@unsyiah.ac.id

**Abstract.** Manganese (Mn) is one of the most abundant metal in the soil. The presence of this metal is essentially needed by living things for the formation of cartilage and bone as well as driving various enzymes. However, in excessive amounts, the metal can cause disruption of the central nervous system. In this study, the removal of Mn was carried out by the adsorption process using zeolite adsorbent. Effects of contact time, pH of sample (2, 3, 4, 5, 6, 7, 8, and 9) and dosage of adsorbent (1; 1,5; 2; 2,5; 3; 3,5; 4; 4,5 and 5 g) were evaluated. The optimal conditions were obtained under operating conditions of 100 minutes contact time, 7 and 9 pH of the sample, and 5 g dosage of adsorbent. The adsorption process was able to reduce Mn metal up to 79.52%. The initial concentration of Mn sample was 5 g/L and the concentration of effluent after the adsorption process reduced to 1,024 g/L. The adsorption process resulted in a significant reduction of Mn level. However, the final result of the process shows the level of Mn that still exceeds the limit permitted by KEPMENKES No. 907/MENKES/SK/VII/2002 of 0.1 mg/L.

## 1. Introduction

Water is the natural resource most needed by all living things on earth. Water is needed by humans for daily use. The kind of water most used by humans is ground water. The problem in groundwater is the heavy metal content that is not expected to exist. One of the heavy metals contained in ground water is manganese (Mn) metal, that can be dangerous for health. Therefore, the government establish the maximum level of Mn in drinking water is 0.1 mg/L, which is regulated in Minister of Health Regulation No. 907/MENKES/SK/VII/2002.

Research on the reduction of metal contained in groundwater has been widely carried out. In general, the method used is the adsorption using various adsorbents from biomass, including candlenut shells [1], crab shells [2], sawdust [3], nanosilica [4], and rice husks [5].

Natural zeolite is a potential adsorbent due to its high adsorption capacity and selectivity, long life, and abundant amount in Indonesia [6]. The adsorption process using natural zeolites has been studied by many researchers and provides results with a good metal ion removal efficiency, because zeolites have many pores and high cation exchange capacity that making it suitable for use as an excellent adsorbent [7]. The important factor in the adsorption process is the surface area [8], therefore in this

[*] To whom any correspondence should be addressed.

study nano zeolites were used to increase the effectiveness of the adsorption of manganese metal. In this study also observed the effect of adsorption operating conditions including contact time, initial sample concentration, pH, and adsorbent dosage, in order to find out the best condition for removing Mn metal in ground water.

## 2. Experimental

### *2.1. Materials*
The material used are Aceh's natural zeolite as an adsorbent and Potassium permanganate ($KMnO_4$) as an artificial solution. $KMnO_4$ was obtained from Sigma Aldrich and used without purification. While the tools used are magnetic stirrer, hot plate, pH meter, ball mill, shive shaker, and spectrophotometer.

### *2.2. Nano zeolite preparation*
Natural zeolites were crushed by using a hammer to make small pebbles, then crushed using a ball mill to form powder. The powder is then filtered using a shive shaker with a filter size of 400 mesh (<37 μm). Nano zeolite was calcined in the furnace at 600°C for 2 hours to remove impurities in the pores of nano zeolite, and then was cooled to room temperature in a desiccator.

### *2.3. Adsorbent characterization*
Nano zeolite has been characterized by using Fourier Transform InfraRed (FTIR) and X-Ray Diffraction (XRD). FTIR analysis was executed to investigate the spectrum of nano zeolites. The analysis was done by Shimadzu Prestige FT-IR 6400 and the spectrum was observed in the range of 2000 to 4000 $cm^{-1}$ wavenumber. The X-ray diffraction analysis was conducted to identify the components contained in nano zeolite. The analysis was performed at angle 2θ within scan range of 10° to 40°, and measurement conditions of 40 kV and 30 mA.

### *2.4. Batch adsorption process*
The adsorption process has been done by contacting nano zeolite adsorbents with $KMnO_4$ solution. The process was done at various contact time (25, 50, 75, 100, 125, and 150 min), concentration of $KMnO_4$ solution (5, 10, 15, and 20 mg/L), pH (3,4, 5, 6, 7, 8 and 9), and the dosage of adsorbent (1; 1.5; 2; 2.5; 3; 3.5; 4; 4.5; and 5 g). The final concentration of $KMnO_4$ solution after adsorption was then analyzed using spectrophotometer every 25 min for 125 min. The removal efficiency of manganese (Mn) metal was calculated using equation 1.

$$E = \frac{Cp}{Cf} x\ 100\% \quad (1)$$

*E*: Mn metal removal efficiency (%)
*Cp*: final concentration of $KMnO_4$ solution (mg/L)
*Cf*: initial concentration of $KMnO_4$ mixture (mg/L)

## 3. Result and Discussion

### *3.1. Nano zeolite characterization*
The results of XRD analysis can be seen in Figure 1. Three dominant diffraction peaks were observed. Diffraction peaks at 25.1° and 25.6° are quartz mineral characteristics ($SiO_2$), while the diffraction peak at 27.9° is the mineral calcite ($CaCO_3$). According to the previous research reported by Gultom [9], the natural nano zeolite peaks were between peaks of 25° to 29° which indicated the Si and Al compounds have been successfully synthesized and contained in natural nano zeolite.

### *3.2. Effect of contact time and initial concentration of the sample*
The effect of contact time on the efficiency of Mn metal adsorption using nano zeolite is observed by making a metal solution with a various concentration of 5, 10, 15, and 20 mg/L into a beaker glass. Then stirring with a speed of 100 rpm, during the contact time of 25, 50, 75, 100, and 125 minutes. The effect of contact time on Mn metal removal efficiency can be seen in Figure 2.

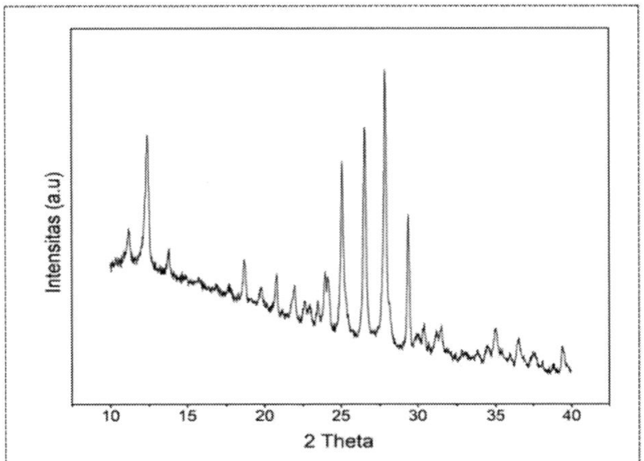


**Figure 1.** XRD peaks of nano zeolite

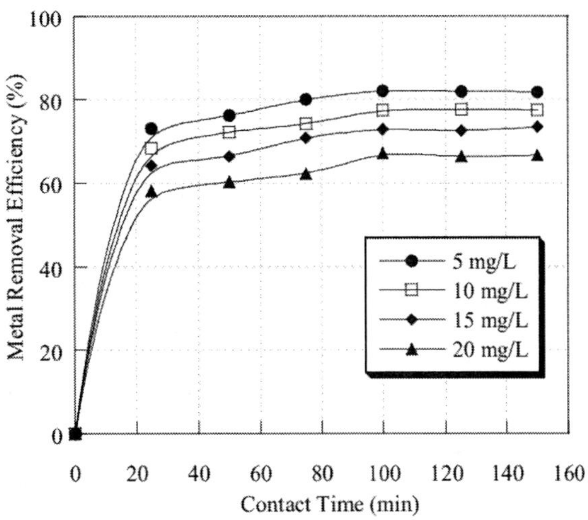


**Figure 2.** Effect of contact time on the Mn metal removal efficiency

Figure 2 shows that the longer the contact time resulted in the increase in metal adsorption efficiency. There is a significant increase in adsorption efficiency during the first 25 min of contact time. This is caused by the availability of contact surfaces of many adsorbents that can adsorb Mn metals well. There is no reasonable increase in efficiency at the next interval of contact time, due to the slow diffusion of cations into the zeolite pores [10]. The Optimum adsorption of Mn metal occurs at 100-minute interval with the efficiency of 79.52%. Based on the figure, it can also be stated that the higher the concentration of the sample solution, the smaller the adsorption efficiency. The decrease in adsorption efficiency is due to the amount of metal ions in the solution is not comparable with the amount of zeolite particles so that the surface of the zeolite reaches the saturation limit in absorbing Mn metal [11].

### *3.3. Effect of pH*

The initial pH of the metal solution is an important factor that must be considered in the adsorption process by nano zeolites. It should be noted that ion exchange depends on the concentration and also the pH of the solution [12]. In this study, the effect of various pH of 2, 3, 4, 5, 6, 7, 8, and 9 with the initial sample concentration of 5 mg/L and contact time absorption of 100 minutes was observed, and the results is presented in Figure 3.

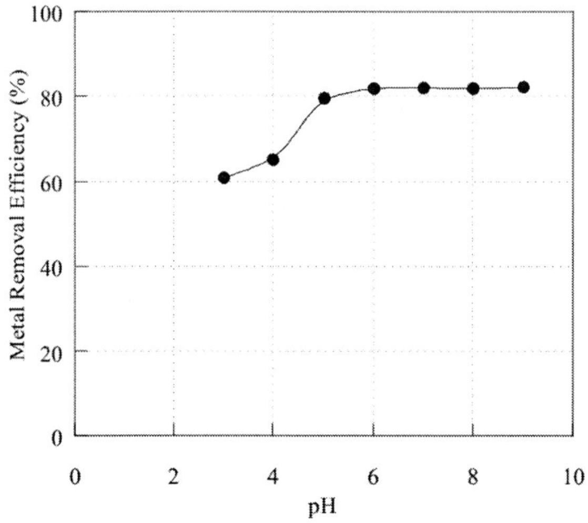


**Figure 3.** Effect of pH on Mn metal removal efficiency

The figure shows that increasing the pH of the solution increases the adsorption efficiency of Mn metal. The amount of heavy metal ions adsorbed increase with an increase in the initial pH value of the solution. This is because the natural zeolite can adsorb $H^+$ ions from the solution into heavy metals. Thus, in more acidic condition, the $H^+$ ion can be adsorbed from the solution [10]. The adsorption efficiency of Mn metal reaches a maximum level at pH 9, that is 79.64%. This result is consistent with the theory that to eliminate Mn metal, the value of the solution is more than 8 [13].

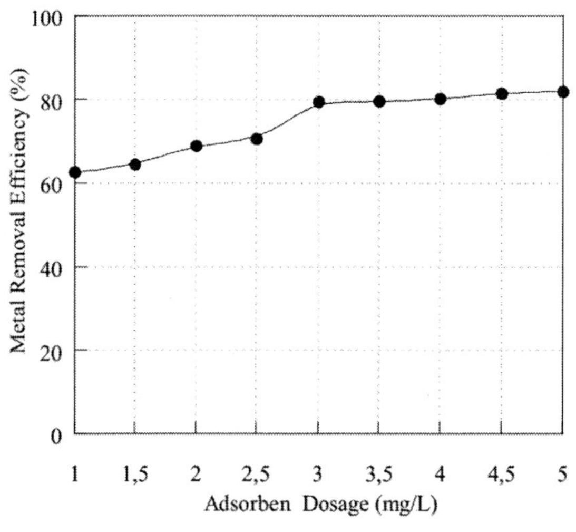


**Figure 4.** Effect of adsorbent dose on Mn metal removal efficiency

*3.4. Effects of Adsorbent Dosage*

The adsorbent dosage is another important parameter that affect the level of metal adsorption efficiency [14-16]. This research studied the effect of various dosage of adsorbent of 1; 1,5; 2; 2.5; 3; 3.5; 4; 4,5;

and 5 g on Mn metal removal efficiency, with a condition of sample concentration of 5 mg/L, pH 7, contact time of 100 min and stirring speed of 100 rpm. The result is shown in Figure 4. The figure revealed that as the adsorbent dosage increase, the adsorption efficiency of Mn metal also increase. Significant increase in adsorption efficiency occurs in the range of adsorbent dosage of 1 to 3.5 g. This is due to the increasing number of particles and surface area available for the process. The maximum adsorption efficiency occurs at a 5-g adsorbent dosage of 79.51%.

## 4. Conclusions

Based on the above results, it can be concluded that:

1. The best condition for the removal of Mn metals using nano zeolite adsorbents was contact time of 100 min and concentration of solution of 5 mg/L, with the efficiency of 79.52%.
2. The maximum pH of the solutions was 9 with the removal efficiency of 79.664%.
3. The best dosage of adsorbent was 5 g with the percentage of Mn metals removal of 79.51%.
4. The adsorption process of Mn metal using nano zeolite with an initial solution concentrasion of 5 mg/L can reach a final concentration of 1.024 mg/L.

## 5. References

[1] Prabarini N and Okayadnya D 2014 *J. Ilmiah Teknik Lingkungan* **5** 33
[2] Pitriani P 2010 *Sintesis dan aplikasi kitosan dari cangkang rajungan* (Jakarta: UIN Syarif Hidayatullah)
[3] Mandasari I and Purnomo A 2016 *J. Teknik ITS* **5** 11
[4] Wibowo E, Hardyanti I, Nurani I, Hardjono D and Rizkita A 2017 *J. Ilmiah Sains* **17** 131
[5] Nurhasni, Hendrawati and Saniyyah N 2014 *J. Valensi* **4** 36
[6] Kundari N A, Susanto A and Prihatiningsih M C 2010 Adsorpsi Fe dan Mn dalam limbah cair dengan zeolit alam *Seminar Nasional IV SDM Teknologi Nuklir*
[7] Panayotova M I 2001 *Waste Manage.* **21** 671
[8] Pujiastuti C and Saputro E A 2007 *Pengaruh ukuran zeolit dan penambahan Na EDTA pada penyerapan logam Zn dalam limbah elektroplating* (Surabaya: UPN Veteran Jawa Timur)
[9] Gultom F, Wirjosentono B, Thamrin, Nainggolan H and Eddiyanto 2015 *J. Chem. Mater. Res.* **7** 20
[10] Motsi T, Rowson N A and Simmons M J H 2011 *Int. J. Miner. Process.* **101** 42
[11] Asih C L, Sudarno and Hadiwidodo M 2014 *J. Teknik Lingkungan* **3** 1
[12] Wang S and Peng Y 2010 *Chem. Eng. J.* **156** 11
[13] Barloková D and Ilavský J 2010 *Pol. J. Environ. Stud.* **19** 1117
[14] Tumin N D, Chuah A L, Zawani Z and Rashid A A 2008 *JESTEC* **3** 180
[15] Iryani D A, Risthy N M, Ginting S Br and Darmansyah D 2019 *J. Rekayasa Kimia & Lingkungan* **14**(1) 81
[16] Muhammad M, Syam A M, Ibrahim I, Nurfarida N and Darmadi D 2019 *J. Rekayasa Kimia & Lingkungan* **14**(1) 72
[17] Fitri R A, Wirakusuma A, Fahrina A, Bilad M R, Arahman N 2019 *Int. J. Eng.* **32**(10) 1372

IOP Conf. Series: Materials Science and Engineering **845** (2020) 012030 doi:10.1088/1757-899X/845/1/012030

# Potential of Edible Oil Production from Rice Bran in Indonesia: A Review

**Y Yunardi[1*], H Meilina[1], U Fathanah[1], R Mahadina[1], A Rinaldi[1] and J Jauharlina[2]**
[1] Department of Chemical Engineering, Universitas Syiah Kuala, Banda Aceh 23111, Indonesia.
[2] Plant Protection Department, Universitas Syiah Kuala, Banda Aceh 23111, Indonesia

E-mail: yunardi@unsyiah.ac.id

**Abstract.** In the year 2018, Indonesia produced 56 million tons of rice paddy. During the paddy milling process, 57-60% rice will be produced, 18-20% husk and 8-10% bran. On this basis, about 4.5 to 5.6 million tons of rice bran is produced every year and unfortunately such massive resource of rice bran is not yet optimally utilized. To this point, rice bran has been used as animal feed. Typically, rice bran contains 15-20 % lipids, 12-16 % protein, 7-11 % crude fibre, 34-52 % carbohydrate and 7-10 % ash. Since it is rich in lipids, rice bran is economically and commercially feasible to be converted into edible oil. With the embargo of Indonesia palm oil by European Union and more portion of palm oil is converted into biodiesel, rice bran oil would have opportunity to supply the country to meet the demand for edible oil. Rice bran oil offers a number of advantages compared to other edible oils in terms high antioxidant contents and high smoke point. This paper will discuss the potential production of rice bran oil from resources and technological point of views, current processing technology, possible methods to increase the recovery, and future prospect for rice bran oil.

## 1. Introduction

Currently, Indonesia is ranked 4th as the country with the largest population in the world, reaching 265,950,668 people with a percentage increase of 1.2% per year [1]. Along with an increase in population, the need for food oil is also increasing. So far, the Indonesian people have only relied on crude palm oil (CPO) to meet the need for food oil. As the largest CPO oil producing country in the world supported by oil palm plantation land area which reaches 12,307,677 hectares [2]. The palm oil industry has indeed helped in improving the economy of the community and the country's foreign exchange. However, the presence of oil palm plantations also provides many problems and negative impacts on the environment, food security, and social. The expansion of oil palm land has caused changes in the forest ecosystem, resulting in the loss of biodiversity, germplasm and loss of water resources. This problem has also penetrated into wetlands thereby increasing the amount of CO2 emissions. In addition, oil palm plantations also cause a reduction in water catchment areas that can trigger natural disasters such as floods and landslides. The negative impacts and problems of land

* To whom any correspondence should be addressed.

ownership from the expansion of oil palm land often lead to conflicts between communities and oil palm companies

Indonesia is one of the biggest rice producing countries in the world. Based on statistical data, the area of paddy fields in Indonesia in 2016 reached 8.19 million hectares [3]. With this vast land, 75.39 million tons of dried paddy is produced annually. Based on research conducted by Hadipernata, Supartono [4], each rice milling process produced rice (67-60%), husks (18-20%) and rice bran (8-10%). This data indicates that Indonesia produces around 6-7.5 million tons of rice bran per year. That is, in addition to millions of tons of rice produced in Indonesia, there is no less abundant bran waste.

The second problem is the low recovery of rice bran oil during the processing of liquid solid extraction generally bran is only used as animal feed, a mixture of natural fertilizer and most of it is burned or left alone. This means that rice bran has not been able to provide added value economically, although it has the potential to be converted into products with higher economic value, such as bioethanol, fibrous foods, cleaners, activated charcoal, and others. Burning is the most popular method used by the community to reduce the volume of rice bran waste. However, this method has a negative impact on the environment and human health. In Thailand, for example, bran is no longer burned but is converted into products that have added value, one of which is cooking oil or vegetable oil. However, the use of rice bran as a raw material for vegetable oil is almost unknown in Indonesia. This is because the government and the community rely more on palm oil as raw material for producing vegetable oil.

There are two problems Indonesia will face if it wants to develop vegetable oil production from rice bran. First, the instability of the free fatty acid content found in the bran, after the grinding process. The content of free fatty acids in bran can increase depending on environmental conditions and storage time. Lipase enzymes help the process of hydrolysis into free fatty acids and glycerol. As a result, bran like this if extracted will produce rancid oil. To avoid rancidity, lipase enzyme inactivation can be done by stabilizing the bran. The second problem is the low recovery of rice bran oil during the processing of liquid solid extraction, which ranges from 6-18% a [5-7]. Until now, the extraction method is still used in the bran oil processing industry as the only way to extract oil from the bran. With this condition, economically the production of bran oil is less attractive because there is still a lot of oil left in the bran. Therefore, efforts are needed to increase the acquisition of bran oil in various ways, such as initial treatment of bran so that oil can easily get out of bran later. This paper will examine the efforts that have been made to stabilize existing bran and the technology that has been developed to increase the yield of bran oil.

## 2. Rice Bran Oil Benefit

Today the interest of society in developed countries against bran oil is increasing because of the benefits that are owned by bran oil. As edible oil, rice bran oil is suitable for frying because of a high smoke point. It is known that the quality of the cooking oil is determined in part by the smoke point, the higher the smoke point the better cooking oil. Rice bran oil has a smoke point of 260 °C, compared with palm oil smoke point that 176 °C. This means that if it is used as cooking oil, palm oil faster than the boiling rice bran oil. From the health side, rice bran oil provides a lot of benefits. Bran oil contains a substance called oryzanol. Oryzanol is an antioxidant that has been known to reduce the content of cholesterol in the human body. Additionally, gamma oryzanol contained in rice bran oil can be a source of power and anti-cancer agents because of the high ability to counteract free radicals. The results also showed that gamma oryzanol beneficial for skin health and beauty. Gamma oryzanol can protect DNA damage of skin cells, increase skin cell regeneration, and prevent damage to collagen and elastin fiber that trigger the occurrence of wrinkles on the skin. In fact, traditionally in Japan, bran is purely used as a mask because of its content of Gamma oryzanol can soften and brighten the skin [8].

In Japan and several Western countries, this oil is known as the "heart of oil" (oil heart), because of the properties for the human heart and has got the category as a healthy food of WHO and FAO. Not only the WHO and FAO, Indian Council of Medical Research also supports the use of rice bran oil is because it is known to have the ability to lower cholesterol, boost the immune system, prevent cancer, improve skin health, accelerate the process of weight loss, relieve symptoms of menopause, as well as prevent heart disease.

From the above, it is clear that rice bran oil has a number of benefits for human health. It is not surprising in developed countries and some Asian countries people are aware of the importance of health switch to using rice bran oil instead of using other vegetable oils. Unfortunately, even though Indonesia is one of the rice barns, turns oil extracted from rice bran is still to be imported from Thailand so the price is relatively expensive.

## 3. Rice Bran Stabilization

Rice bran composition varies depending on the rice variety, weather conditions and methods of rice processing. The chemical composition of the rice bran is strongly influenced by rice varieties. In general, the essential nutrition content of the rice bran in bulk include minerals, vitamins, fiber, protein, amino acids and antioxidants. According to Wibowo [9], the composition of the rice bran in fat and high fiber, as shown in Table 1.

**Table 1.** Composition of rice bran

| Composition | Content (%) |
|---|---|
| Water | 8.33–12.93 |
| Fat | 2.88–5.96 |
| Protein | 8:43 to 13:53 |
| Ash | 4.66–17.96 |
| Crude fiber | 6.74–29.98 |

Although an excellent source of nutrients, bran is not suitable for direct human consumption due to rancidity caused by the presence of lipase. When the bran layer separated from the endosperm during the grinding process, the individual cells burst and consequently lipase in contact with fatty cause triglyceride hydrolysis process into free fatty acids and glycerol [10]. The increase in free fatty acids in the bran occurs within a few the first hours after milling and reaches 5-7% after 24 hours. Dedak free fatty acid containing more than 5% and rice bran oil containing free fatty acids of more than 10% categorized as unfit for human consumption.

**Table 2.** Technology - rice bran stabilization technology [11]

| Mechanical Stabilization | Condition |
|---|---|
| Hot air drying | 100 ° C for 1 hour |
| Steaming | 100 oC for 30 minutes |
| Refrigeration | 2 ° C |
| Sun drying | 47 ° C (maximum) 7 hours per day for 2 days |
| Drying fireside | 84 ° C for 1 hour |
| Spraying chemical stabilization | HCl 1000 ppm |
| Microwave heating | 2450 MHz for 2 minutes |
| Radiation infrared rays | 700 W for 30 minutes |
| Ohmic heating | Humidity 20-40%; 44–72 V / cm voltage gradient |

To prevent rancidity in the bran, the lipase enzyme activity must be stopped at a certain stabilization process immediately after the rice milling process. Various methods have been used to stabilize bran, as shown in Table 2. It should be noted that the purpose of the stabilization process is to inhibit or halt the activity of lipase without damaging the nutrient content in the bran. However, some methods that have been used visible use heat medium so that the operating temperature can reach more than 70 ° C. Methods such as hot air drying, evaporation (steaming), and the methods that can be hot, certainly will damage the protein, or it is called denaturization protein. However, if the protein content of not becoming an important aspect, then the process of heating at high temperature is not a problem.

But in essence, the selection of the appropriate method of stabilization is important because it relates to the effectiveness of the stabilization and operational costs required for stabilization.

How simple and quite effective stabilization developed on the basis of character low lipase activity at low pH. Therefore, the addition of dilute hydrochloric acid so that the pH of the bran changed from 6.9 to 4.0 can reduce lipase enzyme activity. How rice bran stabilization using infrared radiation is also effective for the inactivation of the lipase enzyme that can extend the shelf life of rice bran.

## 4. Technology of Bran Oil Processing

Considering the abundant raw material of rice bran in Indonesia, rice bran is indeed very potential to be converted into food oil. However, efforts to commercialize rice bran oil up to industrial scale such as palm oil still cannot be implemented. This happens because there is no research that produces rice bran oil with a high acquisition. So that both the private sector and the government are still reluctant in commercializing the rice bran oil. However, studies on rice bran production at the research stage have been carried out [6,12]. However, the research that has been done has not been able to have a major impact on oil production from rice bran. At present, research on rice bran oil is still being carried out in order to produce high-yield rice bran oil. If rice bran oil can be commercialized into edible oil, then palm oil can be converted into other products such as biodiesel, soap, cosmetics, and margarine.

At present, to obtain rice bran oil, rice bran oil is processed using a liquid solid extraction process, which is the extraction process using a solvent. According to Susanti, Ardiana [12], the solvents normally used to extract oil from a vegetable material were ethanol, n-hexane, isopropanol, ethyl acetate, acetone and methanol. The use of this solvent to extract edible oil from 100 grams of rice bran yields a succession of 14.70%; 14.94%; 13.30%; 14.26%; 9.30% and 9.16%. It is seen that the acquisition of oil from bran is still very low, which ranges from 9.16% -14.96%. The acquisition of rice bran oil using liquid solid extraction method is still very little. Therefore, the need for pre-treatment of rice bran to increase the recovery of rice bran oil.

Several studies by providing initial treatment of rice bran have also been carried out. Like the use of microwaves (microwave) and far infrared radiation (far infrared), FIR. The use of this method is done as an effort to increase the acquisition of rice bran oil and maintain the stability of rice bran that is easily rancid. Stabilization using microwaves aims to reduce the water content contained in rice bran. The use of microwaves has succeeded in increasing the acquisition of rice bran oil, although not significantly. Based on research conducted by Hadipernata, Supartono [4], the recovery of rice bran oil obtained on the first day after microwaves was 15.48%, while on the 9th day it was 15.79%.

The use of far infrared radiation (far infrared, FIR) before extracting rice bran oil also aims to reduce the water content contained in rice bran and help improve the quality of bran oil. Based on research also conducted by Hadipernata, Supartono (4), we can find out that the use of FIR radiation as a pre-treatment of rice bran affects the rice bran oil produced. where the yield is 15.59% on the first day and 16.24% on the ninth day. The use of sophisticated methods such as supercritical fluid extraction (SFE) has also been tried to increase oil recovery from rice bran. This method uses supercritical $CO_2$ liquid at temperatures between 25-60°C and pressures from 150 to 250 bar. Based on research conducted by Sarmento, Ferreira [13], it was found that this method was quite effective in extracting rice bran oil. Where at a pressure of 250 bar and a temperature of 40°C produces the highest oil yield, which is close to 80%.

### 4.1 Suitability Technology in Indonesia

Related studies have been carried out to obtain rice bran oil with high recovery, so that the commercialization of rice bran oil can be done as one of the edible oils in Indonesia. Comparison of the recovery of rice bran oil that has been obtained from several studies conducted by Hadipernata, Supartono [4], Susanti, Ardiana [12], Sarmento, Ferreira [13] can be seen in Figure 1.

Based on Figure 1, it can be seen that the recovery of rice bran oil using conventional extraction is still very small, only 14.96%. So it needs to be done pre-treatment on rice bran, so that it can increase the recovery of rice bran oil. In the use of microwaves, the increase in yield obtained is relatively small. It can be concluded that this method is not effective for producing edible oil from rice bran. Besides being ineffective the use of microwaves is also not economical because it requires a large amount of energy which will affect production costs. In the use of FIR rays, these results are not much different from the pre-treatment using microwaves, which is 15.79%. So this method is also not effective to be applied in producing bran oil on a large scale.

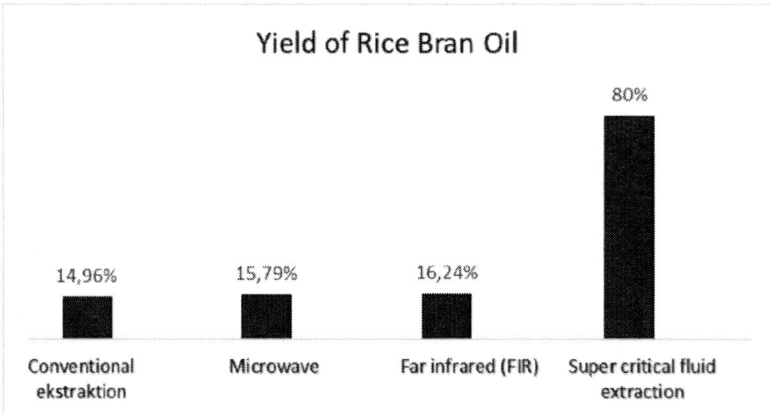


**Figure 1.** Comparison of rice bran oil recovery by various methods

In the use of supercritical fluid extraction methods to produce rice bran oil obtained a high gain, which is 80%. However, the application of this method is constrained by various factors. Among them is this process requires very high pressure for its operation, which is 250 bar. This high pressure will require enormous energy as well. So that it will increase the cost of rice bran oil production. The use of high pressure also requires a high level of safety supervision. So it can be concluded that this method is not appropriate for use in Indonesia.

In addition, there are other methods that are estimated to be used as a pre-treatment to increase the acquisition of rice bran oil, by using fermentation. Fermentation is a way to produce products using microorganisms under certain conditions [14]. According to Ulfa [15], fermentation is divided into two ways of operation, namely liquid phase fermentation and solid phase fermentation. Solid phase fermentation is a pre-treatment method that can be used to increase oil recovery from bran.

Solid phase fermentation has long been known and applied by the people of Indonesia. This process is not only used to produce edible oil, but also has been used to produce everyday foods, such as tempeh, tape, cheese and so forth. Particularly in Aceh, solid state fermentation has been used by people to produce edible oil, coconut cake and dregs from fermented called *pliek u* is used as a flavouring of food for generations. For rice bran fermentation can be carried out in the same manner in a way that is done in traditional solid fermentation in Aceh, where a mixture of microbial cultures are used to increase oil recovery from coconut cake [16], Due to the Indonesian people already familiar with these methods, expected efforts to increase rice bran oil recovery using solid phase fermentation will be more readily accepted by the farmers and people of Indonesia as a method of producing rice bran oil. But these methods to produce rice bran oil has never been done, so it needs to do research related to prove the effectiveness of this method in increasing rice bran oil recovery.

## 5. Conclusions

From the study of literature that has been described above, it can be concluded that the potential of rice bran oil production in Indonesia is very large considering the valley of bran provided a very abundant. The problems with rancidity caused by the lipase enzyme activity can be overcome by a simple method, namely by lowering the pH of the bran to 4.0 by adding hydrochloric acid. Another problem is lack of acquisition can be enhanced by providing early treatment of the bran. Traditional ways that have been made to enhance the coconut oil can be replicated, although a detailed study should be done first. The most important is the willingness of the government to encourage the growth of rice bran oil industry, so that Indonesia does not continuously dependent on palm oil.

## 6. References

[1] Badan Perencanaan Pembangunan Nasional BPS, dan United Nations Population Fund 2013 *Proyeksi Penduduk Indonesia 2010-2035* (Jakarta: Badan Pusat Statistik).
[2] Indonesia KPR 2018 *Statistik Pertanian* (Jakarta: Kementerian Pertanian Republik Indonesia).
[3] BPS 2018 *Luas Panen dan Produksi Beras 2018* (Jakarta: Badan Pusat Statistik)
[4] Hadipernata M, Supartono W and Falah M A F 2012 *J. Aplikasi Tek. Pangan* **1** 103
[5] Mas'ud F and Pabbenteng 2016 *J. INTEK* **3** 82
[6] Nasir S, Fitriyanti and Kamila H 2009 *J. Teknik Kimia* **16** 1
[7] Muis L 2010 *J. Ilmiah Universitas Batanghari Jambi* **10** 62
[8] Patel M and Naik S N 2009 *J. Sci. Ind. Res.* **23** 243
[9] Wibowo A H 2010 *Pendugaan Kandungan Nutrien Dedak Padi Berdasarkan Karakteristik Fisik* (Bogor: Institut Pertanian Bogor)
[10] Malekian F, Rao R M, Prinyawiwatkul W, Marshall W E, Windhauser M and Ahmedna M 2000 *Lipase and Lipoxygenase Activity, Functionality, and Nutrient Loses in Rice Bran During Storage* (Bulletin Lovisiana Agricultural Experiment Station Louisiana State University Agricultural Center)
[11] Gul K, Yousuf B, Singh A K, Singh P and Wani A A 2015 *Bioact. Carbohydr. Diet. Fibre* **6** 24
[12] Susanti A D, Ardiana D, Gumelar G P and Bening Y G 2012 *Polaritas Pelarut Sebagai Pertimbangan dalam Pemilihan Pelarut untuk Ekstraksi Minyak Bekatul dari Bekatulvarietas Ketan (Oriza Sativa Glatinosa)* (Simposium Nasional RAPI XI FT UMS)
[13] Sarmento C M P, Ferreira S R S and Hense H 2006 *Braz. J. Chem. Eng.* **23** 243
[14] John K A and Olusegun O V 2016 *J. Appl. Life Sci. Int.* **5**, 1
[15] Ulfa M 2016 *Penentuan Kadar Glukosamin dari Fermentasi Kulit Udang oleh Mucor Miehei dengan Metode Uji Ninhidrin dan Spektrofotometri UV-VIS* (Bandar Lampung: Universitas Lampung)
[16]. Nurliana 2009 Prospek Makanan Tradisional Aceh Sebagai Makanan Kesehatan: Eksplorasi Senyawa Antimikrob dari Minyak Pliek U dan Pliek U (Bogor: Institut Pertanian Bogor)

# Transesterification of *Pangium Edule Reinw* oil to biodiesel using durian rind ash as heterogeneous catalysts

**D. Yanti[1], H Husin[1,2*], F T Yani[1], A Maulana[2], A Adisalamun[1,2] and H Husaini[3]**
[1] Graduated School of Chemical Engineering, Universitas Syiah Kuala, Banda Aceh, Indonesia, 23111
[2] Departmen of Chemical Engineering, Universitas Syiah Kuala, Banda Aceh, Indonesia, 23111
[3] Department of Mechanical Engineering, Universitas Syiah Kuala, Banda Aceh, Indonesia, 23111

Email: husni_husin@che.unsyiah.ac.id

**Abstract.** The purpose of this research is to know the effect of addition durian rind ash as a catalyst for biodiesel production. Waste of Durian rind ash containing potassium oxide compound can be used as a novel heterogeneous catalyst for the transesterification process. Potassium oxide in durian rind ash catalyst has been synthesized by the simple burning method. In this research, durian rind ash catalyst was potentially for biodiesel production from Pangi (*Pangium Edule Reinw*) oil. The optimum biodiesel yield of 96.24% can be achieved over durian rind ash catalyst of 5 wt.% of catalyst, reaction temperature 65°C at 1 hour, and a methanol-to-Pangi oil ratio of 8:1. The characterization was also conducted using X-ray diffraction (XRD) and Scanning electron microscopy (SEM).

## 1. Introduction

Nowadays fossil fuels are depleting. Great efforts have been done in developing processes for biodiesel generating from renewable sources [1,2]. The first generation biodiesel as an alternative fuel is a renewable fuel with superior process and emissions compared to commercial diesel. Biodiesel has ecological impacts related to potentially sustainable and clean energy for a better future. Indonesia has also participated to produce biodiesel as a motivation to utilize clean energy from emissions of fossil-based energy [3]. Biodiesel which is also known as fatty acid methyl ester (FAME) is generally obtained through the transesterification reaction of vegetable oils or animal fats. This reaction is given by several factors such as alcohol molar ratio to oil, reaction time, reaction temperature, catalyst concentration, stirring speed and water content and free fatty acid content (FFA) [4]. Pangi oil (*Pangium Edule Reinw*) is a type of vegetable oil that potentially converted into biodiesel with an amount of oil weight of 20% of its dry weight. The highest fatty acids composition in Pangi oils are oleic acid and linoleic acid namely 45.2% and 39.3%, respectively [5].

The transesterification reaction is inseparable from the role of the catalyst as a substance that acts to accelerate the reaction. The use of heterogeneous catalysts is now preferred to be developed because it has various advantages over homogeneous catalyst and enzymes such as cheap, easily separated, non-corrosive and reusable [6]. Several types of heterogeneous catalysts used for the production of biodiesel

[*] To whom any correspondence should be addressed.

are zeolite, alkali metals, and the most developed are types of alkaline earth metal oxides [7]. This type of metal oxide catalyst is known to be more active to produce non-toxic products. This catalyst can be obtained from sludges, slags and also ash through thermal treatment above 1200°C [8]. Bennett et al [9] also explain that heterogeneous base catalysts provide a more significant conversion than acid heterogeneous catalysts. Several types of solid catalysts are sourced from several biomass treatments such as snails, eggs, cockle, coconut fiber, oil palm empty fruit bunches and so on [10]. This is closely related to the conservation of natural resources, especially for materials that can rarely be reused. Durian (*Duriozibethinusmurr*) is a plant that grows in Southeast Asian countries such as Indonesia, Philippines, and Malaysia. Lubis et al [11] said that 900,000 tons per year of durian production starting in the next 20 years. Durian has a diameter of about 80 cm. The colour of this fruit is generally green, brownish-green, yellowish-green and yellow. Its rind also rough and has conical spines [12]. Durian fruit consists of 20-30% fruit flesh, 5-15% seeds, and 60-70% rind [13].

In reality, durian rind is produced from durian fruit consumption because the consumed part of the durian is only the flesh, meanwhile, its seeds and rinds are thrown away. The existence of durian rind waste is also disturbing because it has a distinctive aroma that many people do not like [14]. Increased durian production has an effect on increasing durian rind waste. Durian rind is usually reprocessed into adsorbents, pulping and biodegradable films [15-16]. The chemical components of durian rind are $P_2O_5$, MgO, CaO, $SiO_2$, $Fe_2O_3$, $SO_3$, $Na_2O$, $Al_2O_3$, MnO and $K_2O$ as the largest components [13]. The high content of potassium oxide suggests an active phase for a certain reaction. In addition, the results of the proximate analysis also showed that the content of durian rind ash contained was 0.4% of the dry basis [17]. The purpose of this study is to investigate the effect of the ratio of methanol to Pangi oil using a durian skin ash catalyst. The morphology and the composition of the catalyst are also identified by SEM and XRD.

## 2. Methodology

### *2.1 Materials*

The durian ash catalyst was prepared from durian rind including inner skin collected from markets around the city of Banda Aceh, Indonesia. Methanol was purchased from Merck (99.5%), Darmstadt, Germany. Pangi oil was purchase from traditional farmers in the region of Calang Aceh Jaya, Indonesia.

### *2.2 Catalyst Preparation*

Durian rind waste was dried at 32°C. The preparation was then continued by size reduction through cutting then crushed using a mechanical crusher. The sample was then burned in the open air to obtain durian rind ash.

### *2.3 Catalyst Characterizations*

The structural of the samples of durian rind ash were investigated using powder X-ray diffraction (XRD, Shimadzu XRD-6000), having Cu K alpha radiation with 40 kV and 40 mA at scanning rate of 10/min, sampling width 0.02. The morphology of durian rind ash catalyst was characterized using a Scanning Electron Microscopy (SEM, Philips XL-30) apparatus.

### *2.4 Transesterification of Pangi oil*

About 30 grams of *(Pangium Edule Reinw)* oil is put into a three-neck flask. Then methanol was added with volume corresponds to the varied mole ratio of methanol to Pangi oil. Durian rind ash catalyst was added at a constant loading of 5wt.% to oil. The process was carried out with stirring at a constant temperature of 65°C for 60 minutes. After the process was completed, the product was separated using a separating funnel and continued with washing using warm water. The yield of biodiesel was then calculated using equation (1).

$$\%\text{Yield} = \frac{\text{Weight of biodiesel produced}}{\text{weight of oil used}} \times 100\% \qquad (1)$$

## 3. Result and discussion

*3.1 Characterization of durian rind ash catalyst*

A scanning electron microscopy (SEM) image shows that the durian rind ash catalyst has a hollow and irregular morphology. This is one of the effects of the non-uniform distribution of elements. Durian rind ash catalyst also has a micro-size pore ranging of 1-2.6 μm and mesopore with an approximate size of ± 3.7 μm.

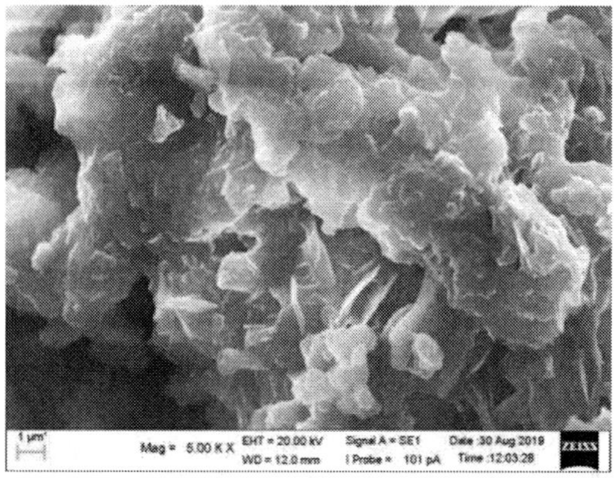


**Figure 1**. SEM image of durian rind ash catalyst

Pore size with the large specific surface area can facilitate the efficiency of reactant contact and its accessibility to active sites [18]. Durian rind ash is also seen to agglomerate as a result of particle interactions during the combustion process [19]. In addition, the wide surface area of the catalyst can also affect the mechanical stability of the durian rind ash catalyst [9].

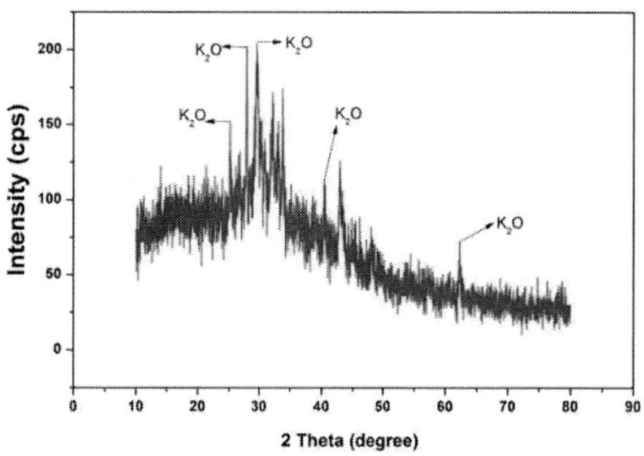


**Figure 2**. XRD analysis of the durian rind ash catalyst

*3.2 XRD Analysis*

The XRD pattern of the durian skin ash catalyst is shown in Figure 2. X-ray diffraction methodology was done to evaluate the composition of durian rind ash. XRD spectrum of durian rind ash can be seen in Figure 2.

Figure 2 reveals that the potassium oxide is the highest component of the catalyst as marked by the appearance of the highest peak with prominent 2θ at 29,6805°, 27,89646°, 40,5767°, 42,893°, and 62,2509°. Potassium oxide is one of the active components that work effectively in the transesterification reaction [20]. Husin et al. also stated that potassium has good catalytic activity and regeneration ability. Potassium can play a role in increasing anion methoxide to break the carbon carbonyl bonds of triglycerides informing biodiesel (Methyl ester) [21].

*3.3 Effect of methanol to oil molar ratio on yield*

The stoichiometry of the transesterification reaction shows that one mole of biodiesel requires 3 moles of alcohol. The reversible nature of the reaction results in the transfer of excess alcohol to the product side [22]. Therefore, a higher molar ratio can be applied to increase miscibility and contact between alcohol molecules with triglycerides [23]. This investigation is needed to minimize the use of methanol in the transesterification reaction. In this study, variations in the molar ratio of methanol to Pangi oil were carried out in 4: 1, 6: 1, 8: 1, 10: 1, 12: 1 and 15: 1. The yield obtained is as presented in Fig. 3.

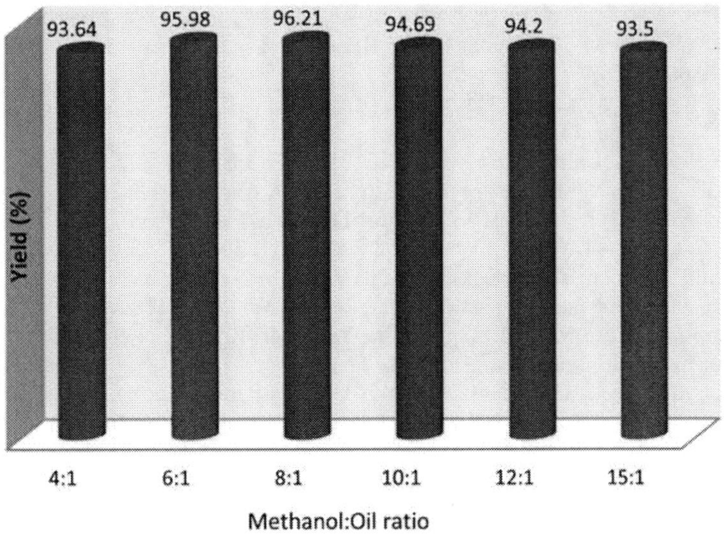


**Figure 3**. Effect of methanol to pangi oil molar ratio (65°C and catalyst 5 wt.%)

Fig. 3 showed that the transesterification reaction with methanol to oil molar ratio variations of 4:1, 6:1 and 8:1 brings about the increment in biodiesel yield. Those are 93.64, 95.98, and 96.21% respectively. For these high yield shows the process of *Pangium Edule Reinw* oil to biodiesel is in a stable polarity so the solubility of glycerol to return become ester phase can be reduced [24]. However, biodiesel yield slightly decrease from 10:1, 12:1 and 15:1 of molar ratio of methanol to oil. It seems, the decreasing of biodiesel yield due to difficulty of product separation from the glycerol layer when the methanol to oil ratio increase. This phenomenon is also reported by other research groups [25]. This acquisition between the two mole ratio variations requires further study related to the role of the diesel fraction as a major component [26].

## 4. Conclusion

The heterogeneous catalyst from durian rind ash has been applied for the transesterification reaction of Pangi oil prepared through a simple combustion method and successfully produced biodiesel with a high yield of 96.21% using 5% catalyst loading at 65°C for 1 hour with 8:1 mole ratio of methanol to oil. The combustion process increased the catalytic activity and mechanical stability of the durian rind ash catalyst. Durian skin ash contains potassium oxide which owns good ability and stability makes it has the potential to be utilized as a material for catalyst biodiesel production.

## 5. Acknowledgments

The author thanks the ministry of research and technology of higher education of Indonesia for financial support through the research project Master Program.

## 6. References

[1] Husin H, Adisalamun, Sy Y, Asnawi T M and Hasfita F 2017 *AIP Conf. Proc.* **1788** 030073
[2] Husin H, Pontas K, Yunardi, Adisalamun, Alam P N and Hasfita F 2017 *ASEAN J. Chem. Eng.* **17** 27
[3] Dharmawan A H, Sudaryanti D A, Prameswari A A, Amalia R and Dermawan A 2018 Pengembangan bioenergi di indonesia *CIFOR* 1-20
[4] Elkady M F, Zaatout A and Balbaa O 2015 *J. Chem.* **2015** 1
[5] Atabani A E, Badruddin I A, Masjuki H H, Chong W T, and Lee K T 2015 *Arabian J. Sci. Eng.* **40** 583.
[6] Husin H, Asnawi T M, Firdaus A, Husaini H, Ibrahim I and Hasfita F 2018 *IOP Conf. Ser. Mater. Sci. Eng.* **358** 012008.
[7] Zaki M, Husin H, Alam P N, Darmadi D, Rosnelly C M and Nurhazanah N 2019 *J. Rekayasa Kimia & Lingkungan* **14** 36
[8] Zuhra Z, Husin H, Hasfita F and Rinaldi W 2015 *Agritech* **35** 69
[9] Bennett J A, K. Wilson K and Lee A F 2016 *J. Mater. Chem. A.* **4** 3617
[10] Syamsuddin Y and Husin H 2010 *J. Rekayasa Kimia & Lingkungan* **7** 112
[11] Lubis R, Saragih S W, Wirjosentono B and Eddyanto E 2018 *AIP Conf. Proc.* **2049** 020069
[12] Yuniastuti E, Nandariyah and Bukka S R 2018 *Caraka Tani: J. Sustain. Agric.* **33** 136
[13] Nurhaeni, Nurakhirawati and Tiaradewi T R 2016 *J. Nat. Sci.* **5** 31
[14] Manickam Y, Nordin M F M, Shameli K, Wannahari R, Iqbal N and Syariff A H M 2015 *J. Optoelectronics and Biomedical Materials* **7** 53
[15] Zhao G, Lyu X, Lee J, Cui X and Chen W 2019 *Food Packag. Shelf Life* **21** 1
[16] Anisuzzaman S M, Joseph C G, Krishnaiah D, Bono A and Ooi L C 2015 *Water Sci. Technol.* **72** 896
[17] Hasan E S, Jahiding M and Arsad J 2017 *J. Aplikasi Fisika* **13** 14
[18] Ferrero G A, Preuss K, Fuertes A B, Sevilla M and Titirici M-M 2016 *J. Mater. Chem. A.* **4** 1
[19] Han J, Yu D, Yu X, Liu F, Wu J, Zeng X, Yu G and Xu M 2019 *Proc. Combust. Inst.* **37** 2733
[20] Fitriana N, Husin H, Yanti D, Pontas K, Alam P N, Ridho M and Iskandar 2018 *IOP Conf. Ser. Mater. Sci. Eng.* **334** 012011
[21] Husin H, Abubakar A, Ramadhani S, Sijabat C F B and Fikri H 2018 *MATEC Web Conf.* **197** 09008
[22] Husin H, Mahidin and Marwan 2011 *Reaktor* **13** 254
[23] Musa I A 2016 *Egypt. J. Pet.* **25** 21
[24] Chuah L F, Yusup S, Aziz A R A, Bokhari A, Klemes J J and Abdullah M Z 2015 *Chem. Eng. Process.* **95** 235
[25] Ayetor G K, Sunnu A and Parbey J 2015 *Alexandria Eng. J.* **54** 1285
[26] Sakdasri W, Sawangkeaw R and Ngamprasertsith S 2015 *Energ. Convers. Manage.* **103** 934

# Landfill Leachate Management in Indonesia: A Review

**N Emalya[1], E Munawar[2], W Rinaldi[2] and Y Yunardi[2,3*],**
[1] Postgraduate Program in Chemical Engineering, Syiah Kuala University, Banda Aceh, Indonesia
[2] Department of Chemical Engineering, Syiah Kuala University, Banda Aceh 23111, Indonesia

E-mail: yunardi@unsyiah.ac.id

**Abstract.** Municipal solid waste has become a major environmental issue in Indonesia, due to the increase of population and quantity of waste, leachate production, lack of management and lack of public awareness. Landfill leachate generation and its treatment have been one of the main focus in the landfill management. Normally, leachate is formed through the seepage of water by external sources such as drainage, rain water and so on through the garbage in the landfill. Leachate contains suspended and dissolved solid pollutants, chemicals both organic and inorganic of high concentrations such as ammonia, nitrates, nitrites, sulfides, heavy metals, nitrogen and others. With such high concentration of pollutants, leachate is highly potential to pollute the groundwater and the environment. This paper presents an overview of landfill leachate management in Indonesia, discussing the characteristics of the leachate from different landfills throughout the country on the basis of the reported literatures, current practice of landfill leachate treatment in the country, and technological development of leachate treatment for possible implementation in the future.

## 1. Introduction

With the increase of population, the changes in consumption patterns and the amount of waste, domestic solid waste in urban area has become one of the main environmental problems in Indonesia. Most of domestic solid wastes will end up at landfills, either managed or unmanaged ones. Wastes in landfills consists of not only solid but also liquid. Liquid waste so-called leachate contains organic and inorganic compounds and some pathogenic bacteria. Leachate is produced when external moisture or water enters the landfill and in contact with the solid waste producing concentrated liquid waste. The quality and quantity of leachate varies greatly, affected by the amount of rainfall, composition, characteristics of waste, landfill age and the operation of the landfill [1]. One important issues of landfilling operation is the treatment of leachate continuously produced with different volumes and compositions depending on the season [2].

Leachate may contains suspended and dissolved solid pollutants, chemicals both organic and inorganic with high concentrations such as ammonia, nitrates, nitrites, sulfides, heavy metals, nitrogen, and others. If it is not managed properly, it could potentially pollute the environment and water bodies [3]. The contamination level at the water bodies caused by leachate discharge depends on the type and

* To whom any correspondence should be addressed.

characteristics of the leachate as well as its impacts in the short and long term [4]. Therefore, leachate management in every landfill becomes crucial to ensure that there will be no contamination to the environment coming from the discharge of untreated leachate.

Leachate management, including its treatment, is not an easy task, since it greatly varies not only in terms of quantity (volume) but also in composition, characteristics and quality [5, 6]. The Government of Indonesia, particularly the regional governments at the provincial and district levels that are responsible for the solid waste management, has not yet put priority on the leachate management. It has been well known that leachate infiltration into the soil could pollute not only the soil, but also contaminate ground water and surface water. Most of landfills in Indonesia currently do not have sufficient landfill leachate treatment facilities, due to many reasons. In case there is landfill leachate treatment facility, the operation of the facility is not run properly, because in many cases the treatment facility is not more than temporary storage tanks. Consequently, in majority the quality of leachate discharged to the water body still exceeds allowable limit set up by the national regulation.

This paper presents an overview of landfill leachate management and also discusses the characteristics of leachate from various landfills in Indonesia based on the reported literature. In addition, current practice of landfill leachate treatment in Indonesia is also presented, followed by issue of possible and available landfill leachate treatment technology that might be implemented in the future.

## 2. Landfill Leachate Regulations and Characteristics

The Government of Indonesia (GOI) has set up a regulation regarding the solid waste management, including landfill leachate management, through Act No. 18/2008. The Act stated that every city or district must have an appropriate landfill that meet the technical and environmental requirements at the latest by 2013 [7]. However, until 2019 there are still many landfills in Indonesia that are operated as an open dumping and do not meet environmental principles. With the open dumping system, it would be impossible to expect that the landfill will have a leachate management system. Consequently, leachate will infiltrate to the soil and ground water without any prior treatment.

Since Indonesia is a tropical country having a rainy season for about six months, between October until March, leachate will be produced from a landfill at higher quantity during this period [8, 9]. Even though the dry season is expected to occur during the mentioned period, in different parts of Indonesia rainy season may occur in different months outside of the mentioned period. In addition, average humidity in Indonesia is quite high even in the dry season. Consequently, leachate from a landfill will be produced throughout the year. In relation to its regulation in the country, the Government of Indonesia issued Ministry of Environment and Forestry Regulation No. P59 2016. According to this regulation, each landfill site has to implement leachate management, stating among others: ensuring the leachate is processed in such a way that it meet the standard quality for discharging to the water body, applying the liners to ensure the leachate will not infiltrate to the environment, and periodically evaluating the quality of the leachate discharged to the water body, at least once a month [10]. Unfortunately, still many landfills are unable to fulfill all requirements set up by the regulation. The main reason is that the solid waste management in a region is under the responsibility of the regency or municipality. Regency or municipality intention to sort out this problem depends on the budget priority, as this issue is not included in high priority. Therefore, leachate produced from the landfill might directly flow to the environment without any prior treatment.

**Table 1**. Leachate quality standards [10]

| Parameter | Highest Level | |
|---|---|---|
| | Value | Unit |
| pH | 6-9 | - |
| BOD | 150 | mg/L |
| COD | 300 | mg/L |
| TSS | 100 | mg/L |
| N total | 60 | mg/L |
| Mercury | 0.005 | mg/L |

| | | |
|---|---|---|
| Cadmium | 0.1 | mg/L |

Table 1 presented the quality standard of leachate that allowable to be discharged to the water body in accordance to the regulation of the Government of Indonesia. Through a survey from a number of literatures, the characteristics of leachates produced from various parts of Indonesia, as depicted in Table 2, are generally high in BOD and COD. Although the pH of the leachates is in the range of allowable standard, the BOD and COD are mostly above the highest recommended levels. It is clear that those leachates have to be treated prior to discharging to the water body.

**Table 2**. Leachate quality data from several landfill Indonesia

| **Landfill, District/City/Region** | **pH** | **BOD (mg/L)** | **COD (mg/L)** | **References** |
|---|---|---|---|---|
| Pekanbaru city landfill | 7.6 | 532.85 | 3750 | [11] |
| Air dingin landfill, Padang | 7,7 | 152.2 | 304.3 | [12] |
| Jatibarang landfill, Semarang | 8.5 | 1600 | 4000 | [6] |
| Supit urang landfill, Malang | ** | 88.18 | 298.67 | [8] |
| Tidal landfill, Banjarmasin | 7.98 | 432 | 6518 | [13] |
| Blondo landfill, Semarang | 8.4 | ** | 1781.4 | [14] |
| Benowo landfill, Surabaya | 7.9 | ** | 5044 | [15] |
| Cipayung landfill, Depok | 7.83 | 3959.6 | 6860 | [16] |
| Sukawinatan landfill, Palembang | 7.45 | 145.7 | 291.1 | [17] |
| Kaliori landfill, Purwokerto | ** | 300 | 670 | [18] |
| Talang gulo landfill, Jambi | ** | 734 | 2800 | [19] |
| Antang tamangapa landfill, Makassar | ** | 729 | 1277 | [20] |
| Klotok landfill, Kediri | 7.89 | 258.8 | 143.5 | [21] |
| Regional Bangli landfill, Bali | ** | 94.55 | 317.44 | [22] |
| Quality standard | 6-9 | 150 | 300 | |

**) Data not available

**Table 3**. Characteristics of landfill leachate [8]

| **Type of leachate** | **Young** | **Intermediate** | **Stabilised** |
|---|---|---|---|
| **Age of landfill (year)** | **<2** | **2-10** | **>10** |
| pH | <6.5 | 6.5 – 7.5 | >7.5 |
| BOD/COD | 0.5-1.0 | 0.1-0.5 | <0.1 |
| COD (mg/L) | >15000 | 3000-15000 | <3000 |
| $NH_3$-N (mg/L) | <400 | NA | >400 |
| TOC/COD | <0.3 | 0.3-0.5 | >0.5 |
| Kjeldahl nitrogen (mg/L) | 100-2000 | NA | NA |
| Heavy metal (mg/L) | >2 | <2 | <2 |

NA = Not Available

The characteristics of leachate generated from a landfill influenced by many factors, including type of solid wastes dumped in the landfill, degree of compaction in the landfill, climatic and hydrogeological condition of the landfill site, pH, age of the landfill, and chemical and biological processes occur during the degradation. Table 3 presented the characteristics of the leachate on the basis of the landfill age. Variations of leachate quality are clearly seen at different stages of landfill operation. At young age, say less than five years, leachate contain high contains high biodegradable organic matter, as indicated by high ratio of BOD/COD. The ratio of BOD/COD can be used as an indicator to determine the age of the solid waste landfill. It has been well known that the BOD levels decrease with age faster than COD because of fast breakdown of the biodegradable refuse. As a consequence, the ratio of BOD/COD of the leachate will decrease with age and can be used to indicate the age of the waste in the landfill. A typical value of BOD/COD ratio of leachate generated from young landfill is between 0.5 to 1.0 as shown in

Table 3. Any leachate or wastewater having a ratio of BOD/COD of greater than 0.5 would be easily treated using biological processes. Therefore, biological means would be appropriate to treat the leachate generated from young landfill. As the ratio of BOD/COD decreases with the age of the landfill, the leachate may contain organic matters that are more difficult to break by biological means. As in the case of medium age landfill, in which the BOD/COD ratio is between 0.1 to 0.5, the treatment of leachate by biological process is still possible utilizing the acclimated microorganisms. In most cases, leachate having the ratio of BOD/COD of more than 0.1 and less than 0.5 still exists in the landfill operated up to 15 years. However, in the case of stabilized landfill having leachate with a ratio of BOD/COD of less than 0.1, biological process is no longer effective to treat such leachate, since it already contains toxic substances which inhibit the microbial activity. Consequently, such leachate would require physical-chemical for its stabilization. Changes in treatment options of leachate on the basis of the landfill age has unfortunately been considered in the design and operation of solid waste landfill in Indonesia.

## 3. Current Practice of Landfill Leachate Treatment in Indonesia

In general, leachate treatment processing facilities in Indonesia have not yet operated in accordance with existing required technical criteria. Lack of funding allocated to operate and maintain the facility is the main issue. Actually, the fund allocated for the solid waste management in a city or regency is very limited, causing the fund for the operation and maintenance of the leachate treatment facility is even smaller. As the responsibility solid waste management, including leachate treatment, is under the local government, up to now such management is not listed in the priority. This is the reason for limited allocation of funding. The next issue is limited competent personnel to operate the facility. In most landfill leachate treatment facilities in the country, the operators do have not background on biological process knowledge while the facility involves with controlling microorganisms to grow optimally at specific conditions. Another problem is that no appropriate control and monitor in the operation of the facility. As a consequence, it would be impossible to get microbes to grow optimally in the system. The most important issue is that there is almost no inspection from the local environmental authority to monitor and evaluate if the operation of the landfill leachate has fulfilled the criteria and the treated leachate has met the quality allowable by the regulation.

In Indonesia, almost all leachate treatment facilities employ purely biological process without considering the characteristics of leachate, very few apply a combination of biological and chemical process. The chemical process is usually an additional unit installed after the operation of the leachate treatment facility and it is not included either in the initial design or construction. Since the chemical unit requires high operational cost, eventually its operation is terminated due to lack of fund. Basically the facility consists of three stages of biological process, namely anaerobic stabilization, facultative and maturation. In addition to these three units, in some facilities another treatment in the forms of biofilter or wetland is added [4,6]. The fundamental reason to install biological process for the treatment of leachate is mainly because it is relatively simple and cheap to operate, which is not always true. This might be right if the quantity and the quality of the incoming leachate are relatively stable throughout the years. Unfortunately, as a tropical country having rainy and dry seasons, it would be very difficult to have a stable leachate composition over time, even though there is possibility to keep the incoming quantity at a constant rate. The anaerobic pond in the facility acts as a pretreatment to reduce the high BOD level and suspended solid of the incoming leachate. The rate of anaerobic biodegradation is faster in the hot region, like Indonesia. The facultative pond represents to further degrade the organic substances in the leachate from the anaerobic pond. The treatment of leachate in this pond involves both anaerobic and aerobic processes. In the upper part of the pond, the aerobic process dominates as the zone is rich in oxygen, while the lower part is regarded as anaerobic zone. The last stage is the maturation pond which functions to treat the incoming leachate from the facultative pond and is usually called as the stabilization pond. In addition to further reduce the suspended solids and BOD, this stage is also expected to remove the pathogenic microbes through rapid changes in the condition and high pH. The degradation is achieved through activities of aerobic microbes and algae [4].

Although the biological processes for the treatment of landfill leachate are relatively simple and inexpensive to operate, they pose a number of drawbacks. The main shortcoming of using such systems is long residence time, besides slow degradation rate and generation of bad smell. An aerobic pond, for example, requires a residence time of 20-50 days with a BOD removal efficiency of 50-85%., while facultative pond needs 5-30 days with 70-80% of BOD removal efficiency and maturation pond takes 7-20 days to remove 60-89% of BOD [23]. As a consequence, the construction of all ponds demands a vast of land area. In most cases, the quality of leachate treated through these processes still can not meet the standard allowable to be discharged to the water body. Therefore, in some cases in the leachate treatment facilities in Indonesia, an additional chemical treatment unit is installed but failed to sustain due to the operational cost constraint.

## 4. Technological Development on Leachate Treatment

Leachate management can be performed in combination of three following ways, namely isolating the landfill to minimize water or liquid from external sources to enter the landfill, recycling the leachate to the landfill, and treating the leachate produced by the landfill [24]. With regard to the landfill leachate treatment, it is also not possible to operate with only single of physical, chemical or biological process to meet the standard for discharge to the water body [25]. Since leachate characteristics vary due to season and age of landfill, among others, such variation in quantity and quality has to be considered in the design and operation of a landfill leachate treatment. Although biological treatment processes (aerobic, anaerobic and anoxic) are widely used in Indonesia to remove biodegradable compounds [26], it has been reported that the effluent quality has not yet met the standard set up by the regulation [27,28,29]. There have been numerous reports describing good practices of landfill leachate treatment that could be implemented in Indonesia.

Recycling leachate back to the landfill has been regarded as a part of a good treatment practice in the landfill management, because it is easy to install and control with reasonable operation cost. Leachate recirculation will preserve the moisture of solid waste in the landfill, promoting the degradation of organic compounds [30] and reducing the level of COD in leachate [31]. In a pilot test trial, COD removal using this method could reach 63-70% [32]. Recirculation leachate back to the landfill not only improves leachate quality, but also shortens the time needed to stabilize landfills [33]. However, if the recirculation rate is too high it will affect the anaerobic degradation of solid waste [31] and will cause problems in the saturation process [34].

Another possible inexpensive treatment is to combine leachate with domestic wastewater [35]. Leachate and domestic wastewater are channeled to the conventional waste treatment units. However, this method becomes ineffective in the case of too high organic compounds in the leachate which can inhibit the activated sludge process and increase the effluent concentration [32]. As an alternative, pre-treatment needs to be done to remove the nitrogen from the leachate and phosphorus from domestic wastewater. The decrease of COD and $NH_3$-N are in line with the increasing volumetric ratio of landfill leachate / domestic wastewater. Using a Sequencing Batch Reactor (SBR) system, with volumetric ratio of leachate and domestic wastewater of 9:1 would be able to eliminate BOD levels up to 95% and nitrogen up to 50% on a daily basis [32].

Leachate treatment using biological process is the most employed method [36]. Biological process of leachate can be performed with three approaches, namely aerobic, anaerobic and anoxic. The biological process treatment of leachate treatment is very effectively used to remove biodegradable compounds. The process is able to remove up to 50% COD from leachate of a new landfill. However, the process is no longer effective for old landfills with a BOD / COD ratio of <0.1 [2].

Research on biological treatment of leachate has been carried out by a number of researchers [23] [43]. The biological process of leachate requires pre-treatment to eliminate the levels of ammonia and organic compounds that may inhibit the microbial growth in the aerobic process [37]. Leachate treatment with anaerobic process using bio-filter of 8 day residence time can eliminate organic pollutants up to 90% [23].

Constructed wetland is another biological method that can be employed for leachate treatment. Such system is designed and arranged involving vegetation, media and microorganisms. Water plants effective for leachate treatment are those having long and stringy roots. *Limnocharis flava, Ipomoaea aquatic*, and *Scirpus validus* plants have been found to be effective to eliminate nitrogen levels up to more than 90% at concentrations of 112 - 610 mg/L. In other study it was shown that *Cyperus haspan, Cyperus alternifolius, Eriocaulon sexangulare* and *Typha domingensis* are effective in reducing BOD and COD up to 60% with the influent BOD concentrations of 514 - 686 mg /L and COD of 923.4 - 1279 mg / L [38].

For several years, flotation has been widely used to reduce the level of colloids, ions, macromolecules, microorganisms and fibers. Research on the use of the flotation method was carried out as a post-treatment of leachate to remove the remaining humic acid, a compound that is difficult to degrade. Under optimized condition, almost 60% of humic acid can be removed from leachate using this method [39].

The use of coagulation and flocculation methods will be effective for the treatment of leachate from old landfills having the BOD/COD of less 0.5 or 0.1 [40]. This method is widely used as a pre-treatment before the biological process or reserve osmosis, or as a final leachate treatment process to remove organic materials that are difficult to degrade. In general, the coagulants usually used in the leachate treatment include aluminum sulfate, ferrous sulfate, ferric chloride and ferric chloride sulphate.

Water stripping is the most common waste treatment method employed to remove high concentration $NH_4^+$-N in leachate. Leachate usually contains high levels of ammonium which can enhance wastewater toxicity [41]. This method will work effectively if the leachate has a high pH value and the used gas is first contacted with H2SO4 or HCL.

Membrane technology has been adopted in various areas of applications, including in the wastewater treatment. membrane is a selective thin barrier allowing relatively free passage of one component while retaining another. Membrane filtration methods utilized in leachate treatment include microfiltration, ultrafiltration, nano-filtration and reserve osmosis. However, membrane filtration methods cannot be used without any pre-treatment or in combination with other membrane processes [42]. Microfiltration is commonly used to retain microorganisms, small particles, large emulsion droplets and colloids [43]. Ultrafiltration is a separation process that utilizes a pressure up to 10 bar [42]. This method can be used to remove macromolecules in the leachate that have the potential to damage the osmotic membrane. Nano filtration can be used to remove organic compounds and heavy metals in leachate [44]. Nano filtration has a looser structure, allowing higher flux and lower operating pressure in the leachate treatment. This method is able to remove particles having weight of more than 300 g/mole through electrostatic interactions between ions and membranes [31].

## 5. Conclusions

On the basis of the literature review on the landfill leachate management in Indonesia, the followings can be concluded:

1. Most of solid waste landfills in Indonesia produce leachates having the strength that are above the standard quality allowable to discharge to the water body, requiring the treatment prior to discharging to the environment.
2. The regulation in Indonesia requires every landfill has to completed with leachate treatment system. However, the design of the leachate treatment system is almost the same throughout the country with lack consideration on the characteristics of the leachate.
3. Although there is a treatment system in most of solid waste landfill in Indonesia, the use of single method of treatment (biological process), poor management, lack of funding, unskilled personnel, and lack of monitoring have caused the failure of the landfill leachate treatment to produce the effluent quality to meet the standard allowable to be discharged to the environment.
4. For future development and implementation, the issue of leachate has to be started from the design by considering various aspect, so the system may consist combination of different approaches, biological, physical and/or chemical processes.

**Acknowledgement**
The authors wish to express their deep gratitude to the Syiah Kuala University for providing financial support for this study through a scholarship given to the first author, under Syiah Kuala University Excellent Research for the Doctoral Acceleration Program.

**References**
[1] Malita Y A, Ratnawulan and Mufit F 2015 *Pillar of Physic.* **5**
[2] Gao J, Oloibiri V, Chys M, Audenaert W, Decostere B, He Y, Langenhove H V, Demeester K and Hulle S W H V 2015 *Rev Environmental Science Biotechnology.* **14**
[3] Vaccari M, Tudor T and Vinti G 2019 *Waste Management.* **95**
[4] Irhamni, Pandia S, Purba E and Hasan W 2017 Tesis. Pascasarjana Unsyiah
[5] Kulikowska D and Klimiuk E 2008 *Bioresource Technology.* **99**
[6] Rezagama A, Hadiwidodo M, Purwono P, Ramadhani N F and Yustika M 2016 *Teknik.* **2**
[7] GoI, Act No. 18 of 2008 of the Republic of Indonesia on Waste Management, Jakarta, Indonesia
[8] Saleh C and Purnomo H 2014 *Jurnal Teknik* Pengairan. **1**
[9] Harjito, Suntoro, Gunawan T and Maskuri M 2018 *IOP Conference Series: Earth and Environmental Science.* **129**
[10] MoE 2016 Ministry of Environment and Forestry Regulation the Republic of Indonesia Act 59 of 2016, Jakarta, Indonesia.
[11] Daryat F, Zul D and Fibriarti B L 2017 *Jurnal Riau Biologia.* **1**
[12] Sari R N and Afdal 2017 *Jurnal Fisika Unand.* **6**
[13] Irfa'i, Yanuwiadi B and Suyadi 2016 *Resources and Environment.* **2**
[14] Rachmad M S, Handayani D S and Rezagama A *Program Studi Teknik Lingkungan Universitas Diponegoro, Semarang*
[15] Kawai M, Pueranti I F, Nagao N, Slamet A, Hermana J and Toda T 2012 *Journal of Environment Management.*
[16] Noerfitriyani E, Hartono D M, Moersidik S S and Gusniani I 2017 *IOP Conference Series: Earth and Environmental Science* **106**
[17] Yusmartini E S, Setiabudidaya D, Ridwan, Marsi and Faizal 2013 *Journal of Physics* **432**
[18] Candra A and Fadlin 2017 *Teknosia* **3**
[19] Adriansyah E, Agustina T E and Arita S 2019 *Indonesia Journal of Fundamental and Applied Chemistry* **4**
[20] Arifin M F, Imran A M, Ramli M and Hatta M P 2015 *International Journal of Engineering and Science Application* **2**
[21] Isnadina D R M, Adi M P and Ardiyanto I 2019 *IOP Conference Series: Earth and Environmental Sciences* **245**
[22] Widyarsana I M W, Damanhuri E, Agustina E and Aulia R N 2019 *International Journal of GEOMATE* **17**
[23] Said N I and Hartaja D R K 2015 *JAI* **8**
[24] Purwanta W 2007 *JAI* **3**
[25] El-Gohary F A, Khater M and Gamal K M *Journal of Applied Sciences* **94**
[26] Ismail T, Tarek D, Mejdi S, Amira B Y, Murano F, Neyla S and Naceur J 2011 *Biosources Technology* **102**
[27] Klauck C R, Rodrigues M A S and Silva L B 2015 *Braz. J. Biol* **75**
[28] Oller I, Altinbas M, Koyuncu I, Arikan O and Gomec-Yangin C 2011 *Science Total Environment* **409**
[29] Zhao R, Novak J T and Goldsmith C D 2012 *Water Res* **46**
[30] Hayati A P, Emalya N, Munawar E, Schwarzbock T, Lederer J and Fellner J 2018 *IOP Conference Series: Materials Science and Engineering* **334**
[31] Abbas A A, Guo J S, Lui Z P, Pan Y Y and Al-Rekabi W S 2009 *American Journal of Applied Sciences* **6**

[32] Renou S, Givaudan J G, Poulain S, Dirassouyan F and Moulin P 2008 *Journal of Hazardous Materials* **150**
[33] Reinhart D R and Al-Yousfi A B 1996 *Waste Management and Research* **14**
[34] Ledakowicz S and Kaczorek K 2004 *Journal of Environmental Sciences and Health* **39**
[35] Ahn W Y, Kang M S, Yim S K and Choi K H 2002 *Desalination* **149**
[36] Said N I, Tenologi Pengolahan Limbah Cair Dengan Proses Biologis.
[37] El-Gohary F A and Kamel G 2016 *Ecological Engineering*
[38] Fajariyah C and Mangkoedihardjo S 2017 *Jurnal Teknik ITS* **6**
[39] Lippi M, Ley M B R G, Mendez G P and Junior R A F C 2018 *Ciencia e Natura, Santa Maria* **40**
[40] Silva A C, Dezotti M and Sant'Anna G L 2004 *Chemosphere* **55**
[41] Marttinen S K, Kettunen R H, Sormunen K M, Soimasuo J A and Rintala 2002 *Chemosphere* **46**
[42] Mojiri A, Aziz H A and Aziz S Q 2013 *International Journal of Sciences and Research: Environmental Sciences* **1**
[43] Ameen E S M, Muyibi SA and Abdulkarim M I 2011 *Environmentalist* **31**
[44] Calli B, Mertoglu B and Inanc B 2005 *Chemosphere* **59**

# Waste Peel of Durian as Solid Cataysts for Biodiesel Production

**F Fitriani[1], H Husin[1,2*], M Marwan[1,2], F Nasution[1], Z Zuhra[2], T M Asnawi[2] and H Hisbullah[2]**
[1] Graduate Program in Chemical Engineering, Universitas Syiah Kuala, Banda Aceh 23111, Indonesia
[2] Department of Chemical Engineering, Universitas Syiah Kuala, Banda Aceh, 23111, Indonesia

E-mail: husni_husin@che.unsyiah.ac.id

**Abstarct.** In this study, peel durian ash was used as a solid catalyst to convert rubber seed oil into biodiesel production via transesterification reaction. The catalyst was fabricated by simple burning and calcined at 600°C for 8 h. The morphology of particles was confirmed by Scanning Electron Microscopy (SEM). The Energy-dispersive X-ray Analysis (EDX) was used to obtain the atomic composition and percentage of the catalyst. The reaction was carried out at 65°C for 1 h. Whereas, the catalyst loading were varied of 1.0%, 2.5%, 5.0%, 10% and 15% wt.%. From EDX data was confirmed that the ash catalyst composes of the potassium oxide ($K_2O$) as the dominant compound. The performance of the catalyst was evaluated from a transesterification reaction for biodiesel production. The highest biodiesel yield of 96.5% was obtained in the reaction using a catalyst of 10 wt.%. The density obtained was 0.865 gr/cm$^3$ confirming the produced biodiesel complies with the SNI standard.

## 1. Introduction

The widespread using of fossil fuels in power plants, transport vehicles, generators, and mining equipment has required higher energy consumption [1]. The demand energy in the transportation sector has hiked with the growth of the population and it is expected to keep on increasing. This has led to a rapid depletion in the world's reserves of fossil fuels. The environmental pollution generated by fossil fuels is currently increasing [2-3]. So that we need to look for other energy sources as alternative fuel which possess all the properties such as renewability, accessibility, sustainable nature, and clean fuel that can meet all the challenges caused by fossil fuels, one of it is biodiesel [4-7].

Biodiesel right now has been used in most countries in the world. Triglycerides that generally exist in vegetable oil (edible or non-edible) or animal fats are reacted with methanol besides either alkali or acid catalyst for industrial-scale manufacturing of biodiesel. The biodiesel can be produced from various natural sources including waste cooking oil, jatropha oil, palm oil, neem oil, soybean oil, rapeseed oil and other vegetable oils. The economics of biodiesel production is highly sensitive feedstock and it comprises a very substantial portion of overall production cost. Hence, the use of non-edible oil sources or waste products of the edible oil industry as biodiesel feedstock to adjust the

* To whom any correspondence should be addressed.

international standards is a recently focused area of research. In this research, the rubber seed oil is used as a raw material for biodiesel production because easy to cultivate [8-17].

To accelerate the transesterification reaction of vegetable oils into biodiesel, various catalysts have been used. The catalyst could be significantly affected the reaction rate and also may favor the formation of the product. There are three types of catalysts that are often used, namely homogeneous, heterogeneous and biocatalyst. The use of heterogeneous catalysts is more preferred than a homogeneous catalyst. The homogeneous catalyst still has a limitation on the separation process that leads to an increase in the total cost of biodiesel production. The heterogeneous catalysts offers more advantages, such as easily separation, recyclable, high glycerol purity without corrosion in the reactor. Several heterogeneous catalysts have been studied for the conversion of vegetable oils into biodiesel, such as SrO, CaO, $TiO_2$-based, $K_2O$, showing relatively high activity [18-22].

Generally, the exploration of solid catalysts based on non-renewable sources. Durian wastes as a renewable source can be used as a solid catalyst due to it contains $K_2O$ compound. When the durian peel was burned at a moderate temperature, it would produce an ash residue. The inorganics parts of the ash residue are $K_2O$ composition as the dominant component. The present work aimed at exploring the possibility of using durian peel as a source of $K_2O$ as a catalyst for biodiesel production. The ash catalyst was characterized by SEM and EDX. The effect of catalyst loading to oil ratio was also evaluated.

## 2. Methodology

### *2.1. Material*
Durian peel is obtained from waste durian, in Banda Aceh City, Indonesia. Methanol was provided by Merck (Germany), and rubber seed oil was purchased from the market. The rubber seed was obtained from a rubber farm in East Aceh, Indonesia. The oil was extracted by pressing the rubber seed using a mechanic press.

### *2.2. Preparation of Catalyst*
Waste durian peel was dried in open area. After drying, durian peel was cut and crushed using a mechanical crusher. The durian peel was put in stainless steel plate and burned until forming ash. The ash of durian peel was calcined at 600 °C for 8 hours. Then the catalyst was cooled at room temperature.

### *2.3. Catalyst Characterization*
Analysis of catalysts was carried out to determine its morphology and chemical composition. The catalysts were analyzed by scanning electron microscopy (SEM). The catalyst element were characterized by energy dispersive X-ray (EDX).

### *2.4. Transesterification Reaction*
The biodiesel production was carried out via transesterification reaction of rubber seed oil, ash catalyst, and methanol. In each experiment the catalyst were varied of 1.0%, 2.5%, 5.0%, 10%, and 15 wt.% to oil. This reaction was conducted in a three-neck round-bottom flask batch reactor equipped with a heater, magnetic stirrer and a water-cooled condense [23]. The rubber seed oil and methanol put into a three-neck flask. Whereas the catalyst was varied of 1.0%, 2.5%, 5.0%, 10%, and 15 wt.% to rubber seed oil. The reaction was kept at a temperature of 65°C for 1 h. The generated biodiesel was separated from the catalyst residue and glycerol using separating funnel and filter paper. After being separated, the biodiesel was then dry-washed in an oven at 80°C for 12 h. The methanol to oil molar ratio was 8:1. The yield of biodiesel was determined using Eq. (1):

$$\%\ \text{yield} = \frac{\text{weight of biodiesel produced}}{\text{weight of oil used}} \times 100\% \quad (1)$$

IOP Conf. Series: Materials Science and Engineering **845** (2020) 012033 doi:10.1088/1757-899X/845/1/012033

## 3. Results and discussion

*3.1. Scanning Electron Microscopy (SEM) and Energy-dispersive X-ray (EDX) Analysis*
The morphology of the durian peel ash catalyst was carried out to determine by SEM analysis, as shown in Figure 1. As can be seen from SEM analysis, the surface morphology of the ash durian peel is inhomogeneous particle size. The shape of the particles of around 300 nm - 2μm. This irregular shape of the ash particle might be due to an agglomeration when burning the durian peel.

**Figure 1.** Morphology of durian peel ash

To identify the compounds in the ash catalyst, EDX was employed to confirm the atomic composition and atomic percentage of the catalyst. The results of the EDX analysis are presented in Figure 2.

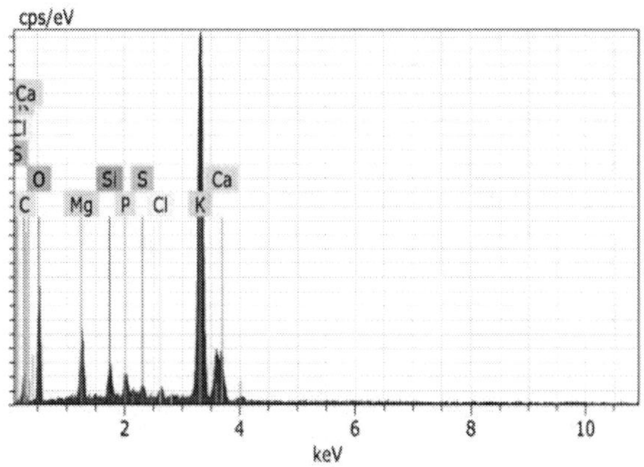


**Figure 2.** EDX analysis of durian peel ash

The spectrums of the EDX were composed of potassium, magnesium, calcium, silica, aluminium, carbon, oxygen, and phosphor, as can be seen from Fig. 1. It appears that the highest peak of the spectrum is potassium. The data analysis reported that the weight percent of potassium is 27.25%.

### 3.2. *Influences of Catalyst Loading*

Catalyst is one factor is greatly affects the effectiveness of biodiesel production [24]. In this experiment, the ratio of the catalyst to the rubber seed oil was varied of 1.0%, 2.5%, 5.0%, 10.0 and 15.0 wt.%. The effect of the weight ratio of the catalyst to the yield of biodiesel is shown in Figure 3.

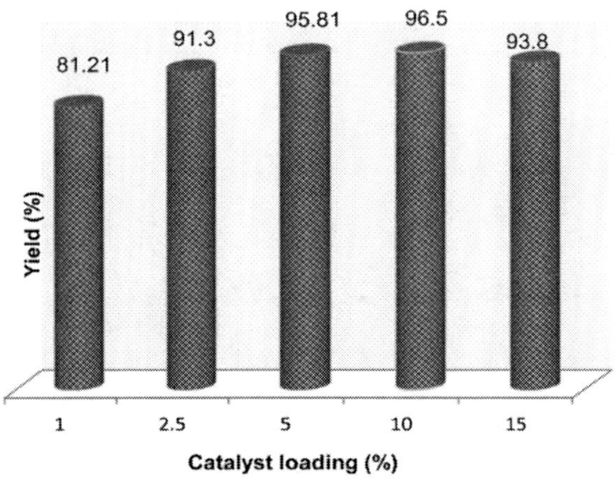


**Figure 3.** Effects of catalyst loading on biodiesel yield

As shown in Fig. 3, the yield of biodiesel tends to increase with an increase in the loading of catalyst. As catalyst loading increased from 1.0 % to 10%, the yield biodiesel also increased from 81.21% to 96.5% %. With further increase in the catalyst loading, above 10 wt%, the biodiesel yield decreased due to emulsification. With an increase in catalyst loading to 15.0 wt.%, the biodiesel yield slightly decreased to 93.8%. The excess catalyst could increase the solubility of glycerol, lead to reduced separation of biodiesel and glycerol and decreased biodiesel yield [25].

### *3.3. Biodiesel Properties*

The properties of biodiesel have been carried out to determine the physical-chemical characteristics of the process using a durian peel catalyst is shown in Table 1.

**Table 1.** Characterization of biodiesel from this study and SNI 7182: 2015

| Parameter | Unit | Result | SNI |
|---|---|---|---|
| Kinematic viscosity | $mm^2/s$ (cp) | 4.42 – 5,90 | 2.3 – 6,0 |
| Density | $kg/m^3$ | 865 - 890 | 850 - 890 |

The properties of biodiesel from ash peel durian catalysts such as kinematic viscosity and density were also analyzed. The kinematic viscosity of biodiesel is ranging from 4.42-5.9 $mm^2/s$ (cp), while, the biodiesel density of biodiesel from this study of 865-890 $kg/m^3$ which were in accordance with SNI standards. Thus, the properties of biodiesel produced in this study using heterogeneous catalysts of durian peels ash are in accordance with Indonesian National Standards. The high fuel properties cause mass injection to be higher in the engine. Because of this, its energy content in the combustion chamber and engine performance are greatly influenced by fuel properties [26-27].

## 4. Conclusion

The heterogeneous catalyst from durian peel ash has been prepared by a simple combustion process. The catalyst has been applied for the transesterification reaction of rubber seed oil. From EDX data was confirmed that the ash catalyst composes of the potassium oxide ($K_2O$) as the dominant compound. The prepared sample has been successfully produced biodiesel with a high yield of 96.5% using 10% catalyst at 65°C for 1 h with methanol to oil ratio of 8:1. Durian peel ash contains potassium oxide which owns good ability and stability makes it has the potential to be utilized as a material for catalyst production.

## 5. Acknowledgement

The author thanks to the ministry of research and technology of higher education of Indonesia for financial support through the research project Master Program.

## 6. References

[1] Liu, Z, Li F, Wang W and Wang B 2019 *J. Energy Inst.* **92** 861
[2] Husin H, Adisalamun, Sy Y, Asnawi T M and Hasfita F 2017 *AIP Conf. Proc.* **1788** 030073
[3] Husin H, Pontas K, Yunardi, Adisalamun, Alam P N and Hasfita F 2017 *ASEAN J. Chem. Eng.* **17** 27
[4] Farooq M, Ramli A, Naeem A, Noman M, Shah L A, Khattak N S and Perveen F 2019 *Int. J. Renew. Energy Res.* **43** 5438
[5] Husin H, Mahidin, Yunardi and Hasfita F 2015 *Key Eng. Mater.* **659** 231
[6] Husin H, Mahidin, Zuhra and Hasfita F 2014 *Bull. Chem. React. Eng. Catal.* **6** 81
[7] Catarino M, Martins S, Paula A, Dias S, Francisco M and Pereira C 2019 *J. Environ. Chem. Eng.* **7** 103099
[8] Reyna-Villanueva L R, Dias J M, Medellín-Castillo N A, Ocampo-Pérez R, Martínez-Rosales J M, Peñaflor-Galindo T and Fuentes G A 2019 *Fuel* **251** 285
[9] Fitriana N, Husin H, Yanti D, Pontas K, Alam P N, Ridho M and Iskandar 2018 *IOP Conf. Ser. Mater. Sci. Eng.* **334** 012011
[10] Syamsuddin Y and Husin H 2010 *J. Rekayasa Kimia & Lingkungan* **7** 112
[11] Arumugam A and Ponnusami V 2019 *Renew. Energy* **131** 459
[12] Yadav M and Sharma Y C 2019 *Energy Convers. Manag.* **198** 111795
[13] dos Santos L K, Hatanaka R R, de Oliveira J E and Flumignan D L 2019 *Renew. Energy* **130** 633
[14] Ayodeji A A, Blessing I E and Sunday F O 2018 *Data Brief* **18** 512
[15] Corro G, Flores A, Pacheco-Aguirre F, Pal U, Bañuelos F, Ramirez A and Zehe A 2019 *Fuel* **250** 17
[16] Fadhil A B, Al-Tikrity E T B and Albadree M A 2017 *Fuel* **210** 721
[17] Onoji S E, Iyuke S E, Igbafe A I and Nkazi D B 2016 *Energy Convers. Manag.* **110** 125
[18] Simões S S, Ribeiro J S, Celante D, Brondani L N and Castilhos F 2020 *Renew. Energy* **146** 719
[19] Vern R, Hua Y, Mubarak N M, Khalid M, Abdullah E C and Nolasco-hipolito C *J. Environ. Chem. Eng.* **7** 103219
[20] Zuhra, Husin H, Hasfita F and Rinaldi W 2015 *Agritech* **35** 69
[21] Husin H, Asnawi T M, Firdaus A, Husaini H, Ibrahim and Hasfita F 2018 *IOP Conf. Ser. Mater. Sci. Eng.* **358** 012008
[22] Zaki M, Husin H, Alam P N, Darmadi, Rosnelly C M and Nurhazanah 2019 *J. Rekayasa Kimia & Lingkungan* **14** 36
[23] Husin H, Marwan and Mahidin 2011 *J. Reaktor* **13** 254
[24] Roschat W, Phewphong S, Khunchalee J and Moonsin P 2018 *Mater. Today-Proc.* **5** 13916
[25] Anwar M, Rasul M and Ashwath N 2018 *Energy Procedia* **160** 268
[26] Mairizal A Q, Awad S, Priadi C R, Hartono D M, Moersidik S S, Tazerout M and Andres Y 2020 *Renew. Energy* **145** 375
[27] Refaat A A 2009 *Int. J. Environ. Sci. Technol.* **6** 677

IOP Conf. Series: Materials Science and Engineering **845** (2020) 012034 doi:10.1088/1757-899X/845/1/012034

# Effect of Polyvinyl Pyrolidone Additive on the Characteristics and Performance of Polyether Sulfone Membrane for Reducement of Fe in Water

**Sofyana***, **Y Syamsuddin, S Mulyati, F Razi**
Departement of Chemical Engineering, Universitas Syiah Kuala, Darussalam, Banda Aceh-23111, Indonesia

*Email: sofyana71@unsyiah.ac.id

**Abstract.** Decreasing the iron (Fe) content in raw water until it reaches the permitted threshold value can be done with filtration using membranes. In this study, the fabrication of membrane for Fe removal was done by using polyethersulfone (PES) as polymer and Polyvinyl Pyrrolidone (PVP) as additive. The addition of PVP aims to increase membrane swelling degree. The asymmetric membrane was prepared by phase inversion method with composition 18 wt% of PES and PVP with a concentration variation of 1; 1.5, and 2 wt%. Variations in the concentration of PVP added to the membrane influence the resulting membrane. Increasing the concentration of PVP shows an increase in the value of permeability, SEM results from modification membranes also show the different morphological structures. The results of the FTIR analysis show that the fabricated membranes have different functional groups. The swelling degree of PES, PES/PVP1, PES/PVP1.5, and PES/PVP2 membranes are increased with a value of 153 to 323%, while the overall porosity value is 62 to 142%. Membrane performance on iron content with 30 ppm feed concentration gives an optimum rejection value of 98.5%.

## 1. Introduction

Metal is a natural substance found in water sources. Metals enter the source of water naturally when rainwater seeps through the mountains where a number of metals will be dissolved into the water. This water will enter large water sources and will be used by the community for various purposes. Iron (Fe) is a metal that is naturally present in water [1]. Water source receive iron either through geogenic sources or through domestic waste deposits or through industrial waste [2]. The sources of iron in surface water are mainly pollution from iron and steel industries, mining and metal corrosion [3]. Aside from surface water, iron is also found in groundwater. Some countries have high iron content in groundwater sources, such as Indonesia, which in some regions has water sources with iron levels above the maximum threshold. Iron content in surface and groundwater varies from 3 to 4 mg/L to 15 mg/L [4]. Limits allowed for drinking water is 0.3 mg/L [5,6]. Water containing iron is generally brownish yellow and stinking, high iron content in the water can be felt when the water is drunk or can be seen from the onset of rust on equipment and clothing [7].

Removal of iron metals from water can be done by various methods both conventional methods such as filtration-precipitation-oxidation, zeolite softening/ion exchange, filter media separation, removal through supercritical fluid, wetland treatment, electrocoagulation/flotation, aeration, and with

* To whom any correspondence should be addressed.

membrane technology [1]. Membrane technology is a process that continues to develop and is widely applied in water treatment containing metals. Membrane separation technology is an applicable technology that is preferred because it does not produce harmful by-products, easy to operate, low energy consumption and does not use significant chemical additives. [8].

Polymeric membranes are often used in the process of Reverse Osmosis, Nanofiltration, Microfiltation Ultrafiltation and Gas Separation. Organic membranes are usually made from synthetic polymers such as polysulfone, polyethersulfone, cellulose acetate, or polyamide [9,10]. Recently polyethersulfone is often used because of some interesting properties, this is indicated by thermal stability, thermal resistance, chemical resistance to acids, petrol, oils and oxidizing agents such as flour and hydrogen peroxide in addition to mechanical strength, tolerance to a wide temperature range, pH, physiological neutrality, low sensitivity to UV [11,12].

The development of materials used in the manufacture of membranes continues to be done to obtain the best membrane in accordance with the application of the membrane. Membrane with a mixed matrix is one membrane that is widely used in the manufacturing process by combining several specific materials with significant characteristics into membrane structures such as titania, alumina, zinc oxide, silica, zirconia , iron oxide , polyvinyl alcohol and polyvinylpyrrolidone [13]. In this study, membranes based on Polyethersulfone polymer are modified with polyvinylpyrrolidone (PVP) as a pore forming agent which is expected to be able to remove iron metal.

## 2. Methodology

### *2.1 Materials*

Materials used in this research included Polyethersulfone was purchased from BASF, Polyvinil Pirolydine and Dymethyl Acetamide was supplied by EMSURE, $FeCl_2$ solution as an artificial feed solution represents the condition of water contaminated with Fe metal, aqua distilled deionized. All chemicals used are analytical grade, except for distilled water.

### *2.2. Preparation of Membrane*

The membrane is prepared using the Phase Inversion technique. Membrane solution is made by mixing PES, PVP, and DMAc, the mixture is stirred for 24 hours with a magnetic stirrer at a suitable speed, the homogeneous solution formed is called a dope solution. The dope solution is left at room temperature for debubling and then cast on a glass plate with a thickness of 175 μm, after casting it is immersed in a coagulation bath containing water as non-solvent until the membrane is separated from the glass plate. The membrane formed is placed in distilled water for 24 hours and then the membrane is dried by placing it between two filter paper.

The membrane is made based on the composition as shown in Table 1, where the PES concentration remains 18% and the PVP concentration is varied, whereas the concentration of the DMAc solvent follows the percentage of PVP. One is a pure PES membrane without the addition of PVP which serves as a comparison. The prepared membrane is then stored for analysis.

### *2.3. Characterization*

The characteristics of the membrane are performed before the membrane is tested for performance, including permeability test, swelling degree, and overall porosity, and also functional groups using Fourier Transform-Infrared (FTIR) (Shimadzu FTIR-Prestige 21 Series). Meanwhile, surface and cross-section morphology are analyzed with Scanning Electron Microscopy (SEM) (Jeol. Jsm - 6510 LA).

#### *2.3.1 Permeability test.*

Permeability testing is carried out using a distilled deionized water feed, by flowing pressure from nitrogen gas as a driving force to drain a disstiled deionized water feed through the membrane. Permeate volume is held at certain intervals until steady conditions are achieved. Permeability testing is carried out at pressures 2,3,4,5 and 6 bar.

**Table 1.** PES, PVP and DMAc composition in membrane preparation

| No | PES (wt%) | PVP (wt%) | DMAc (wt%) | Membrane ID |
|---|---|---|---|---|
| 1 | 18 | 0 | 82 | PES |
| 2 | 18 | 1 | 81 | PES/PVP1 |
| 3 | 18 | 1,5 | 80,5 | PES/PVP1,5 |
| 4 | 18 | 2 | 80 | PES/PVP2 |

*2.3.2 Swelling degree and overall porosity.*

To measure the degree of swelling, membrane samples of a certain size are immersed in water for 24 hours. Then dried by placing between filter paper and then weighed. After that the membrane is dried in an oven at 50 °C for 24 hours and after that the weight is weighed again.

$$\text{Swelling Degree (\%)} = \frac{(Ww-Wd)}{Wd} x\ 100 \quad (1)$$

Where Wd = dry weight (gram), Ww = wet weight (gram)

$$\text{Overall Porosity (\%)} = \frac{Ww-Wd}{A\, x\, l\, x\, dw}\ x\ 100 \quad (2)$$

Where A is effective membrane area ($cm^2$), l is membrane thickness (cm), dw is water density ( $g/cm^3$). To reduce errors, the test was repeated and the average value is shown as the data reported (Ghaemi, 2016).

*2.3.3 Scanning Electron Microscopy*

Cross-section morphology of the prepared membranes was examined by scanning electron microscopy (SEM). Samples were tested using electron beams with kinetic energy of 10 kV. Magnification of the pictures taken is 1000 times for the surface and 800 times for the crossection.

*2.4 Membrane performance testing*

Membrane performance testing using $FeCl_2$ feed solution was carried out in a stainless steel cell, holding the effective area of 10.2 $cm^2$. The experiments were carried out using compressed nitrogen gas and solutions of different concentrations (10, 20 and 30 mg/L) and all measurements were made at 6 bars. The permeate obtained was collected and the flux value was calculated using equation (3) and the permeate obtained was analyzed by Fe content using a spectrophotometer UV-VIS, Shimadzu, 1700. calculation of percentage of rejection according to equation (4) [16].

$$J_w = \frac{V}{A.\Delta t} \quad (3)$$

Where $J_w$ ($L/m^2.h$) is the flux of water, V (L) is the volume of permeate

$$\%\ \text{Removal} = 1 - \frac{Cf - C_p}{Cp} \times 100\% \quad (4)$$

Where $C_f$ and $C_p$ are feed and permeat Fe concentrations, respectively.

The permeate obtained was analyzed for Fe concentration using Spectrophotometer UV-VIS equipment.

## 3. Results and Discussion

*3.1. Permeability, swelling degree and overall porosity of membrane*

A Large porosity of the membrane will make the membrane have pores evenly distributed so that this membrane has the ability to filter unnecessary element in water [14]. It is due to when the water filtration process takes place, the membrane cavity will be filled with Fe molecules, with a large number of membrane pores, so many Fe molecules can be absorbed. And it will be able to reduce Fe content optimally in groundwater.

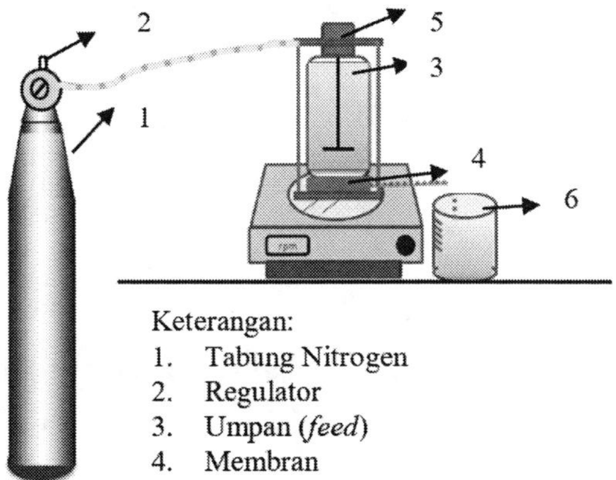


**Figure 1**. The schematic diagram of the membrane equipment

**Table 2**. Permeability of Membrane

| Membrane ID | Lp (L/m$^2$.h.bar) | Swelling degree (%) | Overall porosity (%) |
|---|---|---|---|
| PES | 0,4219 | 153,547 | 62,361 |
| PES/PVP1 | 2,8678 | 221,234 | 120,432 |
| PES/PVP1,5 | 18,817 | 271,102 | 133,680 |
| PES/PVP2 | 19,822 | 342,801 | 148,351 |

From Table 2 it can be seen that the permeability value of pure water (Lp) on the PES membrane is increasing with an increase in PVP concentration. That is because the membrane with the addition of PVP additives has more pore structure and is evenly distributed in the support layer and in the top layer or active layer, so that the feed will easily pass through the membrane compared to the membrane without the addition of PVP [15]. The membrane classified as nanofiltration is a PES/PVP1 membrane with a permeability coefficient value obtained at 2.8678 L/m$^2$.h.bar [16].

*3.2 Functional groups*

As seen in Figure 2, shows that all three membranes have similar functional groups where there are sulfone and ether groups. Sulfon group ($SO_2$) is shown at wavenumber of 1265 - 1289 cm$^{-1}$, ether group (COC) at wavenumber of 1320 - 1137 cm$^{-1}$, CH group at absorption wavenumber of 1465 cm$^{-1}$, and group C = C at group wavenumber of 1600 cm$^{-1}$ .

*3.3. Membrane Morphology*

Surface morphology analysis with SEM exhibits the difference between PES membrane and PES membrane with the addition of PVP at various variations of PVP concentration. Figure 3 (a) and (b) are the results of SEM on the surface of the PES membrane and PES with the addition of PVP, PES / PVP1 membranes have a more flat surface area, this shows a better pore distribution. The appearance of the cross-section structure shows that all membranes have asymmetric membrane characteristics that are fabricated by the phase inversion method, which has a dense top layer structure supported by fine-finger like and macrovoid sub layers. Membranes with the addition of PVP show a different structure, a better pore formation is clearly seen on the support layer (Figure 3(b) to 3(f)). The addition of PVP concentration increases the value of swelling degree and overall porosity. This is in consistent with the cross-sectional structure seen in the results of Scanning Electron Microscopy, which is more pore formed and spread. Pore formation increases membrane permeability.

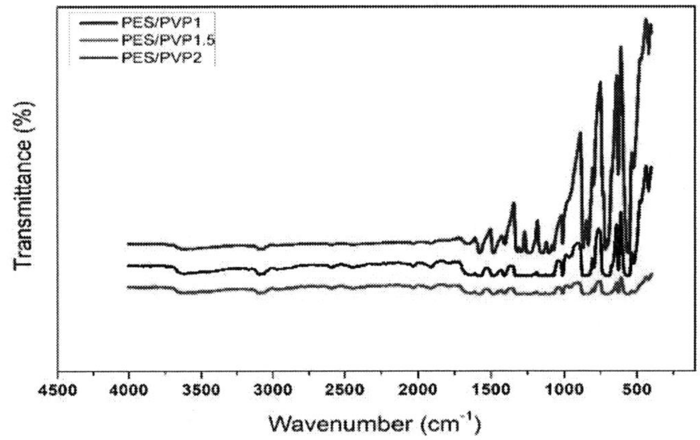


**Figure 2.** FT-IR spectrum of PES/PVP1, PES/PVP1,5 and PES/PVP2

The pure PES membrane has an active layer that is thicker than the PES/PVP membrane, so that the pure PES membrane Lp value is very low at 0.412 L/m$^2$.h.bar, different from the Lp value of the PES/PVP2 membrane which is 19.822 L/m$^2$.h.bar, where the active layer becomes thinner and the support layer forms a larger pore so that the permeability value increases.

### 3.3. *The performance of the membrane in Fe removal*

Flux is one of the most important parameters performed to determine the performance or ability of the membrane in the separation process, the measurement of flux in this study was carried out using $FeCl_2$ solution. $FeCl_2$ solution with a concentration variation of 10; 20 and 30 Mg/L are used as feed conditions in measuring flux values using the optimum pressure from permeability testing which is 6 bar. The flux value is obtained by using equation (4).

Figure 4 shows the difference in flux values between pure PES membrane and membrane after the addition of PVP additives at various concentrations of feed. The results of the flux values obtained indicate that the addition of PVP additives can increase the flux value of the modified membrane compared to the pure PES membrane, where the flux value at the 30 mg/L feed concentration of increased from 5.5 L/m$^2$.h.bar (pure membrane) to 35.1; 49.9 and 60.9 L/m$^2$.h.bar for PES/PVP1, PES/PVP1.5, and PES/PVP2 membranes, respectively. Similarly, in a feed concentration of 20 and 10 mg/L, the flux obtained also increased from 24.1 L/m$^2$.h.bar (pure membrane) into 51.1; 63.2; 69.4 L/m$^2$.h.bar for PES/PVP1, PES/PVP1.5, and PES/PVP2 membranes, respectively and from 8.3 L/m$^2$.h.bar (pure membrane) to 37.1; 59.4; and 82.3 L/m$^2$.h for PES/PVP1, PES/PVP1.5, and PES/PVP2 membranes, respectively. This is because the addition of PVP can increase the pore size and hydrophilicity of the membrane, a larger pore size will make the feed solution easier to pass through the membrane and cause an increase in the flux value.

IOP Conf. Series: Materials Science and Engineering **845** (2020) 012034 doi:10.1088/1757-899X/845/1/012034

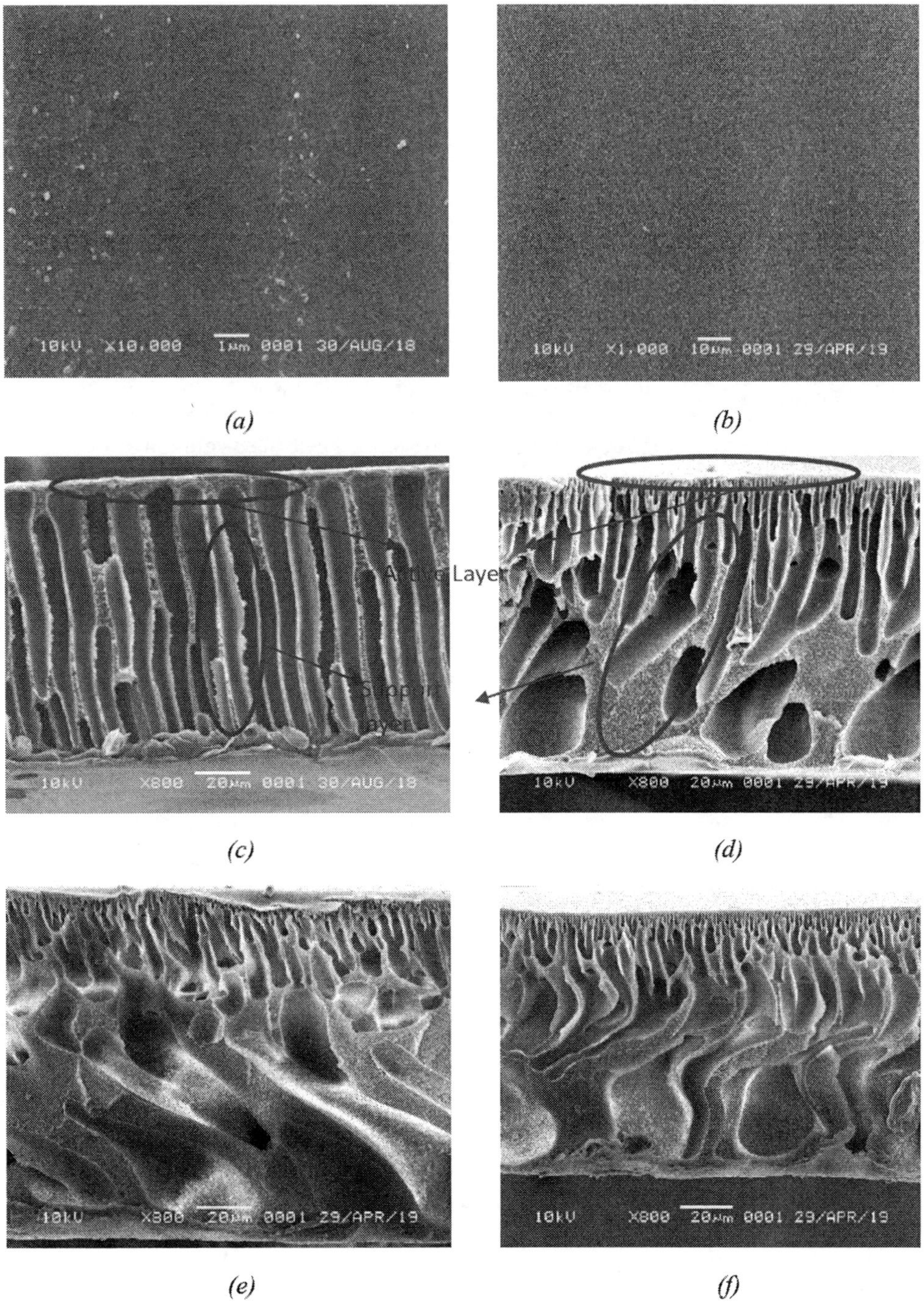


**Figure 3**. SEM image of the surface of membrane: (a) PES and (b) PES / PVP1, cross section of membrane: (c) PES, (d) PES / PVP1, (e) PES / PVP1.5 and (f) PES / PVP2

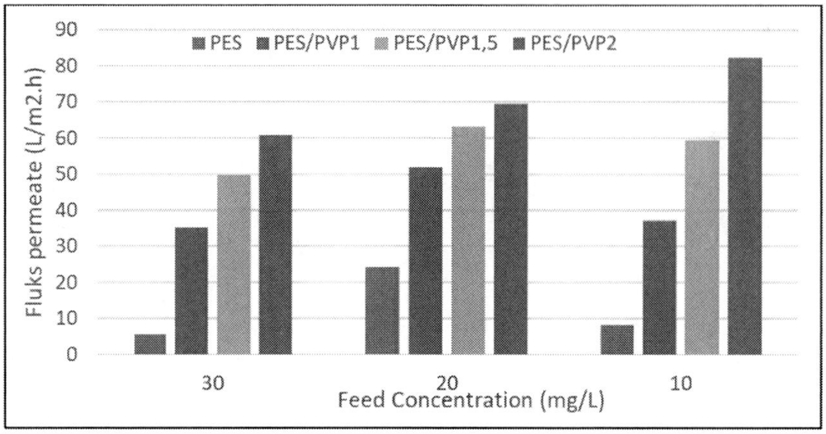


**Figure 4**. Membrane flux to the $FeCl_2$ feed solution

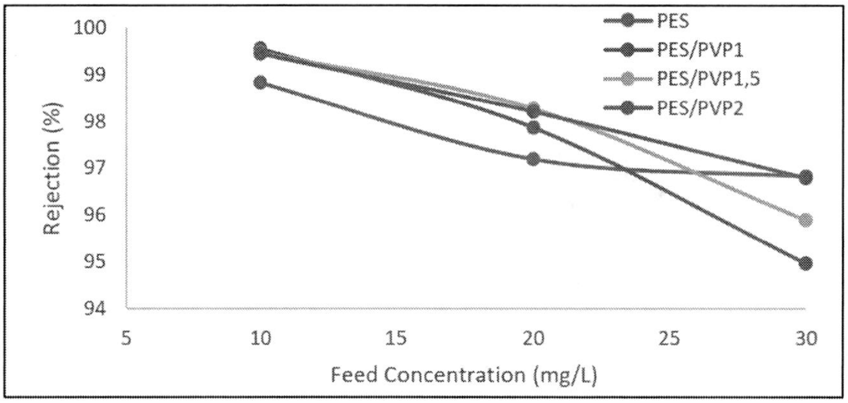


**Figure 5.** Membrane Rejection of $FeCl_2$ feed solution

Figure 5 shows the rejection value obtained after the feed solution is passed on the membrane using pressure as the driving force, the rejection obtained tends to fluctuate but is not very significant at various feed concentrations given. At a feed concentration of 10 mg/L the rejection value obtained reached 99% for the PES membrane with the addition of PVP, here the effect on the percentage of the addition of PVP did not have a significant effect. At a feed concentration of 20 mg/L the percent rejection value obtained reached 98.20% for the PES/PVP2 membrane, as well as 96.8; 94.9; 95.8 and 96.7% at a feed concentration of 30 mg/L respectively for membrane PES, PES/PVP1, PES/PVP1.5 and PES/PVP2. The greater the concentration of feed given, the smaller the rejection value obtained. This occurs because during the testing of different feeds there has been fouling or concentration build-up in the pores of the membrane, causing the value of the percentage of rejection obtained decreases (Liu et al., 2017).

## 4. Conclusion

The addition of Polyvinyl Pyrrolidone as a pore forming agent influences the structure of the Polyether Sulfone membrane by increasing the degree of swelling and porosity of the membrane thereby increasing the permeability value. As the PVP concentration increases, the permeability and flux values of the $FeCl_2$ feed solution increase, the PES membrane has a permeability value of 0.4219 $L/m^2$.h.bar different from the PES/PVP2 membrane that is 19.822 $L/m^2$.h.bar. The PES membrane with the addition of PVP has a morphological structure with a greater number of pores and the support part has a well-distributed finger-like pore. The ability of PES/PVP membranes in Fe metal recovery is quite good, which is in the range of 96 to 98%. The addition of PVP concentration did not have a major effect on the ability to recover Fe metal. Increasing the concentration of the feed solution decreases the rejection value by ± 2% for each increase in the concentration of 10 mg/L.

## 5. Acknowledgement

We would like to express our deep gratitude to Universitas Syiah Kuala for their valuable grant, under PLK program.

## 6. References

[1] Khatri N, Tyagi S and Rawtani D 2017 *J. Water Process Eng.* **19** 291.
[2] Khatri N and Tyagi S 2015 *Front. Life Sci.* **8** 23.
[3] Jusoh A, Cheng W H, Low W M, Nora'aini A, Megat Mohd Noor M J 2005 *Desalination* **182,** 347.
[4] Ellis D, Bouchard C and Lantagne G 2000 *Desalination* **130**·255.
[5] World Health Organization 1996 Iron in drinking water In: Guidelines for drinking-water quality, 2nd ed. Vol 2. Health criteria and other supporting information. Geneva.
[6] Khatri N, Tyagi S and Rawtani D 2016 *Environ. Claims J.* **28** 1.
[7] Febrina L and Ayuna A 2015 *Jurnal Teknologi* **7** 35.
[8] Arahman N, Mulyati S, Lubis M R, Takagi R, Matsuyama H. 2017 *J Water. Proc. Eng.* **20** 173.
[9] Li N N, Fane A G, Winston Ho W S and Matsuura T 2008 *Advance membrane technology and applications* (New Jersey: Wiley & Sons).
[10] Pabby A K, Rizvi S S H and Sastre A M 2009 *Membrane separations – chemical, pharmaceutical food and biotechnological applications* (New York: CRC Press Taylor & Francis Group LLC).
[11] Barth C, Goncalves M C, Pires A T N, Roeder J and Wolf B A 2000 *J. Membr. Sci.* **169** 287.
[12] Zhao W, Huang J, Fang B, Nie S, Yi N, Su B, Li H and Zhao C 2011 *J. Membr. Sci.* **369** 258.
[13] Ghaemi N, Daraei P and Akhlaghi F S 2018 *Carbohydr Polym.* **191** 142
[14] Arahman N, J. 2012 J Rekayasa Kimia & Lingkungan **9** (2) 68.
[15] Ghaemi N 2016 *Appl. Surf. Sci.* **364** 221.
[16] Muchtar S, Wahab M Y, Fang L F, Jeon S, Rajabzadeh S, Takagi R, Mulyati S, Arahman N, Riza M, Matsuyama H. 2019 *J. Appl. Polym. Sci.* **136** 47312.

# Adsorption Behaviour of Hazardous Dye (Methyl Orange) on Cellulose-Acetate Polyurethane Sheets

**M Iqhrammullah[1*], Marlina[1] and S Nur[2]**
[1] Department of Chemistry, Faculty of Mathematics and Natural Sciences Universitas Syiah Kuala, Darussalam 23111, Banda Aceh, Indonesia
[2] Department of Physics, Faculty of Mathematics and Natural Sciences Universitas Syiah Kuala, Darussalam 23111, Banda Aceh, Indonesia

E-mail: m.iqhram@oia.unsyiah.ac.id

**Abstract.** Cellulose Acetate-Polyurethane (CA-PU) copolymer has been successfully prepared by a simple blending method and examined for its adsorption capacity against Methyl Orange (MO). Functional group characterization using Fourier Transform-Infrared (FTIR) confirms the urethane linkage formation as well as the presence of several essential functional groups (such as; O-H, C=O, and aromatic ring). Surface morphology changes are also observed using Scanning Electron Microscopy (SEM), where a smoother and porous structure is obtained in CA-PU. The adsorption isotherm is examined with Langmuir, Freundlich, Halsey, and Harkin-Jura models. MO adsorption on CA-PU sheets fits Freundlich isotherm ($R^2 = 0.9722$) with $K_F = 279.45$ mg/g. Kinetics studies of the adsorption with pseudo first-order and pseudo second-order plots exhibit that chemisorption is the adsorption limiting-rate.

## 1. Introduction

Increase of anthropogenic activities, most of the time, damage the environment, one of which through the release of toxic organic dyes to the environment. Hazardous dyes pollution can affect the environment by interfering the photosynthetic activities through sunlight blocking and increasing chemical oxygen demand (COD). Though it is not particularly harmful, the persistence of these dye compounds leads to long-term exposure, contributing to the mutagenic damage and carcinogenicity for aquatic biota [1]. Azo dyes are commonly used for wide purposes, such in the textile, printing, and the pharmaceutical industries. Upon degradation, most of the azo dyes can produce aromatic amine and benzidine, which are responsible for carcinogenicity [2]. In this research, we take methyl orange as the model for the adsorptive removal of anionic azo dyes.

It is feasible to remove MO from aqueous solutions through adsorptive process. Several adsorbents have been investigated for its capability to remove MO, such as graphene oxide [3], aminated pumpkin seed powder [4], and cross-linked chitosan/β-cyclodextrin composite [5]. On the other hand, cellulose can be used as potential adsorbent material. Cellulose is a well-known abundant biopolymer, which can be obtained from various resources, including biomass. Nevertheless, cellulose performs a low adsorption capacity because of its limited functional groups that contribute to its active sites. To overcome the problem, cellulose is modified to become more functional groups-rich products [6-7].

* To whom any correspondence should be addressed.

Cellulose acetate is the product derivative from cellulose that is used popularly in cigarette filters, textile fibers, pharmaceutical products, and composites. Its ability to adsorb dye pollutants has been investigated by many researchers [8-9]. It has high compatibility, non-toxic, relatively low-cost, and chemical-resistant properties [10]. The biopolymer also has a good interaction with water, while being insoluble. However, most of the time, it requires modifications. It is due to its weak thermal and mechanical properties as well as high swelling degree.

The addition of isocyanates into cellulose acetate mixture allows the formation of cellulose acetate-polyurethane (CA-PU) copolymer. Introduction of urethane linkage can affect its mechanical properties positively [11-12]. Moreover, the modification can balance the hydrophilicity of the cellulose-based material, in which the swelling degree problem has been reported to be solved [13]. Furthermore, the addition of polyurethane can increase the porosity of the material [14], thus can be expected to increase its surface area, resulting in higher adsorption capacity. Therefore, in this research, we prepare CA-PU copolymer by simple blending method with MDI percentage variation, to be applied later for MO adsorption from aqueous solution using a batch method.

## 2. Materials and Method

### *2.1 Materials*
Materials used in this research included cellulose acetate (CA) with 39.3–40.3% acetylation degree, methylene diphenyl diisocyanate (MDI) purchased, 1,4-dioxane, silicon oil commercial as releasing agent purchased from, methyl orange (MO), distilled water, NaOH, acetone. All chemicals used are analytical grade and purchased from Merck, except for silicon oil and distilled water.

### *2.2. Preparation of CA-PU Sheet*
CA-PU sheet was prepared following the procedure reported by G. Li et al. [12] with some modifications and simplifications. The first step was done by pouring 15 mL1,4 dioxane into a beaker glass, heated on a hot plate at 70 °C for 1 minute. CA was weighed for 1 gram and slowly added to the beaker glass filled with 1,4 dioxane while stirred at 250 rpm. The temperature and stirring rate were maintained constant for 5 minutes. After that, MDI was added dropwise as many as the desired percentage (w/w CA) to the mixture. After the addition of MDI, the stirring rate was increased up to 750 rpm for 30 seconds and reduced back to 250 rpm for another 60 seconds to avoid the formation of air bubble. Next, the casting solution was poured onto the 3 x 12 $cm^2$ glass mold that had been coated with silicon oil as the releasing agent. It was left for 12 hours before released from the mold.

**Table 1.** CA-PU copolymer compositions

| No | CA (g) | 1,4 Dioxane (mL) | % MDI | Label |
|---|---|---|---|---|
| 1 | 1 | 15 | 15 | MDI15 |
| 2 | 1 | 15 | 30 | MDI30 |
| 3 | 1 | 15 | 45 | MDI45 |
| 4 | 1 | 15 | 60 | MDI60 |

The obtained cellulose acetate-polyurethane (CA-PU) sheet is cut into small pieces (1 x 0.5 $cm^2$) and washed to remove the unreacted materials by leaving them in 100 mL acetone while stirred for 15 minutes at 500 rpm. After dried in a vacuum oven for 3 hours at 70 °C, CA-PU sheets are washed with NaOH 2% for 60 minutes. They are then neutralized with distilled water for multiple times and dried in a vacuum oven at 70 °C for around 12 h. CA-PU sheets obtained after this process are then stored in a plastic container for further use.

### *2.3. Characterization*
Functional group analysis is carried out with Fourier Transform-Infrared (FTIR) (Shimadzu FTIR-Prestige 21 Series). Meanwhile, surface and cross-section morphology are analyzed with Scanning

Electron Microscopy (SEM) (Jeol. Jsm - 6510 LA). Samples used to compare before and after the copolymerization are CA, and CA-PU prepared with 45% MDI, respectively.

*2.4. Adsorption*
To determine which MDI percentage gives the best adsorption performance against methyl orange, MDI15, MDI30, MDI45, and MDI60 were used. The respective samples were weighed for 0.05 gram and added into Erlenmeyer 100 mL filled with 20 mL MO 20 ppm. The Erlenmeyer was mixed with a rotary shaker at 250 rpm for 3 h. MO solution before and after adsorption was determined for its concentration using UV-Vis spectrophotometer at 460 nm. Each concentration was determined three times to obtain the average value of the measured concentration. The removal percentage is then calculated using the following equation:

$$\%\text{Removal} = \frac{C_t - C_0}{C_0} \times 100\% \qquad (1)$$

Where $C_0$ and $C_t$ are initial and final MO concentrations, respectively.

Sample with the optimum MDI concentration, based on the previous procedure, was used for adsorption with concentration variation (30, 20, and 10 ppm). During each adsorption, 0.4 mL solution was taken periodically to determine the remaining concentration. The adsorption and the concentration determination were done continuously, until it reaches the equilibrium. The results are presented in the form of adsorption capacity, calculated from this following equation:

$$q_t = \frac{(C_0 - C_t)V}{m} \qquad (2)$$

Where $q_t$ is adsorption capacity at *t* minute (mg/L), $m$ is adsorbent weight (g) and $V$ is MO volume (L).

## 3. Results and Discussion

*3.1. Functional groups*
As seen in Fig. 1, broad absorbance band at around 3700-3000 cm$^{-1}$ is a response for hydrogen bond interaction due to the presence of O-H and N-H functional groups. The alcohol hydroxyl group can be observed at 3473 cm$^{-1}$, where the change to narrow band after the copolymerization with PU indicates the formation of N-H groups. It confirms the formation of urethane linkage through hydroxyl and isocyanate reaction. Clearer absorbance peak at 1521 cm$^{-1}$ is an evidence of the addition of aromatic ring from MDI. The absorbance peak at 3372 cm$^{-1}$ and 1736 cm$^{-1}$ correspond to O-H and C=O functional groups, respectively. The presence of functional groups such as C=O, O-H, and aromatic ring, contribute to the intensity of adsorption activities.

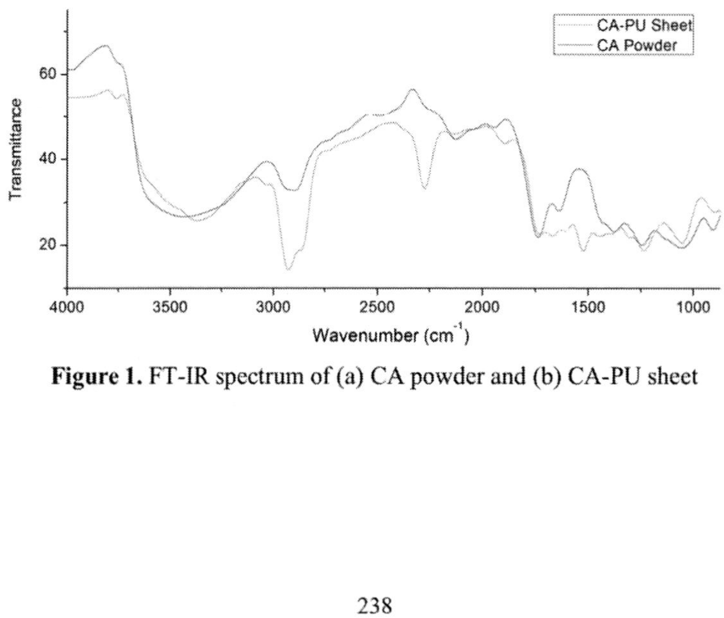


**Figure 1.** FT-IR spectrum of (a) CA powder and (b) CA-PU sheet

*3.2. Morphology Analysis*
Surface morphology analysis with SEM exhibits the difference between CA and CA-PU sheets. Cellulose acetate aggregates can be seen in CA surface, while the smoother surface is observed in CA-PU. It suggests the introduction of urethane linkage between cellulose acetate monomers can disperse the molecule, thus the agglomeration can be avoided. However, heterogenous micro-sized pores are found in the CA-PU surface.

Cross-section morphology analysis exhibits the different pore structure of both materials. While in CA the pore structure is found to be more dense, indicating the formation of porous space after the addition of reactive MDI. This structure is favorable for liquid adsorption due to its capability to give spontaneous diffusion of the liquid-phase adsorbate to the solid surface. As an addition, both CA and CA-PU exhibit the presence of a typical dense layer which is responsible for the adsorption selectivity.

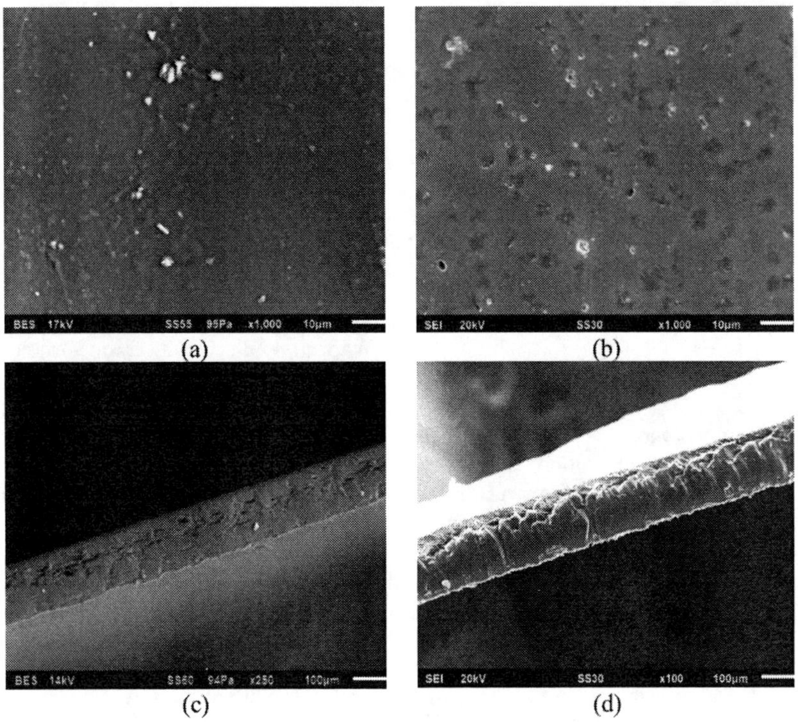


**Figure 2.** SEM image of: (a) pure CA sheet; (b) CA-PU sheet surface; (c) pure CA sheet; and (d) CA-PU sheet cross-section.

*3.3. Adsorption behavior*
*3.3.1 Adsorption by CA-PU prepared with MDI percentage variation*
The highest MO removal percentage (25,54%) is obtained by the CA-PU sheet with the smallest MDI composition (MDI15). By increasing MDI composition, the adsorbent ability to remove MO from the aqueous solution is declining. This can be ascribed to the presence of active sites which is attributed to the functional group of cellulose acetate as observed in the FT-IR spectrum. The aromatic ring and C=O or O-H functional groups, observed in the FT-IR spectrum, enable adsorption of MO on the prepared adsorbent via covalent and $\pi$-$\pi$ stacking interactions. The addition of MDI consequently can reduce the amount of the active functional group per gram adsorbent, which leads to less adsorption capacity.

The addition of isocyanate more than 25% of the total weight is reported to reduce the mechanical properties as reported by J. L. Rivera-Arment et al. [15]. This is due to the increased possibility of uneven distribution of polyurethane in the copolymer blend. However, the copolymerization is necessary to improve pure CA's mechanical, thermal, and chemical resistant properties. T. Riaz et al. [13] reported that the copolymerizing CA and PU increase the mentioned properties. Though it is not investigated in this research, such improvements are required for the application of this adsorbent material. Therefore, to investigate the adsorption behavior of the CA-PU sheet, MDI15 is taken for the adsorption isotherm and kinetics studies.

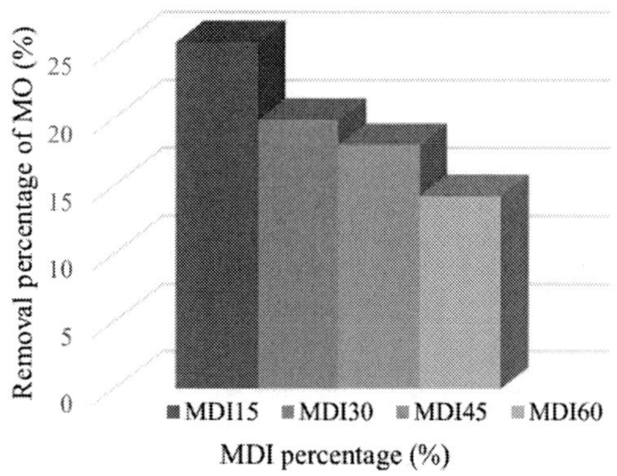


**Figure 3.** Effect of MDI addition on MO removal percentage

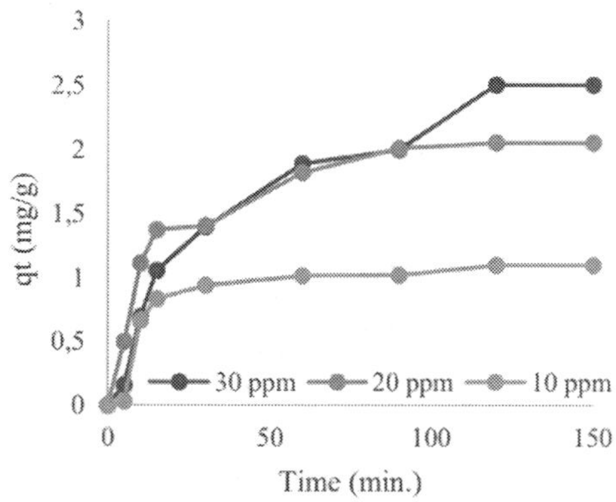


**Figure 4.** Effect of contact time on adsorption capacity with concentration variation of 30, 20 and 10 ppm

*3.3.2 Effect of contact time*

The sharp increase of the adsorption capacity within the initial 30 minutes contact time suggests the diffusion of adsorbates into the pore structure of CA-PU. The rapid diffusion may be ascribed to the hydrophilicity of the adsorbent, where it shows a good affinity with the adsorbate. The diffusion can take form of surface or pore diffusion, where both transport processes can occur at the same time. Given the fact that the adsorbent has a good hydrophilic affinity with the adsorbate, thus surface diffusion may play more important role during the adsorption [16-17]. This is due to the nature of CA that has been known for its high hydrophilicity, in which, however, leads to its high swelling degree. It may later adversely affect the adsorption by reducing the surface area, which determines the adsorption ability of the material. Co-polymerization with PU may balance the hydrophilicity of CA.

As can be observed from fig. 4 adsorption equilibrium is reached after 120 minutes of contact time. During this state, the rate of adsorbate-adsorbent interaction becomes constant, thus no longer affected by the increase of contact time. The adsorption capacity at the equilibrium of each initial concentration is different, where higher initial concentration leads to higher adsorption capacity. This is due to the fact that higher concentration can induce more force to transfer the mass aqueous MO into the CA-PU sheet solid medium.

*3.3.3 Isotherm studies*

To understand the mechanism of MO adsorption onto the CA-PU sheet surface, Langmuir and Freundlich isotherm models are used. Langmuir isotherm models are based on the assumption that adsorption occurs in a single layer, where all the active sites have uniform adsorption energy [18]. For the Langmuir isotherm model, the linear equation is as follow:

$$\frac{C_e}{q_e} = \frac{1}{K_L q_{max}} + \frac{Ce}{q_{max}} \quad (3)$$

Where, $q_e$ and $C_e$ respectively are adsorption capacity and MO concentration at the equilibrium; $Q_{max}$ and $K_L$ respectively are Langmuir constants for adsorption capacity maximum and adsorption energy, respectively. On contrary, the Freundlich isotherm model, derived from Langmuir isotherm, is based on the assumption that the adsorption occurs in multilayers, where the active sites have heterogeneous energy [19]. The linearized form of the Freundlich isotherm model equation can be stated as follow:

$$Logq_e = LogK_f + \frac{1}{n_F}LogC_e \quad (4)$$

Where, $K_f$ (L $g^{-1}$) is Freundlich isotherm constant and $n$ is adsorption intensity-related constant.

The isotherm in this research is initially judged by the correlation coefficient ($R^2$), given by respective Freundlich and Langmuir plots. Both give high correlation coefficients (> 0.75), where they are 0.992 and 0.9722 for Langmuir and Freundlich plots, respectively. Though Freundlich plots give a lower correlation coefficient, $\frac{1}{n}$ value obtained from the slope is not negligible (0.713). $\frac{1}{n}$ explains adsorption linearity, where its value starts from 1 and downward [20].

Due to the stated fact, the adsorption is tested with other isotherm model plots that are based on the assumption of multilayer adsorption; Halsey and Harkin-Jura isotherms [21-22]. Halsey isotherm follows the assumption that adsorbates are distant from the adsorbent's surface. While in Harkin-Jura isotherm, the multilayer adsorption occurs on the adsorbent surface with heterogeneous pore size distribution. The linear equations of both isotherms, respectively are as follows:

$$q_e = \frac{1}{n_H}lnK_H - \frac{1}{n_H}lnC_e \quad (5)$$

$$\frac{1}{{q_e}^2} = \frac{B}{A} - \left(\frac{1}{A}\right)logC_e \quad (6)$$

Where: $K_H$ and $n_H$ are Halsey isotherm constants; while A and B are Harkin Jura isotherm constants.

Correlation coefficients obtained from respective Halsey and Harkin-Jura isotherms are high, indicating their applicability for MO adsorption on CA-PU sheets. These results suggest the adsorption might follow the Freundlich isotherm better than Langmuir isotherm.

**Table 2.** Isotherm constants of Langmuir, Freundlich, Halsey, and Harkin-Jura models for MO adsorption on CA-PU sheets.

| Isotherm model | $R^2$ | Isotherm Constant | | |
|---|---|---|---|---|
| Langmuir | 0.991 | $q_m$ (mg/g) | | 0.185 |
| | | $K_L$ (mg/g) | | 0.786 |
| | | $R_L$ | 10 ppm | 0.113 |
| | | | 20 ppm | 0.060 |
| | | | 30 ppm | 0.041 |
| Freundlich | 0.972 | 1/n | | 0.713 |
| | | n | | 1.402 |
| | | $K_F$ (L/g) | | 279.45 |
| Halsey | 0.972 | n | | 1.402 |
| | | $K_H$ | | 0.167 |
| Harkin-Jura | 0.916 | A (mg/g) | | 0.739 |
| | | B | | 1.430 |

Adsorption fitting both isotherms are common in a liquid phase, especially in low concentration. Low concentration gives less driving adsorption force. Thus, the adsorption is likely to be monolayer. To assess this possibility, dimensionless Langmuir constant ($R_L$) is calculated for each concentration. The formula is as follow:

$$R_L = \frac{1}{1+K_L C_0} \tag{7}$$

$R_L$ value closing to 0 indicates the unlikelihood of Langmuir isotherm model, the otherwise is when the value is close to 1 [22]. As can be seen in Table x, $R_L$ is keep getting close to 0, indicating that the adsorption fitting to Langmuir isotherm is due to the low adsorbate concentration. Therefore, for MO adsorption on CA-PU sheets is suggested to fit the multilayer Freundlich isotherm model.

The value of $n_F$ obtained from the slope of the Freundlich plot equation can be used to indicate adsorption intensity. When the value is more than 1, it indicates the domination of chemisorption taking place. In this research, it is found that the $n_F$ value is 1.402, indicating the possibility of chemisorption-dominant adsorption. In order to determine whether the MO adsorption falls into the chemisorption or physisorption category, kinetics studies are conducted.

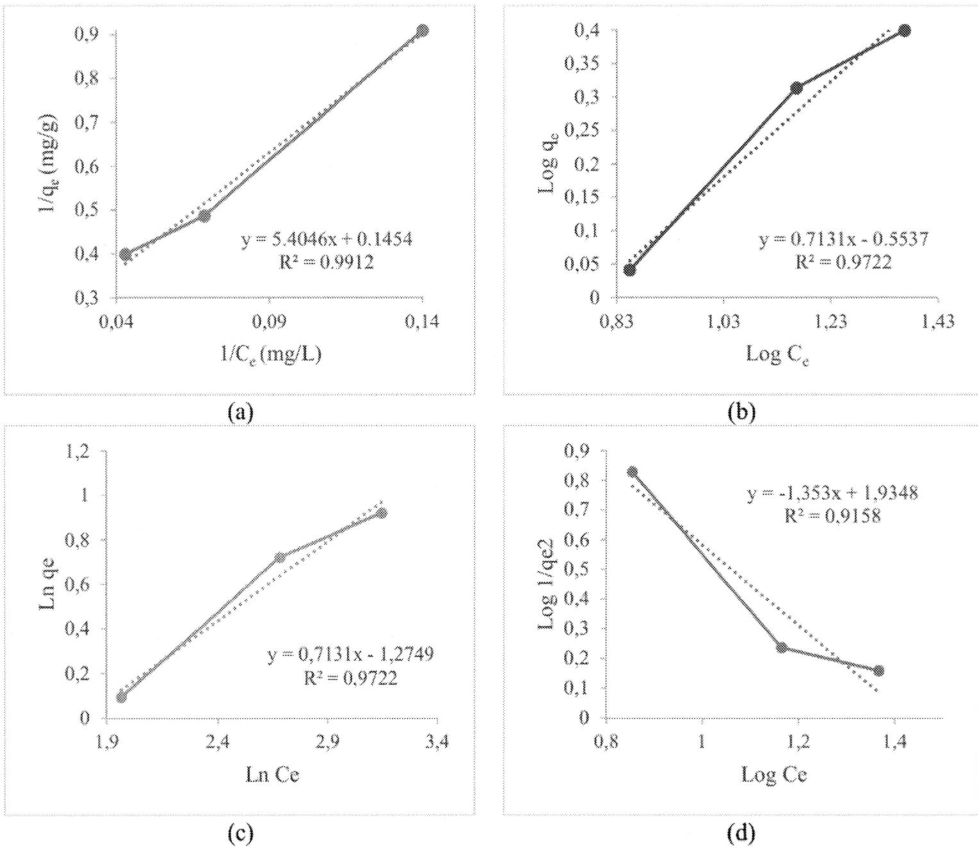


**Figure 5.** Plots of: (a) Langmuir; (b) Freundlich; (c) Halsey; and (d) Harkin-Jura Isotherm models for MO adsorption on CA-PU sheets

*3.3.4 Kinetics studies*

Experimental data of MO adsorption in this research are examined with pseudo-first-order and pseudo-second-order equations. Pseudo first-order modeling is based on the assumption that physical diffusion determines the adsorption rate. While, pseudo-second-order model describes the strong relationship between chemisorption and adsorption rate [23]. The equations of both pseudo-first-order and pseudo second-order can be seen as follows:

$$ln\left(\frac{q_e}{q_e-q_t}\right) = \frac{k_1}{2.303}t \quad (8)$$

$$\frac{t}{q_t} = \frac{t}{q_e} + \frac{1}{k_2 q_e^2} \quad (9)$$

Where, $k_1$ (1/h) and $k_2$ (g/mg h) are the adsorption rate constants for pseudo first-order and pseudo second-order, respectively.

Solid thick lines in Fig. 6 represent the theoretical data while the thin lines with dots represent the experimental data. Pseudo first order plots give no agreement between the theoretical data and experimental data. Meanwhile good agreement between experimental and theoretical data is obtained in pseudo second-order plots. It suggests that chemisorption is the limiting-rate steps for MO adsorption on CA-PU sheets. Therefore, the adsorption is greatly affected by the presence of functional groups interaction of CA-PU surface and MO molecules. The process might involve the valency covalent [23] and π-π stacking forces [12], ascribing to the presence of C=O and aromatic rings of CA-PU sheets, which also contribute to its hydrophilic property.

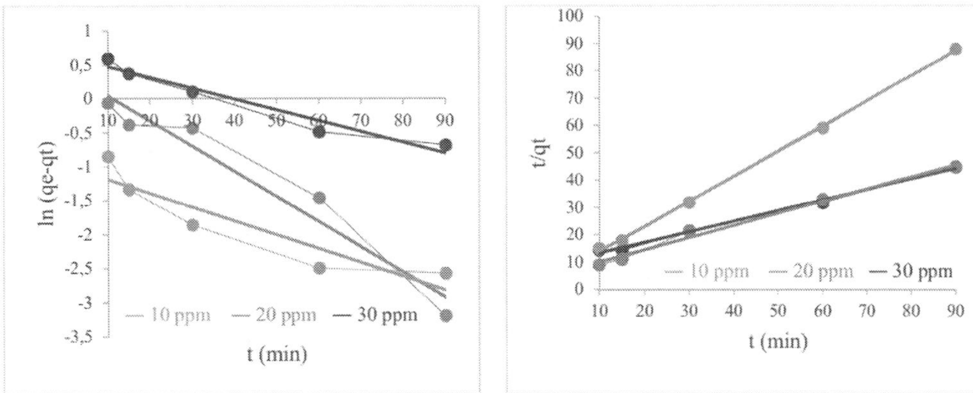


(a) (b)

**Figure 6.** Plots of: (a) pseudo-first-order; and (b) pseudo-second-order models for MO adsorption on CA-PU sheets

## 4. Conclusion

The addition of urethane linkage through copolymerization of PU and CA is suggested to change the chemical properties and morphology of the material. This copolymerization is made possible due to the reaction of hydroxyl groups from CA and diisocyanate from MDI, confirmed by the absorbance peak change and formation in the FTIR spectrum. Surface morphology analysis also proves the improvement of the material properties, where the dispersion of CA molecule and the formation of porous space are found. Nevertheless, the increase of MDI percentage more than 15% adversely affects the adsorption capacity of MO. Evaluation of isotherm models for MO adsorption on CA-PU sheets suggests the applicability of Freundlich isotherm ($R^2$ = 0.9722). This adsorption also fits with pseudo second-order kinetics model, which suggests the chemisorption-dominant adsorption. Therefore, the adsorption of MO on CA-PU sheets is majorly induced by the presence of functional groups that can form covalent and π-π stacking interaction with the adsorbate.

## 5. Acknowledgement

We would like to express our deep gratitude to Universitas Syiah Kuala for their valuable grant, under PRUUPD program Number: 860/UN11/SPK/PNBP/2019.

IOP Conf. Series: Materials Science and Engineering **845** (2020) 012035 doi:10.1088/1757-899X/845/1/012035

## 6. References


[1] Dellamatrice P M, Silva-Stenico M E, Moraes L A B de, Fiore M F and Monteiro R T R 2017 *Brazilian J. Microbiol.* **48** 25

[2] Gičević A, Hindija L and Karačić A 2020 Toxicity of Azo Dyes in Pharmaceutical Industry BT - CMBEBIH 2019 ed A Badnjevic, R Škrbić and L Gurbeta Pokvić (Cham: Springer International Publishing) 581

[3] Robati D, Mirza B, Rajabi M, Moradi O, Tyagi I, Agarwal S and Gupta V K 2016 *Chem. Eng. J.* **284** 687

[4] Subbaiah M V and Kim D-S 2016 *Ecotoxicol. Environ. Saf.* **128** 109

[5] Jiang Y, Liu B, Xu J, Pan K, Hou H, Hu J and Yang J 2018 *Carbohydr. Polym.* **182** 106

[6] Zhang T, Lang J, Liu L, Liu L, Li H, Gu Y, Yan X and Ding X 2017 Effect of carboxylic acid groups on the supercapacitive performance of functional carbon frameworks derived from bacterial cellulose *Chinese Chem. Lett.* **28** 2212

[7] Chen Q, Zheng J, Wen L, Yang C and Zhang L 2019 *Chemosphere* **224** 509

[8] Setyaningsih D, Uju, Muna N, Isroi, Suryawan N B and Nurfauzi A A 2018 *IOP Conf. Ser. Earth Environ. Sci.* **141**

[9] Yang J, Kubota F, Baba Y, Kamiya N and Goto M 2014 *Carbohydr. Polym.* **111** 768

[10] Nazir M S, Wahjoedi B A, Yussof A W and Abdullah M A 2013 *BioRes.* **8**(2) 2161

[11] Zavastin D, Cretescu I, Bezdadea M, Bourceanu M, Drăgan M, Lisa G, Mangalagiu I, Vasić V and Savić J 2010 *Colloids Surf. A Physicochem. Eng. Asp.* **370** 120

[12] Li G, Chai K, Zhou L, Tong Z and Ji H 2019 *Carbohydr. Polym.* **206** 716

[13] Riaz T, Ahmad A, Saleemi S, Adrees M, Jamshed F, Hai A M and Jamil T 2016 *Carbohydr. Polym.* **153** 582

[14] Marcovich N E, Auad M L, Bellesi N E, Nutt S R and Aranguren M I 2006 *J. Mater. Res.* **21** 870

[15] Rivera-Armenta J L, Heinze T and Mendoza-Martínez A M 2004 *Eur. Polym. J.* **40** 2803

[16] Ocampo-Perez R, Leyva-Ramos R, Mendoza-Barron J and Guerrero-Coronado R M 2011 *J. Colloid Interf. Sci.* **364** 195

[17] Zhu Q, Moggridge G D, Ainte M, Mantle M D, Gladden L F and D'Agostino C 2016 *Chem. Eng. J.* **306** 67

[18] Langmuir I 1916 *J. Am. Chem. Soc.* **38** 2221

[19] LeVan M D and Vermeulen T 1981 *J. Phys. Chem.* **85** 3247

[20] Khan T A, Nazir M, Khan E A and Riaz U 2015 *J. Mol. Liq.* **212** 467

[21] Samarghandi M R, Hadi M, Moayedi S and Askari F B 2009 *Iran. J. Environ. Heal. Sci. Eng* **6** 285

[22] Ayawei N, Ebelegi A N and Wankasi D 2017 *J. Chem.* **2017** 1

[23] Zhou L, Yu Q, Cui Y, Xie F, Li W, Li Y and Chen M 2017 *Ecol. Eng.* **102** 443

# Use of hydrodynamic cavitation for esterification of free fatty acids in waste cooking oil

**M Rizal[1], N Suriaini[1], Y Syamsuddin[2] and M D Supardan[2*]**
[1] Graduate Program in Chemical Engineering, Universitas Syiah Kuala, Banda Aceh, Indonesia
[2] Department of Chemical Engineering, Universitas Syiah Kuala, Banda Aceh, Indonesia

E-mail: m.dani.supardan@unsyiah.ac.id

**Abstract**. Esterification followed by transesterification process is widely applied for biodiesel production from oil with high free fatty acids (FFA) content to remove the FFA and increase the yield of biodiesel. The aim of this research was to study the esterification of waste cooking oil using the hydrodynamic cavitation apparatus. Waste cooking oil with initial FFA content 4.33% was esterified with methanol using sulfuric acid 1% -w/w as the catalyst. The reaction condition was kept constant for 120 minutes of total reaction time. At the best operating condition i.e. molar ratio of waste cooking oil to methanol 1:10 and reaction temperature 50 °C, the initial FFA of waste cooking oil 4.33% reduced to 0.76%. The experimental results also showed that the hydrodynamic cavitation had better performance than the conventional process using mechanical stirring.

## 1. Introduction

Vegetable oils undergo several chemical and physical changes during the frying process mainly due to the exposure to high temperatures and the presence of oxygen and water. One of these changes is the increase in acidity of the oil which is provoked by the release of free fatty acids (FFA). Generally, waste cooking oil as the oil produced from the frying process has a high percentage of FFA which is produced from the oxidation and hydrolysis reaction during the frying process. Waste cooking oils are regarded as an alternative to raw vegetable oils for biodiesel production due to the high cost of the vegetable oils and the threat to food security. Recent studies reported that waste cooking oil has been proved as the promising feedstock for biodiesel production because of abundant availability, low cost and have a comparative advantage compared to other feedstock [1-3].

Biodiesel is typically obtained by transesterification process, which involves the reaction of triglycerides with short-chain alcohol to produce fatty acid methyl esters (biodiesel) in the presence of a base catalyst [4]. The presence of FFA cause problems during the base-catalyzed transesterification reaction due to saponification problems. The soaps produced cause problems during the purification of biodiesel, creating an emulsion between biodiesel as the main product and the glycerol as the side product. In order to take advantage of the utilization of waste cooking oil as low-cost feedstocks for biodiesel production, therefore, a method for FFA removal is necessary. Supardan et al. [3] reported

* To whom any correspondence should be addressed.

glycerolysis process to reduce the FFA content of waste cooking oil. Meanwhile, Wang et al. [5] carried out esterification process to reduce the FFA content of oleic acid oil. Currently, the esterification followed by transesterification process is widely applied for the biodiesel production from oil with high FFA content. The esterification process has two purposes i.e. removing the FFA and increasing the yield of biodiesel.

The performance of esterification can be affected by several factors, such as FFA content, water content, molar ratio of alcohol to raw materials, reaction temperature, amount and type of catalyst, reaction time, and the type of alcohol used [6]. Sulfuric acid ($H_2SO_4$) is widely used in the esterification reaction. In esterification reaction, proton $H^+$ can be substituted and binds strongly with oxygen atoms in carbon carbonyl which have a large electronegativity value. Meanwhile, the selection of methanol in the esterification process is based on its reactivity. This is because methanol is an alcohol compound with the shortest carbon chain, in which, the shorter carbon chain has higher reactivity. Many technologies have been developed to produce biodiesel efficiently. Ji et al. [7] proposed hydrodynamic cavitation to intensify transesterification reaction. Meanwhile, the intensification of the esterification reaction using acoustic and hydrodynamic cavitation was reported by Mohod et al. [8]. Hydrodynamic cavitation is a phenomenon of the formation, growth and breakdown of micro bubbles in liquids that occur due to pressure variations in the flowing liquid resulting in changes to the geometry of the flowing system during the esterification process.

The aim of this study is to study the effect of process variables i.e. the molar ratio of waste cooking oil to methanol, temperature and processing time in the esterification of waste cooking oil using hydrodynamic cavitation. The acid value was determined to monitor the progress of the reaction. Furthermore, the performance of the esterification process using hydrodynamic cavitation will be compared with the conventional process using mechanical stirring.

## 2. Materials and method

### *2.1. Materials*

The raw material was waste cooking oil collected around Banda Aceh, Indonesia. The chemicals used were methanol (Merck, Germany), potassium hydroxide (KOH), sodium hydroxide (NaOH), sulfuric acid 98% ($H_2SO_4$), phenolphthalein reagent 99% (Merck, Germany) and ethanol 96% (Merck, Germany).

### *2.2. Esterification of WCO*

The FFA content of waste cooking oil used was 4.33%. The esterification reaction was performed in a 500 mL three-neck flask equipped with a temperature controller and a water-cooled condenser. The stirring speed was kept constant during all experimental runs at 500 rpm. The molar ratios of waste cooking oil to methanol used were 1:4, 1:6, 1:8 and 1:10. The mixture was reacted at the reaction temperature 30, 40 and 50 °C for 120 minutes. The catalyst $H_2SO_4$ concentration 1%-w of oil was used for all experimental runs. The catalyst was mixed into methanol and stirred until homogeneous. Furthermore, this solution was introduced into a three-neck flask that already contains waste cooking oil. After completion of the reaction, the products are transferred to the separating funnel for product separation. The top layer is analyzed to monitor the progress of the reaction. A conventional process using stirrer was also conducted for comparison.

### *2.3. Product analysis*

FFA content of products was analyzed by the Indonesian National Standard (SNI) No. 01-3555-1998. In this method, 5 grams of product is put into the 250 mL erlenmeyer. Then, 50 mL of 96% alcohol is added and then heated for 10 minutes. The solution is then added with 2 drops of phenolphthalein indicator then titrated with 0.1 N KOH. The titration ended when a reddish-brown colour change was observed.

## 3. Results and discussion

### *3.1. Effect process variables*

Figure 1 shows the progress of the reaction in terms of the change in FFA content at different temperature and molar ratio of waste cooking oil to methanol. The experimental results show that the FFA content decreases with increasing reaction time. This is due to the longer reaction time will provide greater contact opportunities between the reaction mixture (waste cooking oil and methanol). Thus, the remaining FFA content will also decrease with increasing reaction time. Sendzikiene et al. [9] reported that the reaction rate changes depending on the reaction duration, especially during the first 15 min. At the end of the reaction time of 120 minutes, the final FFA content obtained was found to be dependent on the process variables. A similar trend was reported by Satriana et al. [10] for glycerolysis reaction.

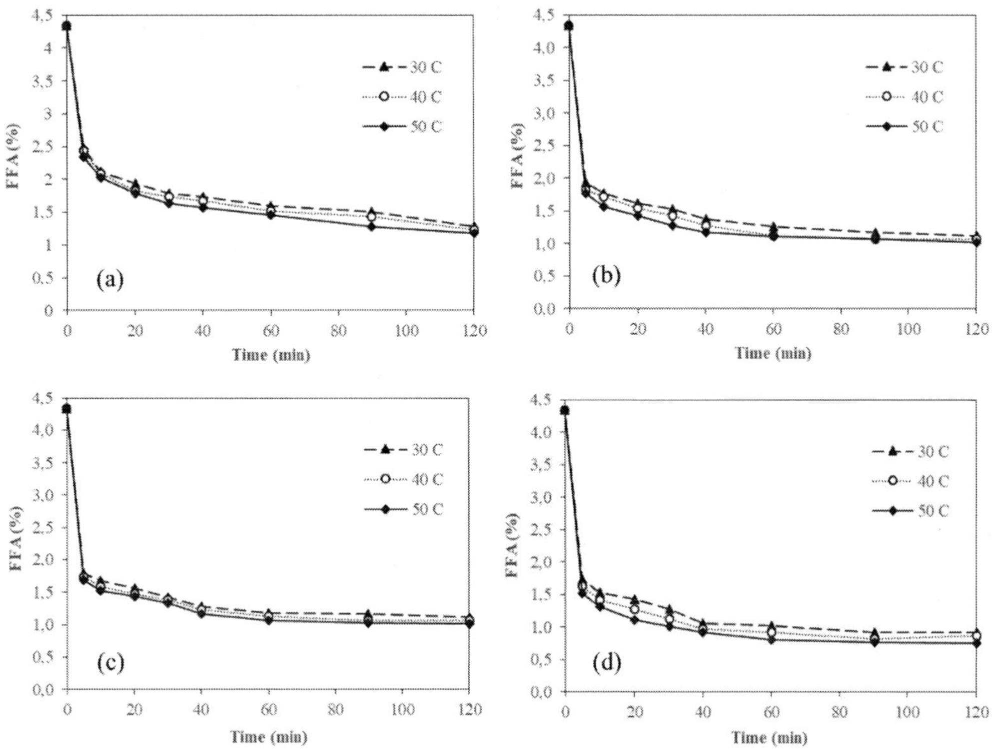


**Figure 1**. The progress of the reaction in terms of the change in FFA content for molar ratio of waste cooking oil to methanol: (a) 1:4; (b) 1:6, (c) 1:8; and (d) 1:10.

It was observed that the FFA content decreased with increasing reaction temperature. This is probably because the heating makes used cooking oil molecules more easily dispersed and distributed to methanol molecules to form reaction products. In addition, an increase in reaction temperature can also increase the movement of oil and methanol molecules so that collisions between particles will increase. The esterification reaction will take place quicker by raising the reaction temperature to near the boiling point of methanol.

Furthermore, the experimental results also showed that FFA content generally decreases from an initial value of 4.33% as the molar ratio of waste cooking oil to methanol increases. For example at reaction temperature 50°C, the FFA decreased from 1.52% to 0.76% and 1.69% to 1.02% at the reaction time between 5 minutes to 120 minutes at a molar ratio of waste cooking oil to methanol 1:10 and 1:8, respectively. Esterification is a type of a reversible reaction. The reaction is carried out in excess of reactants to produce the products at maximum quantity [11]. In this experiment, the use of excess methanol to the reaction mixture will shift the equilibrium to the right side, towards the side of the products. On the other hand, reducing the concentration of any product could also implemented to shift equilibrium to the right side.

### 3.2 *Process comparison*

The comparison of the process performance using hydrodynamic cavitation and conventional process using mechanical stirring at the molar ratio of waste cooking oil to methanol = 1:10 and the catalyst concentration 1% can be seen in Fig. 2. As can be seen in Fig. 2, the FFA content decreases with increasing time and temperature reaction. A similar trend was observed for both the hydrodynamic cavitation process and the conventional process using mechanical stirring. The experimental results showed that the decrease of FFA content for the hydrodynamic cavitation process is faster than the conventional process using mechanical stirring. These results proved that the cavitation produced more intensive contact between oil and methanol (through the phenomenon of formation, growth and destruction of micro bubbles in the liquid) compared to the conventional process. Hydrodynamic cavitation can increase the chemical and physical changes in media through the formation and breakdown of cavitation bubbles. The formation of micro-sized bubbles enlarges the contact surface area between oil and methanol to increase the rate of reaction. According to Ji et al. [7] the use of hydrodynamic cavitation has been shown to accelerate the reaction, reduce the amount of catalyst used and reduce the mole ratio of oil to methanol used and lower energy consumption compared to the process using mechanical stirring.

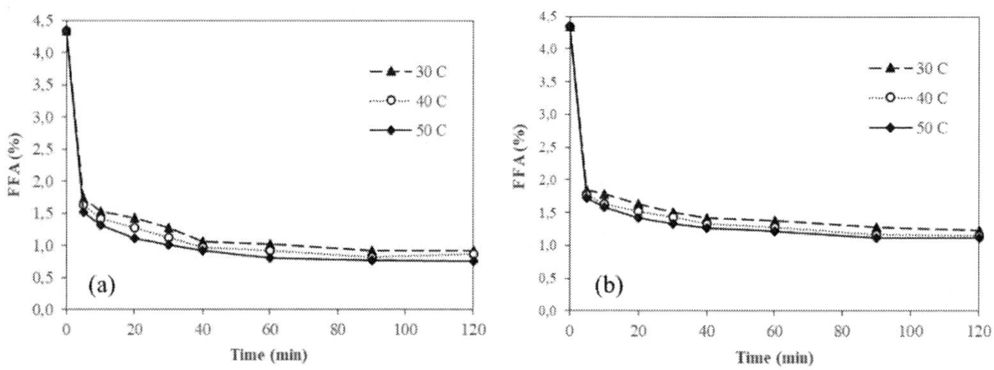


**Figure 2.** Comparison of FFA content for: (a) Hydrodynamic cavitation process; and (b) Mechanical stirring process.

## 4. Conclusion

Esterification using cavitation hydrodynamic was successfully carried out to decrease the FFA content of waste cooking oil. It was observed that the experimental variables studied had a remarkable effect on the FFA content of waste cooking oil. Generally, FFA content decreased as the time of reaction increased. The experiment results showed that a suitable reaction condition should be used to achieve an optimum level of FFA conversion. The best operating condition was obtained at a molar ratio of waste cooking oil to methanol 1:10 with reaction temperature 50 °C in which initial FFA of waste cooking oil 4.33% reduced to 0.76%. It can be concluded that the reaction using hydrodynamic cavitation had a better performance than those using mechanical stirring.

## 5. Acknowledgement

The authors are grateful to Syiah Kuala University for the financial support for this project through the H-Indeks 2019 Research Grant.

## 6. References

[1] Martín J F G, Ruiz J C, García, M T, Feng C H and Mateos P A 2019 *Processes* **7** 1
[2] Suriaini N, Febriana T T, Yulanda A, Adisalamun, Syamsuddin Y and Supardan M D 2019 *J. Rekayasa Kimia & Lingkungan* **14** 155
[3] Supardan M D, Adisalamun, Lubis Y M, Annisa Y. Satriana and Mustapha W A W 2017 *Pertanika J. Sci. Technol.* **25** 1203
[4] Supardan M D, Fahrizal, Moulana R, Safrida D, Satriana and Mustapha W A W 2017 *JESTEC* **12** 847
[5] Wang Y, Fang Z and Zhang F 2019 *Catal. Today* **319** 172
[6] Aranda D A G, Santos R T P, Tapanes N C O, Ramos A L D and Antunes O A C 2008 *Catal. Lett*, **122** 20
[7] Ji J, Wang J, Li Y, Yu Y and Xu Z 2006 *Ultrasonics* **44** e411
[8] Mohod A V, Subudhi A S and Gogate P R 2017 *Ultrason Sonochem.* **36** 309
[9] Sendzikiene E, Makareviciene V, Janulis P and Kitrys S 2004 *Eur. J. Lipid Sci. Technol.* **106** 831
[10] Satriana, Arpi N, Supardan M D, Gustina R T and Mustapha W A W 2018 *AIP Conf. Proceed.* **1940** 020100
[11] Supardan M D and Satriana 2009 *J. Rekayasa Kimia & Lingkungan* **7** 70

# $CO_2$ emission management in palm oil industries in Indonesia: a review

**Novi Sylvia[1,2], Yunardi[3*], Husni Husin[3] and Abrar Muslim[3]**
[1] Doctoral student at Faculty of Engineering, Syiah Kuala University, Banda Aceh 23111, Indonesia
[2] Department of Chemical Engineering, Malikussaleh University, Lhokseumawe, 24351, Indonesia
[3] Department of Chemical Engineering, Universitas Syiah Kuala, Banda Aceh 23111, Indonesia

E-mail: yunardi@unsyiah.ac.id

**Abstract.** Within a few decades, Indonesia has become the largest palm oil-producing country, leaving Malaysia behind. It is undeniable that palm oil has significantly contributed to the economic growth of the country. However, the negative impacts of the palm oil industry are also evident, such as deforestation, environmental pollution, and social conflicts. The palm oil industry has been known as one of the big greenhouse gasses producers, including $CO_2$. Regardless of the pressure from the European Union, the Government of Indonesia (GOI) has committed to sustaining the palm oil industry since there are millions of workforces depend on their livelihood from these industries. Concerning the negative impacts, the GOI requests that each palm oil company has to follow the Indonesian Sustainable Palm Oil (ISPO) certification. So far, around 20% of all palm oil industries in the country have received an ISPO certificate. It is expected by 2023, all palm oil industries have been certified by ISPO since this regulation is mandatory. Implementing all ISPO guidelines in the palm oil industry is not only capable of sustaining the industry but also protecting the environment through the reduction of $CO_2$ and other greenhouse gasses. This paper will review the $CO_2$ management in the palm oil industry in Indonesia in the present situation, connected with ISPO as well as RSPO (Roundtable Sustainable Palm Oil).

## 1. Introduction

Oil palm tree (*Elais quineensis* Jack) is a group of annual tropical plants, originating from Africa. Oil palm trees came to Indonesia for the first time in 1848 and planted in the Bogor Botanical Garden. Several first oil palm plantation experiments were carried out in the southern part of Sumatra, including at Muara Enim (1869), Musi Ulu (1878), and Belitung (1890). The first oil palm plantation was established in 1911 in Tanah Itam Ulu (North Sumatra) by the Olie Palm Culture company and in Raja Island by the Huilleries de Sumatra-RCMA company. It was started with four oil palm seedlings brought by the Dutch from Mauritius in 1848 and then commercially cultivated in 1911. The year 1911 can be regarded as the starting point of the era of the people and the Indonesian economy's dependence on oil palm [1,2]. In 1919, oil palm companies began exporting the first palm oil, Crude Palm Oil (CPO) to European countries from Tanah Deli Sumatra and Tamiang Aceh, at the amount of

* To whom any correspondence should be addressed.

IOP Conf. Series: Materials Science and Engineering **845** (2020) 012037 doi:10.1088/1757-899X/845/1/012037

576 tons. In 1923, as much as 850 tons of Palm Kernel Oil (PKO) began to be exported to Europe. In 1980, the area of Palm Oil had reached 294,560 ha with CPO production of 721,172 tons [3].

Indonesia produced 3 million tons of PKO and became the most significant crude palm oil (CPO) producer in the world, with a total production of 32 million tons or equivalent to around 46.6% of the world's total CPO production. The world market demand for CPO continues to increase and is estimated to reach 95.7 million tons in 2020. Meanwhile, the development of the Indonesian palm oil industry shows a positive trend, with an increase in production volume every year. This condition shows that the future of Indonesia's palm oil industry continues to grow to meet the world's CPO demand.

It is undeniable that the rapid development of the palm oil industry in Indonesia has had a positive impact on the country's economy. One indication is a decrease in poverty and an increase in the welfare of the people/community involving with the industry. The palm oil industry supports the lives of around 4.5 million farmers owning the oil palm plantation and million other people who work in this sector [4]. The palm oil industry also triggers a multiplier effect; for each increase in the final demand of CPO, there will be an increase in the output of the national economy, which eventually will provide a significant contribution to the regional development.

## 2. The palm oil industry in Indonesia

Although received many critics from the developed country, particularly from the European Union, the palm oil industry has become the expectation not only from the Government side but also from the village communities. The industry possesses the significant potential to increase economic and social development in Indonesia. As the demand for palm oil increases, more and more new lands for palm oil plantation are opened to back up the production. It is, therefore, not surprising if Indonesia produces 47 % of the world's palm oil from the plantation area of 7.9 million hectares. Also, the oil palm industry absorbs around 16.2 million workers, of which 4.2 million are direct ones, and the remaining are indirect ones. Since the industry support the lives of million peoples both in urban and rural areas, the government and the people work together to sustain the industry even develop further [4].

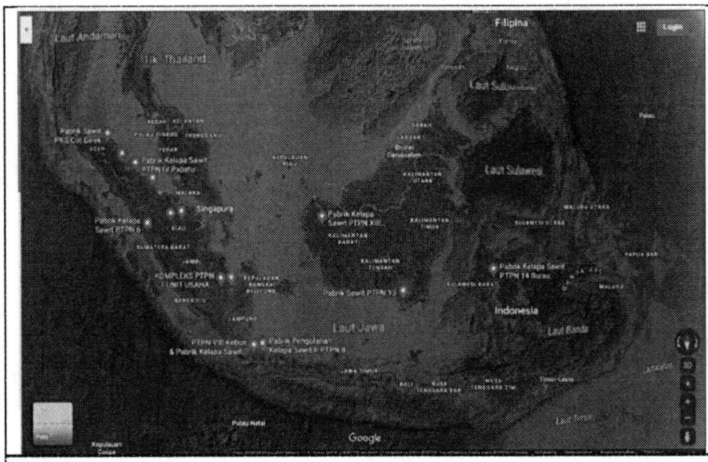


**Figure 1.** Map of palm oil plantation distribution in Indonesia.

Figure 1 presented the map showing the distribution areas of palm oil plantation in Indonesia. Data from the Indonesian Central Bureau of Statistics (BPS) recorded that there are 1,592 large oil palm plantation companies in 2016, having a common area of oil palm plantations of 7,726 million ha out of the total oil palm plantation of 11.94 million hectares (ha). The remaining 4.66 million ha belong to individuals, cooperatives, communities, or small scale plantation companies. With this vast of palm oil plantation, yearly oil production could reach up to 33.23 million tons, of which about 10.87 million tons produced by smaller scale plantations such as cooperative and communities, while a bigger portion of 22.36 million tons coming from big scale plantation companies. In terms of plantation location, most of oil palm plantations in Indonesia are located in the island of Sumatra, with the most significant area being in the Riau Province, having 146 units of Palm Oil Mill (POM). The other growing areas of oil palm plantations are in Kalimantan and Sulawesi.

### *2.1. Palm oil production process*

Extracting palm oil from fresh fruit bunches (FFB) involves many sequential units of operations. There are two types of oils produced from the palm tree, namely, crude palm oil (CPO) and kernel oil. The CPO is the palm oil extracted from the fibrous mesocarp, and the latter is lauric oil produced from the palm kernel. Figure 2 presents the process flowsheet for the manufacture of palm oil from the FFB.

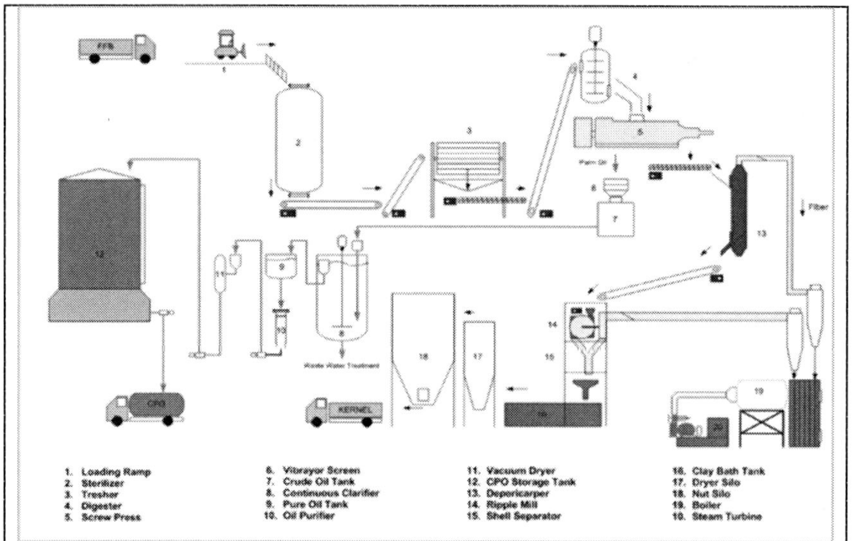


**Figure 2.** Typical production process of palm oil in Indonesia.

**Sterilization**: Sterilization process comes first in the crude palm oil extraction. For 75-90 minutes, the fresh fruit bunches (FFB) are sterilized inside autoclave using steam at a temperature of 140°C, aiming at deactivating hydrolytic enzymes responsible for the breakdown of oil to free fatty acids and loosening the fruits from bunches. Oil cell breakdown will, then, coagulate the mucilage.

**Stripping or threshing:** Upon the completion of the sterilization process, fruits are stripped from the bunches and separated in a rotary drum stripper by lifting and dropping them through the stripper in order to be knocked out of the bunches. Bucket conveyor then collects the detached fruits falling through the space between the bars on the stripper and discharged into a digester. Empty fruit bunches (EFB) resulted from the stripping operation, after all, their fruits being detached, will be wasted.

**Digestion:** To digest the fruits, they are pulverized under the steam heated condition in a vertical cylindrical digester fitted with rotating arms. This process is carried out to break the mesocarp oil-bearing cells. To facilitate the flow of the oils, hot water is added in a digester and no waste residue resulting from this step.

**Palm oil extraction:** The vertical cylindrical digester released oil mash, which then passes through a screw press and a vibrating screen, a hydrocyclone, and decanters, to separate the fine solids and water. The oil is purified by using the centrifuge and vacuum drier before the transfer to the storage tank, where the temperature is maintained at 60°C with steam coil heating. The crude oil slurry is then fed into a clarification system for further separation and purification. The fiber and nuts are separated by passing through the depericarper [5]. At this stage, the crude palm oil (CPO) constitute a mixture of palm oil (35-45%), water (45-55%), and the variable portion of fibrous materials. Having passed the screw presser, the CPO will be pumped to a horizontal or vertical clarification tank in order to skim the oil from the top of the clarification tank. The CPO is then passed through a high speed, centrifuge, and a vacuum dryer before being transferred to storage tanks. This process produces waste in the form of decanter wastewater and decanter cake.

### *2.2. Emissions in palm oil mills (POM)*

The continuous development of the palm oil industry in the country will have an impact on increasing solid, gas, and liquid wastes resulted from FFB processing. Solid waste coming out of POM includes empty bunches amounting to about 23% of FFB, boiler ash (about 0.5% of FFB), fiber (around 13.5% of FFB) and shells (around 5.5% of FFB). In the extraction process, every 100 tons of fresh fruit bunches processed will produce 20 tons of shells, 7 tons of fiber, and 25 tons of empty bunches [5,6]. Solid waste coming out of the POM generally does not require complicated handling. Solid waste can be used again as fuel, fertilizer and animal feed, and can be sold to generate additional income. Fiber, shells, and FFB can be used as fuel, while boiler ash can be applied directly as a potassium fertilizer. Also, empty bunches can be converted into fertilizer by making mulch and composting; and core pulp can be utilized used as animal feed.

#### *2.2.1.Emissions from transportation vehicles and agriculture machine*

In general, the fuel used inside the plantation area is in the form of fossil fuels, especially diesel oil. Diesel fuel is used for agricultural machinery in the sections of maintenance, harvesting, harvest collection, and internal transportation between oil palm plantations and factories.

#### *2.2.2. Emissions from combustion in boilers*

The boiler is the primary source for the production of the energy required in the POM because the electricity used in the POM is not from the state electric company of Indonesia. The fuel used to start the boiler is fiber and palm kernel shells. The boiler produces steam, which in turn is used to drive turbines to run the generator. All processes and operations in the POM required electricity supported by this internal generator.

**Figure 3.** Palm fiber and kernel shell residues.

In general, all palm oil mill in Indonesia utilizes solid wastes resulted from the processing of palm oil as fuel in the forms of fiber and kernel shells, as shown in figure 3. The fiber and shells obtained from the palm oil extraction process are used as the primary fuel in the palm oil mill boiler (POM) [6]. Based on several studies, it shows that the heating value of the waste can cover the need for steam used to drive a steam turbine. The electricity produced is used to meet all the electricity needs in the mill [7]. The kernel shell is reported to have a calorific value of 16.9 MJ /kg, while the fiber has a calorific value of 9.2 MJ / kg [8]. Table 1 presented the heating value for each type of solid wastes produced in the palm oil mill [9].

**Tabel 1.** The heating value for each type of solid wastes in the palm oil mill.

| Biomass waste | Caloric value (MJ $kg^{-1}$) | Element Content (%) | | | |
|---|---|---|---|---|---|
| | | C | H | N | O |
| Empty fruit bunch (EFB) | 18.54 | 44.09 | 6.18 | 1.62 | 48.11 |
| Fiber | 18.12 | 45.19 | 6.62 | 1.44 | 46.75 |
| Shell | 18.98 | 47.64 | 6.57 | 1.08 | 44.71 |

To decrease the dependence of fossil fuels, the Government of Indonesia encourages every sector to utilize renewable energy resources, such as solar, wind, and biomass. It is in line with the government policy that the palm oil industry utilizes fiber and shells to produce energy needs within the industry or even share the excess energy to the national electricity company. However, it should be noted that biomass combustion is frequently regarded as one of the primary sources of particulates and pollutant gas emissions. Every combustion process is expected to occur in a complete mode, so the result of the combustion will produce energy, carbon dioxide, and water. The same case is also expected in the combustion biomass in the boiler of palm oil mill, as shown in reaction 1.

$$CH_2O + O_2 \longrightarrow CO_2 + H_2O \quad (1)$$

In equation (1), the $CH_2O$ represents the average composition of cellulosic biomass materials. As incomplete combustion will most probably occur, other unwanted species that the above mentioned in equation (1) may be formed, such as carbon monoxide (CO), methane ($CH_4$), non-metal hydrocarbons (NMHC) and particulate matter (PM). Besides, nitrogen and sulfur species (especially NOx and SOx) are also formed, which may also pollute the environment. Unburned fuel will come out together with exhaust gas through the chimney, identified by the appearance of black smoke during the burning process, as shown in Figure 4.

**Figure 4.** The smoke of a chimney resulted from biomass burning in the boiler of palm oil mill.

Pollutants, of any kind, not only threaten human health but also affect agriculture crops, forest species, and ecosystems. Referring to the regulation by ministerial of environment article 7 of 2007 [12] as shown in table 2, Indonesia has set the air emission-quality standard from stationary combustion sources. As for the palm oil industry, the primary concern of this regulation is to control the dark smoke and dust emissions from combustion waste, which is a mixture of fiber and shells. The article also stated that the maximum permissible dust concentration level in $CO_2$ gases is at 12%.

**Tabel 2.** Standard of air emission.

| Standard Parameter | Quality |
|---|---|
| Particulate | 300 mg $m^{-3}$ |
| Sulphur Dioxide ($SO_2$) | 600 mg $m^{-3}$ |
| Nitrogen Oxide ($NO_2$) | 800 mg $m^{-3}$ |
| Hydrogen Chloride (HCl) | 5 mg $m^{-3}$ |
| Chlorine Gas ($Cl_2$) | 5 mg $m^{-3}$ |
| Ammonia ($NH_3$) | 1 mg $m^{-3}$ |
| Hydrogen Fluoride (HF) | 8 mg $m^{-3}$ |
| Opacity | 30% |

*2.2.3. Emissions from liquid waste*

Palm Oil Mills (POM) produce liquid waste, better known as POME (palm oil mill effluent). POME is waste-water produced by palm oil mills mainly from boiled condensate, hydro-cyclone water, and sludge separator. For every ton of FFB treated, about 0.6 to 1 m3 of POME is formed. POME is rich in organic carbon with a COD value of more than 40 g/L and nitrogen of about 0.2 and 0.5 g/L ammonia nitrogen and total nitrogen. The characteristics of POME are shown in table 1. POME sources come from different processing units: 100% total POME, 60% comes from the clarification station, 36% comes from the cooking station, and the remaining 4% comes from the core station.

Generally, POME is treated by a pool system. Naturally, the oxygen available in the treatment pond system is not able to meet the aerobic decomposition needs of organic matter contained in liquid waste, so the decomposition process runs anaerobically and produces biogas, which later is released into the atmosphere. Most of the biogas produced from POME decomposition are methane (CH4), convertible to CO2 emissions, which is a source of GHG emissions.

## 3. Management of $CO_2$ emissions in Indonesia

Export of Crude Palm Oil from Indonesia to several European Union countries has been rejected from 2012-2014. This happens because of competition with other products that produce renewable energy. The European Union requires the use of renewable energy to reduce GHG emissions by a minimum of 35%, while Indonesia's CPO is only 19% [13, 14]. In 2011, the European Commission recognized ISCC as one of the first certification schemes to demonstrate compliance with the requirements of the European Union Renewable Energy Directive (EU-RED) (International Sustainability & Carbon Certification (ISCC)).

Meanwhile, according to a study by the United States Environmental Protection Agency (EPA) conducted in early 2012, by applying the Notice of Data Availability (NODA), GHG emissions from renewable diesel palm oil and Indonesian palm oil biodiesel can only meet 11-17 percent of the GHG emission reduction limits. This value is obtained using an emission factor from peatlands of 95 tons CO2 per ha/year. In the Renewable Fuel Standard (RFS) program implemented in the United States, renewable energy requirements must meet the minimum 20 percent threshold for reducing greenhouse gas (GHG) emissions [15,16].

The Presidential Regulation No. 61 of 2011 concerning the National Action Plan for Reducing Greenhouse Gas Emissions and Presidential Regulation Number 62 of 2013 concerning the Management Agency for Reducing Greenhouse Gas Emissions from Deforestation, Forest Degradation, and Peatlands are Indonesian government's policies that support emissions reduction. The government also requires palm oil plantations and palm oil mills to implement Indonesian Sustainable Palm Oil (ISPO) with the issuance of Permentan No. 11/Permentan/OT.140/ 3/2015 concerning the Indonesian Mandatory Sustainable Palm Oil Certification System (ISPO). Also, there is a voluntary certification, namely RSPO (Roundtable Sustainable Palm Oil). The application of ISPO and RSPO is intended to ensure the sustainability of oil palm plantations [17, 18, 19, 20]. To meet the criteria, below are a joint list of efforts undertaken by ISPO and RSPO:

- Limiting land clearing in areas with high carbon content such as primary forests and peatlands.
- Paying attention to soil and water conservation when opening / adding new land.
- Using fertilizers according to fertilizer recommendations and the use of empty bunches for fertilization.
- Controlling the use of pesticides.
- Making methane capture facilities in palm oil mills effluent.
- Managing of water level in peat areas.
- Utilizing shells and fiber for boiler fuel as renewable energy.
- Making efficient use of fossil fuels and electricity.

## 4. Best management practice through RSPO

Figure 5 illustrates the waste produced from FFB processing into CPO and the kernel, as well as how it is treated. The green box represents lower GHG treatment in the plant area itself compared to the orange box.

This section discusses the Best Management Practice (BMP), which directly and indirectly reduces GHG emissions in factories. Some BMPs such as methane capture and joint composting were included in previous RSPO BMPs. Although it is effective in reducing GHG emissions, it has not yet become a common practice in the palm oil industry. Methane capture and composting have the potential to be applied to reach zero waste in the palm oil mills. POME processed in methane capture can produce biogas, which can be used further for renewable energy production. In the case of renewable energy from biogas, it is used to replace existing biomass-based electricity production combined with increased efficiency in Combined Heat and Power (CHP).

Meanwhile, excess biomass available for fuel can be used for other production (e.g., paper production). POME can also be used in co-composting facilities to produce organic fertilizer. BMP aimed at reducing GHG emissions in factories and plantations by reducing the application of chemical fertilizers. This section also covers emerging technologies such as improved biogas for transportation fuels, which have been proven in other sectors, but has yet been adopted in the palm oil sector [21, 22]

Implementation of BMP in factories can reduce GHG emissions elsewhere or contribute to a broader context. For example, biogas produced by the capture of methane that is converted to electricity can replace fossil fuel-based power plants. Implementing proper air pollution devices as well as adequate height of stack would lead to a better air quality in the surrounding areas of the palm oil mills [23]. An increase in CHP can lead to biomass savings. Excess biomass can be sold or transported elsewhere for renewable energy generation [24].

Biomass can also be used as material to produce paper, thus avoiding the cutting down of trees. On the other hand, joint composting using EFB and POME can reduce chemical fertilizers on plantations. Besides, the factory implements several BMPs for optimal GHG emission reductions in a reasonable economic investment. Some BMP implementations are not only motivated by GHG reduction but also the goal of sustainability, efficiency, and process improvement. Most factories with internal demand for electricity or those close to the network, choose the capture of methane and biogas for electricity generation [25].

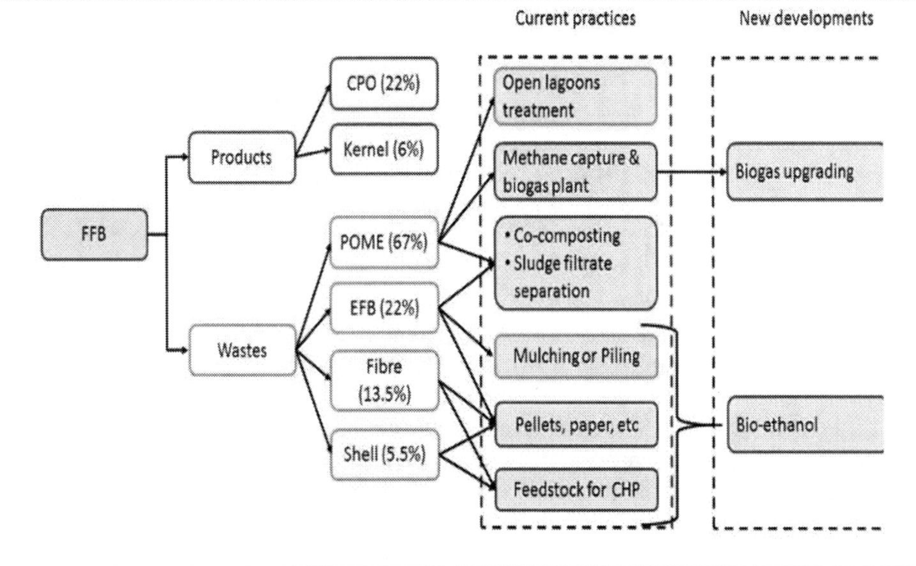


**Figure 5.** Waste utilization in palm oil mills.

## 5. Conclusion

The oil palm industry actively applies BMP to improve its plantation development and management practices in order to reduce emissions. Emission reduction is made through various ways including engagement with smallholders to reduce deforestation, adopting practices without fuel, replacing synthetic fertilizers with factory by-products or organic fertilizers, thus, reducing the use of fossil fuels, and other methods.

## Acknowledgments

The authors express their sincere gratitude to Malikussaleh University through the Ministry of Research, Technology, and Higher Education for the financial support provided to the first author to conduct this research under the doctoral program at Faculty of Engineering, Syiah Kuala University, Banda Aceh, Indonesia.

## References

[1] Badan Pusat Statistik 2015 *Statistik Kelapa Sawit* (Jakarta: Badan Pusat Statistik) p 56
[2] I Gusti Putu W, Sudrajat and Hermanto S 2018 *Pembangunan Perkebunan Kelapa Sawit Berkelanjutan Dengan Pendekatan Dinamis* (Bogor: Idemedia Pustaka Utama) p 29-50
[3] Noviyani P I A and Aryo S 2017 *Jom. F. Tek* **4** 1
[4] Rashid M, Chong W C, Ramli M, Zainura Z N, and Norruwaida J 2013 *Sains Malaysiana* **42** 1289
[5] Rupani P F, Singh R P, Ibrahim M H and Esa N 2010 *World Appl. Sci. J.* **11** 70
[6] Muhammad A N, Tjahjono H and Meta R 2014 *Energy Procedia* **47** 166
[7] Muhamad F O, Sulastri S, and Mohd F M B 2017 *IOP Conf. Ser. Mater. Sci. Eng.* **226** 012003
[8] Najmi W M W A, Rosli A N and Izat M S S 2006 *ICEE* 1
[9] Zhang H, Qiang T C, Huang Y, Wu Q, Wang Y, and Zeng G 2017 *Int. J. Hydrogen Energ.* **42** 23871
[10] Demirbas A 2008 *Energy Sources* **30** 170
[11] Nussbaumer T 2003 *Energ. Fuel.* **17** 1510

IOP Conf. Series: Materials Science and Engineering **845** (2020) 012037 doi:10.1088/1757-899X/845/1/012037

[12] Roundtable on Sustainable Palm Oil 2018 *RSPO Management System Requirements and Guidance for Group Certification of FFB Production* (Kuala Lumpur: RSPO Secretariat Sdn Bhd) p 62
[13] Roundtable on Sustainable Palm Oil 2018 *Compilation of Best Management Practices to Reduce Total Emissions from Palm Oil Production* (Kuala Lumpur: RSPO Secretariat Sdn Bhd) p 157
[14] Kun Y C and Abdullah A M 2018 *Int. J. Energy Environ. Eng.* **4**
[15] Nur M R M, Hanafi H, Mimi and Hassim 2016 *J. Eng. Sci. Technol* **3** 101
[16] Kavalek M, Havrland B, Pecen J, Ivanova T, and Hutla P 2013 *Agron. Res* **11** 183
[17] Kementerian Pertanian Republik Indonesia 2015 *Studi Bersama Persamaan dan Perbedaan Sistem Sertifikasi ISPO dan RSPO* (Jakarta: Kementerian Pertanian Republik Indonesia Sekretariat Komisi Indonesian Sustainable Palm Oil (ISPO)) 1-56
[18] ISPO Matrix 2018 Penerima Sertifikat ISPO (http://www.ispo-org.or.id)
[19] RSPO 2019 Certified Growers Are Producers Of Palm Oil Whose Operations Have Been Certified Against The RSPO Principles And Criteria (https://rspo.org/certification)
[20] Salman F, Najib M, and Djohar S 2017 *Indonesia. J. Bus. Entrep.* **3** 219
[21] Moreno-Peñaranda R, Gasparatos A, Stromberg P, Suwa A, Pandyaswargo A H and Puppim de Oliveira J A 2015 *Sustain. Prod. Consum.* **4** 16
[22] Tampubolon N and Pasaribu M 2017 *IOP Conf. Ser. Mater. Sci. Eng.* **237** 012036
[23] Regina A. A., I. Mohammad Halim Shah, 2010, Jurnal Rekayasa Kimia & Lingkungan., **7**, 176
[24] Rukaiyah R D E H, 2018 *J. Pembang. Berkelanjutan* **1** 105
[25] Rusli Anwar W, Santun R.P Sitorus, Anas Miftah Fauzi, and Machfud 2016 *J. Littri* **22** 11

IOP Conf. Series: Materials Science and Engineering **845** (2020) 012038 doi:10.1088/1757-899X/845/1/012038

# Functional beverage instant ginger powder (*zingiber officinale)* with addition of betel extraction (*piper bettle)*

**F Arifan[1*] S Winarni[2], Wahyuningsih[1], I Pudjihastuti[1] and R T D Wisnu Broto[1]**
[1] Chemical Engineering Diploma, Diponegoro University, Semarang, Indonesia
[2] Department of Biostastics and Population Study, Diponegoro University, Semarang, Indonesia

E-mail: fahmiarifan80@gmail.com

**Abstract.** Ginger is one of the herbal plant species that is used as cook seasoning because it can add a distinctive flavor to foods. Ginger also has so many substance contents that is needed for the human body, such as essential oil (0,5-5,6%), zingiberon, zingeberin, zingibetol, barneol, kamfer, folandren, sineol, gingerin, vitamin (A, B1, and C), carbohydrate (20-60%), resin and organic acid. This research aims to find the value of antioxidant and quality requirements of traditional beverage powder from ginger powder. The tools which are used in the making of ginger powder with the addition of betel leaf extract are: knife, pan, filter, blender, stove, and spatula. While the tools for analyzing are beaker glass, measuring cylinder, test tube, measuring flask, oven, petri dish, Erlenmeyer. The ingredients which are used in the making of ginger powder are : ginger, betel leaf, sugar and water. The result of this research found that water contain of instant ginger betel is 2,080% while ash content of instant ginger betel is 0,533%. This research also analyzes the level of metal contamination, where the metal that is tested is Pb, Cu and Zn metal. The result of the Cu metal contamination is equal to 0,198 ppm, Zn metal is 0,527 ppm, while the Pb metal contamination is undetected. The research is also analyzing the content of antioxidant and found that the content of antioxidant is 41,632%. The conclusion is the content of metal contamination is Cu, Zn and Pb, while water content and ash content is still the same with SNI 01-4320-1996.

## 1. Introduction

Indonesia has abundant natural resources, including types of herbal plants. From various herbal plants, some types of them have a health benefit. The one plant that is known to have health benefit is Ginger. Ginger is one type of medicinal plant that is usually used as a spice because it can add the flavor to food. Ginger can also be relied as a non-oil and gas export commodity in the form of fresh ginger, dried ginger, essential oil, and oleoresin. Ginger has been widely known and used for various purposes, such as a mixture of food, beverage, cosmetics, and perfume. Many research results proved various benefit of ginger rhizomes, including anti-diarrheal, antioxidant, anti-hepatotoxic and antipyretic activities[1], [2]

Ginger has a functional contains as an antimicrobial and antioxidant. One of inhibiting growth component of E. coli bacteria is gingerine and gingerol, while antioxidant ability comes from gingerol and shogaol contain [3]. Based on previous research, the major constituents in ginger are

[*] To whom any correspondence should be addressed.

carbohydrates (50–70%), lipids (3–8%), terpenes (zingiberene, β-bisabolene, α-farnesene, β-sesquiphellandrene, and α-curcumene), and phenolic compounds (gingerol, paradols, and shogaol).[4]

The purpose of making ginger powder is to increase the sale value of ginger and make ginger easily to consume, then the addition of betel leaf extract to increase the benefit of the ginger powder. One of the benefits of betel leaf extract is to cure asthma and help in menstruation. The purpose of this study was to determine antioxidant activity, as well as the quality requirements (water content, ash content, and content of heavy metal contamination) from traditional beverage powder, namely ginger powder in addition to betel leaf extract.

## 2. Research methodology

### *2.1. Research material*

The tools used for making ginger powder with the addition of betel leaf extract are: knives, pan, filter, blender, stove, and wood stirrer. The tools for analysis are beaker glass, measuring cup, spatula, test tube, measuring flask, oven, petri dish, erlenmeyer, and funnel. The ingredients used for making ginger powder are ginger, betel leaves, sugar, and water. The chemicals used are supporting materials for the analysis of antioxidant DPPH (1,1-diphenyl-¬2¬pcrylhydrazil).

Antioxidant testing using the DPPH method was carried out using 0.1 ml of sample extract diluted with various concentrations. The extract was then reacted with 3.9 ml of DPPH solution (2.4 mg DPPH / 100 ml methanol) and homogenized. After 30 minutes incubated at room temperature followed by absorbance measurements with a spectrophotometer at a wavelength of 516 nm. 0.1 ml of methanol (as blanco) was reacted with 3.9 ml of DPPH in the 0th minute. For each test, three replications were carried out, and the results were delivered in an average form. Antioxidant activity is calculated by the equation:

$$\%\ inhibition = \frac{(Abs\ blanco - Abs\ Sample)}{Abs\ blanco} \times 100\ \% \qquad (1)$$

Abs : absorption value

Testing of moisture content using the gravimetric method. The gravimetric method is carried out by means of drying using an oven at 105oC for 4-5 hours (W0) weighed and dried to a constant mass. Then the petri is weighed with a sample and dried to a constant mass (W1) until the weight of the water evaporates with the weight of the wet sample.

### *2.2. Research implementation*

Material preparation includes making betel leaf extract, which is betel leaf collecting, weighing, blending it to a pulp, soaking by using water, filtering, and producing filtrate (betel leaf extract).

The process of making ginger powder begins with ginger collection, sorting, stripping, washing, and weighing of ginger and water. Then shredded, then filtered to obtain the filtrate or ginger juice. Ginger and betel leaf extract are mixed in a pan which is then heated to boiling. Fast stirring until crystalized. Ginger powder that produced is mashed using a blender, then sieved with a sieve to produce ginger instant powder.

Instant ginger powder will be analyzed for nutritional value, antioxidant, and traditional drinks powder standard quality from ginger powder.

## 3. Results and discussion

### *3.1. Water content*

Moisture is the percentage of the water content of a material. Water in food ingredients is one factor that affects metabolic activity, such as a microbial activity that can affect nutritional quality[5]. The water content in the powder material also determines the durability of raw material. In this study, a sample of instant ginger produced in Tlogowungu Village, Kaloran, Temanggung. The result of the water content analysis of instant betel ginger is 2.080%, where the moisture content matches into the standard of traditional beverage powder quality of SNI 01-4320-1996. Where in the SNI, the water content is allowed up to 3.00%.[6] Drying method that can be used to make powder drinks such as ginger powder, among others, using spray drying. The obstacle when using this method is that it is very expensive in terms of cost so it is not suitable for medium businesses or small businesses. Another method of drying is drying with an oven with a temperature of <100oC. High temperatures adversely affect nutrient content. The temperature of the oven used is too low (<50oC) causing a long drying process. Therefore, the drying process is better at a temperature of 60-80 °C[7], [8].

The benefits of water in food as one of the factors that affect the activity of enzymes, microba, chemical processes, which cause changes in organoleptic properties, as well as nutritional value. Water in food is expressed as a percentage of water content, AW, or RH. Water content is the percentage of the water content of a material can be expressed in wet weight (wet basis) or dry weight (dry basis). Water activity or water activity (AW) is the amount of free or unbound water in a system that can support biological and chemical reactions. Relative humidity (RH) is defined as the ratio between the water vapor partial pressure to a certain temperature saturated vapor pressure [9], [10].

### *3.2. Ash content*

Ash is the residual inorganic material from the complete combustion process at 600 °C for a few times. The amount of ash content of food products depends on the amount of mineral content of the ingredient used. Food ingredient is at least 96% containing organic and water. The rest consists of mineral elements or ash content.[5] According to SNI 01-4320-1996 the ash content allowed in traditional drinks powder is up to 1.5%. The result of ash content from instant betel ginger obtained at 0.533% where the ash content was still allowed by SNI 01-4320-1996. Determination of total ash content can be used to determine whether or not processing of food ingredients, knowing the type of material used, and as a determinant of nutritional value parameters.

### *3.3. Metal contamination content*

In this study, the metal contamination content analysis of instant betel ginger sample. The metal contamination content analysis was carried out because if the content of metal contamination in the sample exceeded into the limit and consumed by a human could cause spontaneous interference. The metal contamination analysis in this sample are metal contamination content of Lead (Pb), metal contamination content of Copper (Cu) metal contamination content of, and metal contamination of Zinc (Zn). The method used in measuring metal contamination is the Atomic Absorption Spectrophotometer (AAS). The result of the analysis is that Lead content (Pb) in instant betel ginger is unknown, the content of Copper (Cu) in instant betel ginger is 0.198 ppm and the content of Zinc (Zn) for instant betel ginger is 0.527 ppm. Where the metal contamination content still allowed into SNI 01-4320-1996. Where the maximum content of Lead (Pb), Copper (Cu), and Zinc (Zn) are 0.2 ppm; 2.0 ppm and 50 ppm.[6]. Food is often polluted by inorganic components, including various dangerous heavy metals. Heavy metals are high molecular weight metal elements. Low-grade heavy metals are generally toxic to plants, animals, and humans.

### *3.4. Antioxidant content*

The determination of antioxidant activity is using DPPH method. The DPPH method is chosen because it is simpler, easier, and faster. Another advantage is the DPPH method requires only a small sample to be tested so it can be used widely to test the ability of compounds as electron donor. DPPH testing method is a conventional method and has long been used to determine the activity of the antioxidant compounds.[11] The test is carried out at a wave of 516 nm because it is the maximum wavelength of DPPH.[12] The test method using DPPH is based on a decrease in absorbance due to discoloration solution color in DPPH color, where DPPH reacts to hydrogen atom from free radical reducer compound to form a more stable DPPH-Hydrazine. DPPH reagent that reacts to antioxidants will cause discoloration from purple to yellow. Color intensity depends on the ability of the antioxidant.[13]

The principle of this method is the measurement of DPPH fishing arrests with antioxidant activity compounds using spectrophotometry so the value of the antioxidant activity is known. From the picture, it can be seen that the highest value of the antioxidant activity of instant ginger is 41.632%, and in instant betel ginger is 34.990%. This difference is because instant ginger only consists of ginger, where the value of antioxidant activity from ginger is very high. While instant betel ginger consists of two mixes, betel leaf, and ginger extract, in which the mixture of these two ingredients can increase the value of the antioxidant activity. The antioxidant strength of ginger powder mixed with betel leaf extract increased 6.642% and included in the strong category (IC50 value is between 10-50 μg/mL)[14].

## 4. Conclusion

Ginger is a rhizome plant that can be used for various kinds of processed products. One of the commodities in Tlogowungu Village is ginger. This ginger is used as a raw material for making instant ginger in Tlogowungu Village. The produced instant ginger met the standard of traditional beverage powder quality (SNI 01-4320-1996). The combination of ginger powder and betel leaf extract can increase antioxidant strength by 6.642% from its original value. This drink is categorized as containing strong antioxidant categories.

## Acknowledgment

We thank the farmers involved in this study, this study was funded by Diponegoro University and the Ministry of Research, Technology and Higher Education of the Republic of Indonesia.

## References

[1] Lete I and Allué J 2016 *Integr. Med. Insights* **11**
[2] Ygonum P O L and Ra O D O 2011 *Benha Vet. Med. J.* **22** 9
[3] Ko M J, Nam H H M and Chung M S 2019 *Food Chem.* **270** 149
[4] Prasad S and Tyagi A K 2015 *Gastroenterol. Res. Pract.* **2015**
[5] Cabral J P S 2010 *Int. J. Environ. Res. Public Health* **7** 3657
[6] BSN, Traditional Beverage Powder 1996 *Indonesian National Standard.*
[7] Clement A, Olatunde M, Patrick O and Joyce O 2017 *World J. Food Sci. Technol.* **1** 93
[8] Yusufe M and Mohammed A 2017 *Ann. Food Sci. Technol.* **18** 145
[9] Leclercq-Perlat M N, Sicard M, Trelea I C, Picque D, Corrieu G 2012 *J. Dairy Sci.* **95** 4666
[10] Onilude A A, Igbinadolor R O and Wakil S M 2010 *African J. Microbiol. Res.* **4** 1939
[11] Simamura T, Sumikura Y, Yamakazi T, Tada A, Kashiwagi T, Ishikawa H, Matsui T, Sugimoto N, Akiyama H and Ukeda H 2014 *Anal. Sci.* **30** 717
[12] Kasangana P , Haddad P and Stevanovic T 2015 *Antioxidants* **4** 410
[13] Molyneux P 2003 *J. Sci. Technol.* **13** 1
[14] Phongpaichit S, Nikom J, Rungjindamai N, Sakayaroj J, Hutadilok-Towatana N, Rukachaisirul V and Kirtikara K 2007 *FEMS Immunol. Med. Microbiol.* **51** 517

# Refrigeration system based-dehumidifier

**D Ariwibowo[1*], Indartono[1] and S Darmanto[1]**
Mechanical Engineering Dept. Vocational School, Diponegoro University. Semarang, Indonesia

E-mail: didik.ariwibowo@live.undip.ac.id

**Abstract.** This research aimed to develop a dehumidifier prototype based on R-404C vapor compression refrigeration system and observed its performance. Condenser and evaporator became a unit to condense the moisture content of the air and heat it to an elevated temperature. Condenser and evaporator were arranged in line with airflow to be treated. The air velocity drawn through the unit of condenser-evaporator arrangement was set at 4.5, 5.0, 5.5 m/s by fan current input controller. Evaporator and condenser are a fin-tube type with areas of 0.6 m$^2$. Refrigerant inlet temperature to evaporator was set at -20 °C. Temperature and Relative humidity (RH) of air at the inlet and outlet of the unit were measured by thermohygrometer. The humidifier performance was indicated by a ratio between moisture condensed (kg) and system energy input (kW). The heat released in the cooling and condensation process was 10.1 kW and 17.8 kW, whereas heat consumed in the heating process was 24.8 kW. The best performance of the dehumidifier was 16.7 kg/kWh at air velocity, outlet temperature and RH were 5 m/s, 42 °C and 17 %, subsequently.

## 1. Introduction

In the food industry, drying process development for heat sensitive raw materials becomes crucial. Drying atmosphere affects to properties and qualities of dried product. Air at elevated temperature and lower relative humidity (RH) is needed to dry a temperature-sensitive product, such as herbal-based product. The drying temperature for the product is in the range of 30 °C – 40 °C [1;2]. The temperature not only avoids nutritional damage but also case hardening of the product [3].

At room temperature, water diffusion in herb-fiber is drawn slower to achieve product surface so that the drying process becomes longer. An effort to solve the problem is to expose the product in a lower air relative humidity. In this environment, the drying process will be faster [4;5]. The dehumidification process can lower relative humidity. One of the dehumidification process methods is by lowering air temperature until its dew point. At dew point, a part of moisture content in the air will be condensed. The rest portion will be as dried-air with lower relative humidity.

One of the dehumidification process methods is by lowering air temperature until its dew point. At dew point, a part of moisture content in the air will be condensed. The rest portion will be as dried-air with lower relative humidity.

The equipment to utilize the dehumidification process namely Air Conditioning (AC) unit. The evaporator in AC unit will condense air drawing through it. But, the temperature of the air is relatively low. To increase the air temperature, the heater is installed [6;7]. In this scheme of dehumidification process, air flowing out from the equipment about 30 °C – 57 °C [8] with relatively low RH .

* To whom any correspondence should be addressed.

IOP Conf. Series: Materials Science and Engineering **845** (2020) 012039 doi:10.1088/1757-899X/845/1/012039

Dehumidifier using AC and heater affects to higher operational cost, because AC is designed for room conditioning, not for obtaining dried-air. Therefore, modification of evaporator-condenser arrangement and their size could be a prospect design.

Modification of the evaporator-condenser arrangement and their size in a refrigeration system will affect the vapor compression system cycle balance [9]. For this preliminary research, the evaporator and condenser were in the same size. The cycle balance and air flowing out from the equipment were observed.

## 2. Literature review

The extraction rate of moisture takes into account an observation of a dehumidifier system. The drying load will decrease as the product dries out, affect the property of the process air. The change in air property will disturb the refrigeration cycle. The air property, together with an inter-dependence of the air and refrigerant, results in the dynamic behavior of the dehumidifier system. This leads to different behavior in each period of the dehumidification process. During all differing dehumidification periods, adjustment to the main influencing parameters will achieve an optimal performance of the dehumidifier. The parameter includes compressor power and the air flow rate. The regulation for a batch-type dehumidifying process is therefore difficult. Steady-state operation of dehumidifier by using a refrigeration system or sorption systems to be a better choice, if the process dynamics are avoided.

The aim of dehumidifier design is how to use energy efficiently [9]. Evaporator and condenser take a role in the amount of internal heat loss effects. Most refrigeration system based-dehumidifier perform far away from efficient points. The varies design condition of the dehumidifier will lead to varies optimum points. Inappropriate dimensioning of the heat exchanger component leads to this inefficient design conditions. Heat exchanger components size and arrangement will be one of the most critical challenges of a refrigeration based-dehumidifier design.

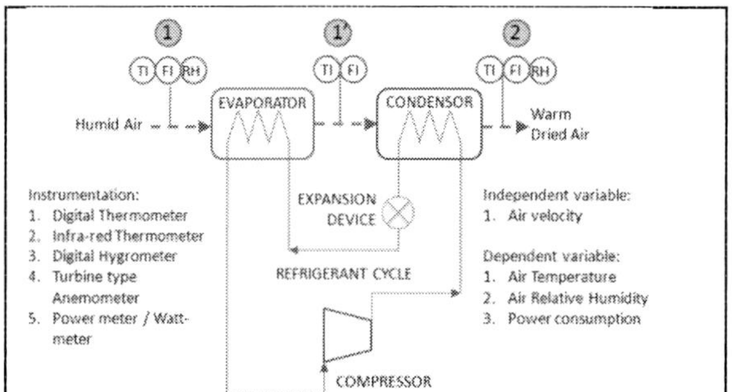


**Figure 1.** Scheme of the dehumidifier prototype.

The capacity of refrigeration or evaporation is proportional to the specific enthalpy difference of air. And this enthalpy difference is proportional to the temperature difference of the evaporator's inlet and surface temperature, with considering of slope of the saturation line. It is possible to calculate the optimum temperature difference, with a given relative humidity (RH) in a continuous process and given dehumidifying air temperature. This optimum temperature difference is for a minimum evaporation capacity.

At this point, a certain amount of water will be discharged. On the other hand, it is possible to calculate the optimum temperature difference related to the influence-parameters mentioned above to achieve a maximum moisture extraction. There is no correspondence between maximum moisture extraction and the minimum evaporation capacity. This is due to the fact that there are different optimal enthalpy differences.

## 3. Materials and methods

Equipment used in this research were the dehumidifier prototype built from modified refrigeration equipment. The observed parameter was measured using thermohygrometer, power mete,r and pressure gauge. Data collected from the measurement were analyzed and shown as graphs.

### *3.1. Equipment*

The dehumidifier prototype was built based on the scheme shown in Fig. 1. There were two elements of the prototype. Element 1 consists of the main equipment of the refrigeration system; those are a compressor, condenser, expansion device, and evaporator. Whereas element 2 consists of a duct and fan that draws air through evaporator and condenser arrangement. Evaporator and condenser were arranged in line with airflow so that air experiences cooling and heating processes. The Condensor placed immediately downstream of the evaporator.

Kulthorn WJ 2450ZK-Pyang was the compressor for the refrigeration system. Evaporator and condenser have a heat transfer area of 0.6 square meters ($m^2$), whereas the fan used were LTF3E-300-50, 80 watts, 1380 revolution per minute. The refrigerant used in the refrigeration system was R-404C.

### *3.2. Data analysis*

The parameter measured in this research were air velocity (m/s), air temperature (°C), air relative humidity (RH), power input to the refrigeration system and fan (Watt).

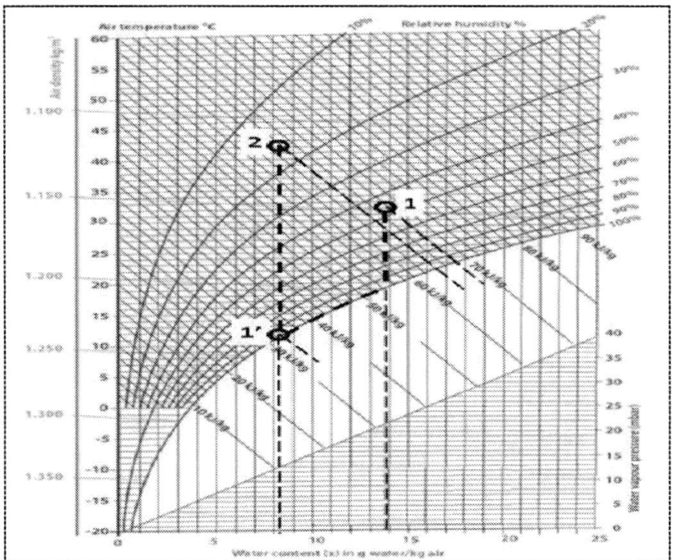


**Figure 2.** Cooling, condensation, and heating processes of air flowing through evaporator-condensor arrangement.

The temperature was measured at inlet (1), intermediate (1'), and outlet point (2) of the dehumidifier. Data collected from the measurement were analyzed. The performance of the dehumidifier was calculated as the ratio between moisture condensed and system energy input (kg/kWh). The air velocity was varied at 5.5 m/s, 5 m/s, 4.5 m/s.

## 4. Results and discussion

Air that experienced cooling, condensation, and heating processes when flowing through the evaporator-condenser arrangement was described in Mollier chart shown in Fig. 2. Air, from initial temperature 30 °C and 50% RH (point 1), was drawn into the dehumidifier, and indicated at point 2 when out from dehumidifier.

The temperature of air out from evaporator (point 1') was 12 °C, whereas air out from condenser (point 2) had temperature and relative humidity of 42 °C and 17% RH. Fig. 2 shows that moisture decreasing was about 5.8 g/kg air.

At the best point, the air mass flow rate was 2,800 kg/h. This condition would condense moisture from the air about

$$\dot{w} = 2{,}800\ \frac{kg}{h}\ x\ 5.8\frac{g}{kg}\ =\ 16.24\ kg/h$$

The total power input to the system was 0.97 kW. Therefore, the ratio between moisture condensed and power input was

$$\text{Moisture to power input ratio} = \frac{16.24\ kg/h}{0.97\ kW} = 16.7\ \text{kg/kWh}$$

The air experienced a cooling and condensation processes when contacted with the evaporator surface. At this time, there was an amount of heat released in the form of sensible heat and latent heat. The sensible heat is the heat released from the entry point of the evaporator to the dew point (cooling process), whereas latent heat is the heat released when condensation occurs. The amount of the heat was calculated by specific enthalpy and mass flow rate of air.

**Table 1**. Enthalpy, Temperatur and RH at inlet, intermediate, and outlet point.

| Air velocity, Vair (m/s) | | Inlet (Ti) | Intermediate (Tint) | Outlet (To) |
|---|---|---|---|---|
| 5.5 | T (°C) | 30 | 21 | 35 |
| | RH (%) | 50 | | 34 |
| 5 | T (°C) | 30 | 12 | 42 |
| | RH (%) | 50 | | 17 |
| 4.5 | T (°C) | 30 | 9 | 42 |
| | RH (%) | 50 | | 11 |

Refer to Figure 2, specific enthalpy at the entry point of the evaporator was 68 kJ/kg. The specific enthalpy at dew point and out from evaporator was 55 kJ/kg and 32 kJ/kg, subsequently. Therefore, the sensible heat and latent heat were determined by the following calculation.

$$\text{Sensible heat} = \frac{2{,}800\frac{kg}{h}\ x\ (68\frac{kJ}{kg} - 55\frac{kJ}{kg})}{3{,}600}$$

$$= 10.1\ \text{kW}$$

$$\text{Latent heat} = \frac{2{,}800\frac{kg}{h}\ x\ (55\frac{kJ}{kg} - 32\frac{kJ}{kg})}{3{,}600}$$

$$= 17.8\ \text{kW}$$

The heat consumed in heating process determined by the following calculation.

$$\text{Heating} = \frac{2{,}800\frac{kg}{h}\ x\ (64\frac{kJ}{kg} - 32\frac{kJ}{kg})}{3{,}600}$$

$$= 24.8\ \text{kW}$$

The temperature measured at inlet, intermediate, and outlet points for the various velocity of air is shown in Table 1. Table 1 shows that the intermediate temperature of the air was decreased as air velocity decreased. It is reasonable that the lower air velocity affects heat transfer contact time longer, so that tends to decrease the final temperature with decreasing of air velocity. This longer contact time also affects to heating process at the condenser.

The ratio between air heating effect at the condenser and air cooling effect at evaporator at air velocity 5.5 m/s, 5 m/s, and 4.5 m/s were 1.5, 1.44, 1.47, subsequently. This ratio depends on the pressure ratio of the compressor.

The cooling process of air flowing through evaporator released an amount of heat termed as the heat of air, in kilowatt (kW) unit. The relation between the heat of air, the mass flow rate of air and log mean temperature difference (LMTD) was figured at Fig. 3. This LMTD was the temperature difference between air temperature flowing through the evaporator and surface temperature of the evaporator (Tse). The surface temperature of the evaporator was constant from inlet to outlet, and closed to refrigerant temperature (Te) of -20 °C.

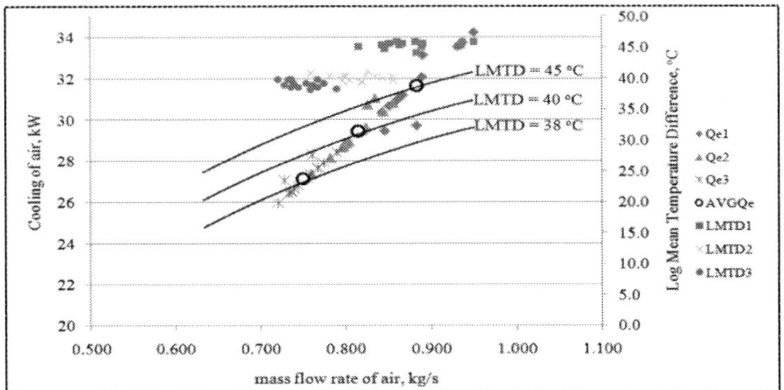


**Figure 3.** Cooling of air with a mass flow rate of air and LMTD evaporator.

The cooling of air was obtained from this equation,

$$Q_{air,e} = \dot{w}_{air} \; x \; Cp_{air} x \; (T_{int} - T_i) \; kW \tag{1}$$

whereas LMTD was calculated from this equation

$$LMTD = \frac{(T_i - Ts_e) - (T_o - Ts_e)}{Ln\frac{(T_i - Ts_e)}{(T_o - Ts_e)})} \tag{2}$$

Figure 3 shows that the cooling of air was increased when air mass flow rate and LMTD increased. Similar to the evaporator performance described by Stoecker [10], this description of Fig. 3 is a rationale. It could be said in other terms that the air side heat transfer coefficient hair increases as the airflow rate increases as a result, the overall heat transfer coefficient of the evaporator increases.

The observation was done to the whole system when operated. There was an abnormality of the compressor when air velocity was lower than 4.5 m/s. In this condition, the temperature of the compressor increased, and the compressor current was rising close to its upper limit. Until the upper limit was reached, the system was turned off. It was caused that air cooling capacity to condenser was lower. This lower cooling capacity affected to increase the temperature of the refrigerant flowing in condenser and evaporator tube.

There was not frozen condensate on the evaporator tube surface. It was indicated that the evaporator was operated at normal condition.

The best performance of the dehumidifier was 79.77 at air velocity, outlet temperature and RH were 5 m/s, 37 °C and 17 %, subsequently. At the best performance, the enthalpy change of air flew through the evaporator and condenser.

## 5. Conclusion
The evaporator-condenser arrangement in the refrigeration system could be a prospect design for dehumidification. The design could lower the relative humidity of air from 50% to 17% with air velocity of 5 m/s. The outlet temperature was 42 °C, the suggested temperature for heat sensitive products.

## Acknowledgment
This research was supported by Vocational School Diponegoro University. We thank our colleagues from Mechanical engineering who provided insight and expertise that greatly assisted the research.

## References
[1] Qing-guo H, Min Z, Mujumdar A S, Wei-hua D and Jin-cai S 2006 *Dry. Technol.* **24** 1025
[2] Kuntjoko A 1989 Paripurna, Jakarta
[3] D Ariwibowo and M E Yulianto 2013 *Majalah Gema Tekologi* **22** 31
[4] Hawlader M N A, Perera C O and Tian M 2006 *J. Food Eng.* **74** 392
[5] D Ariwibowo and Yulianto M E 2012 *METANA* **6** 7
[6] Perera C O and Rahman M S 1997 *Trends Food Sci Tech.* **8** 75
[7] Minea V 2015 *Dry. Technol.* **33** 515
[8] Strumillo C 2006 *Dry. Technol. 24* 1059
[9] Kumar R and Sharma K K 2018 *Natl J. Multidiscip. Res. Dev.* **3** 491

# Physical characteristic manag'ement and palatability of processed mung bean sprout for duck

**Retno Iswarin Pujaningsih**[1*] **and Istna Mangisah**[1]
Department of Animal Agriculture, Diponegoro University, Semarang, Indonesia

E-mail: retno.marwoto@gmail.com

**Abstract.** This research was conducted to observe three forms of processed mung bean sprouts on physical characteristic and palatability in 6 weeks male ducks. Mung bean sprouts were feed in the form of fine, fresh and fermented. Completely Randomized Design was used in this study as experimental design with 3 treatments and 4 replicates. Water activity, water absorption and palatability were measured as variables. The outcomes of this study showed that treatments significantly influenced water activity and mung bean sprouts palatability ($P<0.05$), respectively. High significantly difference was found among treatments in water absorption ($P<0.01$). It can be inferred from the research that to improve palatability, duck feed must contain water content between 11.5-14.46%. Mung bean sprout waste can be provided in the form of fresh or even fermented for the duck. According to its water activity, diet consist of mung bean sprout waste in the form of dried mash is recommended to be stored for a relatively longer period of time compare to the other treatments.

## 1. Introduction

Naturally, ducks have a higher tolerance for fiber-containing feed than other birds. Duck farmers usually utilize agricultural and plantation waste as an alternative feed ingredient, because ducks are able to digest crude fiber better than other poultry up to ± 10%. High feed costs encourage farmers to try to use alternative feed ingredients that are cheap and have good nutrient content to reduce production costs. One of agricultural waste that can be used as an alternative feed ingredient is mung bean sprouts waste. Mung bean sprout waste is a part that has not been used again in the production process. It contents 88.5% dry matter, 13.56% crude protein, 33.07% crude fiber, 0.22% crude fat and 64.58% of Total Digestible Nutrient [1].

Grains contain a variety of Non Starch Polysaccharides (NSP), which consist mainly of arabinoxylans (pentosane), ß-glucan and cellulose [2]. The type of non-starch polysaccharides in mung bean sprout waste flour is arabinose. Arabinose is a pentose sugar, in the class of polysaccharides or non-soluble fibers (NSP) so that in utilizing waste flour mung bean sprouts require the help of digestive tract microbes. Non-starch polysaccharides can be used optimally as a prebiotic source through a fermentation process. Fermented mung bean sprouts can increase the population of Lactic Acid Bacteria so that short chain fatty acids (SCFA) and lactic acid increase [3].

Mung Bean Sprout can be processed as feed for duck. However, improper feeding management can affect duck productivity. From an economic point of view, converting it to cheap animal feed is the right effort, it does not compete with human needs. Birds must be grown and developed with enough

[*] To whom any correspondence should be addressed.

nutrients since they kept in a close housing cage all the time. Feed drying technology can compose feed product into powder form. Processing the bean sprout into powder form allows it to be easily mixed into the feed as a complete feed. Complete feeds are recommended to be given to ducks that are intensively maintained [4].

Furthermore, the use of mung bean by using feed processing technology may reduce environmental pollution. A simple and cheap processing technology is needed to make the mung bean sprouts wastes more durable, easy to store and to give. A technology process is required because this mung bean sprout wastes are perishable and voluminous (bulky). It's also needed to overcome feed scarcity by simple, cheap and applicable processing technology to increase feed quality and storage. Therefore, the aim of this experiment is to find out the physical characteristics (water activity and water absorption) and palatability of the bean sprout waste in formed of fresh, fine and fermented.

## 2. Materials and method

### *2.1. Materials*

Six weeks of male ducks used for the study were choosed in weight's range of 900 ± 110.15 g. The ration made up of yellow corn, rice bran, soybean meal, fishmeal, mineral mix and mung bean sprout waste. Waste of mung bean sprouts is served in fine, fresh and fermented. *Trichoderma harzianum* was used to mung bean sprout fermentation process. The composition and nutritional content of the research ration are listed in Table 1, referred on the results of the analysis in the Nutrition and Feed Science laboratory.

### *2.2. Preparations of mung bean sprouts waste*

Making sprouts is done by washing the mung beans and then soaking them for 8 hours before draining them. Mung bean incubation is carried out at room temperature for 24 days. During the germination process, watering is done every 12 hours. Mung beans are separated between bean sprouts and mung bean sprouts waste. The wastes sprouts are set aside for the treatment of feeding fresh bean sprouts. While feeding fine treatment is done by drying the waste in the sun. Afterward, milling is done with the aim of reducing the size of the feed material so that it is easily digested.

The fermentation mung bean sprout waste processing begins by conditioning the wastes at 60-70% moisture content. Then Trichoderma harzianum was dissolved with 6% water of dry matter content. Trichoderma harzianum solution added by 5% molasses of dry matter content and then mixed with mung bean sprout wastes and stirred until homogeneous. At that point, it is put into a plastic that has been perforated and fermented for 6 days. Next 6 days, it aerated to stop the fermentation process [11]. Then the available feed ingredients are arranged in accordance with the formulations that have been previously calculated.

### *2.3. Reseacrh conducting*

The experimental design used in this research was Completely Randomized Design with 3 treatments and 4 replicates. The treatments were feeding T0: fresh mung bean sprouts waste; T1: fine mung bean sprouts waste; T2: fermented mung bean sprouts waste. Data were analyzed using ANOVA and followed by Duncan Test [5]. Variables measured were (1). Physical characteristic of mung bean sprouts waste included water absorption and water activity. (2). The palatability of mung bean sprouts waste.

## 3. Results and discussions

### *3.1. Chemical composition*

The nutrient composition of complete feeds which content mung bean sprouts wastes were presented in Table 1. Refer on proximate analysis, the lowest water content (9.42%) was found in diet contained fine mung bean sprouts waste. The highest water content (14.46%) was noticed in the ratio contained of fermented bean sprouts. The level of water content is affected by water content of raw materials and relative humidity (RH). Reference [6] stated that an increase in pellet water content in the range of

9.9% -17.25% still meets the maximum water quality feed standard for poultry rations, namely 14%. The microorganism could be inactivity if water content of wafer was approximately 12% -14% [7]. At this level the feed will not be effortless to be moldy and decayed. Water content of approximately 8-12% is necessary to get an ideal bounding [8].

**Table 1.** Nutritional Composition and Content of Diet's Study

| Feed Ingredients | Diet's Study | | |
|---|---|---|---|
| | T0 | T1 | T2 |
| | ---------------- | % ---------------- | |
| Corn | 39.6 | 40 | 40.7 |
| Rice bran | 18.4 | 18.5 | 18.3 |
| Soy bean meal | 19.2 | 19 | 18.9 |
| Fish meal | 6.8 | 6.5 | 6.1 |
| Fresh mung bean sprout waste | 15 | - | - |
| Mineral mix | 1 | 1 | 1 |
| Fine mung bean sprout waste | - | 15 | - |
| Fermented mung bean sprout waste | - | - | 15 |
| Total | 100 | 100 | 100 |
| Nutritional content (%) | ---------------- | % ---------------- | |
| Metabolize energy (kkal/g) | 2832.73 | 2843.55 | 2845.87 |
| Dry matter | 88.5 | 90.58 | 85.54 |
| Crude fiber | 13.22 | 13.61 | 13.18 |
| Fat | 4.23 | 4.16 | 4.14 |
| Crude protein | 19.11 | 19.01 | 19.06 |

### *3.2. Water absorption*

Table 2 presented physical characteristic of ratio contained mung bean sprout waste includes water absorption. Water absorption is defined as the capability of materials to absorb water from the air to link with material's particle [12].

T2 was found as lowest water absorption in diet contained of fermented mung bean sprout waste (109.75 ± 53.41). The highest water content was discovered in diet consisted of fermented mung bean sprout waste (14.46%). In diet contained with fine mung bean sprout waste (137.79 ± 32.48), the highest water absorption was found. This was because T1 include more fibers with more air hollow to absorb extra water. The water absorbing volume was inversely proportional with compactness. The higher the bulk the lower water absorbing capacity will occur [6].

**Table 2.** Physical characteristic of ratio contained mung bean sprout waste

| Treatments | Variables | |
|---|---|---|
| | Water absorption (%) | Water activity (%) |
| T0 | $121.53 \pm 36.14^{A}$ | $0.91 \pm 0.01^{a}$ |
| T1 | $137.79 \pm 32.48^{B}$ | $0.82 \pm 0.11^{b}$ |
| T2 | $109.75 \pm 53.41^{C}$ | $0.91 \pm 0.01^{a}$ |

### *3.3. Water activity*

One of the important chemical compound for life is water. Microorganism capable to grow at water activity around 0.70. Water activity is the sum of free water that microorganisms rely on for growth [9]. When water activity outside the cell becomes low enough, it causes osmotic stress: the cell cannot take up water and becomes dormant. The microorganisms are not eliminated, they just become unable to grow enough to cause infection [10].

The water activity which found in diet consisted of fresh mung bean sprouts waste (T0) and fine mung bean sprout waste (T1) was not significantly difference (Table 2). The lowest water activity was found in ratio contained fermented mung bean sprouts waste (0.82±0.11). The higher the Aw value (near 1), the more microbes can grow [9]. Addition of molasses to fermented mung bean sprouts waste decreases the water activity in accordance with the hygroscopic nature of sugar which has the ability to form hydrogen bonds with water. The presence of hydrogen bonds between water and sugar causes a decrease in the amount of free water and a decrease in the Aw value, so that water cannot be used for microbial growth. Reference [10] stated that the process of drying, evaporation, adding sugar, adding additional ingredients that are hygroscopic or adding salt is a way to reduce the value of Aw. The water activity ratio can be stored. The water activity of ratio could be concerned by water content in raw materials or environment temperature where the diets was kept [6].

### *3.4. Palatability*

In the current study, the palatability of ratio contained mung bean sprout waste was consumed as indicator to identify how male ducks choose to consume. The palatability of feed ingredients might be indicated by the organoleptic such as aroma and perception (salty, sweet, and bitter), consistency and temperature that draw livestock to eat. Palatability is defined as the reaction provided by the feed given to the livestock. The ability of livestock to digest feed deeply persuaded by the performance and the physical condition of the feed stuff processed. Livestock will be easier to received and attracted to the food particles that are easier to consumed and digest. Improved digestibility will lead to increased consumption [13]. This palatability test was conducted in 3 days presented in Table 3.

**Table 3.** Palatability of diet contained mung bean sprout waste (g/head/day)

| Replication | Treatments | | |
|---|---|---|---|
| | T0 | T1 | T2 |
| 1 | 122.50 | 89.70 | 118.10 |
| 2 | 125.10 | 80.50 | 116.90 |
| 3 | 121.60 | 86.80 | 118.55 |
| 4 | 120.50 | 97.20 | 120.20 |
| Mean | 122.43$^{a}$ | 88.55$^{b}$ | 118.44$^{a}$ |

Ratio contained fresh mung bean sprout waste showed the highest palatability. Since, this diet contains some mung bean sprout waste which had particle size that is easy to be consumed. The results in Table 3 presented that ratio contained fermented mung bean sprout waste (T2) is not significantly different with diet content of fresh mung bean sprout waste (T0). Besides, T2 had the highest water content that make it easy to swallow. Good diets should have good palatability. This remains crucial because it is an outcome of a combination of aspects that animals were stimulated by eyesight, aroma, toching base and taste [14].

From the three types of feed given, the treatment of fine mung bean sprout waste containing rations shows the lowest average yield (88.55 g / head / day). Reference [15] clarified that duck should provide with a wet diet instead of a fine dried diet. Because the duck's beak is wide and jagged. It is difficult for the duck to take smooth and dry feed. However, if they are forced to eat dry food, some of the feed will stick to the beak. Feed that has entered the oral cavity will be attached to the palate of the oral cavity, tongue and throat. This situation will stimulate the duck to drink so that it is easy to swallow the feed. But when the duck is drinking, some of the feed in the beak and oral cavity will be left in the drinking water and will be wasted. Based on study observations, the food scattered in drinking water reaches 20-22% of the given.

## 4. Conclusion

It can be inferred from the research that to improve palatability, duck feed must contain water content between 11.5-14.46%. Furthermore, mung bean sprout can be provided in the form of fresh or even fermented for the duck. According to its water activity, diet consist of mung bean sprout waste in the form of dried mash is recommended to be stored for a relatively longer period of time compare to the other treatments.

**Acknowledgment**
We thank our colleagues and students from "the ducks and mung bean team" who delivered insight and expertise that seriously assisted the study, although they may not agree with all of the clarifications/conclusions of this paper. We would also like to present our gratitude to the academic staff of Laboratory of Feed Technology and Laboratory of Feed Nutrition Science at Faculty of Animal Agriculture, Diponegoro University for sharing their pearls of wisdom with us during the course of this research.

**References**

[1] Yulianto J 2010. *Pengaruh Penggunaan Kulit Kecambah Kacang Hijau dalam Ransum Terhadap Kecernaan Bahan Kering dan Organik pada Kelinci Keturunan vlaams reus jantan.* Fakultas Pertanian Universitas Sebelas Maret, Surakarta
[2] Caprita R, Caprita A and Julean C 2010 *Anim Sci Biotechnol.* **43** 1 368
[3] Purwati D, Nyoman S and Istna M 2018 *J. Agromedia* **36** 1
[4] Fardana D H and Wahyono F *2018 Prosiding Seminar Nasional Kebangkitan Peternakan III, Semarang 3 Mei 2018*: *"Hilirisasi Teknologi Peternakan pada Era Revolusi Industri 4.0"*
[5] Steel R G D and Torrie J H 1993 *Prinsip dan Prosedur Statistika* Gramedia Pustaka Umum Jakarta
[6] Retnani Y, Hasanah N, Rahmayeni and Herawati L 2010 *Agripet* **10** 13
[7] Yuliastanti T 2013 Jurnal Kebidanan 2 Desember
[8] Verma A K, Mehra U R, Dass R S and Singh A 1996 *Anim. Feed Science and Technol* **59** 255
[9] López E M, García H S and Malo A L 2012 *Food Res. Int.* **45** 713
10] Ray Bibek 2005 Fundamental food microbiology. -3rd ed. CRC Press LLC, 2000 N.W. Corporate Blvd., Boca Raton, Florida 33431. ISBN 0-203-99825-1
[11] Pangestu G A, Pujaningsih R I and Mangisah I 2018 *J. Ilmiah Peternakan Terpadu* **6** 77
[12] Jayusmar E, Trisyulianti and Jacja J 2002 *Media Peternakan* **24** 76
[13] Pujaningsih R I F, Fani and Pambudi AW 2016. *J. Animal Prod.* **18**(1) 8
[14] Lawrence T L J 1990 In: J Wiseman and PJA Cole (Editor). 1990 *Feedstuff Evaluation* University Press. Cambridge 115
[15] Daud M, Mulyadi and Zahrul F 2016. *Agripet* **16** 1

# Utilization of durian seeds as a substitute for corn in the laying hen feed on consumption and nutrient digestibility for improving local

**I Mangisah[1,2*], V D Y B Ismadi[1], F Wahyono[1], N Suthama[1], B Sukamto[1], and R I Pujaningsih[1]**
[1]Animal Science Departement, Diponegoro University, Semarang, Indonesia
[2]Vocational School, Diponegoro University, Semarang, Indonesia

E-mail: istnamangisah@yahoo.co.id

**Abstract.** The objective of this study was to evaluate the effects of durian seed meal as a substitute for corn in the ration on feed and nutrient consumption, egg weight, and production. The study was conducted for six weeks. The material used is laying hens strain Lohmann Brown age 42 weeks as much as 120 birds. The study used Completely Randomized Design (RAL) with four treatments and five replications; each replication consisted of 6 birds. The treatments were the use of durian seed meal as a substitute for corn as much as 0, 3, 6, and 9%. Data were analyzed by ANOVA. The results showed that durian seed meal as a corn substitution significantly affects the parameter observed except feed and protein consumption. In conclusion, corn can be replaced with durian seed meal maximally by 6% without decreasing productive performance of laying hens

## 1. Introduction

Yellow corn is the primary source of energy in formulating laying hen feed. The use of yellow corn in poultry rations in general and laying hen in particular due to its high energy content, as about 3.313 kcal / kg, and contains lots of starch in the form of amylopectin, and high digestibility. Although yellow corn contains low protein (8%), it's as rich in a yellow pigment called xanthophyll on average 5 ppm and carotene 0.5 ppm. Rations containing yellow corn in large quantities affect the color of the yolk. Domestic corn production is lower than demand, so imports should fulfill it. Based on the Central Bureau of Statistics [1], the amount of corn imported in 2014 was about 3,253,616 tons and increased in 2015 to be 3,267,694 tons. One of the fruit seeds that contain high carbohydrates similar to corn is durian seeds. Durian fruit was composed of flesh of the fruit (20-35%), seeds (5-15%) and rind (60-75%). Durian production in Indonesia is highly available because of the average harvesting time for eight months per year [2]. According to data of Central Bureau of Statistics (BPS), Indonesia is high durian producing country, it was reported in 2012 durian production reached 6,917 tons/year and increased in 2015 to be 10,173 tons/year [3] In general, durian rind and seeds were just thrown away into waste, whereas durian seeds can be processed and utilized for further purpose [4]. Nutrient content of durian seeds was carbohydrate 67.40%; crude protein 6.43%; crude fat 1.48%; crude fiber 6.15%; and calcium 0.92% [5]. Besides, durian seeds also contain polysaccharides gels and antioxidants, which are useful for disease resistance and health [6,7,8]. On the other hand, durian seed

* To whom any correspondence should be addressed.

contain anti-nutritional factors in the form of cyclopropene acid and oxalic acid [9] which can negatively affect productivity if excessive feeding level and possibly can be toxic [2]. Other anti-nutritional substances in durian seeds were crude gum [10]. Crude gum is a water-soluble polysaccharide that included nonstarch polysaccharides that have high water holding capacity when dissolved in water, high viscosity, and stability so that the particles in the durian seeds were difficult to be absorbed [8]. The presence of crude gum in durian seeds can affect the digestion process of nutrients and the number of nutrients absorbed. Previous studies have shown that durian seeds can be used in quail rations as much as 5%, and if their use was more than 5%, it caused a decrease in consumption and productivity [11]

Feeding durian seeds in broiler chickens was not more than 10% [5] Utilization of processed durian seeds as laying hen feed has not been previously carried out. Based on the background described above, the present research was conducted to evaluate the effect of feeding durian seed meal to substitute yellow corn on consumption and nutrient digestibility on laying hen.

## 2. Materials and methods

The study used 120 laying hens of 52 weeks and reared in individual cages (battery). The experimental ration was Experimental ration composition and nutrient content are presented in Table 1. The study was assigned in a completely randomized design 4x5. The treatment was the level of feeding durian seed meal, namely 0, 3,6, and 9%. The parameters observed in the present study were as follows:

- Feed consumption = the amount of ration given - the amount of residual ration
- Protein or crude fiber consumption was calculated by multiplying feed consumption with a nutrient percentage in the ration
- Nutrient digestibility was calculated using the formula of [12] namely : Nutrient digestibility = (nutrient consumption – nutrient in excreta) : nutrient consumption x 100%"

Data were analyzed of variance and continued to Duncan test [13].

## 3. Results and discussion

### *3.1. Feed and nutrient consumptions*

Feeding durian seed meal did not significantly affect on the feed and protein consumption in laying hens (Table 2). This was caused by the dietary energy content given that was relatively the same, ranging between 2858.73 - 2879.60 kcal/kg. Substitution of corn with durian seeds up to 9% did not cause any change in dietary energy content. The metabolizable energy (ME) content of durian seeds meal was 2,780 kcal/kg, slightly lower than yellow corn energy (3,167.98 kcal/kg), but in this study, a slight decreased in energy levels did not reduce feed consumption. Feed consumption depends on dietary metabolizable energy because chickens consume rations to meet energy needs. Base on reported by [14] that feed intake is inversely proportional to the concentration of energy in the diet [15]. Lower feed consumption in chickens depends on crude protein (CP) and metabolizable energy content; if ME is low, then feed consumption is higher.

Anti-nutritional factors in the form of cyclopropene fatty acids and oxalic acid in durian seed meal did not affect feed consumption, because the level was low. (T0 = 0%, T1 = 0.0199, T2 = 0.038% and T3 = 0.057%). Oxalic acid content of durian seed meal was also low, namely T0 = 0%, T1 = 0.011%, T2 = 0.023% and T3 = 0.034%. The higher feeding durian seed meal in the present study resulted in the higher fatty acid of cyclopropene and oxalic acid. Anti-nutritional substances of oxalic acid and cyclopropene in the ration did not significantly reduce feed consumption, and also did not affect palatability. The results of the present study were different from that of Hasnawati et al. [16] that 5% of boiled durian seed meals given in male local duck reduced feed consumption by 5,069 g/bird compared to control group.

**Table 1**. Composition and nutrient content of the experimental ration.

| Ingredients | Treatment | | | |
|---|---|---|---|---|
| | T0 | T1 | T2 | T3 |
| | ---------- (%) ---------- | | | |
| Yellow corn | 43 | 40 | 37 | 34 |
| Rice bran | 18 | 18 | 18 | 18 |
| Palm oil | 0.5 | 0.5 | 0.5 | 0.5 |
| Durian seeds | 0 | 3 | 6 | 9 |
| Soybean meal | 19 | 19 | 19 | 19 |
| Fish meal | 10 | 10 | 10 | 10 |
| $CaCO_3$ | 3,5 | 3,5 | 3,5 | 3,5 |
| Oyster shell | 5 | 5 | 5 | 5 |
| Mineral mix | 1 | 1 | 1 | 1 |
| TOTAL | 100 | 100 | 100 | 100 |
| Nutrient content *(%) | | | | |
| Metabolizable energy**(kcal/kg) | 2733,84 | 2609,66 | 2513,73 | 2417,80 |
| Crude protein | 17,78 | 17,58 | 17,36 | 17,14 |
| Crude Fiber | 4,44 | 4,53 | 4,25 | 4,15 |
| Ether extract | 6,32 | 6,19 | 6,07 | 5,94 |
| Calcium | 3,31 | 3,30 | 3,29 | 3,28 |
| Phosphorus | 0,63 | 0,68 | 0,66 | 0,65 |

* Chemical analysis at the laboratory of Animal Nutrition and Feed Science, Faculty Animal and Agricultural Sciences, Diponegoro University

** Calculated base on the value of metabolizable energy = 70% x Gross Energy

Protein consumption also indicated a similar pattern with feed consumption. However, the result of crude fiber consumption increased significantly by the feeding level of durian seed meal at the level of 6 and 9% (T2 and T3) (Table 2). The higher level of feeding durian seed meal resulted higher consumption of crude fiber as well, although feed consumption was not significantly different. The higher crude fiber in the ration, also caused higher crude fiber consumption, and it could have an impact on decreasing nutrient digestibility [17].

**Table 2.** Nutrient consumption and nutrient digestibility of rations contain durian seeds.

| Parameter | Perlakuan | | | |
|---|---|---|---|---|
| | T0 | T1 | T2 | T3 |
| Feed consumption (g/bird/28 d} | 3,202.2 | 3,146.4 | 3,039.6 | 3,082.5 |
| Protein consumption (g/bird/day) | 18,88 | 18,57 | 17,95 | 18,21 |
| Crude fiber consumption (g/bird/day) | 4.72 $^{a}$ | 5.17 $^{a}$ | 5.50 $^{b}$ | 6.10 $^{b}$ |
| Protein digestibility (%) | 71,84$^{a}$ | 70,43$^{a}$ | 68,12$^{b}$ | 60,69$^{b}$ |
| Crude fiber digestibility (%) | 25,56$^{a}$ | 15,16$^{b}$ | 15,08$^{b}$ | 14,93$^{b}$ |

### *3.2. Nutrient digestibility*

Feeding durian seeds meal at 3% resulted in the same protein digestibility as that of control group. However, level 6% and 9% durian seed meal, significantly reduced protein digestibility. The digestibility of crude fiber at T0 was significantly higher than that of T1, T2, and T3. This was due to the presence of anti-nutritional substances content in the form of cyclopropene and oxalic acids in durian seed meal. Hasnawati et al. [16] reported that durian seed meal contains anti-nutritional substances that can affect nutrient digestibility.

The content of oxalic acid in durian seeds can bind calcium to be calcium oxalate that cannot be absorbed by the animal so that it reduces the process of calcium absorption and causes the damage to the intestine, thereby reducing the process of protein absorption in the digestive tract [18].

Durian seeds have a slippery water-soluble polysaccharide called crude gum. The compound can interfere with the process of nutrient absorption, so that nutrients released through the excreta. The amount of nutrient release together along with the excreta when the feeding durian seed meal increase. Gum derived from durian seeds had a high water holding capacity [10]. The high water holding capacity causes the digest to become thick and affects the rate of passage. The decrease in protein digestibility and crude fiber digestibility due to the feeding durian seed meal lowered the absorption of amino acids and decrease in egg production.

## 4. Conclusion

In conclusion, corn can be replaced with durian seed meals maximally by 6% without decreasing the productive performance of laying hens.

## References

[1] Central Bureau of Statistics 2016 Import of Corn 2016 Lowest in Last 5 Years
[2] Djaeni and Prasetyaningrum 2010 *Riptek.* **4** 37
[3] Central Bureau of Statistics 2017 Fruit Production by Province (Tons)
[4] Nathanael R, Efendi R and Rahmayuni 2016 *J JOM Faperta* **3** 1
[5] Suhaidi I 2004 University of Northern Sumatra Indonesia.
[6] Thunyakitpisal P, Saladyanant T, Hongprasong N and Apinhasmit W 2010 *J. Investig. Clin. Dent.* **1** 120·
[7] Haruenkit R S, Poovarodom, Vearasilp S and Gorinstein S. 2010 *Food chem.* **118** 540
[8] Amin A M, Ahmad S A, Yin Y Y, Yahya N and Ibrahim N 2007 *J. Food Hydrocolloids* **21** 273
[9] Ho L and Bhat R 2014 *Food chem.* **168** 80
[10] Mirhosseini H and Amid B T 2012 *Molecules.* **17** 6465
[11] Rozako M R, Sudjarwo E and Hamiyanti A A 2015 *Faculty of Animal Husbandry*
[12] Meluzzi A, Sirri F, Tallarico N, and Franchini A 2001 *Br Poult Sci.* **42** 213
[13] Steel R G D and Torrie J H 1981 *Statistical Principles and Procedures for a Geometric Approach* PT Gramedia, Jakarta.
[14] Perween S, Kumar K, Candramoni Kumar S, Singh P K, Kumar M and Dey A 2016 *Vet World* **9** 893
[15] Zeng Q F, Cherry P, Doster A, Murdoch R, Adeola O and Applegate T J 2015. *Poult Sci* **94** 384
[16] Hasnawati, Hafsah, and Sugiarto 2016 *J. Agrisains.* **17** 62
[17] Has H, Napirah A and Indi A 2014 *Ilmu dan Teknologi Peternakan Tropis* **1** 63
[18] Jiang Z L, Zhao M X, Liao L X 1996 *J. Analytical Chemical.* 139

# Optimization of biogas production in Indonesian region by liquid anaerobic digestion (L-AD) method from rice husk using response surface methodology (RSM)

**W D Nugraha[1*], Syafrudin[1], V N Kusumastuti[1], H H A Matin[2], Budiyono[3]**
[1] Environmental Engineering Department, Engineering Faculty, Diponegoro University, Semarang, Indonesia
[2] Environmental Science Study Program, Faculty of Science and Mathematic, Sebelas Maret University, Surakarta, Indonesia
[3] Chemical Engineering Department, Engineering Faculty, Diponegoro University, Semarang, Indonesia

E-mail: winardi_punk@yahoo.com

**Abstract.** Rice husk is one of agricultural waste that is potentially converted to biogas. In this research, rice husk is used as the material to produce biogas by liquid anaerobic digestion (L-AD) with TS concentration below 15/5. The purposes of this research are finding the optimum value of total solid (TS), C/N ratio, and microbial consortium (MC) for biogas production and then analyzing the optimum biogas production rate from rice husk. Because of the high lignin content, rice husk must be through the pretreatment stage using NaOH 3%. The variations of the observed factor are 5%, 10%, 15% for TS concentration; 20, 35, 50 for C/N ratio; and 3%, 6%, 9% for MC concentration. There are 16 reactors in which the variation for each reactor was determined by Central Composite Design. Data obtained from the research will be optimized using Response Surface Methodology (RSM) in Statistica 10. The result shows that the optimum condition of biogas production from rice husk is TS in between 1%-2.5%, C/N ration 33.45, and MC in between 5.5%-7%. Reactor having a combination of value factors that close to the optimum value will be analyzed using Polymath 6.0, which is reactor 10. The kinetics constant of biogas production rate (u) is 3.625 ml/grTS.day, the maximum biogas production (A) is 277.374 ml/gr TS, and the minimum time needed to produce biogas ($\lambda$) is 3.878 days.

## 1. Introduction

Rice husk is one of lignocellulose biomass that can be used as an alternative energy source that is not optimally used. In Central Java, the rice production in the year of 2015 is 11.006.570 tons [1] so that the amount of rice husk produced is around 2.202.314 tons with the assumption of 20% rice production [2]. Biogas, one of a method to transform biomass into energy, is a gas mixture produced from the decomposition of organic compounds by methanogenic bacteria in anaerobic conditions [3]. Before used as the material to produce biogas, lignocellulose material generally will be going through the pretreatment stage to increase the material digestibility [4].

* To whom any correspondence should be addressed.

Various research has been conducted by the former researcher about the effect of TS and C/N ratio on biogas production. Based on the former study, the highest biogas production from rice husk will be obtained if the C/N ratio 35 [5] and a total solid 9% [6] for L-AD method. The microbial consortium consists of Streptomyces sp., Trichoderma sp., and Geobacillus sp. which can help the degradation process of cellulose, hemicellulose, and lignin [7,8,9].

Based on those researches, further research is needed to find the optimum value of the factor to gain optimum biogas production. In this research, the optimization of the biogas production will use response surface methodology (RSM) which the factors are total solid, C/N ratio, and microbial consortium. This research intends to determine the optimum value of total solid, C/N ratio, and microbial consortium for biogas production from rice husk with L-AD method.

## 2. Materials and method

This research is divided into 2 stages. The first stage is a preliminary test, and it is to know the characteristic of the materials used. For determining total solid of rice husk and cow manure, we used *NREL/TP-510-42621* about *Determination of Total Solids in Biomass, and Total Dissolved Solids in Liquid Process Samples* and the value of C-organic was determined by walkey and Black method, while for the value of N-total, we used spectrophotometry method.

The reactor used in this research is a 1-liter plastic bottle equipped with rubber as the bottle cap. The reactor is connected to a 3/16 inch plastic hose used for gas transferred in the measurement. Materials used in the research are rice husk collected from a granary in Rowosari, Tembalang district, rumen cow fluid as inoculum obtained from Penggaron slaughterhouse in Pedurungan District, Semarang, the microbial consortium is in Decoprima packing powder, and urea obtained from the farm shop in Karangrejo, Banyumanik district. All the materials used are the same as the former researcher [5,6].

There are 3 factors in this research, which are total solid with variations of 5%, 10%, 15%; C/N ratio with variations of 20, 35, 50; and microbial consortium with variations of 3%, 6%, 9%. Optimization of the 3 factors was conducted using response surface methodology (RSM) in central composite design (CCD). Based on data processing by RSM, there are 16 reactors used in the research with a variation of the factors in every reactor will be showed in table 1.

The study was conducted for 60 days with the gas measurement was every two days. It aimed to keep the reactor in the anaerobic condition and there was no significant biogas production every day. Biogas volume was observed by transferring the gas into the measuring glass and the amount of gas produced was calculated from the difference of the water volume in the measuring glass.

After 60 days of biogas observation, data obtained from the observation was analyzed using response surface methodology (RSM) in Statistica 10 to determine the optimum value of each factors. Besides that, the reactor with the highest biogas production was analyzed using Polymath 6.0 to determine the rate of biogas production.

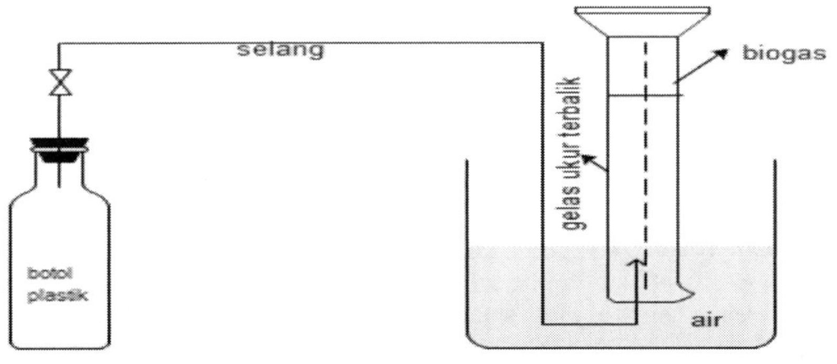


**Figure 1.** Schematic diagram of series laboratory batch assessment of L-AD.

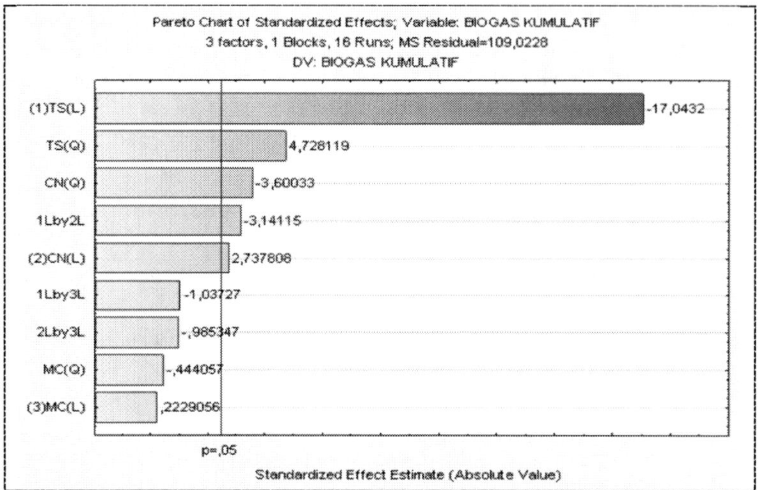


**Figure 2.** Pareto chart result of RSM analysis.

**Table 1**. Observation result of yield biogas in 60 days.

| No. | Factors | | | Biogas Production (ml/grTS) |
|---|---|---|---|---|
| | Total Solid (%) | C/N Ratio | Microbial Consortium (%) | |
| 1 | 5.0 | 20.0 | 3.0 | 69.8 |
| 2 | 5.0 | 20.0 | 9.0 | 94.8 |
| 3 | 5.0 | 50.0 | 3.0 | 122.8 |
| 4 | 5.0 | 50.0 | 9.0 | 129.4 |
| 5 | 15.0 | 20.0 | 3.0 | 13.967 |
| 6 | 15.0 | 20.0 | 9.0 | 19.8 |
| 7 | 15.0 | 50.0 | 3.0 | 16.733 |
| 8 | 15.0 | 50.0 | 9.0 | 11.867 |
| 9 | 10.0 | 35.0 | 6.0 | 50 |
| 10 | 1.591 | 35.0 | 6.0 | 192.453 |
| 11 | 18.409 | 35.0 | 6.0 | 12.167 |
| 12 | 10.0 | 9.773 | 6.0 | 14.6 |
| 13 | 10.0 | 60.227 | 6.0 | 28.4 |
| 14 | 10.0 | 35.0 | 0.955 | 59.25 |
| 15 | 10.0 | 35.0 | 11.045 | 45 |
| 16 | 10.0 | 35.0 | 6.0 | 57.2 |

## 3. Results and discussion

### *3.1. Optimization of total solid, C/N ratio, and microbial consortium*

The result of 60 days observation can be seen in Table 1. As the data shown in the table above, each reactors has a different amount of biogas produced. The highest biogas production was conducted in reactor 10 with 192.453 ml/grTS. These observation result was analyzed with RSM in Statistica 10 resulting an equation as follows

$$Y = 57.0878 - 48.1543x1 + 16.2198x12 + 7.7355x2 - 12.3509x22 + 0.6298x3 - 1.5233x32 - 11.5958x1x2 - 3.8292x1x3 - 3.6375x2x3 \quad (1)$$

Y = Biogas Production (mL/gr TS)
X1 = Concentration of Total Solid (%)
X2 = C/N Ratio
X3 = Concentration of microbial consortium (%)

Based on the equation, the value with the positive sign means that the increasing amount of variables both linear (L) and quadratic (Q) will give a positive impact or increase biogas production. Meanwhile, the value with the negative sign both linear (L) and quadratic (Q) will decrease biogas production. Therefore, the coefficient of X1 (L) is in negative sign, it means that the increase of total solid linearly will decrease the biogas production but the increasing of total solid quadratically will increase the biogas production because the coefficient of X1 (Q) has a positive sign. The second variable is X2 which X2 (L) is in a positive sign that means the increasing C/N ratio linearly will increase the biogas production but the increasing C/N ratio quadratically will decrease the biogas production. The last variable is X3 which X3 (L) is in positive sign and X3 (Q) is in negative sign, so that the increase of microbial consortium linearly will increase the biogas production.

Figure 2 shows the level of significance for each variable and the interaction between two variables. The significance level for all variables is 98.41%. The red line in the Pareto chart is the value of $P = 0.05$. In Figure 2, the variable significantly affects if the variable chart passes the red line which the value of $P<0.05$. On the contrary, if the chart does not pass the red line, it means the variable does not significantly affect the biogas production with the value of $P>0.05$. The following figure is the Pareto chart.

Based on the Pareto chart, the variables affecting the biogas production are total solid (linearly and quadratically), C/N ratio (linearly and quadratically) and the interaction between the variable of total solid and C/N ratio linearly because the chart passes the red line ($P<0.05$). The other variables like microbial consortium either linearly or quadratically and the interaction between total solid and microbial consortium also between C/N ratio and microbial consortium does not significantly affect the biogas production cause the chart does not pass the red line or the $P>0.05$.

The optimum condition of total solid, C/N ratio, and microbial consortium can be determined by the response surface analysis in the surface plot (3 dimensions) and contour plot chart (2 dimensions). Either the surface plot or the contour plot chart, it consists of 2 independent variables and one dependent variable, cumulative biogas production (ml/grTS), so that the other one variable will be a constant value. The two diagrams show the same analysis, but it is easier to determine the optimum range of variables in the contour plot chart (2 dimensions). In the chart, there is various colors showing the amount of biogas production. The more red the color the greater the amount of biogas production. It can be seen in figure 3 until figure 5.

The following figures 4 and 5 show the surface plot and contour plot between total solid and C/N ratio on 6% MC concentration and between total solid and MC concentration on C/N ratio 35. Based on Figures 3 and 4, it can be seen that the optimum value for the variable total solid is under 10%. This result is the same as the previous research stating that the optimum amount of total solid is 5% [10]. Another reference state that excess substrate value fed into the reactor will increasingly activate and accelerate the growth of acidogenic and acetogenic bacteria leading to an imbalance stage of acidogenesis and acetogenesis [11]. Meanwhile, the organic material that has a low total solid content will be degraded faster leading to un-optimal performances of methanogenic bacteria on producing biogas because of low pH.

Figures 4 and 6 show the surface plot and contour plot between total solid and C/N ratio on 6% MC concentration and between C/N ratio and MC concentration on 10% total solids. Based on the diagram, the optimum value of C/N ratio is in the range of 30 – 40, with the specific value is 33.45. As well as the former research stating that the optimum value of C/N ratio is 35[5]. The other research also stating if value of C/N ratio more than 43, it will inhibit the bacteria performance [12]. If the value of C/N ratio is too high, nitrogen will increase bacterial growth reacting to carbon that gives a negative effect on biogas production [13]. On the contrary, the inadequate value of C/N ratio will lead to nitrogen accumulation in the form of ammonia, affecting pH increasing [13].

Figures 5 and 6 show the interaction between total solid and MC concentration on C/N ratio 35 and the interaction between C/N ratio and the microbial consortium on 10% total solid. It shows that the optimum value of MC concentration is in the range of 5.5% - 7%. Based on the reference, Trichoderma sp. and Geobacillus sp. can produce cellulase enzyme but Geobacillus sp. is thermophilic bacteria [7,9] and Streptomyces sp. can produce ligninase enzyme [8].

Therefore based on data analyzing, it can be concluded that the optimum value for total solid is under 10%, for C/N ratio is 33.45, and the last for MC concentration is in the range of 5.5% - 7%. Reactor 10 shows the highest biogas production and has the closest value with the optimum value with TS 1.54%, C/N ratio 35, and MC 6%.

### *3.2. Biogas production optimum rate from rice husk with L-AD method*

Reactor 10 has the closest value to the optimum variable value with 1.54% total solid, C/N ratio 35, and 6% MC concentration. Cumulative biogas production of reactor 10 then calculated its kinetic constant of biogas production rate (U) (ml/gr.TS.day), maximum biogas production (A) (ml/gr.TS), and time needed to produce biogas (λ) (day) using Polymath 6.0. From the calculation, it is obtained the following results.

**Table 2.** Kinetic constants for optimum biogas production.

| Constants | Value |
|---|---|
| U (ml/gr.TS.day) | 3.625 |
| A (ml/gr.TS) | 277.374 |
| λ (day) | 3.878 |

The constant values then input to the equation 2 to determine the biogas production with t filled with time (day) to be calculated.

$$P = 277,374\ \exp\left\{-\exp\left[\frac{3,625\ x\ 2,178}{277,374}(3,878 - t) + 1\right]\right\} \quad (2)$$

P = cumulative biogas production (ml/grTS)
t = time cumulative for biogas production (day)

Figure 3 below is the result of Polymath 6.0 analysis until 184 days, which can be seen that the highest biogas production is achieved on 184th days with the result is 277.355 ml/grTS. After 184 days, biogas is still produced in the reactor but the volume produced only ±0.002 ml/grTS per day, so it does not significantly affect the accumulation of yield biogas. It also can be seen in the graph the $R^2$ value is 0.9892, which means that the calculation data from Polymath is not significantly different with the data obtained from the research.

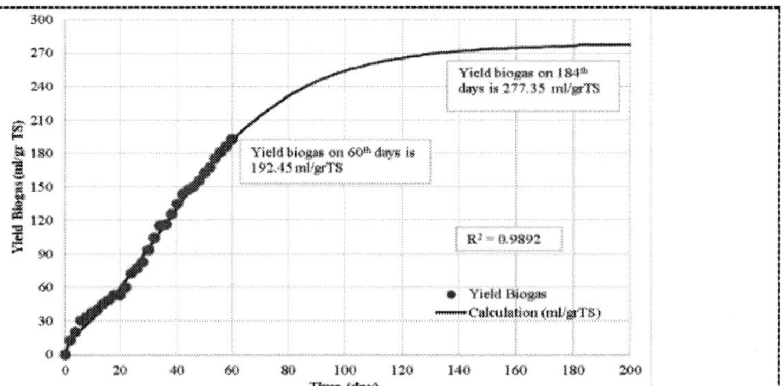


**Figure 3.** Relationship between trial data and polymath 6.0 calculation results on optimum biogas production.

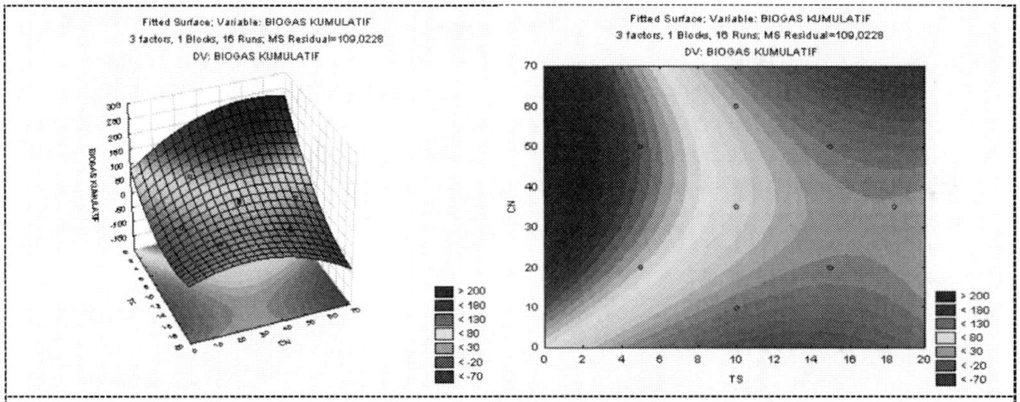


**Figure 4.** Three dimensional / surface plot (left) and two dimensional / contour plot (right) graphs for variable TS and C/N ratio at 6% concentration of microbial consortium from RSM analysis.

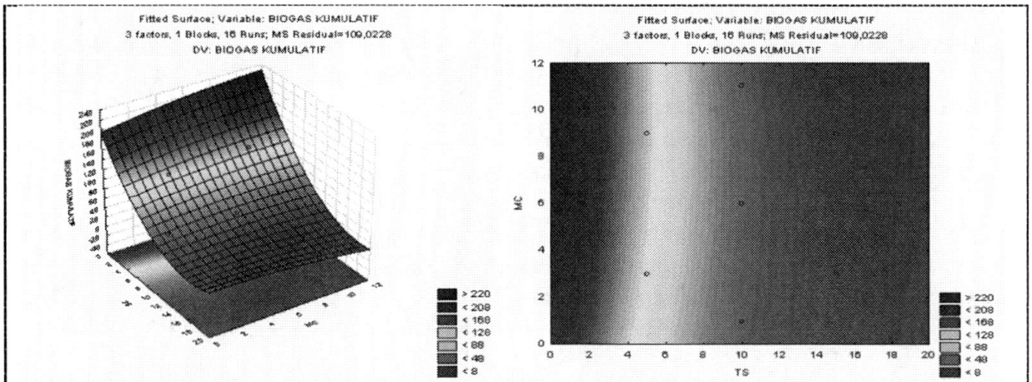


**Figure 5.** Three dimensional / surface plot (left) and two dimensional / contour plot (right) graphs for variable TS and C/N ratio at 6% concentration of microbial consortium from RSM analysis.

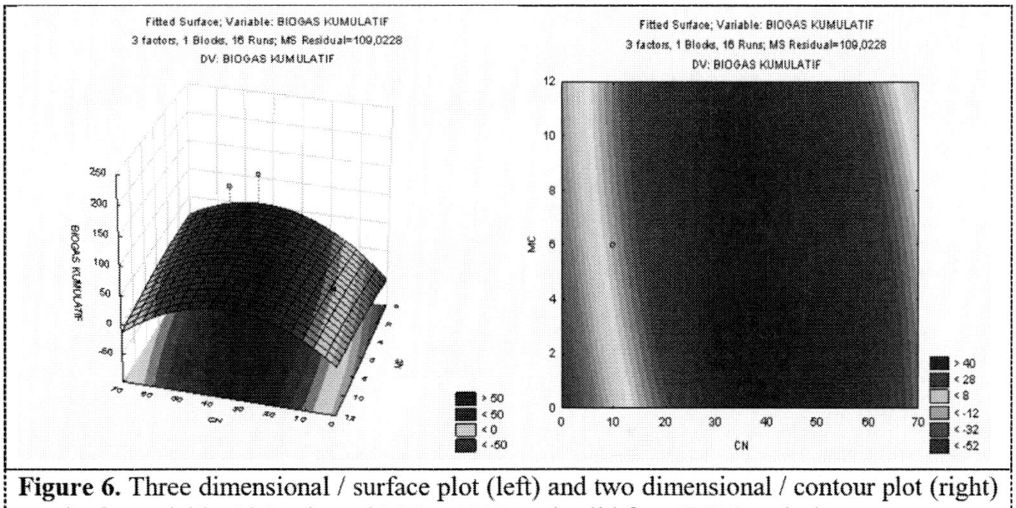


**Figure 6.** Three dimensional / surface plot (left) and two dimensional / contour plot (right) graphs for variable C/N ratio and MC at 10% total solid from RSM analysis.

## 4. Conclusions

The maximum biogas production can be obtained at the optimum condition of variables: (1) TS concentration is below 10%, (2) C/N ratio of 33.45, and (3) microbial consortium in the range of 5.5% - 7%. The maximum biogas production rate in the reactor that has the closest value to the optimum variable value is reactor 10 with 1.54% TS, C/N ratio 35, and 6% concentration of MC producing biogas yield of 277.355 ml/grTS. The kinetics constant rate of biogas production per day (U) is 3.625ml/grTS, maximum biogas production rate (A) is 277.374 ml/grTS, and minimum time needed to produce biogas is 3.878 days.

## Acknowledgment

We highly thank and appreciate to Diponegoro University through the Research Professorship Program (RPP) 2018 due to funding this research.

## References

[1] Statistik Nasional Indonesia 2017
[2] Waries A 2006 Jakarta *Gramedia Pustaka*
[3] Mara I 2012 *Dinamika Teknik Mesin* **1**
[4] Al Seadi, Rutz D and Prassl H 2008 Biogas Denmark
[5] Syafrudin, Winardi D N and Hasfhi H A M 2017 *Adv Sci.* **101** 2204
[6] Syafrudin, Winardi D N and Erica S S 2017 *Jurnal Teknik Lingkungan* **6** 1
[7] Bardant T B, Abimanyu H and Epriyani P L 2013 *Jurnal Kimia Terapan Indonesia* **15** 35
[8] Gawande P V and Kamat M Y 1999 *J. Appl Microbiol.* **87** 511
[9] Potprommanee L, Wang X Q and Han Y J 2017 *PLoS One* **2** 1
[10] Sawasdee V 2014 *Energy Procedia* **61** 1129
[11] Soeprijanto, Ismail T and Lastuti M D 2010 *Jurnal Purifikasi* **11**
[12] Dennis A and Burke P E 2001 *Olympia: Environmental Energy Company*
[13] Budiyono G, Kaerunnisa and Rahmawati I 2013 *Jurnal Teknologi Kimia* **11** 1

IOP Conf. Series: Materials Science and Engineering **845** (2020) 012043 doi:10.1088/1757-899X/845/1/012043

# Analysis the cause of leakage and repair of floor plates diesel fuel tank 20-D-2 at PT BADAK NGL Indonesia

**I Isbah[1,2*] and W Mangestiyono[1]**
[1]Vocational School of Mechanical Engineering
Diponegoro University
[2]PT BADAK NGL, Bontang, Indonesia

E-mail: ishaisbah@gmail.com

**Abstract.** PT. BADAK Natural Gas Liquefaction or better known as PT. BADAK NGL is a liquefied natural gas (LNG) company in Indonesia with installed capacity 22.5 MPTA of LNG. PT BADAK NGL has LNG, LPG, condensate, and diesel fuel storage tanks. There are two diesel fuel tanks available for supporting diesel equipment such as turbine diesel, marine boats, and vehicles. One of the diesel fuel tank, named 20-D-2, was constructed in 1975 based on standard API 650-Welded Tanks for Oil Storage. In 1996, after being operated for 21 years, plate replacement was needed on the tank floor and had been done as well. But in 2017, during five years of periodic inspection, severe corrosion was found again on the plate of the tank floor. In this paper, the effort to investigate the failure and plate repairment was reported. Visual observation at internal and external side on the plate of the tank floor was conducted based on NDT method (Non Destructive Testing) testing with MFL (Magnetic Flux Leakage). Plate thickness measurement of tank floor was also conducted with UT (Ultrasonic Testing). In addition, chemical analysis of residual fuel on the tank was performed to investigate sulphur concentration that potentially corrodes the plate. Scanning Electron Microscopy (SEM) was also performed to investigate the plate surface that has been corroded. Based on the analysis above, it was known that the leakage of 20-D-2 tank floor was caused by pitting corrosion, general corrosion and crevice corrosion due to rainwater that accumulated on the bottom of tank floor. Direct contact among rain water-soil-plate of tank floor was identified as the main cause of the leakage of tank floor. After leakage mechanism was identified as well, an appropriate method could be formulated to repair the tank to avoid the leakage.

## 1. Introduction

The operating license certificate knows as SKPP (Sertifikat Kelayakan Penggunaan Peralatan) of hydrocarbon storage tanks in PT Badak NGL had been ended in March 2017. As a recertification program, all hydrocarbon storage tanks (LNG, LPG, condensate, and diesel fuel tank) were inspected in 2017. As a part of this recertification program, Diesel Fuel Tank 20-D-2 was opened for internal inspection purposes. It was inspected externally by visual examination and UT (Ultrasonic Thickness) measurement. Internally, it was inspected by visual examination, Magnetic Flux Leakage (MFL) examination, UT measurement, and pit depth measurement. From external inspection, it showed no major non-conformity found. However, the internal inspection showed a major non-conformity on the floor plates of the tank. Metal loss and scattered pitting corrosion was found on the floor plates of the storage tank. Therefore, proper repair shall be conducted to restore the condition of the tank and to extend the SKPP.

This research was conducted to analyze the causes of corrosion that occur on 20-D-2 tank floor plates by carrying out several testing methods to find the root cause of the problem so that it can make an appropriate repair to this 20-D-2 tank leak.

* To whom any correspondence should be addressed.

## 2. Experimental design

In this paper, the authors examined the causes of corrosion occurred at the bottom plate of the 20-D-2 diesel fuel storage tank. The study began by studying the historical data from the Inspection Section about a 20-D-2 tank. Then the bottom plate sample was taken by cutting the corroded plate of bottom tank 20-D-2 for further analysis. Material and environmental aspects of 20-D-2 tank both visually, by performing non-destructive testing and using laboratory tests. For material, NDT methods like MFL and UT have been performed. In addition, Scanning Electron Microscopy or known as SEM (XRF) and Metal Analizer test were also performed to see the actual metal composition of it. The result of the analysis will be compared with a material composition that conforms to standard specification. Apart from it, the sulfur content of diesel fuel in the tank and the pH of the soil where the 20-D-2 tank stands were checked by laboratory tests.

## 3. Results and discussion

### *3.1. Visual observation*

Visual inspection was carried out by the Inspection Section through internal inspection when the tank was offline by focusing on the welded parts and floor plates on the tank. From the result of internal inspections, the coating/lining of the bottom plate of the tank floor was peeled off. Pitting corrosion, general corrosion and crevice corrosion was found. Pitting corrosion occurred almost throughout the surface of bottom plate of the tank floor.

In addition, visual inspection was also carried out on the external side of the tank floor plate. From the results were found that the floor plate was seated on the ground with no concrete foundation. This can cause floor plate corrosion due to environmental exposure under the floor.

When performing the tests, the bottom plate of the tank floor was divided into ten parts with a particular configuration. The visual examination showed many pitting corrosions scattered on almost the entire surface of the tank floor. Visually, the plates at numbers 1/2 and 1/3 had more concentrations of pitting corrosion compared to other number plates. Therefore, a measurement of pitting depth (pit depth measurement) was used to determine the remaining thickness of the bottom plate.

**Figure 1.** The layer of paint peeled off.

**Figure 2.** Pitting corrosion throughout the surface of the bottom plate.

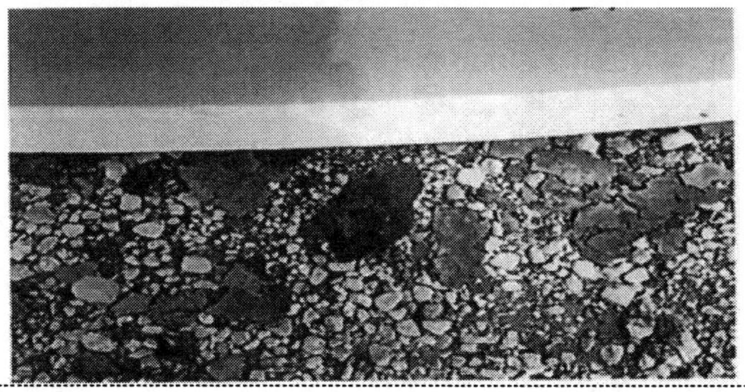

**Figure 3.** Direct contact between floor tank and soil.

*3.2. Non-destructive testing*

Non-destructive testing was performed to find out the remaining thickness of the defect on the floor plate. Based on API 653, joints. Identification of defects is carried out using Magnetic Flux Leakage (MFL), while for thickness measurements is done by Ultrasonic Testing (UT).

The result of MFL could be seen in Figure 5. It could be seen that defects were observed on all plates. However, not all defects observed from the MFL were in a critical state, only some of them were in a critical state.

Pit depth measurement was also performed on all plates to determine the deepest possible pitting. After measuring, the deepest pitting is on plate number 2/1, which had a depth of 3.94 mm.

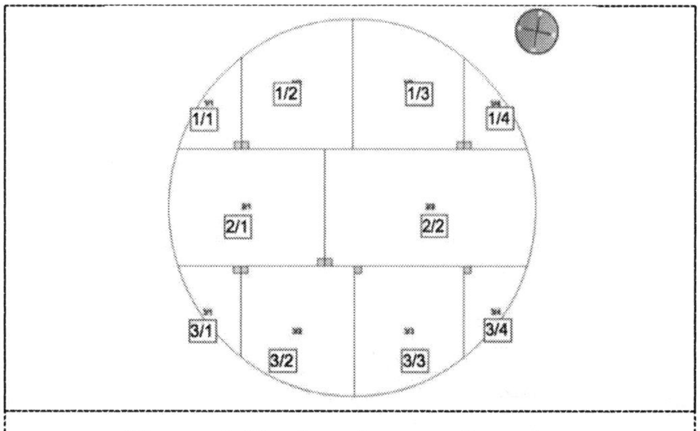


**Figure 4.** Number plates configuration.

Pitting corrosion was expected to occur due to contact with water accumulation on the internal side of the tank. Accumulation of water was guessed to had been there since the last time the tank was cleaned in 2006. Because the tank bottom plate was not coated at the last inspection resulting in pitting corrosion.

In Figure. 5. it can be seen that the blue dot represents the defect on the underside of the tank floor plate. Then from the results of the MFL thickness measurements were made on each plate number using the UT method. Measurements were made on blue dots to determine the minimum thickness remaining on the plate. With the results of these measurements can be known thickness reduction on each plate by reducing the thickness of the initial plate known from the datasheet was 6.5 mm, with the UT reading on the bottom side of the tank floor plate.

After knowing the remaining thickness of the tank floor plate, then we can calculate the required thickness (Minimum Required Thickness) so that we can compare whether the remaining plate thickness is still in accordance with the Standard. Based on API Standard 653, calculating the required thickness minimum can use the formula in Sub-Chapter 2.6.3 regarding the minimum thickness for tank floor plates.

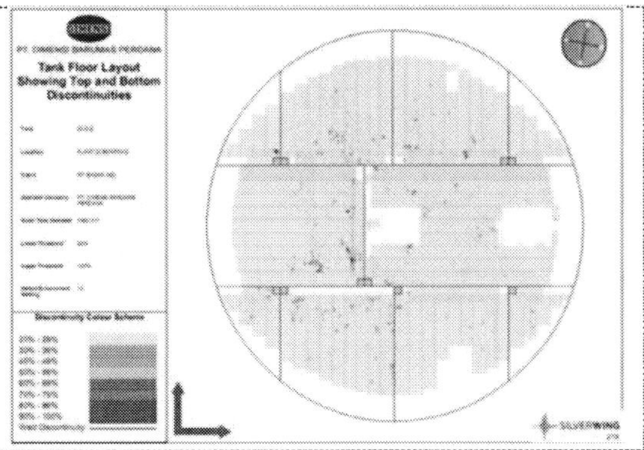

**Figure 5 a.** MFL result.

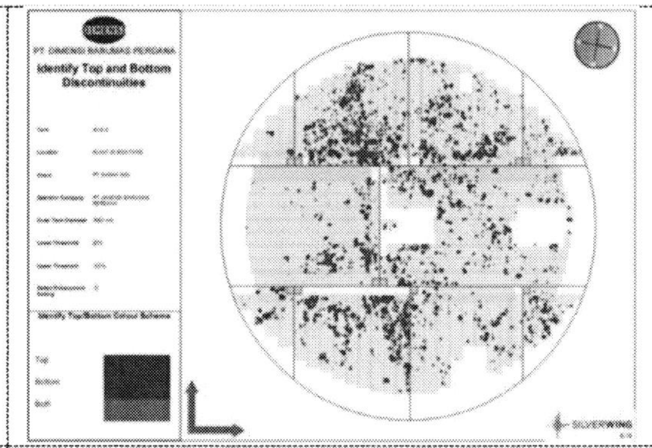

**Figure 5 b.** MFL result.

According to the API 653 Standard about the minimum floor plate thickness, if the tank built did not have a leak detection system or a leak-retaining foundation, the minimum thickness of the floor plate was 0.1 inch. Tank 20-D-2 (MRT) did not have a foundation that can detect leakage, therefore the minimum thickness of the floor slabs allowed for subsequent inspections was 0.1 inch or 2.54 mm.

$$MRT = (Minimum\ of\ RT_{bc}\ or\ RT_{ip}) - Or\ (StP_r + UP_r) \quad (1)$$

The value of Or is based on the next inspection schedule which is 5 years due to the latest conditions that indicate a defect in the previous inspection. Then the value of Or to be filled in Equation (1) was 5 years. StPr and UPr were the maximum corrosion rates on the upper and lower sides of the tank floor plate. The corrosion rate was obtained by dividing the thickness reduction by the total years since the tank was first built. While thickness reduction can be obtained from the maximum pit depth and minimum thickness. Meanwhile, the total year since the 20-D-2 tank was built since 1996 is 21 years.

All of the above data descriptions were used to find the RTbc or RTip value. To make it easier, we can define the minimum remaining allowable minimum (minimum thickness allowed) as RT. The results of these calculations show that many plates are still acceptable for the minimum thickness for the next inspection interval. Since the inspection was carried out with the help of the MFL method, it can be seen that the highest thickness reduction occurs in plate number 1/2, which was 50% and the lowest thickness reduction occurs in plate 2/2, which was 35%.

### *3.3. Material composition testing*

Material composition testing aimed to see the composition of the actual material composition used for 20-D-2 tank floor plate. In the datasheet, it was known that the material used was ASTM A36. Based on ASTM A36, the composition for A36 material is shown in Table 1.

**Tabel 1.** ASTM A 36 material composition.

| Product | Shapes[A] | Plates[B] | | | | |
|---|---|---|---|---|---|---|
| Thickness, in. [mm] | All | To ¾ [20], incl | Over ¾ to 1½ [20 to 40], incl | Over 1½ to 2½ [40 to 65], incl | Over 2½ to 4 [65 to 100], incl | Over 4 [100] |
| Carbon, max, % | 0.26 | 0.25 | 0.25 | 0.26 | 0.27 | 0.29 |
| Manganese, % | ... | ... | 0.80–1.20 | 0.80–1.20 | 0.85–1.20 | 0.85–1.20 |
| Phosphorus, max, % | 0.04 | 0.04 | 0.04 | 0.04 | 0.04 | 0.04 |
| Sulfur, max, % | 0.05 | 0.05 | 0.05 | 0.05 | 0.05 | 0.05 |
| Silicon, % | 0.40 max | 0.40 max | 0.40 max | 0.15–0.40 | 0.15–0.40 | 0.15–0.40 |
| Copper, min, % when copper steel is specified | 0.20 | 0.20 | 0.20 | 0.20 | 0.20 | 0.20 |

Then to determine the composition of the current tank flooring material, it was tested using a Metal Alloy Analyzer and SEM (Scanning Electron Microscopy) testing using the XRF method. The test was carried out by shooting at four points using Metal Alloy Analyzer.

**Figure 6.** Corroded sample from top side of 20-D-2 tank floor plate.

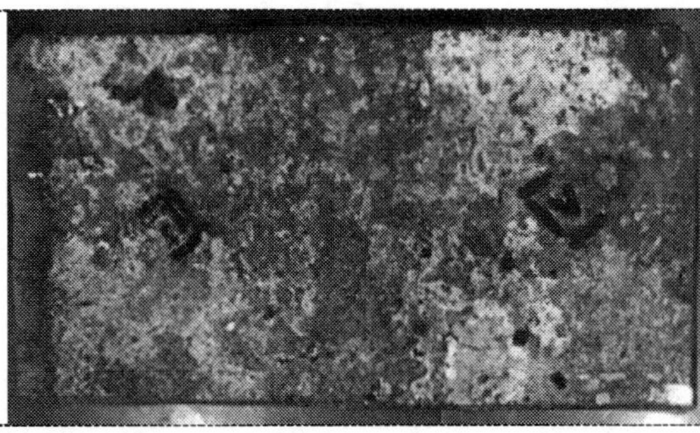

**Figure 7.** Corroded sample from bottom side of 20-D-2 tank floor plate.

From the metal alloy analyzer result in Tabel 1, it can be seen and compared with the A36 material standard in Table 2. Based on Tabel 2, the reference for the thickness plate was ¾ inch. The following comparison can be seen in Table 3. From the results of reading Table 3, it can be seen that Silicon and Sulfur products had levels that exceed the maximum limit of the ASTM A36 Standard.

**Tabel 2.** Metal alloy analyzer result.

| Element | % | | | |
|---|---|---|---|---|
| | Point-1 | Point-2 | Point-3 | Point -4 |
| Zr | 0.019 | 0.036 | 0.005 | - |
| Ti | 0.054 | 0.043 | 0.034 | 0.124 |
| Si | 8.49 | 8.01 | 3.21 | 3.45 |
| Mo | 0.007 | - | - | - |
| Mn | 0.681 | 0.685 | 0.644 | 0.703 |
| Fe | 90.61 | 90.93 | 95.84 | 95.5 |
| Cu | 0.015 | 0.009 | 0.048 | - |
| Co | 0.13 | 0.223 | 0.115 | - |
| Zn | - | 0.044 | 0.03 | 0.067 |
| Nb | - | 0.006 | - | - |
| Cr | - | 0.018 | 0.02 | 0.016 |
| S | - | - | 0.052 | 0.112 |
| W | - | - | - | 0.028 |

**Tabel 3.** Comparison of metal alloy metalizer results with ASTM A36 standard on 20-D-2 tank floor plate material.

| Product | Standard ASTM A36 | Metal Alloy Metalizer pada Corroded Sample | Metal Alloy Metalizer pada Furnished Sample |
|---|---|---|---|
| Cr, max (%) | 0.25 | 0.018 | - |
| Mn, max (%) | - | 0.703 | 0.731 |
| P, max (%) | 0.04 | - | - |
| S, max (%) | 0.05 | 0.112 | - |
| Si, (%) | 0.4 | 8.49 | 0.6 |
| Cu, min (%) | 0.2 | 0.048 | 0.021 |

### *3.4. Chemical analysis for sulphur content*

The diesel fuel inside of 20-D-2 tank was taken as a sample for chemical analysis. The diesel fuel was analyzed by sending the sample to PT PERTAMINA RU V Laboratory to analyze whether there was sulfur content in it. The results of sulfur content analysis were the sulfur content of diesel fuel found 0.14% m/m (1400 ppm) from the maximum limit of 0.30% m/m (3000 ppm)

### *3.5. pH testing*

pH testing was performed by taking soil where the location of the 20-D-2 storage tank is built as a sample. pH testing of soil was carried out in order to find out whether the soil was acidic, alkaline or neutral. Acidic soil tends to be more corrosive than neutral or alkaline soil. After testing the pH using litmus paper, it was found that the soil where the 20-D-2 location was built on alkaline soil with 8 pH values.

Based on the results of the explanation of several tests above, it can be seen in the results of visual inspection that 20-D-2 tank had no tank foundation so that the tank was seated right on the ground which causes accumulation of rainwater entering the gap between the tank floor and the ground resulting corrosion on the bottom plate of tank floor. In addition, from the results of material testing analysis with Metal Alloy Analyzer and XRF method, it can be concluded that the 20-D-2 tank floor material didn't use standard material accordance with Standard ASTM A36 so the material could be said out-of-spec.

## 4. Alternative repair method

After knowing the cause of corrosion on the 20-D-2 tank floor plate, the next step was to determine the correct repair method for the 20-D-2 tank floor plate and in accordance with the criteria in API 653 Section 9. Some alternative repair methods that can be applied to tank 20 -D-2 was as follows:

*A. Repair by installing patch plates*

The alternative to the first repair method was to patch the tank floor plate with a thin plate. Based on API 653 Section 9, the plates used for patching can be rectangular, square or circular. To do this patching, there are criteria that must be fulfilled, among others, there was a minimum patching distance between plates and a maximum plate thickness when the plates were stacked in three layers due to the patching plate. From the results of the MFL reading it can be seen that the distance between corrosion was too close to each plate so that the alternative repair method with partial patching on the 20-D-2 tank floor plate was not in accordance with the provisions in API 653 Section 9.

*B. Repair by installing proper patch plates*

The second alternative repair method was to install patch plates (additional plates) that meet all the rules in the API 653 Section 9. To accommodate all the rules, larger patch plates must be made and had a very large size which covers about half of all floor plates. Although this configuration was in accordance with Standard API 653 Section 9, there were obstacles if an alternative method of repair was carried out because the size of the patching plate was large and quite impossible to be inserted into the tank through the available manway tank. Therefore, the second alternative repair method was more complicated to be done than the first alternative.

*C. Repair by replacing entire floor plates*

The third alternative repair was by replacing all floor plates. This repair method was acceptable to be done based on API 653. Since the condition of floor plates was severe based on inspection result and calculation, and then it was possible to repair floor plates by replacing all floor plates. However, this alternative was only done if there were no other better alternatives.

*D. Additional modification*

Referring to the previous leakage in 2006, the 20-D-2 tank floor plate leaked within 10 years from the first replacement in 1996. If repairs were made only by replacing the entire tank slab as it was done in 1996, it was estimated the same case of leakage can occur again in the next 10 years after repairs. To prevent this in the future, additional modifications based on the mechanism of the tank floor damage must be accommodated.

Based on the damage and leakage analysis described above, two findings are the dominant cause of tank leakage, so the follow-up of these findings is necessary. Some modifications can be made to the 20-D-2 tank as a follow-up as well as an improvement in the 20-D-2 tank so that similar cases do not recur. Follow-up carried out included the following:
1)Foundation Installation on 20-D-2 Tank: Installation of new storage tank foundation shall refer to Section 7 of API 650 – Welded Tanks for Oil Storage. There are four types of foundation mentioned in the standard. These four types of foundation are listed below :

a)Earth foundation without a ringwall.

b)Earth foundation with a concrete ringwall.

c)Earth foundation with a crushed stone and gravel ringwall.

d)Concrete slab foundation

Except for the concrete slab foundation type, all these foundations are not directly categorized as a release prevention barrier. Release prevention barrier (RPB) means that if a leak seeps through floor plates, it could be detected immediately from the RPB (foundation) and not contaminates soil at that time. Since the main intention is to eliminate the contact between soil and bottom side of floor plates so corrosion will not occur there, then the concrete slab type foundation is the only type of foundation that could overcome the problem. However, installing the foundation could not directly prevent corrosion on the bottom side of the floor plates if there is still a gap between floor plates and the foundation. This gap could entrap water and lead to crevice corrosion. Therefore, a proper seal shall be installed between floor plates and the foundation or at the tip of floor plates, to prevent water entrap on the bottom of the floor plates.

2)Use of standard materials: From the results of material testing carried out with a metal alloy analyzer and XRF test, it can be seen that the composition of the floor plate material was not in accordance with ASTM A36 Standards so it can be said that the 20-D-2 floor plate material was out-of-spec. For this reason, if 20-D-2 tank floor plate replacement was carried out, it must be ensured that the plate material has an original certificate from the manufacturer concerned so that the material could be guaranteed according to the standard.

## 5. Conclusion

•Damage mechanisms on the top side of floor plates were pitting and general corrosion due to the presence of water at the bottom of diesel fuel. Damage mechanisms on the bottom side of floor plates were general, pitting, and crevice corrosion due to direct contact between floor plates and soil
•The thickness reduction or thinning allowed in floor plates were varied from 35% to 50% for each plate. All floor plates have metal loss more than the mentioned percentage. Thus, all floor plates shall be repaired.

## 6. Recommendations

•To conduct a repair on floor plates Diesel Fuel Tank 20-D-2 by replacing all plates.
•F&PE Section is requested to design concrete slab as foundation referring to API 650.
•Use standard material in accordance.

## Acknowledgment

The author thanks to First, Mr. Wiji Mangestiyono, ST, MT, and Mr. Vicky Indrafusa, ST, MT, as my assigned lecturers who always provide guidance and direction both in lessons and report making. Second, my family, especially my mom, who still supports me all the time. Third, Aan and Niken as my fun time. Fourth, Hani Rahmawati, who is my role model and always encourages me to get an

MSc. Fifth, Ayu Primastuti, who helps to build my confidence to do this Final Project in less than 2 months.

**References**

[1] American Petroleum Institute. "*API 620 : Design and construction of large, welded, low-pressure storage tanks*". Tenth edition USA 2002
[2] American Society for Testing and Materials. "ASTM A 36 : Standard spesification for carbon *structural steel*". USA 2004
[3] American Society for Testing and Materials. "*ASTM D 1266 : Standard test method for sulfur in petroleum products*". USA 2004
[4] Pagirik, Ayub. " *Signification of diesel oil testing : Signifikasi pengujian minyak solar.*" Kompas.com. Indonesia 2013
[5] Carucci, Vincent A , Delahunt and John F 2002 Corrosion New Jersey
[6] MFE Enterprises 2015 *What is MFL?* . http://www.mfeenterprises.com/what-is-mfl
[7] Rizal and Syaiful 2015 *Practice Report : NDT (Non destructive test)* Madura, Indonesia
[8] Prasetyo and Yudi 2011. *Scanning electron microscope (SEM) and optical emission spectroscopy (OES).* https://yudiprasetyo53.wordpress.com/2011/11/07/scanning-electron-microscope-sem-dan-optical-emission-spectroscope-oes/

IOP Conf. Series: Materials Science and Engineering **845** (2020) 012044 doi:10.1088/1757-899X/845/1/012044

# Differences in precision of wax molds made of portland cement and silicone rubber in investment casting process for economic efficacy

**I Kusyairi[1*], M A Choiron[2], H M Himawan[1] and Y S Irawan[2]**
[1]Mechatronic Engineering, Politeknik Kota Malang, POLTEKOM, Malang, Indonesia.
[2]Mechanical Engineering, Brawijaya University, UB, Malang, Indonesia,

E-mail: kusyairi1408@gmail.com

**Abstract**—The investment casting method has several advantages, namely the ability to mold complex objects, create products smooth surface and high precision. This is closely related to the wax pattern used, if the wax already has a precision geometry with a precise shrinkage estimation during the casting process, the product will be in accordance with the plan, and vice versa. This study compares the level of roughness and precision of wax molds made of Portland cement and Silicon Rubber. The first step was making a product design with CAD, the product pattern to be made was the origami crash box with a thickness of 3 mm. The second step, converting the CAD file to .stp/.step then printing the design with 3D Printing. The third step, duplicating the 3D printing results with Portland cement material on the first mold and silicone rubber on the second mold. The first mold was designed with the assembly method, the assembly will be carried out after the first part and the second part have been finished. The fourth step was pouring wax that has been mixed with dammar gum (Shorea javanica), onto each gate that has been designed. The study found that the percentage difference between design and cement was 14.7% while the difference between design and RTV Silicon Rubber was 2.7%. This proves that RTV Silicon Rubber can form more precise patterns than cement. RTV Silicon Rubber mold is able to form complex patterns. The surface roughness adjusts to the imitated pattern, if the pattern is smooth then the surface produced by RTV Silicon Rubber mold is smooth, and vice versa.

## 1. Introduction

The investment casting method is the most widely used method for making parts that require high precision. Investment casting is able to make complex, thin and smooth structures. The investment casting method has several stages, (1) making wax molds, generally using metal, (2) pouring heated wax onto wax molds, can be poured manually or by injection method, (3) making ceramic slurry, in this process, the powder used for ceramic slurry was divided into two, namely fine powder and coarse powder (4) Coating ceramic slurry, wax pattern was coated with ceramic slurry (5) dewaxing, heating the coated wax pattern, this stage was carried out so that the wax came out of the mold (6) Pouring, casting metal into molds.

To achieve the desired precision, various factors need to be considered in the six processes, starting from making wax dies to finishing. A lot of research has been done on the investment casting method, proving that the investment casting method can be used in molding complex objects and producing desired thicknesses [1,2], research also leads to pouring methods, which influence the quality of wax patterns. Therefore, researchers designed a low-pressure injection device to get the maximum wax pattern [3].

* To whom any correspondence should be addressed.

Research on investment casting method also leads to gate design, so that the incoming liquid can fill every space [4,5], on the issue of smoothness, research on investment casting method develops the composition of ceramic slurry, adds some refractory materials that have a certain smoothness that can produce desired smooth surface [6,7] research on wax pouring temperature concluded that, the high pouring temperature gave more shrinkage, the casting of wax onto the mold was carried out between the pour temperatures of 50° - 60 °C. Apart from the casting temperature, shrinkage also happens due to the geometry factor [8].

**Figure 1.** Wax mold made of cement.

Making wax molds is generally done with metal-based molds, the results obtained from metal molds are able to precisely mold complicated patterns by including 2% wax shrinkage value from the pattern design. Since the costs incurred are too high for some circles, there are several solutions to overcome it, one of them is the development of RTV Silicon Rubber mold [9,10,11] which was predicted to form complicated patterns. Household-scale industries also using cement as a wax mold, as shown in Figure 1, this research will compare two alternative substitutes for metal molds, in the form of precision comparison between cement mold and RTV Silicon Rubber mold.

## 2. Method

### *2.1. Cad design*

The first step was to design the pattern for metal casting. In this research, the origami Crash Box design which was split into two parts was chosen, as shown in Figure 2 [12]. With a thickness of 3 mm, the wall of the origami crash box consisted of 3 indentations. The challenge in the process of pouring wax and metal was to ensure that the liquid flows, passes, and fills the angles and indentations of the origami crash box.

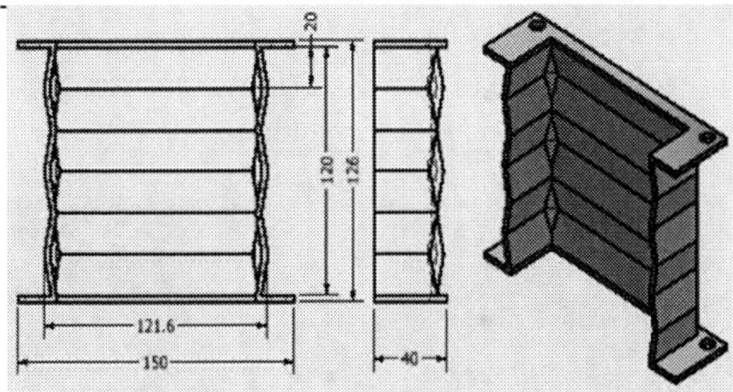


**Figure 2.** Origami crash box.

IOP Conf. Series: Materials Science and Engineering **845** (2020) 012044 doi:10.1088/1757-899X/845/1/012044

### *2.2. Design printing*

After the modeling process in CAD software is complete, then CAD files are converted into .stp / .step files, and run on 3D Printing machines that are operated with Cura software. The material made to print the parts was PLA filament with a diameter of 1.75 mm, the temperature of 240°C, nozzle size of 0.4 mm, speed of 30 mm/s, processing for 17 hours 14 minutes. Modeling the images shown in Figure 3a work using buffer (blue) and object in red. The results of 3D printing work as shown in Figure 3b, the buffer is still attached to the part and needs to be finished. Figure 3c is the result of finishing and the surface is still not smooth.

**Figure 3a.** 3D Printing modelling.

**Figure 3b.** Result of 3D printing.

**Figure 3c.** 3D Printing process.

### *2.3. Mold making*

In making a wax mold from cement, the result of 3D Printing is pressed with cement from the front and back. After the cement dries, the next process is to smooth the mold surface. The result of the cement mold processing was divided into 3 parts as shown in Figure 4a. Figure 4b is the picture of mold parts put together, consisting of 2 (two) gates which are the entry points for the wax.

As for the manufacture of the wax mold with RTV Silicon Rubber by mixing RTV Silicon Rubber with Hardener. The stirring process then followed by pressing the parts into the mixture, let stand until the liquid hardens. The mixture hardening duration depends on the type of hardener. The results of RTV Silicon Rubber mold can be seen in Figure 5a, which consists of two parts. The Silicon Rubber mold design is supported by cement so that Silicon Rubber does not change shape like in Figure 5b.

**Figure 4a.** Divided into 3 Parts.

**Figure 4b.** Cement mold parts put together.

**Figure 5a.** Outer and inner molds.

**Figure 5b.** Mold parts put together.

### *2.4. Pouring*

Paraffin wax, which has the chemical formula $C_nH_{2n}{}^{+2}$, will harden at room temperature and will melt at a temperature of around 37 °C (99 °F) [13] was used in this research. Paraffin, which has been cut into pieces, then mixed with dammar gum in a heated furnace. The ratio of paraffin wax to dammar gum was 1:4, as shown in Figure 6a. Before being put into the furnace, dammar gum was ground into a powder to avoid the explosion that might happen if it is burned in the form of a lump. The combustion process took up to 120 minutes until the paraffin and dammar gum was mixed and turned brown, as shown in Figure 6b. Figure 6c is the picture of solid paraffin. The pouring temperature was in the range of 50°C then allowed to stand for 5 minutes so that the wax hardened. Before pouring, cement mold and RTV rubber mold were coated with non-sticky liquid, so that the wax can be easily removed from the mold.

**Figure 6a.** Paraffin and dammar gum.

**Figure 6b.** Mixture.

**Figure 6c.** Solid.

### *2.5. Measurement*

Printing with molds made of cement and silicon rubber is carried out three times. The printout is measured in the plane of height, length, thickness, and width. The mold measurement area is as shown in Figure 7. After the measurement process, the percentage of dimensional distortion in each field is calculated.

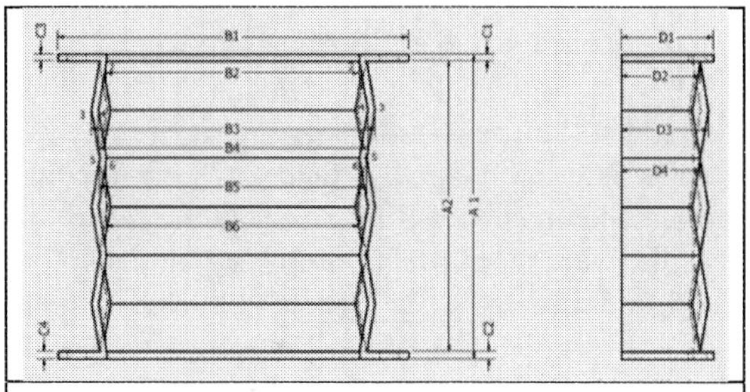


**Figure 7.** Measurement area.

A 1 = the Outer Height
A2 = the Inner Height
B1 = the Outer Length
B2 = The Length of point 2 to 2
B3 = The Length of point 3 to 3
B4 = The Length of point 4 to 4
B5 = The Length of point 5 to 5
B6 = The Length of point 6 to 6
C1=C2=C3=C4 = The Difference of A1 and A2
D1 = width of Origami Pattern

D2 = Width of the edge to the middle diamond
D3 = Width of the edge to the outer angle
D4 = Width of the edge to the Inner angle

## 3. Results and discussion

The result of cement mold can be seen in Figure 8.a and the result of RTV Silicon Rubber mold can be seen in Figure 8.b. Both of them has succeeded in making the origami crash box. From table 1, the differences in design dimensions of cement mold and RTV Silicon Rubber mold can be seen. It shows a significant difference between cement and RTV Silicon Rubber materials. in area A was obtained the percentage data of the added dimensions of 1.6 percent on Portland cement molds and 0.7 percent on silicon rubber molds, for area B added 2.6 percent percentage occurs in cement molds and 1.9 percent on silicon rubber molds. Significant differences are found in area C, the percentage of the added dimensions on cement molds is 43.9% and on silicon rubber rtv molds is 4.4%, while in the area D the percentage of added molds are made from cement and silicon rubber rtv are 10.2% and 3.25%, respectively, as shown on Figure 9.

a) Cement b) RTV Silicon Rubber

**Figure 8.** Results of wax molds.

The biggest percentage of added dimension in the overall is area C, which is an uneven thickness, this is because the workmanship is done manually, then the surface could not be precision between one point and another point. In silicon rubber RTV, it takes the release of silicon rubber RTV from the holder every time we print the wax, when it is reassembled on the holder, RTV silicon rubber is no longer precise with the holder, and affects the 2nd and 3rd molds. The percentage difference between design and cement was 14.7% while the difference between design and RTV Silicon Rubber was 2.7%. This proves that RTV Silicon Rubber can form more precise patterns than cement. RTV Silicon Rubber structure that has good formability, can mold patterns precisely, the pattern of casting wax into silicon rubber mold has not changed the geometry of silicon rubber because the wax pouring temperature, which is 50°C, is still below the Maximum Service Temperature, which is 226,5°C. However, further research needs to be done if RTV silicon rubber mold is used in a large scale.

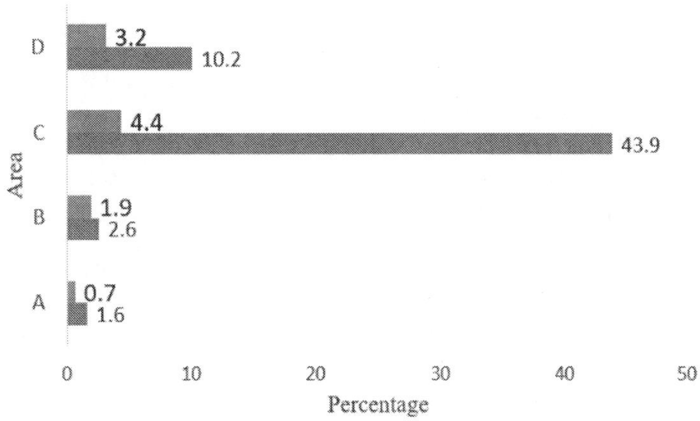


Figure 9 Average of dimensional percentage area.

As for roughness, through observation and touching on the surface, cement-based mold is smoother than RTV Silicon-based mold because cement mold is processed twice, the first process is modeling and the second process is smoothing. This is different from RTV Silicon Rubber mold which only takes one process. If the pattern is smooth then the mold will also be smooth, and vice versa. In Figure 8.b the result of mold follows the 3D printing pattern. The execution of cement pattern in the second workmanship was done manually, so it affects the precision of the mold. The more thorough the manual work is, the precision level will also increase.

**Table 1**. Comparison of the dimensional percentage of dimensional distortion.

| Dimension | Cement | | Silicon Rubber | |
|---|---|---|---|---|
| | Average (mm) | Percentage (%) | Average (mm) | Percentage (%) |
| A1 | 124.6 | 1.1 | 125.20 | 0.6 |
| A2 | 117.4 | 2.1 | 119.17 | 0.7 |
| B1 | 151.3 | 0.8 | 146.00 | 2.7 |
| B2 | 114.0 | 5.2 | 110.33 | 1.8 |
| B3 | 123.0 | 1.2 | 119.67 | 1.5 |
| B4 | 116.0 | 0.5 | 115.87 | 0.4 |
| B5 | 120.0 | 4.8 | 116.33 | 1.6 |
| B6 | 111.8 | 3.1 | 112.40 | 3.7 |
| C1 | 4.2 | 38.9 | 3.03 | 1.1 |
| C2 | 4.5 | 50.0 | 3.07 | 2.2 |
| C3 | 4.5 | 48.9 | 3.20 | 6.7 |
| C4 | 4.1 | 37.8 | 3.23 | 7.8 |
| D1 | 39.4 | 1.5 | 41.00 | 2.5 |
| D2 | 24.0 | 29.9 | 35.67 | 4.1 |
| D3 | 38.1 | 0.8 | 38.80 | 2.7 |
| D4 | 31.4 | 8.4 | 33.07 | 3.5 |
| Average | | 14.7 | | **2.7** |

IOP Conf. Series: Materials Science and Engineering **845** (2020) 012044 doi:10.1088/1757-899X/845/1/012044

## 4. Conclusion

a. The percentage difference between design and cement was 14.7% while the difference between design and RTV Silicon Rubber was 2.7%.
b. RTV Silicon Rubber mold can be more precise than cement mold. RTV Silicon Rubber mold is able to form complex patterns. The surface roughness adjusts to the imitated pattern, if the pattern is smooth then the surface produced by RTV Silicon Rubber mold is smooth, and vice versa.
c. Cement mold tends to be smoother than RTV Silicon Rubber mold because cement mold can be processed and smoothed again after the pattern is pressed.
d. Printing results from 3D Printing are still rough. It affects the mold products, especially RTV Silicon Rubber mold. Therefore, further finishing is required, both evaporating and coating of the product.

**Acknowledgment**

The researchers would like to express sincere gratitude to the Ministry of Research, Technology and Higher Education, the Directorate General of Strengthening Research and Development, and the Directorate of Research and Community Service that has financed the entire research (Grant Number: 0045/E3/LL/2018).

## References

[1] Liu C, Jin S, Lai X, He B and Li F 2015 *Int.J.of Adv.Manufac Tech.* **77** 1191
[2] Jiang R, Wang W, Zhang D and Wang Z 2016 *Int.J.of Adv.Manufac Tech.* **83** 949
[3] Chil-Chyuan K and Ren-Lung T 2015 *Polymers & Polymer Composites* **23** 647
[4] Huang P and Lin C 2015 *Int.J. of Adv.Manufac Tech.* **79** 997
[5] Biernacki R, Haratym R, Bałkowiec A, Wawulska-Marek P, Matysiak H, Zdunek J. and Kurzydłowski K J 2015 *Archives of Metallurgy & Materials* **60** 345
[6] Jin S, Liu C, Lai X, Li F and He B 2017 *Int.J. of Adv.Manufac Tech.* **88** 663
[7] Małek M, Wiśniewski P, Matysiak H, Zagórska M and Kurzydłowski K 2014 *Archives of Metallurgy & Materials* **59** 1059
[8] Chaudhari B Y, Kanoongo N, Sulakhe N and Pathak S D 2013 *IOSR-JMCE.* **25** 2320
[9] Rahmati S, Akbari J and Barati E 2007 *Rapid Prototyp. J.* **13**
[10] Valentan B, Pilipovic A and Drstvensek I 2012 *Adv. in Prod. Eng. & Man.* **7** 255
[11] Chil-Chyuan K and Ren-Lung T 2015 *Polym. Polym. Compos.* **23** 647
[12] Kusyairi I 2017 *J.of Energy Mech. Eng. 2*
[13] Sandhu E C S and Sharma E A 2012 *IOSR-JMCE* **3** 1

# Drying kinetic modelling of dried black potato (*Plectranthus rotundifolius*) cultivated in Indonesia

**V Paramita[1*], H Kusumayanti[1], M E Yulianto[1], D A Rachmawati[1], I Hartati[2], P R Ardi[3]**
[1]Department of Industrial Technology, Diponegoro University, Semarang, Indonesia
[2]Department of Chemical Engineering Wahid Hasyim University, Semarang, Indonesia
[3]Department of Management Economic, STIE Asian East College Semarang, Indonesia

E-mail: vparamita@live.undip.ac.id

**Abstarct**. The drying kinetic modelling of dried black potato (Plectranthus rotundifolius) was studied by employing electrical cabinet oven. The parameter investigated was moisture content, regarding to the drying temperature (70, 80 and 90 °C) and slice thickness (2, 3 and 4 mm). The lowest sum square error of drying kinetic model (0.0014) was obtained by applying logarithmic model of thin layer drying. This value performed by 4 mm of slice thickness of black potato with 80 °C of drying temperature during 195 min of drying time. This result indicated that the logarithmic model is the most suitable model for interpreting the thin layer drying kinetic of black potato.

## 1. Introduction

Black potatoes are a type of tuber in the same group as potatoes and are classified in the lamiceae family. It provides a potential food crop as an alternative food source, but cultivation in the community actually decreases increasingly scarce. *Plectranthus rotundifolius* is an erect, 30 cm tall of bushy, and semi-succulent stem of annual herb with a thick leaf. Plant grows well in poor or sandy soil with direct sunlight, moreover it is also tolerant of drought and rainfall [1]. The tubers can be harvested after 4 to 5 months flowering and the aerial parts of plant have died [2]. The selection of black potatoes as objects in this study is in line with the national strategic planning in the field of diversification of local resource-based food consumption with exploration of the utilization of non-conventional food crops, including tubers [3].

Black potato tubers contain more carbohydrates (33.7 gr/100 g) compared to yellow potatoes (13.4 gr/100 g). Energy in black potato tubers (400 cal) is even 6 times that of yellow potatoes (64 cal) [4]. Moreover, [5] also mentioned that black potato is rich in essential minerals and vitamins, in addition to low glycemic index of carbohydrate source. This carbohydrate source potentially developed as a nutritious functional food, due to the high content of ascorbic acid in the tuber [6].

Crispy snack derived from potato possibly served by applying fresh, slice or dough taste of potatoes. For obtaining potato chips, the process preparation includes the potato harvesting, selecting, storage before processing, peeling, slicing, frying or drying, applying flavour and packaging [7]. In order to

* To whom any correspondence should be addressed.

result the identical colour of chips, therefore the tubers should be consistency sliced in thickness. Before drying or frying, any starch or other materials resulted from peeling and slicing fed into running water in order to clean the slices. Most of dry crisp snack are mass production, and therefore their initial and final structure and also texture are very concerned [8].

Potato chips with thin slices are mostly losing their moisture content more than 75 % and reaching around 2 % during processing. However, during the hot temperature contact of oil frying, unavoidably causing a substantial absorption of oil around 35 % and mostly located on the surface of the chip. It is indicated that there was poor hot oil penetration during frying and stuck in the surface at the end of frying [9]. The application of hot air during drying can be avoid this problem and resulted non-oily dried product. Therefore, this study investigated the drying process and the kinetic of black potato at 70, 80 and 90 °C, regarding the slice thickness of 4 mm.

## 2. Methodology

### *2.1. Drying treatment*

An oven universal UN 110 produced by Memmert (Schwabach, Germany) was applied for determination of drying kinetics of black potatoes. The hot air produced by heating coil in the dryer cabinet was employed to determine the dry sample mass with temperature 70, 80 and 90 °C. The material thickness was 4 mm. The dried materials were weighed subsequently every 15 min until constants weight by using Ohaus PA 214 (NJ, USA).

### *2.2. Analysis of drying kinetics*

The drying kinetics of different permeable materials are frequently observed by using thin-layer equations, which possibly divided into three groups (i.e. theoretical, semi-theoretical and empirical. The theoretical discuss about internal resistance to moisture transfer only, while the others discuss about the external resistance to moisture transfer between the product and the air only [10]. Among the semi-theoretical thin-layer drying models, the Lewis model (2), the Henderson and Pabis model (3) and the Two-term model, are used widely.

The Fick's second law applied by assuming that there is evenly distributed resistance of the moisture flow along with the homogeneous material and avoid the diffusion coefficient, D, of the local moisture content. If the decreasing of volume is neglected, the Fick's second law possibly re-writes as:

$$\frac{\partial M}{\partial t} = D\nabla^2 M \quad (1)$$

$$MR = \frac{M - M_e}{M_0 - M_e} = \exp(-kt) \quad (2)$$

$$MR = \frac{M - M_e}{M_0 - M_e} = a\exp(-kt) \quad (3)$$

The empirical models explained a direct relationship between the moisture content obtained by experimental and the drying time. They avoid the basic theory of the drying process and there is no meaning on its parameter. Therefore, they gave an un-clear and inaccurate critical point of view during drying, although it describes the drying curve for the conditions of the experiment better than the semi-theoretical ones [11]. The empirical models studied in this work are including Logarithmic model (4), Two-term exponential model (5), Diffusion Approach model (6), and Modified Henderson and Pabis model (7).

$$MR = a\exp(-kt) + c \quad (4)$$

$$MR = a\exp(-kt) + (1-a)\exp(-kat) \quad (5)$$

$$MR = a\exp(-kt) + (1-a)\exp(-kbt) \quad (6)$$

IOP Conf. Series: Materials Science and Engineering **845** (2020) 012045 doi:10.1088/1757-899X/845/1/012045

$$MR= a \exp(-kt) + b \exp(-gt) + c \exp(-ht) \quad (7)$$

All of the kinetic data obtained by applying solver program included in Microsoft Excel 2010 Software.

## 3. Results and discussion

Figure The drying kinetics of black potato tubers were studied by plotting the experimental data into the semi-theoretical thin-layer drying models and the empirical models. The semi-theoretical thin-layer drying models discussed are Lewis model, the Henderson and Pabis model and the Two-term model, the empirical models are Logarithmic model, Two-term exponential model, Diffusion Approach model, and Modified Henderson and Pabis model.

### *3.1. Modeling of the drying kinetics*

Figure 1 presents the concept of drying processing of thin layer. $M_0$ indicated the initial mass of single chips, while $M_e$ presented the equilibrium mass of single chips after t min of drying with x mm of thickness. $T_s$ showed the surrounding temperature affected the chips and resulted to the heat (Q) absorbed by the chips. $H_2O$ is indicating the amount of $H_2O$ vaporized from the wet chips.

The assumptions applied during drying are (i) the chip is uniform; (ii) no alteration with the chip's characteristics and regardless to the shrinkage may occur; (iii) condition is isobaric; (iv) avoid the diffusion of water from internal porous and regarding only to the surface evaporation; (v) evenly the initial moisture contents of the chips and exactly similar parts facing each other while drying is occurring; (vi) the moisture equilibrium arises on the surface due to the finishing of surface diffusion; (vii) surrounding temperature is equals to the ambient drying air temperature, (viii) the heat transfer is done by conduction within the product, and by convection outside of the product; (ix) effective moisture diffusivity is constant versus moisture content during drying [12].

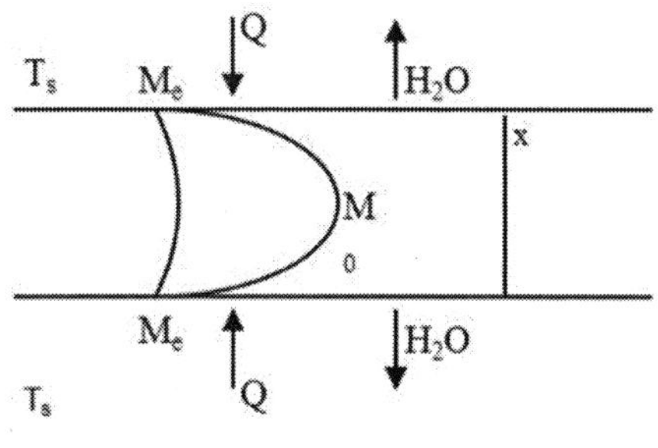


**Figure 1.** Drying mechanism of thin layer [12].

### *3.2. Comparison of experimental and calculated MR*

The empirical value of each models showed on the table 1. Fig. 2 presents the effect of drying time on the moisture content of experimental data (symbol) and calculated data (line) of drying kinetics on 4 mm slice thickness of black potato, regarding to the drying temperature. It shows that Logarithmic (empirical approach) model and Henderson & Pabis (semi-theoretical approach) model gave excellent fit on the experimental data (less than 0.01 of SSE (sum square error) with exception of 90 °C of Henderson & Pabis). The lower value of SSE provided the better the fitting models of experimental data. The others model shows the value of SSE more than 0.01 which indicated the poor fitting experimental data on the model. The Logarithmic model provided better modelling than the Henderson & Pabis model, with the minimum value of SSE were obtained (less than 0.005). Since there were many negligible points of view on the empirical model, it shows that Henderson & Pabis provided more accounted calculated data for drying kinetic approach. The constant of c on the Logarithmic model provided the better accuracy on the fitting model while comparing to the Henderson & Pabis model.

Akpinar [19] found that other empirical model of Midilli or Thompson model [20] well explained the drying potato chips at 60–80 °C. Higher drying temperature provided the faster the lower moisture content obtained, however higher temperature also leads to the case hardening of the chips and therefore the browning colour of the surface.

**Table 1.** Drying kinetics of black potato on 4 mm slice thickness of chips.

| Thin Layer Drying Model | **Temperature (°C)** | **k** | **a** | **g** | **B** | **h** | **c** | **SSE** |
|---|---|---|---|---|---|---|---|---|
| Lewis [13] | 70 | 0.0176 | - | - | - | - | - | 0.1199 |
| | 80 | 0.0295 | - | - | - | - | - | 0.0501 |
| | 90 | 0.0285 | - | - | - | - | - | 0.0596 |
| Henderson & Pabis [14] | 70 | 0.0131 | 0.7447 | - | - | - | - | 0.0074 |
| | 80 | 0.0242 | 0.8098 | - | - | - | - | 0.0038 |
| | 90 | 0.0235 | 0.8117 | - | - | - | - | 0.0129 |
| Logarithmic [15] | 70 | 0.0112 | 0.7667 | - | - | - | -0.0393 | 0.0020 |
| | 80 | 0.0220 | 0.8264 | - | - | - | -0.0261 | 0.0014 |
| | 90 | 0.0179 | 0.8796 | - | - | - | -0.0898 | 0.0039 |
| Two-term exponential [16] | 70 | 0.0176 | 1.000 | - | - | - | - | 0.1199 |
| | 80 | 0.0295 | 1.000 | - | - | - | - | 0.0501 |
| | 90 | 0.0285 | 1.000 | - | - | - | - | 0.0596 |
| Diffusion approach [17] | 70 | 0.0176 | 1.000 | - | 1.000 | - | - | 0.1199 |
| | 80 | 0.0295 | 1.000 | - | 1.000 | - | - | 0.0501 |
| | 90 | 0.0285 | 1.000 | - | 1.000 | - | - | 00596 |
| Modified Henderson & Pabis [18] | 70 | 0.0131 | 0.2482 | 0.0131 | 0.2482 | 0.0131 | 0.2482 | 0.0074 |
| | 80 | 0.0242 | 0.2699 | 0.0242 | 0.2699 | 0.0242 | 0.2699 | 0.0038 |
| | 90 | 0.0235 | 0.2706 | 0.0235 | 0.2706 | 0.0235 | 0.2706 | 0.0129 |

### *3.3. Effective diffusivity*

The effective moisture diffusivity, $D_e$, is used to explain the moisture diffusion process during drying and calculated by using the method of Fick's second law. Its solution (presents as the diffusion of liquid in a slab shaped solid in terms of the dry basis moisture content) can be written as follows [21]:

$$MR = \frac{8}{\pi^2}\exp\left[-D_e\left(\frac{\pi}{Ht}\right)^2 t\right] \tag{8}$$

Rizvi [22] mentioned that the effective diffusivities are regarding to the uniformity of the material and the air temperature during drying, which commonly described by using the Arrhenius-type relationship [23, 24]. The natural logarithm of $D_e$ as a function of the absolute temperature is plotted to show a linear relationship between (ln De) and (1/T) as shown at Fig. 3, leading to an Arrhenius-type relationship:

$$D_e = D_o \exp\left(\frac{-E_a}{RT}\right) \tag{9}$$

IOP Conf. Series: Materials Science and Engineering **845** (2020) 012045 doi:10.1088/1757-899X/845/1/012045

The slope of the straight line is (-Ea/R) from which the activation energy Ea is calculated for the drying process. The diffusivity constant $D_0$ is also calculated from the intercept of the straight line. The value of $D_0$ obtained is 1.258E-11 m2/s and the value of Ea is 3.23E-08 J/mol.

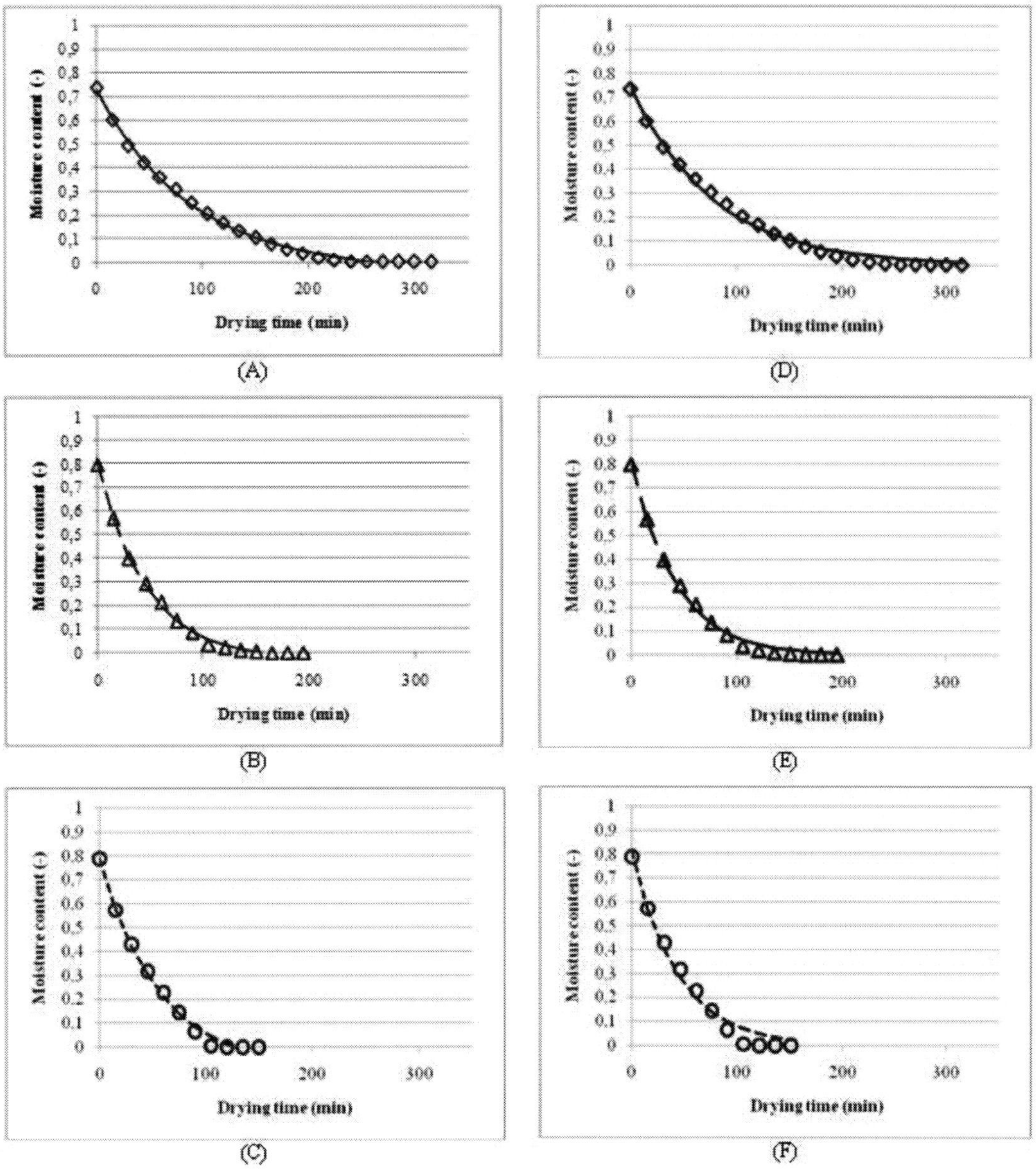


**Figure 2.** Plotting the drying time against the moisture content of experimental data (symbol) and calculated data (line) of drying kinetics on 4 mm slice thickness of black potato; Logarithmic model (A, B, C) and Henderson & Pabis model (D, E, F). Drying temperature: 70 (A, D), 80 (B, E) and 90 (C, F) °C.

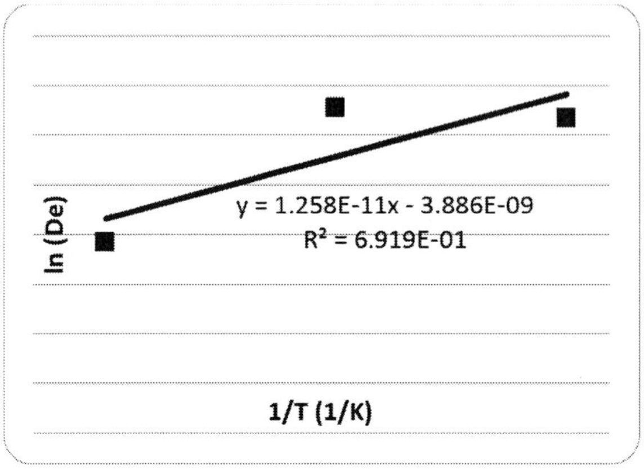


**Figure 3.** Arrhenius plot of effective diffusivity and temperature.

## 4. Conclusion

The lowest sum square error of drying kinetic model (0.14) was obtained by applying logarithmic model of thin layer drying. This value performed by 4 mm of slice thickness of black potato with 80 °C of drying temperature during 195 min of drying time. This result indicated that the logarithmic model is the most suitable model for interpreting the thin layer drying kinetic of black potato.

## 5. Nomenclature

| | |
|---|---|
| MR | moisture content (-) |
| t | drying time (min, for MR and s, for D) |
| M | mass (gr) |
| k | empirical coefficient of drying rate (-) |
| a | empirical coefficient of constants (-) |
| g | empirical coefficient of drying rate (-) |
| b | empirical coefficient of constants (-) |
| h | empirical coefficient of drying rate (-) |
| c | empirical coefficient of constants (-) |
| D | diffusivity (m2/s) |
| Ht | thickness of the chip (m) |
| Ea | activation energy (J/mol) |
| R | gas constant (8.3145 J/mol K) |
| 0 (subscript) | initial |
| 1 (subscript) | continuation |
| e (subscript) | equilibrium |

## 6. References

[1] Lewis W H 1986 *Econ. Bot.* **40** 176
[2] Priya M H and Anbuselvi S 2013 *J. Chem. Pharm. Res.* **5** 12
[3] Nugraheni M, Santoso U, Suparmo and Wuryastuti H 2013 *J. Teknol. Ind. Pangan* **24** 138
[4] Anbuselvi S and Priya M H 2013 *Int. J. Pharm. Sci. Rev. Res.* **22** 213
[5] Lestari P, Utami N W and Setyowati N 2015 *Bot. Garden Bull.* **18** 59
[6] Kishorekumar A, Jaleel C A, Manivannan P, Sankar B, Sridharan R, Murali P V and Panneerselvam R 2008 *Colloids Surf. B Biointerfaces* **62** 307
[7] Bawa A S and Sidhu J S 2003 *Encyclopedia of Food Sciences and Nutrition* (London: Academic)

[8] Duizer L M 2013 *Instrumental Assessment of Food Sensory Quality*, ed Kilcast D (Cambridge: Elsevier) p 403
[9] Pedreschi F, Mery D, Bunger A and Yanez V 2010 *J. Food Process Eng.* **34** 1
[10] Kucuk H, Kilic A and Midilli A 2014 Progress in Exergy, Energy, and the Environment, ed Dincer I, Midilli A and Kucuk H (Switzerland: Springer) p 669
[11] Onwude D I, hashim M, Janius R B, Nawi N M and Abdan K 2016 *Compr. Rev. Food. Sci. F.* **15** 599
[12] Erbay Z and Icier F 2010 *Crit. Rev. Food Sci. Nutr.* **50** 441
[13] Lewis W K 1921 *Ind. Eng. Chem.* **13** 427
[14] Henderson S M and Pabis S 1961 *J. Agric. Eng. Res.* **6** 169
[15] Lopes D C, Neto A J S and Santiago J K 2014 *Biosyst. Eng.* **118** 105
[16] Roman F and Hensel O 2011 *Agric. Eng. Int. CIGR J.* **13** 1
[17] Yaldiz O and Ertekin C 2001 *Dry. Technol.* **19** 583
[18] Karathanos V T 1999 *J. Food Eng.* **39** 337
[19] Akpinar E K 2006 *J. Food Eng.* **73** 75
[20] Thompson T L, Peart P M and Foster G H 1968 *Trans. ASAE.* **11** 582
[21] Aguerre R, Suarez C and Viollaz P E 1982 *J. Food Technol.* **17** 679
[22] Rizvi S S H 1986 *Engineering properties of foods* ed Rao M A and Rizvi S S H (New York: Marcel Dekker)
[23] Panchariya P C, Popovic D and Sharma A L 2002 *J. Food Eng.* **52** 349
[24] Minaei S, Motevali A, Najafi G and Seyedi S R M 2011 *Aust. J. Crop Sci.* **6** 548

# Optimization of biogas production from water hyacinth by liquid anaerobic digestion (L-AD) using response surface methodology

**Syafrudin[1*], W D Nugraha[1], H H A Matin[2], F Margaretha[1], and Budiyono[3]**
[1]Department of Environmental Engineering, Diponegoro University, Semarang, Indonesia
[2]Environmental Science, Sebelas Maret University, Surakarta, Indonesia
[3]Department of Chemical Engineering, Diponegoro University, Semarang, Indonesia

E-mail: udin_syaf@yahoo.com

**Abstract.** In this biogas research from water hyacinth, we are using Central Composite Design and full factorial design. Enzyme (X1), C/N ratio (X2) and total solid (X3) is three experimental variables were selected as controlled factors. Performed on a laboratory scale and at room temperature with 16 runs for 60 days measured every two days. The variables were set at 3-9% for Enzyme concentration, C/N at between 20-50, and concentration of TS between 5-15% (LAD condition). Based on the research that has been done, the optimum conditions for the biogas production process from water hyacinth in Liquid State Anaerobic Digestion (LAD) conditions use Response Surface Methodology (RSM) in the variable range are the C/N ratio 30-40, 6% enzyme, and for TS was not significant factor. Following the run in the central composite experimental design, these values correspond to run number 10 with variable C/N ratio 35, enzyme 6% and TS 1.59% with biogas production 202.51 mL/g TS.

## 1. Introduction

Biogas have a significant potential as a source of energy latest because the womb methane $CH_4$ his own they received a heat engine 50 mJ/kg. Methane $CH_4$ that has one carbon in every chain, can produce combustion that is environmental friendly than the serial fuel long carbon. This is because the amount of $CO_2$ put which is produced during the combustion of fuel serial short carbon is more a little. Biogas is flammable gases (flammable) resulting from the process of fermentation organic materials are bacterium by anaerobic derived from an agricultural waste, animal husbandry and of human waste. The material that can be used as raw materials of manufacture for biogas that is biodegradable material like biomass, dirt, the results of the activity of garbage solid urban and others [1].

One of material used in this research is a waste product the hyacinth. Biogas obtained from the results of the decipherment of water hyacinth without oxygen are anaerobic. Water hyacinth that containing a great the water level in its body which is approximately 90 % is a profit in making use of as a result of biogas through a yeast process fermentation with methane formulation bacteria besides figures to be womb a compound of carbon dioxide and nitrogen high between 30-35. Breed of this fast causing a plant water hyacinth it has turned into a weed water. A very fast the water hyacinth growth but also that it caused many problems, among others speed up silting in the rivers or lakes, decreasing

* To whom any correspondence should be addressed.

the production fish, has created a difficulty for an irrigation channel, and cause the evaporation of water to 3 to 7 times greater than the evaporation of water out on the open water [2].

In the process of biogas there are renewable biomass material as of aquatic plants used for cleansing of waste water or invasive species in lakes and the rivers can be harvested and implemented in the process of L-AD. L-AD can be operating at concentrations Total Solids which are different. The process comparison between of L-AD with TS fewer than 15 % classified as Liquid State Anaerobic Digestion (L-AD), are suitable for waste by high water content include cesspool mud and waste flood. Therefore, that water L-AD need a lot to handle waste low water as biomass containing lignocellulose. On the other experiment that discuss, a Solid State Anaerobic Digestion are (SS-AD) operating on Total Solid more than 15 % so it is more suitable to digest lignocellulose on biomass [3], [4].

Microorganism's activity that play a role during the process of fermentation depends for C/N Ratio. On anaerobic are the ratio of c / n optimal about 20-30. The ratio of c / n high on the material organic will causing the production of methane that is low. The problem is that of material by C/N Ratio are only containing nitrogen in a low. The increase has caused consumption nitrogen that is fast by bacteria methanogen and result in the yield applied for biogas down because of nitrogen lack available for the growth bacteria. That is why to increase the level of nitrogen on the material, organic matter has accompanied containing nitrogen that is as high as the excrement of an animal. While, if the C/N ratio very low to cause of ammonia accumulating and could raise its pH digester amounted to 8.5 who is poisonous against bacteria methanogen [5]. This is a reason why not need other material containing carbon or high fibre, as grass, straw, the husks and foliage. The trick is to mixing organic matter with material containing carbon until achieved the C/N ratio term.

Water content known as one of important parameter anaerobic process are. Hence, the womb total solid used define two types of them are anaerobic process, wet are Liquid Anaerobic Digestion are to the womb TS 15 % and dry are Solid-State Anaerobic are to TS 15 %.

If temperature between 4-60°C and temperature kept constant gas will be produced. The high temperature was expected to fast but are becoming less of bacteria would the temperature had an influence on the process of digestion anaerobic of organic matter and the production of gas. Digestion lasted well at a temperature of 30-40°c for the condition of mesophilic and at a temperature of 45 to 60°C for the condition are thermophilic. The speed of fermentation down in the 2008 second temperature under 20°C. Most bacteria mesophilic optimal temperature reached at 35°C, but to bacteria are thermophilic at a temperature of 55°c. At a temperature of are thermophilic the activity of microorganisms resulted higher 25 to 50 % to produce more biogas yield [6].

The range are the condition of the neutral pH 7-7.5 who is favourite field most microorganisms. The process of biogas there are types of microorganisms and pH different for each microorganism. An optimal growth pH between 6.8 - 7.6 are the condition for the optimal bacteria methanogen and at pH under 6.6 the growth of bacteria methanogen would decrease, while bacteria with hydrolysis process and acetogenesis will work at between pH 5.5-6.6. If pH values dropped below 6.5, so the production of an organic acid by bacteria hydrolytic can result decrease in more noticeably pH and the process of fermentation will stop [7].

## 2. Materials and Methods

Biogas research has been conducted for 60 days located in The Engineering Laboratory Waste Processing Chemical Engineering and Environmental Engineering Laboratory, the building in college together of Engineering Faculty, Diponegoro University. Namely by preparing in the implementation stage research reactor in accordance with draught which has been made (Fig. 1 and Fig. 2) drum-shaped plastic volume 2 liters a number of seventeen reactor (Duplo) [8]. By controlling the reactor in the extent of the contents of or reactor by determining variable free the C/N ratio: 20, 35, 50; Enzyme: 3%, 6%, 9% and Total Solid 5%, 10%, 15%, and the rumen: water ratio is 1:1, for rumen cow fluid as inoculum was obtained from slaughter house in Pedurungan, Semarang [9].

This research was done on a laboratory scale. Biogas production process was operated on bio digester with volume of 1.5 liters. Bio digester made from a polyethylene bottle plugged in tightly rubber plug and fitted with a valve on the top of bio digester to measure the biogas.
Water hyacinth that used on the reactor up to 16 is water hyacinth that not through pre-treatment, while 17 reactors is control reactor that uses water hyacinth by concentration of the TS 5 %. The research was done for 60 days with the method of measurement in the form of a method of displacement of water every once in two days.

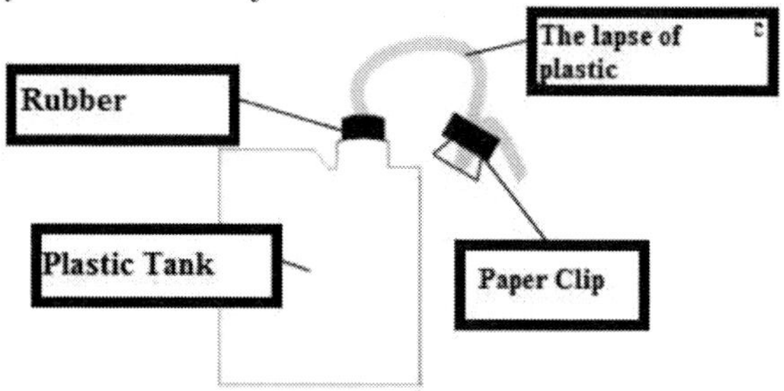


**Figure 1.** Detailed reactor design.

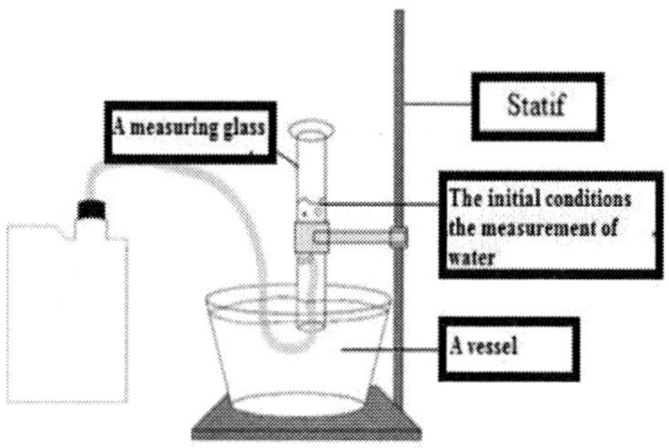


**Figure 2.** Schematic diagram of biogas measurement in laboratory.

*2.1. Biogas analysis*

The research studies were used water hyacinth as the principal ingredient of making biogas. The implementation of this research with the processing organic waste removal of water hyacinth (bio waste) and rumen fluid. Then examined the composition of TS, water content, and C/N Ratio. After that composition are calculation and determination of enzyme variation, C/N Ratio and TS content and preparation of an anaerobic a batch or reactor. The substrate will be mixed with inoculum and water according to the TS concentration as well as the technical urea to adjust the C/N Ratio variation and the Enzyme added accordingly. Other preparing samples can be incorporating into the reactor, seal tightly for anaerobic conditions, and ready for operation. In this super-special stage, we observed the structure groups of water hyacinth before and after of them using. During the processes, observed the volume of biogas produced each time intervals of 2 (two) days. Water hyacinth that used on the reactor up to 16 is water hyacinth that not through pre-treatment, while 17 reactors is control reactor that uses water hyacinth by concentration of the TS 5 %. The research was done for 60 days with the method of measurement in the form of a method of displacement of water every once in two days.
The biogas volume is calculating by passing the gas into a water-filling measuring cup by utilizing the gas properties of pressing in all directions from the or reactor (Boyle' s Law), so that when the valve will be open, or gas from reactor directly moves into the measuring cup to observe the difference volume. The biogas the volume of observation scheme was presented in Fig. 2. Observations will stop after no more biogas is forming. After that, data analysis using Response Surface Method to obtain optimum position in biogas production in cardiac condition.

### *2.2. Response surface methodology*
Variables are set between 3-6% enzyme concentration variation, C/N ratio 20-50, and TS 5-15%. The parameters are optimize using a technique called Response Surface Methodology. A Central Composite Design with full factorial design employ in this regarding. Three independent variables namely, enzyme (X1); C/N ratio (X2); and TS content (X3) as control factors and Yi was set at response variable, with Equation 2. The lower, upper, and center point of the design were code as -1, 1, 0 and α, where + 1 denotes high level, -1 low level, α = 2 n/4 (n=number of variables or factors) is the star point, and 0 corresponds to the center point. The result points of design to provide curvature estimation for the model. From experimental design used, 16 experiments are need and data will be resulting. Statistical analysis of the model was performed in the form of analysis of variance (ANOVA). For each variable, which the quadratic models were represented as contour plots (2D) and surface plots (3D) [10].

$$Yi = \beta0+\beta1X1+\beta2X2+\beta3X3+\beta12X1X2+\beta13X1X3+\beta23X2X3+\beta11X12+\beta22X22+\beta33X32 \quad (1)$$

Did calculation material is in accordance with predetermined variables and prepare us bio digester needed. The substrate was mixing with inoculum, water, enzyme, and technical urea to adjust the C/N Ratio variation and Total Solid. After the sample was preparing, can be inserting into the bio digester, then sealing tightly to get the cardiac condition without oxygen or anaerobic, and ready for operation. During the operation, biogas production was measured every two days. Furthermore, the observation was stopped after the biogas was no longer formed.

**Table 1.** Central composite experimental design for optimizing enzyme, C/N ratio, and TS content on biogas production factorial.

| Run | Factors | | | Biogas production |
|---|---|---|---|---|
| | Enzyme (%) | C/N ratio | TS (%) | (mL/g TS) |
| 1 | 3 | 20 | 5 | 120.8 |
| 2 | 9 | 20 | 5 | 111.3 |
| 3 | 3 | 50 | 5 | 104.1 |
| 4 | 9 | 50 | 5 | 100.9 |
| 5 | 3 | 20 | 15 | 21.66 |
| 6 | 9 | 20 | 15 | 15.56 |
| 7 | 9 | 50 | 15 | 24 |
| 8 | 3 | 50 | 15 | 22.2 |
| 9 | 6 | 35 | 10 | 52.3 |
| **10** | **6** | **35** | **1.59** | **202.51** |
| 11 | 6 | 35 | 18.41 | 16.53 |
| 12 | 3 | 9.77 | 10 | 46.125 |
| 13 | 6 | 60.2 | 10 | 44.05 |
| 14 | 1 | 35 | 10 | 42 |
| 15 | 11 | 35 | 10 | 3.425 |
| 16 | 6 | 35 | 27.5 | 42.275 |

## 3. Results and discussions
### *3.1. Optimize the TS, C/N ratio, and enzyme against the biogas from water hyacinth use Response Surface Methodology (RSM) in LAD*
This research having aims to determine the ratio TS, C/N ratio and the enzyme use RSM. TS concentration (Total Solid) 5-15 % [11], the ratio of C/N 20-50 % and enzyme 3-9%. At this stage, a model quadratic served as meanwhile the contours of (contour of the plot) two dimension (2D) .The production of biogas maximum indicate by the surface area of the smaller in the diagram contour. The

IOP Conf. Series: Materials Science and Engineering **845** (2020) 012046 doi:10.1088/1757-899X/845/1/012046

level of significance the influence of factors MC, C/N Ratio, and TS presented in the form of the chart of pare to chart as are presented in Fig. 3.

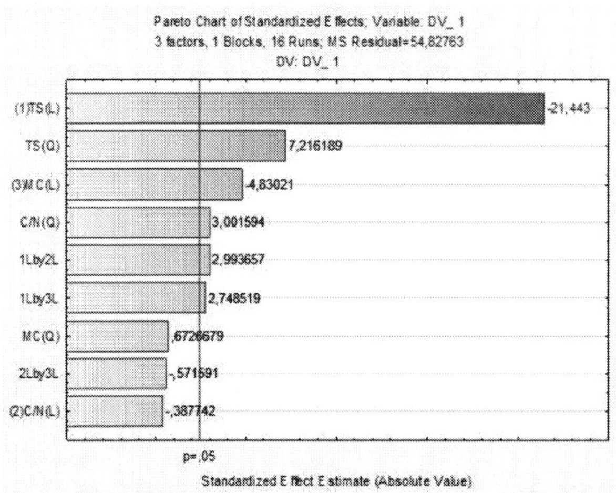


**Figure 3.** Significant factor of enzyme, C/N ratio, and total solid against the production of biogas.

The pare to principle (also known as a rule restore us) said that for many the incident, around 80% than the effect caused by 20% of the cause. From the Fig. 4 pare to chart shows the significance of the influence of TS is directly proportion to the results of production of biogas, microbial consortium the said / enzyme is directly proportional to the results of the production of biogas, and the C/N Ratio is directly proportional to the production of biogas.

A picture pares to chart indicating the level of significance for each variable whether it is being linearly and quadratic, as well as the relations between 2 variables. The total solid are linearly and quadratic as well as the C/N Ratio are linearly influence significantly to the production of biogas ($p<0.05$).While to the ratio of C/N having the character of quadratic and the concentration of microbial consortium said (mc) be it the is linear and quadratic did not influence on the biogas production ($p>0.05$) [10]. In addition, diagrams on Fig. 4. Above can be seen that factors TS having the character of linear having influence the highest against the production of biogas.

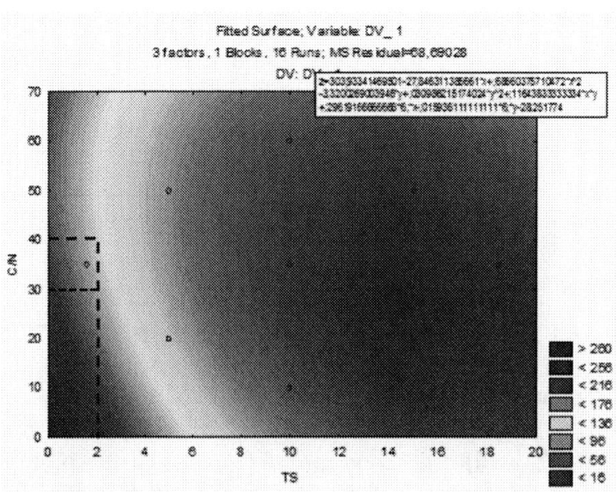


**Figure 4(a).** The 2D contours of TS and C/N ratio to biogas production at concentrations MC.

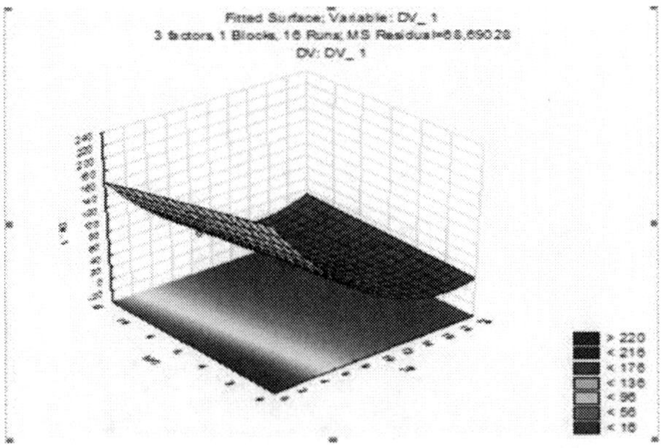


**Figure 4(b).** The 3D contours of TS and MC against biogas production at concentrations the C/N ratio.

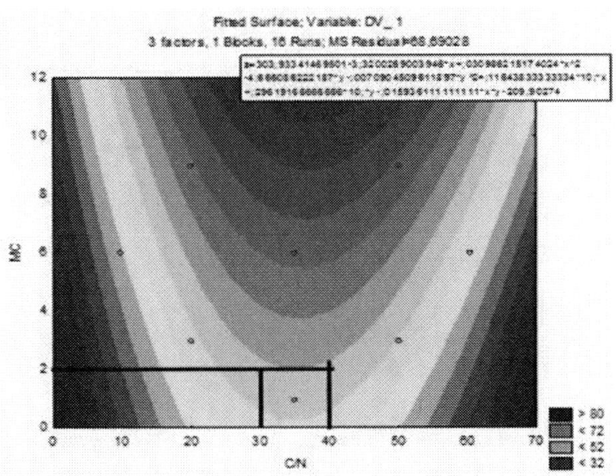


**Figure 5(a).** The 2D Contours of C/N ratio and MC against biogas production at concentrations TS.

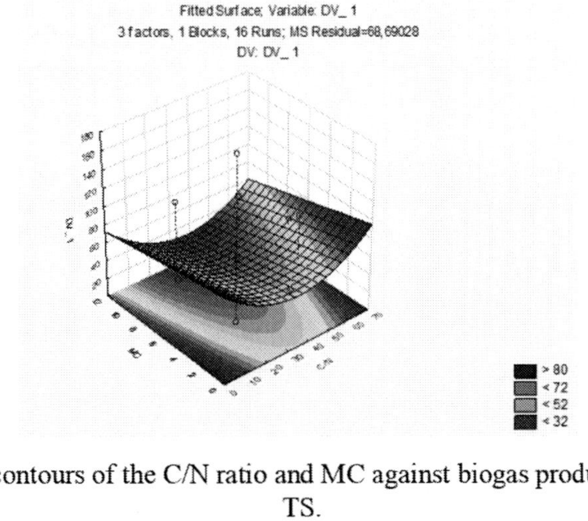


**Figure 5(b).** The 3D contours of the C/N ratio and MC against biogas production at concentrations TS.

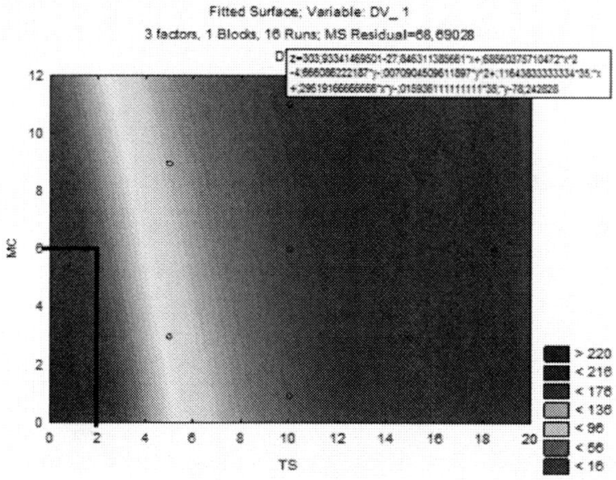


**Figure 6(a).** The 2D contours of TS and MC against biogas production at concentrations the C/N ratio.

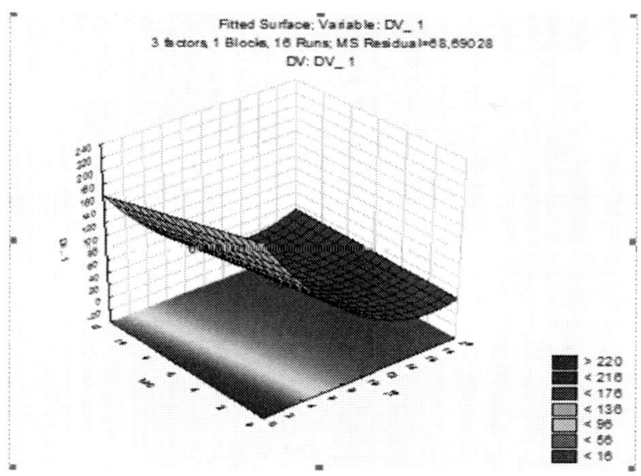


**Figure 6(b).** The 3D contours of TS and MC against biogas production at concentrations the C/N ratio

In Fig. 4 shows that the concentration of the ratio of C/N optimum position for the production of biogas is between 30-40. On other illustration namely in Fig. 5, the concentration of enzyme optimum position for the production of biogas was 6%. In Fig. 6 also shows that the C/N Ratio optimum position for the production of biogas is between 30-40. On other illustration namely in Fig. 5, the concentration of TS optimum position for the production of biogas is around 2%. Based on the response the surface that was displayed on a Fig. 4 and Fig. 5 found the condition of optimum position to the factors of TS not significant on biogas research of water hyacinth in lad this. Of a number of things that has been related is in accordance with the outcome of several study to show that the concentration of TS for the production of biogas optimum position at concentrations below 10 %, the experimental work on this is an L-AD water hyacinth on the condition of the concentration of TS among defeat at 2-0 %.

## 4. Conclusion

Based on the research that has been done, the optimum conditions for the biogas production process from water hyacinth in Liquid State Anaerobic Digestion (LAD) conditions use Response Surface Methodology (RSM) with variable ranges of 5-15% TS, C/N ratio of 20-50, and enzyme 3-9%.

Optimum conditions in the variable range are the C/N ratio 30-40, 6% enzyme, and for TS was not significant factor. Following the run in the central composite experimental design, these values correspond to run number 10 with variable C/N ratio 35, enzyme 6% and TS 1.59% with biogas production 202.51 mL/g TS.

## 5. Acknowledgment

We highly thank and appreciate to Diponegoro University through Research Professorship Program (RPP) 2018 due to the funding of this research.

## 6. References

[1] F Monlau, E Latrille, A C Da Costa, J P Steyer and Carrere H 2013 *Appl. Energy* **102** 1105
[2] J Gao, L Chen, Z Yan and Wang L 2013 *Bioresour. Technol.* **132** 361
[3] K Michalska, K Miazek, L Krzystek and Ledakowicz S 2012 *Bioresour. Technol.* **119** 78
[4] J Zhu, C Wan and Li Y 2010 *Bioresour. Technol.* **101** 7523
[5] A Teghammar, K Karimi, I Sarvari Horvath and Taherzadeh M J 2012 *Biomass and Bioenergy* **36** 116
[6] D Elango, M Pulikesi, P Baskaralingam, V Ramamurthi and S Sivanesan 2007 *J. Hazard. Mater.* **141** 301
[7] H Hawali Abdul Matin and Hadiyanto H 2018 *E3S Web Conf.* **31** 02007
[8] Budiyono, I Syaichurrozi and Sumardiono S 2013 *World Appl. Sci. J.* **26** 1464
[9] W D Nugraha, Syafrudin, A T Senduk, H H A Matin and Budiyono 2018 *E3S Web Conf.* **73** 01016
[10] H H A Matin and Hadiyanto H 2018 *J. Environ. Sci. Technol.* **11** 147
[11] Syafrudin, W Dwi Nugraha, H Hawali Abdul Matin and Budiyono B 2017 *MATEC Web Conf.* **101** 02016

# A review: biogas production from tofu liquid waste

**B Budiyono[1,2*] and I Syaichurrozi[3]**
[1]Vocational School, University of Diponegoro, Semarang, Indonesia
[2]Department of Chemical Engineering, University of Diponegoro, Semarang, Indonesia
[3]Department of Chemical Engineering, University of Sultan Ageng Tirtayasa, Cilegon-42435, Indonesia

E-mail: budiyono@live.undip.ac.id

**Abstract.** The tofu consumption will increase in the future. As consequent, the tofu production increases significantly. In other side, the amount of tofu wastes also increases. The solid waste can be utilized as solid compost but the liquid waste is discharged directly to the waster bodies at the moment. The latter is potential to be used as biogas feedstock. Some important factors affecting biogas rate need to be considered such as power Hydrogen (pH), Chemical Oxygen Demand (COD) to Nitrogen ratio, Total solid (TS) content, biological agent addition, initial concentration and two-stage anaerobic digestion. Furthermore, purification of the biogas formed by digestion of the tofu liquid waste is needed to be conducted to increase the methane content and decrease carbon dioxide ($CO_2$) and hydrogensulfide ($H_2S$) content. The $H_2S$ has to be removed prior to $CO_2$ removal. The $H_2S$ can be removed by biofiltration and $CO_2$ can be removed by biological absorption with help of microalgae. The effluent of anaerobic digester (digestate) may still contain high pollutant so it can be used as nutrient sources for microalgae. The zero waste system for tofu industries is successfully proposed in this paper. Future trend in treating tofu liquid waste in not only for biogas production butalso for biohydrogen production.

## 1. Introduction

Tofu is a traditional food made from soybean [1]. It is very useful for human body as a protein source. In addition, its price is low. Hence, many Indonesian people consume it [2]. Tofu consumption in households in Indonesia per year is shown in Fig. 1. Development of the tofu consumption for the year 2002-2018 is fluctuation with average value of 7.41. Furthermore, the tofu consumption is predicted to be 8.38, 8.52, 8.67 in year of 2019, 2020, 2021 respectively [3]. In other word, the demand of tofu is predicted to become increasing for the future years. According to BPS [4], as protein source, tofu is more desirable than the others because consumption (kg/capita/week) for tofu is higher (0.158) than ras chicken meat (0.121) or beef (0.009). More than 84,000 tofu industry unitsare operated in Indonesia and total tofu production per year is more than 2.56 million tons [5].

* To whom any correspondence should be addressed.

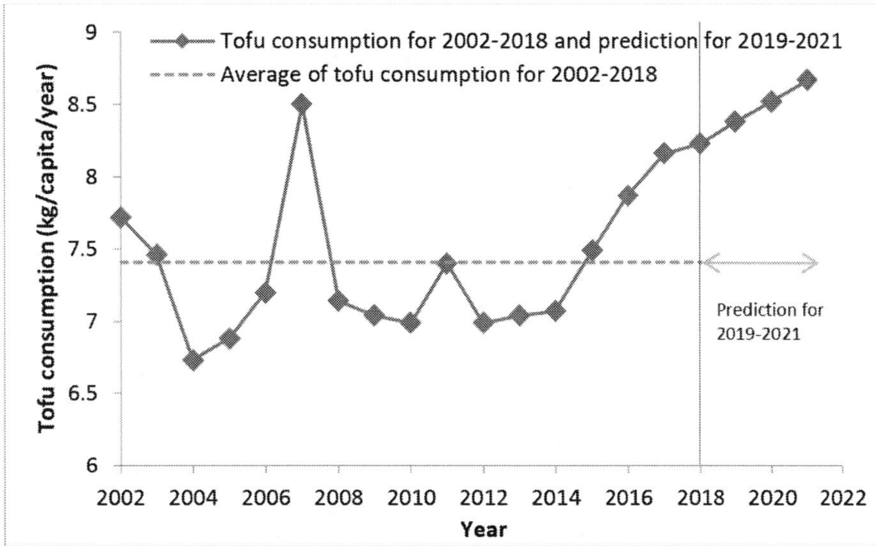


**Figure 1.** Development and prediction of tofu consumption in households in Indonesia (Adapted from [3]).

The shape of tofu in Indonesia is shown in Fig. 2. To produce the tofu traditionally, some process are needed [6-7].

a) The soybeans are washed and then are soaked in clean water for 8-12 hours.
b) The soaked soybeans are drained and then grinded to be a slurry.
c) The slurry is mixed with water (water:beans of 10:1) and then It is cooked at 100-110°C for 10 min.
d) It is filtered or centrifuged to separate the soymilk from solid waste (okara).
e) Next step is coagulation in which coagulants (usually calcium chloride or magnesium chloride) are added to allow for curdling.
f) Then, the supernatant is separated and the curds are pressed to form blocks of tofu. In this step, the liquid waste (whey) is generated
g) Finally, it is cut and packaged

The tofu production scheme is shown in Fig. 3.

**Figure 2.** Tofu in Indonesia [8].

## 2. Tofu wastewater characteristic and its environmental impacts

Based on Fig. 3, in tofu production, industries generate two kind wastes i.e. solid waste (okara) and liquid waste (whey). To produce tofu as much as 80 kg, the industries need soybeans of 60 kg and water of 2,700 kg. While, the solid and liquid wastes are generated as much as 70 kg and 2,610 kg respectively [9]. The tofu solid waste has a very little impact on the environment because it is mostly used as fodder for cattle animals or biofertilizer for crops [2]. Chemical content of tofu solid waste and solid compost is shown in table 1. In other side, usually, tofu liquid waste is discharged directly to the water bodies. However, this activity is forbidden because its characteristics are not below the quality standard. Comparison between chemical characteristics of fresh tofu liquid waste with quality standard of tofu liquid waste that can be discharged to the water bodies is presented in table 2.

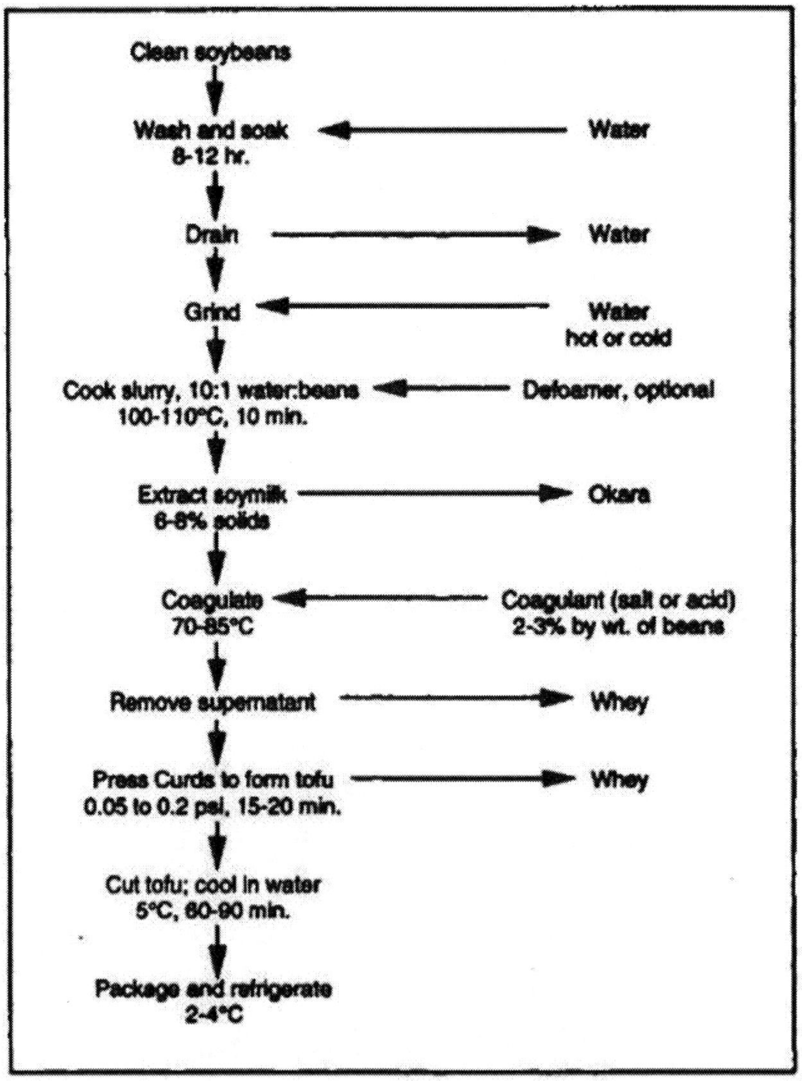


**Figure 3.** Tofu production process[6-7].

**Table 1.** Comparison of chemical content of tofu solid waste and solid compost [2]

| Parameters | Tofu solid waste | Solid compost |
|---|---|---|
| N (%) | 1.24 | 1.44 |
| $P_2O_5$ (mg/L) | 5.54 | 2.37 |
| $K_2O$ | 1.34 | 3.03 |
| Protein (%) | 7.72 | - |

If the waste is discharged directly to the water bodies, some problems will occur. The high organic compound (high COD and BOD) will decrease the dissolved oxygen, so it will disturb the aquatic organisms such as fish. The high nitrogen content also results bad odor because of ammonia gas. Its murky color will decrease the water quality and decrease the sunlight penetration to the water plants in which it cannot do photosynthesis process easily. Furthermore, when the water is used for human daily activity, the human can get skin diseases, stomachache, etc [2,10-11].

Based on information above, a treatment is needed to solve the problems. The one of potential methods is anaerobic digestion (AD) because (1) it is very good to treat wastes having high COD content, (2) it has low cost, (3) it converts a waste to biogas (renewable energy), (4) it is easy to be operated.

**Table 2.** Quality standard of tofu liquid waste and characteristics of fresh tofu liquid waste

| Parameters | Values | |
|---|---|---|
| | Quality standard for tofu liquid waste (Based on PP No. 82 year of 2001) | Fresh tofu liquid waste [10,12] |
| pH | 5-9 | 4.82-5.5 |
| Chemical oxygen demand (COD) (mg/L) | 500 | 5,000-8,640 |
| Biological oxygen demand (BOD) (mg/L) | 100 | 3,575.5-6,586 |
| Phosphate (mg/L) | 5 | 0.97-95.5 |
| Nitrogen (mg/L) | 20 | 297.5 |
| Nitrate (mg/L) | 20 | 25.355 |
| TSS (mg/L) | 400 | 2,350 |

## 3. Biogas production

### *3.1. Biogas production process*

Anaerobic digestion is one of interesting methods to treat wastes. During digestion, the organic compound in the tofu liquid waste is converted to be biogas with help of anaerobic bacterial activity. Biogas contains methane ($CH_4$) as main product, carbon dioxide ($CO_2$) as main by products, and other gas in little amount (such as carbon monoxide (CO), hydrogen ($H_2$), ammonia ($NH_3$), hydrogen sulfide ($H_2S$), etc.) [13-14]. The percentage of the gasses in biogas depends on the biogas feedstocks. Some phases occur in biogas production i.e. hydrolysis, acidogenesis, acetogenesis, methanogenesis [1,15]. In hydrolysis phase, complex organic compounds (such as polysaccharides, lipids, protein) are degraded to become simple organic compounds (such as glucose, long-chain fatty acids, amino acids). In acidogenesis phase, the simple organic compounds are degraded to become volatile fatty acids (acetic acid, butyric acid, propionic acid, hydrogen, carbon dioxide, methanol, ethanol). In acetogenesis phase, propionic acid, butyric acid and ethanol are converted to become acetic acid. In methanogenesis, the methane is formed in which there are three mechanisms i.e. (1) hydrogenotropic methanogenesis: reaction between carbon dioxide and hydrogen results methane and water (equation 1), (2) acetoclastic methanogenesis: acetic acid is converted to methane and carbon dioxide (equation 2), (3) methyltropic methanogenesis: methanol is converted to methane, carbon dioxide and water (equation 3). Scheme of biogas production is presented in Fig. 4.

$$2H_2 + CO_2 \rightarrow CH_4 + 2H_2O \quad (1)$$

$$CH_3COOH \rightarrow CH_4 + CO_2 \quad (2)$$

$$4CH_3OH \rightarrow 3CH_4 + CO_2 + 2H_2O \quad (3)$$

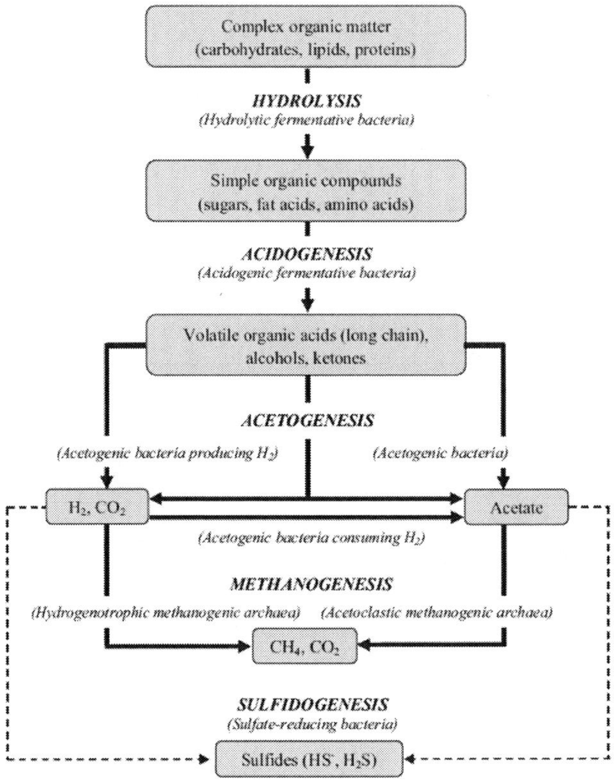


**Figure 4.** Phases in anaerobic digestion [15].

*3.2. Factor affecting biogas production*

*3.2.1. Power of hydrogen level.* The fundamental factor for microbial growth is power of hydrogen (pH) level [16]. Syaichurrozi et al. [1] has studied the effect of initial pH on biogas production from tofu liquid waste with variation of initial pH of 5, 6, 7, 8, 9. The results are shown in Fig. 5. Based on Fig. 5, the best initial pH is 8. Lay et al. [17] also has a conclusion in which the optimum initial pH range for biogas from tofu wastewater is 7.5-8. The pH of tofu liquid waste is 4.82-5.5 (table 2), so the base compounds (such as NaOH) can be used to increase the pH level prior the waste is treated in anaerobic digestion.

*3.2.2. Total solid content.* Tofu liquid waste contains low total solid (TS) content (below 7%). According to Budiyono et al. [18], the optimum range of TS in AD is 7.4-9.2%. Hence, Rahmat et al. [11] have added some solid wastes (sheep dung, paddy straw, bamboo litter). The results showed that biogas increased drastically. If TS content more than the optimum range, organic materials is overload in digester. In other side, TS content less than the optimum range, fermentation process is unstable so biogas production rate is low [19].

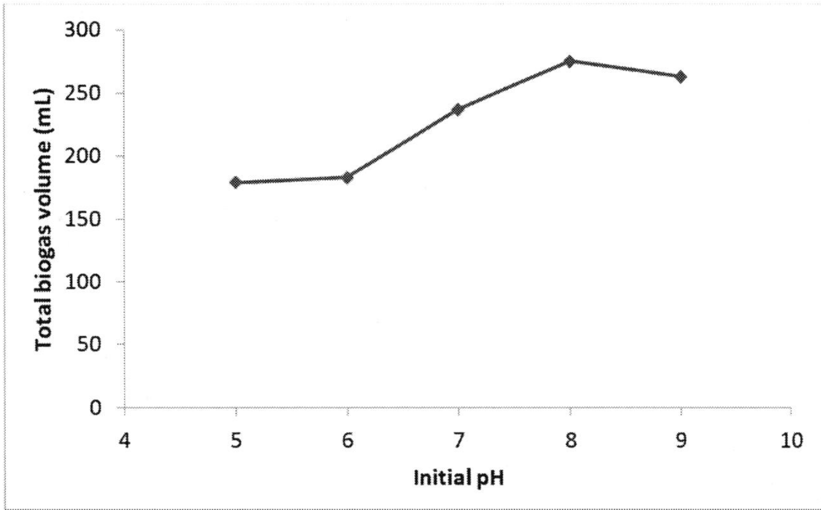


**Figure 5.** The effect of initial pH on total biogas volume in AD of tofu wastewater (adapted from [1]).

*3.2.3. COD to nitrogen ratio.* Ratio between COD to nitrogen (COD/N) affects the success of biogas production. Anaerobic bacteria can grow in substrates containing COD/N ratio with range of 350/7 to 1000/7 [20]. If the ratio is higher than the optimum range, volatile fatty acids (VFAs) is produced in a large amount. The abundant of the VFAs decreases the substrate pH, so bacterial activity will be disturbed and biogas volume is low [21]. In other side, if the ratio is lower than the optimum range, ammonia and ammonium are easy to be formed. The high concentration of ammonia (more than 150 mg/L) and ammonium (more than 30,000 mg/L) is toxic for anaerobic bacteria. The ammonia is more toxic than ammonium [22]. According to Syaichurrozi et al. [21], tofu liquid waste contains COD/N ratio of 270/7. This value is lower than the optimum range. Biogas yield produced from AD of tofu liquid waste is 77.01 mL/g COD. To increase the COD/N ratio, carbon source such as glucose can be added to the tofu liquid waste. Besides that, a waste containing high carbon content also can be used as carbon source. This concept is called as co-digestion. Syaichurrozi et al. [21] utilized vinasse as co-partner for tofu wastewater because it contains high COD/N ratio (1436/7). The best mixture of vinasse and tofu wastewater to produce high biogas yield (153.7 mL/g COD) is 20:80 (volume basis) with COD/N of 1042/7.

*3.2.4. Initial concentration.* Widyarani et al. [23] stated that initial concentration of substrate affects biogas production from tofu liquid waste. The waste concentration is varied to be 20, 50, 100%. The concentration of 100% (without dilution) results the higher biogas volume than the two others. It shows that the organic materials (total solid) in tofu wastewater is needed to be increased not to be decreased. Thus, addition of the other waste is better to get suitable total solid content (correlation with point 3.2.2).

*3.2.5. Biological agent addition.* Addition of a biological agent also can increase biogas production. Addition of yeast of *Saccharomyces cerevisiae* succesfully increases total biogas volume of tofu wastewater in which the total biogas increases 22.91-81.97% compared to without yeast addition [1]. In fermentation process, the yeast helps anaerobic bacteria to convert glucose to become acetic acid, butyric acid and ethanol. The composition of these compounds depends on pH condition [24]. Thus, optimum pH level is needed to be obtained to get optimum condition for growth of both the yeast and the anaerobic bacteria in order to biogas is produced maximally. The lower the pH level, the higher the ethanol concentration presents in system so it disturbs the anaerobic bacteria activity. In other side, the higher the pH level, the lower the ethanol concentration is generated but too high pH level is also not good for anaerobic bacteria. The best initial pH for collaboration of the yeast and anaerobic bacteria is 8-9 [1]. A previous study predicts the scheme of biogas production when the yeast is added and rumen fluid is used as inoculum (Fig. 6). The rumen fluid contains bacteria of *Clostridium sp.*, *Clostridium sporogenes*, *Clostridium butyricum* and methanogenic bacteria [1].

*3.2.6. Two-stage anaerobic digestion.* This concept is proposed to separate the hydrolysis-acid formations (acidogenesis and acetogenesis) and methane formations (methanogenesis). Commonly, biogas production process occurs in one-stage anaerobic digestion where all phases occurs in one system. However, this technique is not optimal because acidogenic and acetogenic bacteria requires a acid acidic condition and methanogenic bacteria needs a neutral condition. The scheme of the two-stage anaerobic digestion is shown in Fig. 7. Hence, the two-stage anaerobic digestion is proposed to treat tofu liquid waste (Iriani et al., 2018). The pH is adjusted to be 5.3-5.7 in first reactor and 7 in second reactor [25]. The hydrogen ($H_2$) and carbon dioxide ($CO_2$) is generated in first digester and then the methane ($CH_4$) and carbon dioxide ($CO_2$) is generated in second digester [26]. Purification to decrease the $CO_2$ can be applied to get $H_2$ and $CH_4$ gas in high purity. The $CO_2$ removal in biogas will be discussed in section of 4.1.

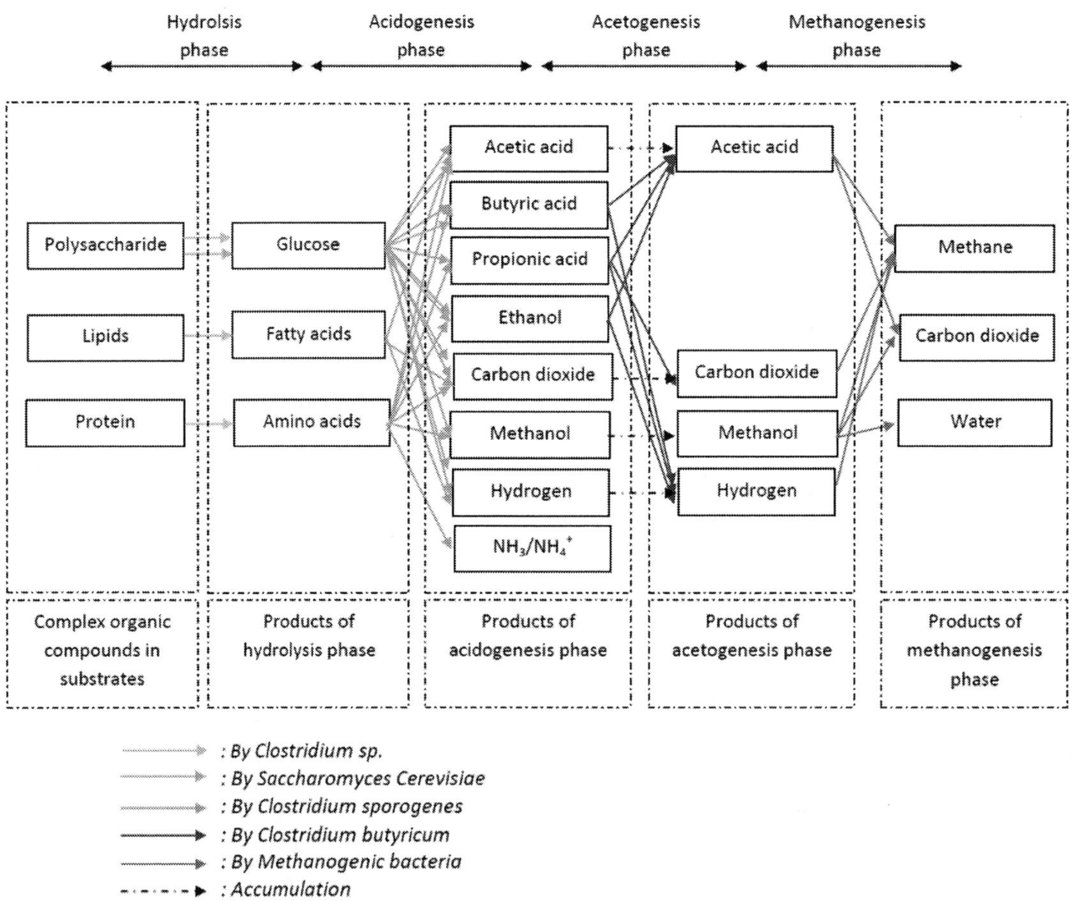


**Figure 6.** Prediction of biogas production process with help of anaerobic bacteria from rumen fluid and yeast *Saccharomyces Cerevisiae*[1]

## 4. Biogas Purification

Based on Harihastuti et al. [27], biogas generated from AD of tofu liquid waste contains 45.89% of $CH_4$, 29.69% of $CO_2$, 0.33% of $H_2S$, 0.0006% of $NH_3$ , and 24.10% of $H_2O$. Hence, it is needed to be purified to increase the biogas quality.

### *4.1. Removal of $CO_2$*

The $CO_2$ is a main by product in biogas. The presence of $CO_2$ can decrease caloric value. Hence, purification methods are needed to remove the $CO_2$ in biogas.

*4.1.1. Physical absorption.* The principle of this method is based on different solubility of gas components in liquid absorbent. Commonly, waster is used as the absorbent. The solubility of $CO_2$ is 26 times higher than that of $CH_4$. However, the solubility of $H_2S$ is higher than $CO_2$. The $H_2S$ will be dissolved and disturb the operation because it has corrosive characteristic. Thus, it is better to remove $H_2S$ in biogas prior the $CO_2$ removal process is conducted [28]. The advantages of this method are (1) high efficiency (more than 97% $CH_4$), (2) adjustable capacity by changing temperature or pressure, (3) low $CH_4$ loss (less than 2%). The disadvantages of it are (1) high cost in investment and operation, (2) clogging by bacterial growth, (3) foaming [28].

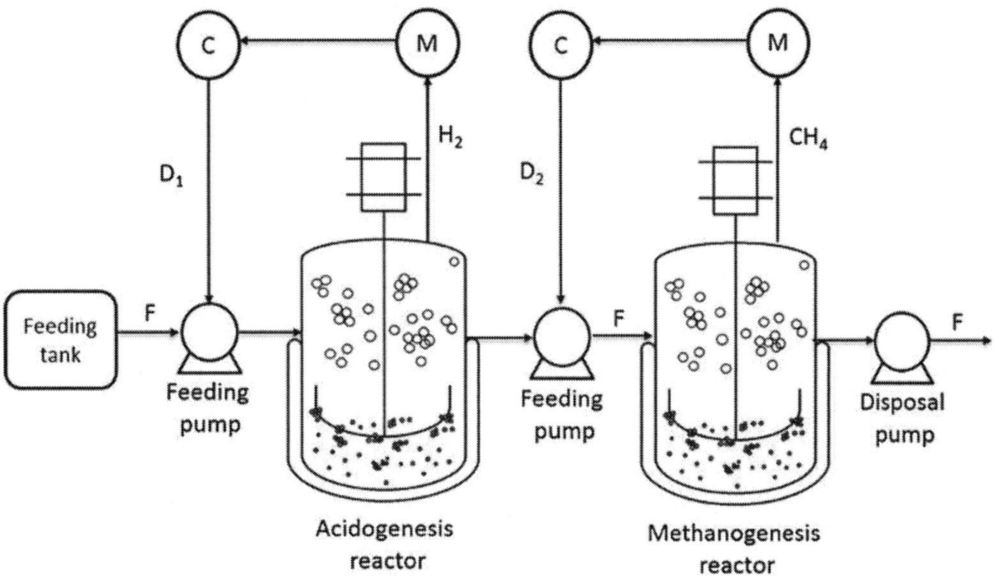


**Figure 7.** Two stage anaerobic digestion[26].

*4.1.2. Chemical absorption.* In this method, a solvent is used as liquid absorbent. The common solvents applied are mono ethanol amine (MEA) and di-methyl ethanol amine (DMEA). The $H_2S$ is removed prior to chemical absorption because it causes an amine poisoning [28]. The advantages of this method are (1) high efficiency (more than 97% $CH_4$), (2) low operating cost, (3) low $CH_4$ loss (less than 0.1%), (4) power cost of 0.05-0.25 kWh/m$^3$ of gas. The disadvantages of it are (1) Expensive investment, (2) heating for regeneration, (3) poisoning of the amine [28].

*4.1.3. Biological absorption.* Microalgae can be applied to remove $CO_2$ in biogas. The microalgae will uptake it as direct carbon source to do photosynthesis. This method is very interesting because biogas purification and biomass production occur in same time. The advantages of this method are (1) high efficiency, (2) low operating cost, (3) microalgal biomass production, (4) decreasing cultivation cost.

The disadvantages of it are (1) drop in pH due to too high $CO_2$ content in biogas or too high the flow rate, (2) light energy is needed to do photosynthesis[29]. Some microalgae can be applied in this method but the most interesting is *Spirulina* because (1) it has big biomass size so the biomass is easy to be harvested and (2) it has fast growth rate.

*4.1.4. Cryogenic separation.* This method utilizes the different boiling point of $CO_2$ (-78°C) and $CH_4$ (-160°C). The high pressure is applied (200 bar) [28]. The advantages of this method are (1) high efficiency, (2) low extra energy cost to reach liquid biomethane. The disadvantages of it are (1) expensive investment and operation, (2) high energy for cooling, (3) high pressure [28].

*4.2. Removal of $H_2O$*

The high $H_2O$ content in biogas gives some disadvantages because (1) it causes a corrosion in metal burner equipments or compressors and (2) it decrease the calorific value. The $H_2O$ can react with $H_2S$, $NH_3$, $CO_2$ to form acids [28].

*4.2.1. Physical separation (condensation).* This method includes demisters, cyclone separators, moisture traps and water traps. However, the ways are low efficient [28].In previous study, the $H_2O$ in biogas generated from tofu liquid waste can be removed by condensation [27].

*4.2.2. Chemical separation.* The principle of this method is absorption of water in glycol (a drying agent). The glycol has a binding component decreasing the dew point from −5 until −15 °C. Furthermore, the silica gel, magnesium oxide, activated carbon and alumina also can be used because they have a binding component decreasing the dew point until −40 °C at pressure of 6–10 bar [28].

*4.3. Removal of $H_2S$*

Some methods to remove $H_2S$ in biogas are absorption by water, chemical absorption, membrane, adsorption with activated carbon [28]. However, they need high investment and operating cost. In other side, biofiltration needs cheap investment and operating costs. Furthermore, it can be operated at low pressure and does not produce further waste [30]. Biofiltration has a multi-phase system in which $H_2S$ gas is dissolved and adsorbed in the biofilm, and then degraded by microorganisms that are immobilized on a packing material forming a thin layer (biofilm). The packing material can be made from natural and synthetic materials. The packing material must have (1) high porosity as well as high water retention capacity, (2) nutrient availability for microorganisms, (3) high surface area, high permeability and high adsorbance, (4) low price. Some materials such as soil, compost, wood chips, leaves, sawdust, bagasse ans salak seed can be used. There are several microorganisms that could be utilized in sulfur removal. Bacteria of Thiobacillus thioxidans, Thiobacillus feroxidans, Chromatiaceae (Chlorobioceae) can be chosen in biofiltration method [30].

**5. Treatment of AD effluent (digestate)**

In AD, effluent of digester still contains organic compound. Hence, the effluent cannot be discharged directly since it has high COD content. Some methods have to be applied to utilize the effluent.

*5.1. Coagulation*

This method utilize coagulants of iron hydroxide to adsorb the COD and then settled as sludge. The sludge is easy to be separated through sedimentation or filtration. There are two kinds of coagulation i.e. chemical coagulation and electrocoagulation. In chemical coagulation (CC), chemical compounds such as $FeCl_2$, $FeCl_3$, $Fe(SO_4)$, $Fe_2(SO_4)_3$ is added to the AD effluent. Furthermore, the compounds react with water to form coagulants ($Fe(OH)_2$ or $Fe(OH)_3$). Meanwhile, in electrocoagulation (EC), the $Fe^{2+}$ or $Fe^{3+}$ ion is generated from electrolysis. In EC, iron electrodes (anode and cathode) are immersed to the AD effluent and then the current is flowing, so the anode is oxidized to be $Fe^{2+}$ or $Fe^{3+}$ ion. In other side, the water is reduced to be $OH^-$ ion and $H_2$ gas in cathode. The $Fe^{2+}$ or $Fe^{3+}$ ion will react with $OH^-$

to form coagulants. The CC is high cost because it needs chemical compounds. The EC needs no chemical compounds but it needs electrical sources.Application of CC and EC on effluent of AD (substrate of AD is not tofu liquid waste) is successfully conducted by Rodrigues et al. [31] and Liu et al. [32] respectively. Based on our literature study, application of CC and EC on anaerobically treated tofu liquid waste has not reported yet.

### *5.2. Utilization as biofertilizer*

The AD effluent also can be used directly as biofertilizer since its characteristic is suitable with quality standard of compost. Tofu liquid waste contains high nitrogen sources so the AD effluent may contain high ammonia or ammonium. Hence, some further treatments are needed to decrease the ammonia or ammonium level prior the effluent is used as biofertilizer. Koszel and Lorencowicz [33] reports that digestate from agricultural biogas plants can be used as a fertilizer directly.

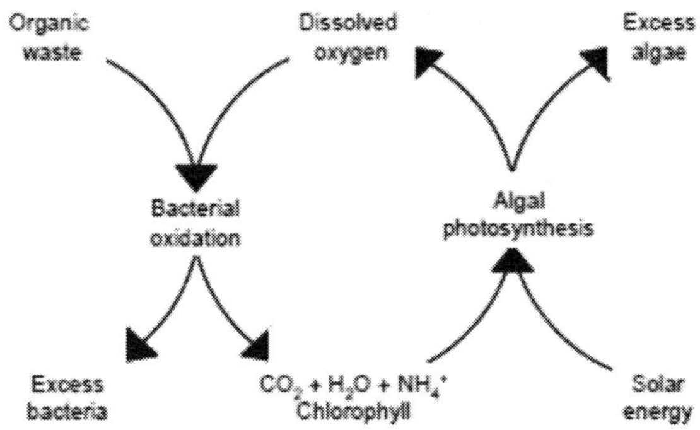


**Figure 8.** Photosynthetic oxygenation principle [34].

### *5.3. Utilization as organic nutrient for microalgae*

The AD effluent can be used as nutrient source for microalgae because it still contains organic carbon and nitrogen sources. This concept is very interesting because some advantages can be reached at same time i.e. wastetreatment and biomass production. The AD effluent may contain oxidation bacteria. When it is applied as nutrient sources for microalgae, the oxidation bacteria and microalgae present in the system. According to Oswald and Gotaas [34], during cultivation, there are collaboration between oxidation bacteria and microalgae that may occur (Fig. 8). At the first time of cultivation, the oxidation bacteria consumes organic compounds and produces energy and $CO_2$ via respiration and then the $CO_2$ reacts with water to form carbonate, so pH substrate to be low. After that, the carbonate is consumed by microalgae via photosynthetic process and released $OH^-$. Photosynthetic activity needs $CO_2$ as primary carbon source and light as primary energy source. Accumulation of $OH^-$ increases substrate pH. The microalgae can also consume organic carbon as source of carbon and generate energy and $CO_2$ via respiration. This growth is called as a heterotrophic growth. The $CO_2$ is applied as carbon source in photosynthesis. This growth is called as photoautotrophic growth. The microalgae that can grow through either photoautotrophic or heterotrophic processes is called as mixotrophic bacteria [35].

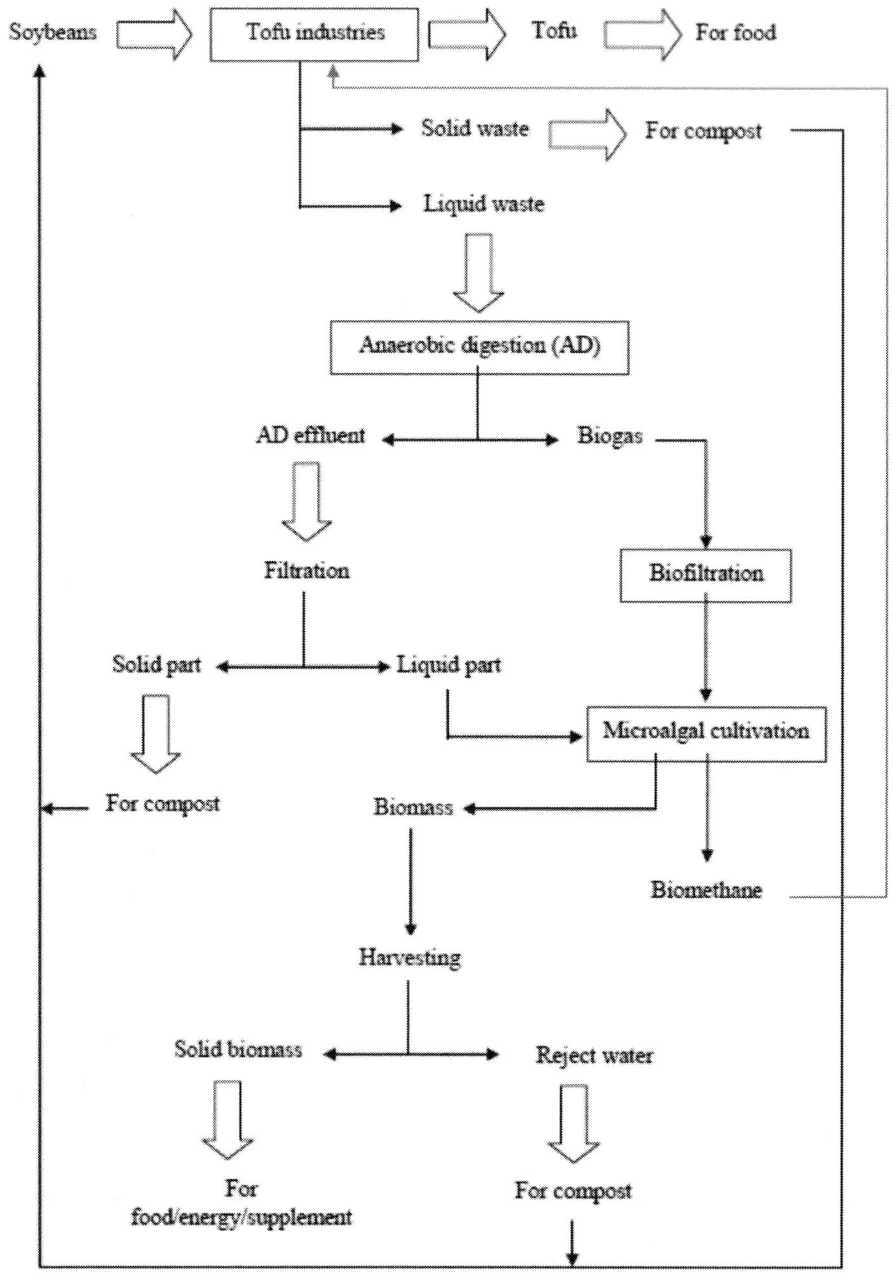


**Figure 9.** Scheme of zero waste.

## 6. Scheme of zero waste system

A scheme that can be applied by tofu industries with zero waste can be seen in Fig. 9. Tofu industries produce tofu as a main product which is consumed by human as protein source. The solid waste is utilized as biofertilizer (as compost). The compost is applied in fields to increase soybean production. In other side, the liquid waste is brought to anaerobic digesters. In anaerobic digesters, the organic compounds in the liquid waste are converted to biogas by help of anaerobic bacteria. The biogas still contains impurities ($H_2S$ and $CO_2$), so purification methods are applied to decrease these gas components. The $H_2S$ and $CO_2$ gasses are removed by biofiltration and biological absorption (microalgae) respectively so the biomethane is obtained. The $H_2S$ in low concentration also can be removed by microalgae but it in high concentration is not suggested because it is toxic. Hence, biofiltration is better to be conducted prior the biogas is purified through microalgae. The biomethane can be used as energy source in tofu production. The other product of anaerobic digestion is effluent. The effluent is separated to obtain the solid part and liquid part. The solid part is used as solid compost and the liquid part is interesting to be applied as organic nutrient sources in microalgal cultivation. After cultivation, the biomass of microalgae is separated by filtration (harvesting). The biomass can be applied as a source of food, energy, and supplements.

## 7. Conclusion

Tofu liquid waste can be treated by anaerobic digestion and produce biogas as top product and effluent as bottom product. Some factor affecting biogas production are important to be considered i.e. pH, COD to N ratio, TS content, biological agent, initial concentration and two-stage anaerobic digestion. The biogas formed still contains high $CO_2$and $H_2S$. Hence purification methods are applied to decrease these gas components. The $H_2S$ is removed by biofiltration. The $CO_2$ can be removed by biological absorption with help of microalgae. In other side, the effluent can be used as organic nutrient sources for microalgae. The zero waste system is successfully proposed in this paper.

## 8. References

[1] Syaichurrozi I, Rusdi R, Hidayat T and Bustomi A 2016 *Int. J. Eng. Transactions B: Applications* **29** 1037
[2] Faisal M, Gani A, Mulana F and Daimon H 2016 *Asian J. Chem.* **28** 501
[3] Buletin Konsumsi Pangan 2019 Http://Epublikasi.Setjen.Pertanian.Go.Id/Epublikasi/Buletin/Konsumsi/2019/Buletin%20konsumsi%20vol%2010%20no%201%202019.Htm (Online 20 September 2019)
[4] BPS 2019 Https://Www.Bps.Go.Id/Statictable/2014/09/08/950/Rata-Rata-Konsumsi-Per-Kapita-Seminggu-Beberapa-Macam-Bahan-Makanan-Penting-2007-2018.Html (Online 20 September 2019)
[5] Adisasmito S, Rasrendra C B, Chandra H and Gunartono M A 2018 *Int. J. Eng. Technol.* **7** 30
[6] Hui T K 2013 https://www.livewelltoday.info/overview-of-soybean-processing-and-production (Online 27 September 2019)
[7] Shurtleff W and Aoyagi A 1979 *Tofu & Soymilk Production*, The Book of Tofu vol 2 (Soyfoods Center, Lafayette, CA.
[8] Rahmawati F 2013 Http://Staffnew.Uny.Ac.Id/Upload/132296048/Pengabdian/Teknologi-Proses-Pengolahan-Tahu-Dan-Pemanfaatan-Limbahnya.Pdf (Online 20 September 2019)
[9] Arinto D J, Paramastri H P and Soetrisnanto D 2013 *Jurnal Teknologi Kimia Dan Industri* **2** 233
[10] Faisal M, Mulana F, Gani A and Daimon H 2015 *Res. J. Pharm. Biol. Chem. Sci.* **6** 1053
[11] Rahmat B, Hartoyo T and Sunarya Y 2014 *Am. J. Agric Biol. Sci.* **9** 226
[12] Myrasandri P and Syafila M 2012 http://www.ftsl.itb.ac.id/wp-content/uploads/sites/8/2012/07/15308036-Putri-Myrasandri.pdf (Online 25 September 2019)
[13] Juanga J P, Visvanathan C and Trankler J 2007 *Waste Manage. Res.* **25** 30
[14] Karellas S B 2010 *Renew. Sust. Energ. Rev.* **14** 1273
[15] Moraes B S, Zaiat M and Bonomi A 2015 *Renew. Sust. Energ. Rev.* **44** 888
[16] Kafle G K, Kim S H and Sung K I 2013 *Bioresour. Technol.* **127** 326
[17] Lay C H, Sen B, Huang S C, Chen C C and Lin C Y 2013 *Renew. Energ.* **58** 60
[18] Budiyono, Widiasa I N, Johari S and Sunarso 2010 *Int. J. Basic Appl. Sci.* **10** 68
[19] Syaichurrozi I 2016 *Waste Technol.* **4** 16
[20] Speece R E 1996 (USA: Archae Press)
[21] Syaichurrozi I, Rusdi, Dwicahyanto S and Toron Y S 2016 *Int. J. Renew. Energ. Res.* **6**
[22] Deublein D and Steinhauser A 2008 Wiley-Vch Verlag, Weinheim

[23] Widyarani, Victor Y, Sriwuryandari L, Priantoro E A, Sembiring T and Sintawardani N 2018 *IOP Conf. Series.: Earth and Environmental Science* **160**
[24] Lin Y, Zhang W, Li C, Sakakibara K, Tanaka S and Kong H 2012 *Biomass Bioenergy* **47** 395
[25] Iriani P, Utami S and Suprianti Y 2018 *AIP Conf. Proc.* **2021**
[26] Gurubel K J, Sanchez E N, Coronado M A, Zuniga G V, Sulbaran R B and Breton D L 2019 *Optim. Contr. Appl. Met.* **1**
[27] Harihastuti N, Purwanto P and Istadi I 2015 *AIP Conf. Proc.* **1699**
[28] Awe O W, Zhao Y, Nzihou A, Minh D P and Lyczko N 2017 *Waste Biomass Valorization* **8** 267
[29] Sumardiono S, Budiyono, Syaichurrozi I and Sasongko S B 2014 *Cur. Res. J. Biol. Sci.* **6** 53
[30] Lestari R A S, Sediawan W B, Syamsiah S, Sarto and Teixeira J A 2016 *J. Environ. Chem. Eng.* **4** 2370
[31] Rodrigues I J, Fuess L T, Biondo L, Santesso C A and Garcia M L 2014 *Desalination Water Treat.* **52** 22
[32] Liu Z, Stromberg D, Liu X, Liao W and Liu Y 2015 *J Hazard. Mater.* **21** 483
[33] Koszela M and Lorencowicza E 2015 *Agric. Agric. Sci. Proc.* **7** 119
[34] Oswald W J and Gotaas H B 1957 *Trans. Am. Soc. Civ. Eng.* **122** 73
[35] Budiyono, Syaichurrozi I, Sumardiono S and Sasongko S B 2014 *Trends Appl. Sci. Res.* **9** 93

# Morphology characterization of lignocellulosic materials in ozone based degradation

**A P Siswanto[1*], M E Yulianto[1], D Handayani[1], E F Sapatra[1], I Yuniarti[1], R A S Imamsyah[1]**
[1]Department of Industrial Technology, Diponegoro University, Semarang, Indonesia

E-mail: anggun.siswanto@live.undip.ac.id

**Abstract.** Oil-rich plants are widely cultivated as raw materials for the biofuel industry in developing countries. However, this plant is mainly also a potential source for food supply. Therefore, the supply of food crops and oil competes, making a solution to this conflict very desirable. Technological developments aimed at contributing to an increase in food and fuel supply to meet market demand is the next step in increasing existing food and oil production. Renewable energy is renewable energy and, if managed properly, these resources will not be used up. Renewable energy types include biomass, geothermal, solar energy, water energy, wind energy, and ocean energy. This method promises an economical and environmentally friendly process. Ozone is a strong oxidant for potential use in the degradation process of lignocellulosic materials. This can be produced by air flow through a plasma micro-reactor. Utilization of water as a solvent will produce a solution that is rich in ozone. This research aims to develop a degradation scheme for lignocellulosic materials using an ozonolysis reactor. The novelty of this research is the use of ozone in dissolved water for the degradation of lignocellulosic materials. Water is used as a solvent for ozone dissolving where water molecules with dissolved ozone can be separated into OH radicals which are very reactive oxidants.

## 1. Introduction

The energy needs of petroleum fuels in various countries in the world in recent years have increased sharply, not only in developed countries, but also in developing countries including Indonesia. Petroleum fuels in the future, is currently developing new and renewable energy sources that are environmentally friendly. Renewable energy is energy that can be approved and approved to be managed well, the resources will not be used up. Renewable energy types Updated biomass, geothermal, solar energy, air energy, wind energy, and ocean energy. Ethanol is a biofuel, and has good prospects as a substitute for liquid and gasohol fuels with raw materials that can be supportive, environmentally friendly, and very beneficial for the microeconomics of a farmer-based community.

Ozone that cannot be revised will re-form molecules [1]. This method promises economic and environmentally friendly processes. Ozone is a strong oxidant for potential use in the degradation process of lignocellulosic material. This can be done by the flow of air through a plasma micro-reactor. Utilization of water as a solvent will produce a solution that is rich in ozone. Plasma has promising advantages because of its low energy consumption, conversion in atmospheric pressure, and high conversion [2]. Maintenance using a plasma micro-reactor is usually carried out at atmospheric pressure

* To whom any correspondence should be addressed.

which contributes to economic efficiency. Does not require process under vacuum conditions. Because of this, it is easier to process a reaction.

This research develops lignocellulose degradation materials using an ozonolysis reactor. The novelty of this research is the use of ozone in dissolved water for the degradation of lignocellulosic material. Water is used as a solvent for ozone dissolving where water molecules with dissolved ozone can become OH radicals which are very reactive oxidants. These molecules are expected to attack carbon double bonds in the structure of the lignocellulose molecule. This process is similar to the Criegee interaction which explains the completion of esters at the ozone junction. The benefit of this new technology is that there are no by-products left after ozonolysis treatment.

## 2. Methodology

### 2.1. Ozonolysis treatment

Raw materials in the form of lignocellulosic materials were pre-treated using the size reduction method. The raw material was processed first then treated by ozonolysis. The ozone concentration is adjusted to the capacity of the ozone generator it has. Variables included: ratio of distilled water and weight of lignocellulosic materials was 1:20, room temperature and ambient pressure, duration ozonolysis (5; 10; 15; 20; 25; 30 minutes).

### 2.2. Morphology analysis

Morphological test of lignocellulosic materials after ozonolysis using the Scanning Electron Microscope (SEM) method. This test aims to determine qualitatively lignocellulosic materials after the ozonolysis process. Ozonolysis of carbon double bonds is well explained by the Criegee mechanism, shown in Fig. 1. The reaction takes place in three different stages namely (Reaction I) the formation of "primary ozonide", (Reaction II) "primary ozonide" decomposition into becarbonyl compounds and "carbonyl oxides", (Reaction III) insertion of carbonyl oxides into the structure of carbonyl compounds. It was reported that during the decomposition stage (Reaction II), selective fission occurs with chemical bonds. Oxygen bonds (O-O) and carbon bonds (C = C) split where the stronger bond (C-O) remains intact [3].

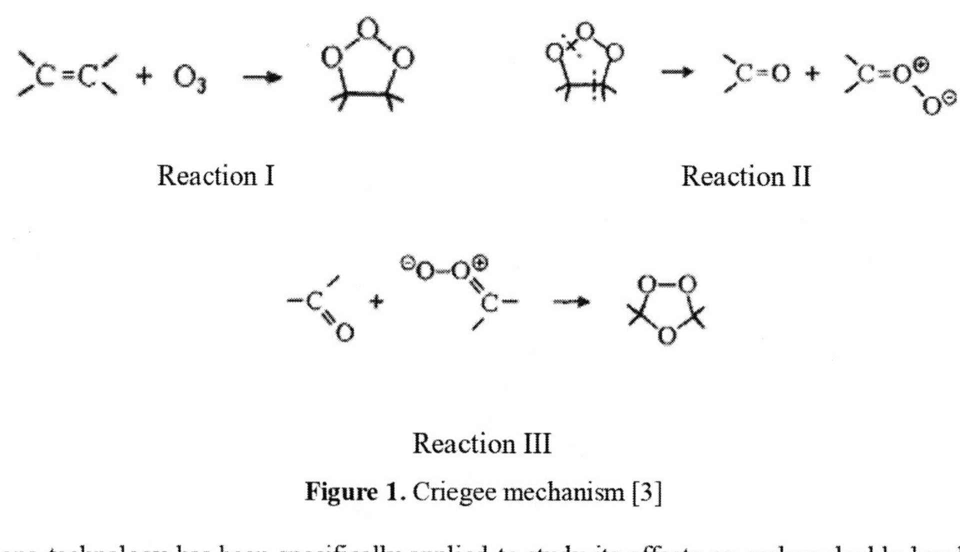


**Figure 1.** Criegee mechanism [3]

Ozone technology has been specifically applied to study its effects on carbon double bonds from biodiesel [1] and in the field of detoxification of phorbol ester [4, 5]. Ozone is expected to attack carbon double bonds (C = C) in the phorbol ester structure which can cause molecular recombination. The division of a double bond is predicted to modify the strength of the configuration molecule. As a result, the modified phorbol ester is produced by changing the activation power as a tumor promoter after ozonolysis.

## 3. Result and discussion

Investigation of the effects of ozonolysis in biodiesel, especially on carbon double bonds under extreme operating conditions, has been reported [1]. Methanol and dichloromethane are used as solvents with triethylamine as a catalyst where ozonolysis is carried out at -75 0C. This study aims to analyze the molecular structure of soy methyl as biodiesel samples after ozonolysis. Fourier Transform Infrared Spectroscopy (FTIR), Gas Chromatography (GC) and Gas Chromatography-Mass Spectrometry (GC-MS) are used for structural analysis. A 90% reduction in the number of double bonds in the mixture is produced after 2 hours of ozonolysis. A theoretically predicted ozonolysis ester product was found. This shows that chemical-based reactions in combination with ozone treatment affect the configuration of the double bond. Further experiments to the ambient temperature are recommended to meet the requirements of industrial processes.

### *3.1. Plasma application in ozone production*

Plasma refers to an electrically neutral ionized gas, the same amount of positively and negatively charged ions, and is often defined as the fourth state of matter. Ionized gas is produced from oscillations which are considered as compressed electric waves, similar to sound waves [6]. Langmuir also stated that: "Except near the electrode, where there is a casing that contains very few electrons, ionized gas contains approximately the same number of ions and electrons so that the resulting space charge is very small. We will use the name plasma to describe this region containing a balanced charge of ions and electrons "Plasma envelope is defined as the transition region separating plasma and its boundaries. Unlike plasma and its boundaries, the characteristics of the plasma envelope are typical which give impetus to the movement of charged particles" [7]. Fig. 2 shows the plasma phase compared to the phases of other substances.

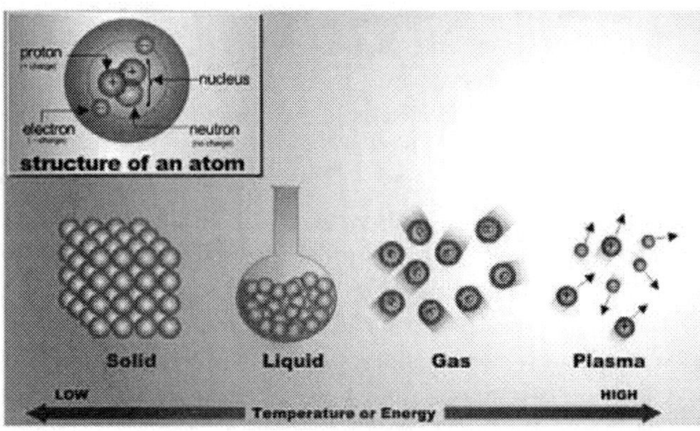


**Figure 2.** Various state of matters [8]

Nehra et al [9] have summarized various types of non-thermal plasma reactors. Corona release, plasma jets of atmospheric pressure and dielectric barrier discharge are described. Corona release is the first type of plasma reactor. Typically, this reactor release is produced from a pair of asymmetrical electrodes with a DC voltage that is continuously lit or pulsed. The homogeneous electrode configuration is surrounded by the electric field that is generated. Atmospheric pressure plasma jets create radio frequency interference during ignition release. This speeds up the free electrons from the central electrode so that energetic collisions in the feed gas occur quickly.

### *3.2. Dielectric barrier discharge plasma reactor*

Another class of plasma reactors, Dielectric barrier discharge (DBD) is also non-thermal. DBD is called a silent discharge, because a typical discharge is generated from AC power, and operates under atmospheric pressure and moderate gas temperatures. Fig. 3 shows the configuration of a non-thermal plasma reactor.

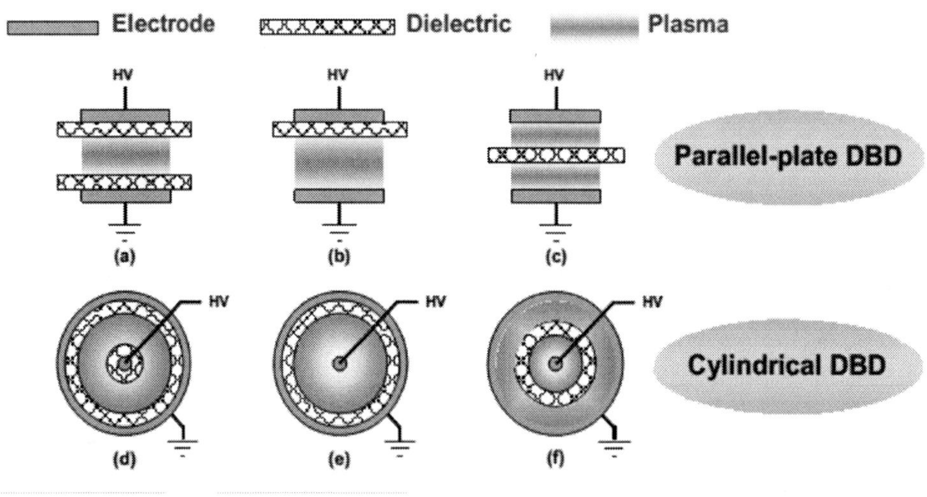


**Figure 3.** Configuration of non-thermal plasma reactor [9, 10].

Typically, DBD reactors consist of two metal parallel electrodes separated by a small gap in between. To ignite plasma, gas is passed through the electrodes and induced by an electric current. Various gases can be used to produce plasma such as helium, neon, argon and air. DBD plasma can consist of independent current filaments, which are micro-discharges [11]. The occurrence of micro-release in the DBD configuration develops because it reaches an electric field strong enough in the reactor gap to exceed the point of damage. This micro-release is homogeneously distributed on the dielectric surface in nanosecond duration [12].

### *3.3. Ozonolysis treatment in lignocellulosic materials*

Ozone is an inorganic compound consisting of three oxygen atoms arranged in a linear structure. It is colourless, partially soluble in water and usually conductive in the form of gas. It occurs naturally in the Earth's atmosphere when oxygen is activated by UV light and is also produced in storms [13, 14]. In addition, ozone produced artificially can be found in copiers, electric motors, etc. Ozone is known as a strong oxidant with an oxidation potential of 2.07 V [14, 15]. Table 1 presents the oxidation potential of various oxidant agents.

**Table 1.** Oxidation Potential of Various Oxidant [13]

| Oxidant | Oxidation Potential (V) |
|---|---|
| Fluorine | 3.06 |
| Hydroxyl free radicals | 2.80 |
| Atomic oxygen | 2.42 |
| Ozone | 2.07 |
| Permanganate | 1.67 |
| Chlorine | 1.36 |
| Oxygen | 1.23 |
| Hypochlorite | 0.94 |

Lignin is recognized as an outer layer of cell structure that is polymerized into hemicellulose and cellulose. Agricultural waste, food waste and organic industrial waste mainly contain lignocellulosic material [16, 17]. Hemicellulose is cellulose which can be used for many industrial products such as enzymes, biofuels, organic acids and food additives. This compound is known for its sugar-based chemical composition, and therefore can be used as a source of economic value [18, 16]. Fig. 4 illustrates the biological structure of lignocellulosic material.

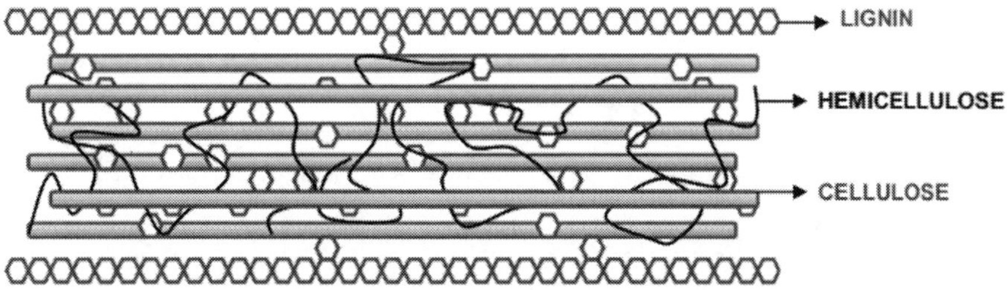


**Figure 4.** Cross sectional view of lignocellulosic structure [18].

The process of lignin degradation is highly desirable to biologically access hemicellulose and cellulose layers. The presence of ether bonds in the lignin structure limits this process. This results in lignin insusceptibility during conventional biodegradation approaches such as enzymatic and chemical treatments [17-20]. Fig. 5 shows the images of Scanning Electromagnetic Microscope (SEM) on the lignocellulosic materials.

**Figure 5.** SEM images of lignocellulosic samples.

Studies on the use of ozonolysis in the treatment of lignin compounds have been conducted for decades [21-23]. Vidal and Molinier [23] have studied lignin ozonolysis to improve the digestion of sawdust. They found that the accessibility of enzymes to the experimental substrate was improved during ozonolysis. Unfortunately, a high voltage input, 15-33 kV is needed during ozone generation. These high-power supply requirements have associated high processing costs. Subsequently, laboratory-scale experiments were carried out in 350 ml of a gas washer by Wu et al [21] with a modern ozone generator. Ozone is used in the pre-treatment stage of hydrogen production using barley straw as a raw material.

## 4. Conclusion

Ozone is known by its high oxidation potential which make it as one of the strongest oxidant. There are several ways of producing ozone in which Dielectric Barrier Discharge Plasma Reactor was applied in this research. The application of ozone in lignocellulosic materials was studied and showed that the lignocellulosic materials is susceptible to ozone.

## 5. References

[1] Baber T M 2005 *Biomacromolecules* **6** 1334
[2] Zimmerman W B, Lozano P J H and Bandulasena H H 2010 *J. Sewerage & Water*
[3] Criegee R 1975 *Angew. Chem. Internat. Edit.* 14 745
[4] Diwani G I E, S A E Rafei and Hawash S I 2011 *Adv. Applied Sci. Res.* **2** 221
[5] S Kongmany, H Matsuura, M Furuta, S Okuda, K Imamura, Y Maeda 2013 *J. Phys.: Conf. Ser.* **441** 012006
[6] Langmuir I 1928 *Proc. Natl. Acad. Sci. USA,* **14** 628
[7] Fridman A 2008 *Cambridge University Press*
[8] Cheng H H, C S Shing, W Y Chi and H D Lit 2007 *J. Environ. Eng. Manage.* **17** 427
[9] Nehra V, A Kumar and H K Dwivedi 2008 *Int. J. Eng.* **2**
[10] Wu Z, Z Pengyan, T Lin, Z Duo, W Ailing and G Xingang 2012 *Adv. in Biomed Eng.*
[11] Kogelschatz U, Eliasson B and Egli W 1997 *J. Phys.* **IV** 7
[12] Wagner H E, Brandenburg R, Kozlov K V, Sonnenfeld A, Michel P and Behnke J F 2003 *Vacuum* **71** 417
[13] Barlow P J 1994 *Watertec Engineering Pty Ltd*
[14] Franken L 2005 *Food Safety and Security at Kansas State University*
[15] Babikov D, Kendrick B K, Walker R B and Pack R T 2003 *J. Chem. Phys.* **119**
[16] Chuck C J, Parker H J, Jenkins R W and Donnelly J 2013 *Bioresour. Technol.* **143** 549
[17] Palmqvist E and Hahn H B 2000 *Bioresour. Technol.* **74** 25
[18] Mussatto S I and Teixeira J A 2010 *Formatex Research Center* **2**
[19] Malgorzata M, Jadwiga S O and Anna M 2012 *Fibres Text. in East. Eur.* **20** 191
[20] Wu J, Upreti S and Ein M F 2013b *Int. J. Hydrog. Energy* **38** 10270
[21] Wu J, Ein M F and Upreti S 2013a *Bioresour. Technol.* **144** 344
[22] Nouwezem R, Borredon M E, Parisi J P and Gaset A 1993 *Bioresour. Tech.* **45** 43
[23] Vidal P F and Molinier J 1988 *Biomass* **16** 1

IOP Conf. Series: Materials Science and Engineering **845** (2020) 012049 doi:10.1088/1757-899X/845/1/012049

# The effect of pH on Synthesis of ZnO-Natural Zeolite Nanocomposite by Co-Precipitation Method

**T E Agustina[1*], A U Prajawita[1], and S A S Sinaga[1]**
[1]Chemical Engineering Department, Universitas Sriwijaya, Palembang-30662, South Sumatera, Indonesia

E-mail: tuty_agustina@unsri.ac.id

**Abstract.** Currently, the method being developed to overcome environmental pollution generated by synthetic dyes produced from the textile industry wastewater is by photocatalysis using photocatalyst materials. Photocatalysis is a relatively inexpensive and easy method to implement. Photocatalysts that have received major attention and widely studied are metal oxide semiconductors such as ZnO. Photocatalyst activity can be increased through the development of supporting materials such as zeolites. This study aims to synthesize ZnO-natural zeolite photocatalysts nanocomposite by co-precipitation method in order to take advantage of the adsorbent and catalyst properties in the natural zeolites and the photocatalyst properties in the ZnO. To study the effect of pH, ZnO-natural zeolite nanocomposites are synthesized with different pH under alkaline conditions (pH of 8-11). Nanocomposite materials obtained then tested its photocatalyst activity in degrading of 50 ppm procion red synthetic dyes. The ZnO-natural zeolite nanocomposites produced were characterized by using XRD, SEM, and BET. The synthesized ZnO-natural zeolite nanocomposite showed the best results at pH of 10, with the ZnO crystallite size of 20.99 nm, the surface area of 134.35 $m^2/g$, and the Zn contained of 10.05%. Procion red degradation of 75.54% was attained by using ZnO-natural zeolite nanocomposite photocatalyst after 120 minutes of UV irradiation.

## 1. Introduction

Many industries in Indonesia use dyes in the production processes. The manufacture of furniture, paints, varnish, housewares, and textiles need dyes to further improve the production yield. Among these industries, the textile industry is the industry that uses the greatest number of dyes in the production process. Synthetic dyes often used in the textile industry are non-biodegradable in nature, such as procion, erionyl, and auramin [1]. However, the textile industry wastewater is a hazard in the form of toxins and carcinogenic in nature, not only a danger to human health but also a bad impact on the ecosystem. Most traditional water treatment methods such as membrane filtration, adsorption, and chemical treatment are not effective for removing dyes properly. Conversely, this method also produces secondary pollutants such as poison gas and sludge as solid waste that require further treatment.

* To whom any correspondence should be addressed.

Another alternative method for treating wastewater that is cheap, relatively fast, and capable of being used in the long term is photocatalysis [2]. Photocatalysis is the process of decomposition of chemical compounds with the help of photon energy, where oxidation and reduction reactions occur on the surface of semiconductors, for example ZnO, $TiO_2$, CdS, and $Fe_2O_3$. When the semiconductor is affected by ultra violet (UV) light, the material can decompose organic compounds through photocatalytic reactions [3]. One of the good semiconductors used is ZnO. ZnO semiconductors are more profitable than $TiO_2$ because they are more able to absorb the spectrum of the sun and quantum rays than $TiO_2$ [4].

The weak adsorption capacity is a problem for the photocatalytic process. To cover up the deficiencies, photocatalyst material needs to be combined with an adsorbent as a matrix or dopant [5]. Zeolite is a good solid support for dispersing ZnO semiconductors because it has a number of surface area, pore volume, and relatively uniform pore and channel sizes [6]. Therefore, in this research, ZnO nanocomposite together with natural zeolite will be made, to maximize the photodegradation process accompanied by adsorption of solid particles in wastewater. The selected photocatalyst synthesis method is the co-precipitation method, which has some advantages such as the use of inexpensive raw materials, easy handling, ambient processing temperature and pressure, and a large-scale production when compared to other synthesis methods [7].

The pH is an important factor in making nanocomposite photocatalysts by co-precipitation method. Hence, the effect of pH on synthesis of ZnO-natural zeolite nanocomposite photocatalyst was studied in this research. In addition, the ability of nanocomposite photocatalyst ZnO-natural zeolite in the process of photodegradation of procion red synthetic dyes were examined.

## 2. Methodology

### *2.1. Chemicals and materials*

The materials used in this study were aquadest, natural zeolite, deionized water, $Zn(CH_3COO)_2.2H_2O$, ZnO solids, 0.1 M NaOH, and procion red synthetic dyes. All the chemicals was supplied from Merck, while natural zeolite and procion red were purchased from Fajar Kimia chemical store in Jakarta. The apparatus used is a reactor equipped with UV Evaco 15W 254 nm lamp, pH meter and EMCLAB-61 UV Spectrophotometer.

### *2.2. Procedures*

#### *2.2.1. Synthesis of ZnO-natural zeolite nanocomposite photocatalyst*

A 5 grams of activated natural zeolite are dissolved in 100 ml of 0.1 M $Zn(CH_3COO)_2.2H_2O$ solution in 100 ml of deionized water. The mixture is stirred and heated at 80°C in a reflux flask for 4 hours. The mixture then washed with deionized water to remove residual acetate and dried at 60°C for 5 hours. After that 0.1 M NaOH was added to the mixture until a pH reached 8 and stirred for 2 hours with a magnetic stirrer. Then the mixture is washed by deionized water to remove sodium and dried at 60°C for 5 hours. The product is calcined for 2 hours at 450°C [8]. The procedure is repeated for the pH of 9, 10, and 11. The nanocomposite of ZnO-natural zeolite photocatalyst is characterized by using XRD, SEM-EDX and BET.

#### *2.2.2. Photodegradation of Procion Red Synthetic Dyes*

The ZnO-natural zeolite photocatalysts will be tested for the degradation of procion red synthetic dyes. To test the photocatalyst activity, photodegradation of procion red is carried out using a UV lamp reactor. As a comparison of effectiveness, the degradation was also tested using zeolite only and using ZnO only. The ZnO-natural zeolite weighed 100 mg then mixed with synthetic dye with a concentration of 50 ppm as much as 25 ml in a shaker and placed in a reactor equipped with a UV lamp. When the UV lamp was turned on, the time were recorded. The samples were taken every 30 minutes for 120

IOP Conf. Series: Materials Science and Engineering **845** (2020) 012049 doi:10.1088/1757-899X/845/1/012049

minutes of reaction. After the sample is filtered, the procion red color degradation was analysed by using UV Spectrophotometer. The procedures were repeated using only ZnO and only natural zeolite.

## 3. Results and Discussion

### *3.1. Synthesis and characterization of nanocomposite ZnO-natural zeolite*

The process of synthesis of ZnO- natural zeolite nanocomposites for the degradation of synthetic dyes is carried out through several stages of the processes, which includes preparation through activation of natural zeolites, the process of co-precipication of the production of ZnO-natural zeolite nanocomposites, calcination and catalyst characterization. This research was conducted to determine the effect of pH on the product of synthesis of ZnO-natural zeolite nanocomposite by co-precipication method through the characterization and degradation of procion red synthetic dye.

The synthesized samples with variations in pH were characterized by BET (Brunauer Emmet Teller) analysis to determine surface area, pore area, and the ability of adsorption of nanocomposite material. Also the nanocomposite samples produced were analyzed by Scanning Electron Microscope - Energy Dispersive X-ray Spectroscopy (SEM-EDX) to determine the composition of atoms and morphology of the sample material. The composites were also characterized by X-Ray Diffraction (XRD) analysis to determine the crystal size of the sample. This will be compared to find out which conditions are the optimal one.

Natural zeolite was activated by using HCl solution. This acid treatment purposes to dissolve and remove the metal oxides that are absorbed and covered the surface of the natural zeolite, so that it is become more porous and the surface of the contact area becomes larger [9]. Activation in natural zeolites aims as a preparation process before co-precipitation process is carried out. The size of crystallites is generally determined by XRD. The crystallite size of natural zeolite before and after activation were listed in table 1. The size of the crystallites increases after activation. The size of crystallites in commercial ZnO is greater than in natural zeolite.

There is a difference between zeolites before and after the activation process; this proves that the activation process can change the surface morphology of natural zeolites, as can be seen from SEM images in Fig. 1 and Fig. 2. After the activation, natural zeolite surface looks smoother and flatter. This is caused by the loss of impurity in the activation process. In addition, the crystallite size of natural zeolites increased after the activation.

**Table 1.** Crystallites size of natural zeolite and ZnO.

| Sample | Crystallite size (nm) |
|---|---|
| Natural zeolite before activation | 42.54 |
| Natural zeolite after activation | 43.89 |
| ZnO | 384.90 |

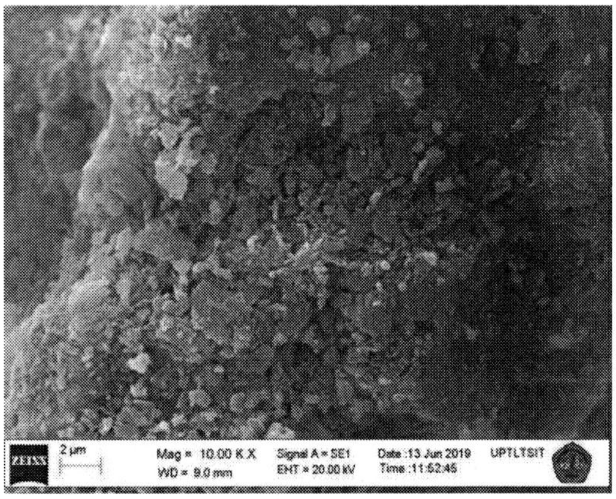

**Figure 1.** SEM images of natural zeolite before activation

**Figure 2.** SEM images of natural zeolite after activation

Figure 3 shows the morphological results of commercial ZnO. The surface morphology of ZnO shows homogeneous shape and distribution and has a hexagonal phase. Figure 4 demonstrates the XRD pattern of ZnO. The diffraction pattern of ZnO at 2θ (degree) with the peak of 31,950;32,707; 34,614; 36,438; 47,709; 56,755; 63,019; 66,536; 68,10; 69,248; 72,711; 76,61; and 77,130; correspond to the reflection from 100, 002, 101, 102, 110, 103, 200, 112, 201, 004, and 202. The XRD pattern is identical to the hexagonal phase with the Wurtzite structure [10]. The crystallites size of commercial ZnO is listed in table 1.

**Figure 3.** SEM image of ZnO

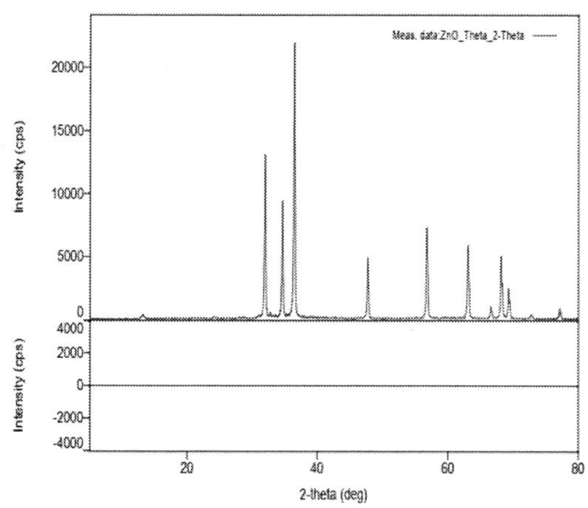


**Figure 4.** The XRD pattern of ZnO

The process of synthesize of ZnO-natural zeolite nanocomposites consists of a number of following steps; activation, precipitation, filtration, washing, forming, drying, calcining [11]. The natural zeolite was activated by HCl as a preliminary step. Then, co-precipitation consists of three main steps, namely, the mixing of the liquid acetate precursor with natural zeolite, the ion exchange between $Zn^{2+}$ and natural zeolite, nucleation and crystal growth to form primary particles, and aggregation of primary particles. Table 2 shows the crystallite size obtained from the XRD analysis of the ZnO-Zeolite nanocomposite. It has been proven that the synthesized ZnO is in nano size. In this study, the largest ZnO crystallite size occurs at pH 8, and the smallest crystallite size occurs at pH 10. Further increasing the concentration of OH− from pH 8 reduces the ZnO crystallite size. The crystallite size decreases at pH 9 and 10, but at pH 11, it increases. It can be seen that at pH 11 the co-precipication process is no longer optimal. This is because the amount of dissolved OH− was larger during synthesis of ZnO at pH > 10. When ZnO react with too much OH−, the dissolution of ZnO occurs. The dissolution made the crystallites became smaller and agglomerated [12].

**Table 2.** Crystallites size of nanocomposites ZnO-natural zeolite

| Sample pH | Crystallite size (nm) | |
|---|---|---|
| | ZnO | Zeolite |
| 8 | 67.82 | 107 |
| 9 | 21.64 | 111 |
| 10 | 20.99 | 140.8 |
| 11 | 33.59 | 71.25 |

*3.2. Degradation of procion red using nanocomposite photocatalyst ZnO-natural zeolite*

To determine the effect of pH on nanocomposite synthesis, the nanocomposites are made using different alkaline pH from 8 to 11. Nanocomposites produced from this different pH are then tested for their ability to degrade procion red dyes. A procion red concentration of 50 ppm was used as a model pollutant in this study. The degradation percentage at different pH value was illustrated in Figure 5. As shown, the highest degradation percentage of 75.54% was found when using nanocomposites that synthesized at pH 10, after 120 minutes of mixing time under UV irradiation. Degradation of rocion red in the term of C/Co (C is procion red consentration at t; Co is initial procion red concentration) can be depicted in Figure 6. The lowest C/Co is also found when using the nanocomposite when synthesize at pH 10. This can be explained as follow; at pH 10 the nanocomposites have the smallest crystallite size. In addition, based on the BET test, the largest surface area of 134.35 $m^2/g$ obtained where nanocomposites was synthesized at pH 10. According to [12] the crystallite size of ZnO is relevant to the particle size. It can be considered that the smaller the particle size and the greater the surface area, the better the adsorption process. That is why the highest degradation was achieved at pH of 10. Based on SEM-EDX results, the highest content of Zn in the composites of 10.05% was also found at pH of 10.

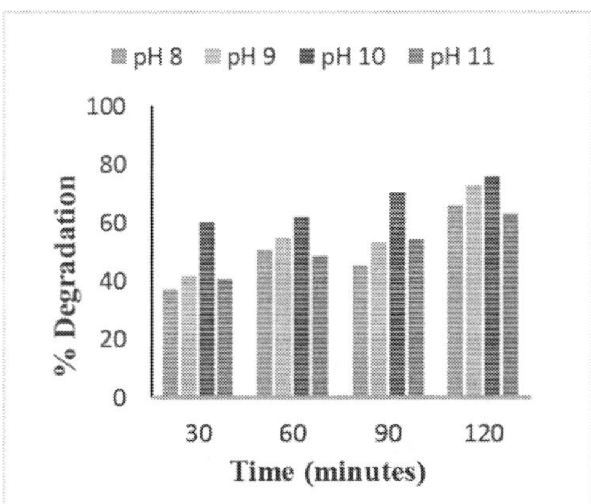


**Figure 5.** Effect pH on degradation of procion red

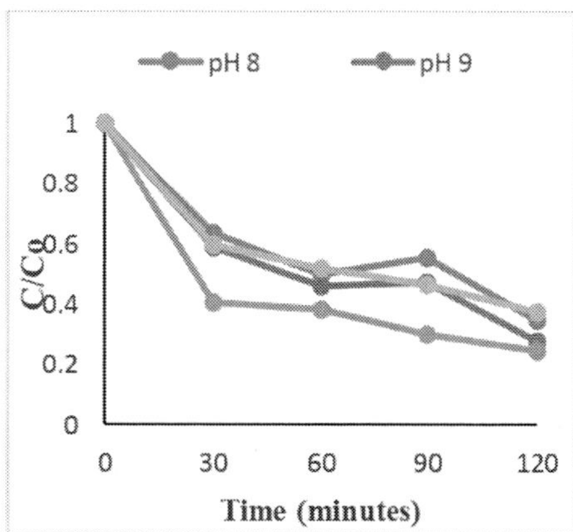


**Figure 6.** Effect of mixing time on C/Co of

procion red

### *3.3. Comparison of degradation procion red using zeolite, ZnO, and nanocomposite ZnO-natural zeolite*

Figure 7 shows the comparison of degradation percentage of procion red by using three types of materials, in this case natural zeolite, ZnO, and ZnO-natural zeolite nanocomposites (best results of pH 10) at 30 minutes of mixing time. The ZnO-natural zeolite nanocomposite produces the highest degradation compared to ZnO and natural zeolite alone. However, the longer the contact time the ability of natural ZnO-natural zeolite nanocomposites has decreased compared to ZnO, which is still able to further degrade. The specific density of ZnO-natural zeolite composite is greater than ZnO, causes this material not spread evenly during the process, so that degradation process is not optimal. Moreover, the ZnO used here is a commercial ZnO, which has a high purity and uniformity in size, so that the degradation process can be more optimal.

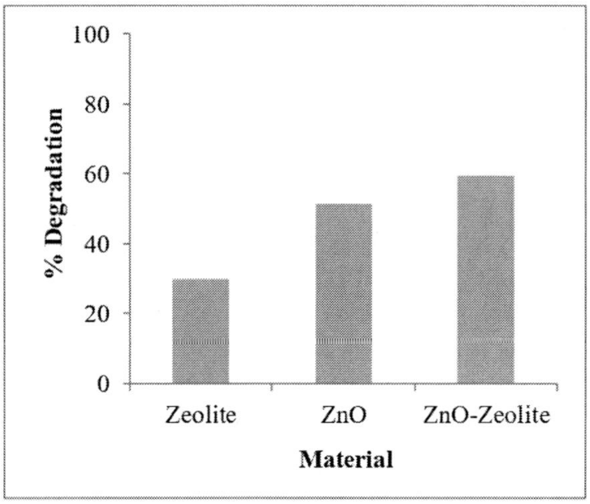


**Figure 7.** Degradation percentage of procion red by using natural zeolites, ZnO, and ZnO-natural zeolites (procion red concentration of 50 ppm)

## 4. Conclusion

In this research, ZnO-natural zeolite composite material was synthesized by using co-precipitation method. The material obtained was characterized by XRD, SEM-EDS, and BET. The analysis shows that ZnO and natural zeolite have been successfully synthesized to produce nano-sized composites. The pH effect significantly on the nanocomposites properties. The best physical characters were found when the nanocomposite synthesized at pH 10. At this pH the crystallite size of 20.99 nm, the surface area of 134.35 $m^2/g$, and the Zn content of 10.05% in the nanocomposite were obtained. The highest degradation percentage of 75.54% was reached when using nanocomposites photocatalyst, after 120 minutes of mixing time under UV irradiation.

**Acknowledgments**

The authors wishing to acknowledge the financial support from LPPM Universitas Sriwijaya through Hibah Unggulan Kompetitif 2019. The authors also thank the Integrated Research Laboratory of the Postgraduate Program of Universitas Sriwijaya and the Laboratory of Waste Treatment Technology of Chemical Engineering Department for laboratory assistances.

IOP Conf. Series: Materials Science and Engineering **845** (2020) 012049 doi:10.1088/1757-899X/845/1/012049

## References


[1] Salam A, Agustina T E, and Mohadi R 2018 *International Journal of Scientific & Technology Research* **7** 54-9

[2] Dhamayanti Y, Wijaya K, and Tahir I 2005 *Proseding Seminar Nasional DIES ke 50 FMIPA UGM* 22–29

[3] Perdana D N, Wardhani S, and Khunur M 2014 *Kimia Student Journal* **2** 576-82

[4] Hutabarat R 2012 *Sintesis dan Karakteristik Fotokatalis $Fe^{2+}$-ZnO Berbasis Zeolit Alam* (Depok: Universitas Indonesia)

[5] Ruliza M, Agustina T E, and Mohadi R 2014 *Earth and Environmental Science* **10** 1-7

[6] Hartini, E 2011 *Modifikasi Zeolit Alam dengan ZnO Untuk Degradasi Fotokatalis Zat Warna* (Depok: Universitas Indonesia)

[7] Chanu L A, Singh W J, Singh K J, and Devi K N 2019 *Results in Physics***12** 1230-37

[8] Fereshteh Z, Loghman M R, Estarki, Razavi R S and Taheran M 2013 *Mat Sci in Semiconductor* **16** 547-43

[9] Ngapa Y D 2017 *Jurnal Kimia dan Pendidikan Kimia* **2** 90-6

[10] Hamedani N F and Farzaneh F 2006 *J Sci Islamic Republic of Iran* **17** 231-34

[11] Jong K P 2009 *Synthesis of Solid Catalyst* (Germany: Wiley-VCH)

[12] Alias S S, Ismail A B, Mohamad A A 2010 *J Alloys and Comp* **499** 231-37

# Effects of cup lump natural rubber as an additive on the characteristics of asphalt-rubber products

**Bahruddin[1*], Arya W[1], Yanny S[1] and Alfian M[2]**
[1]Department of Chemical Engineering, Engineering Faculty, University of Riau, Pekanbaru, Indonesia
[2]Department of Civil Engineering, Engineering Faculty, University of Riau, Pekanbaru, Indonesia

Corresponding author : bahruddin@lecturer.unri.ac.id

**Abstract.** Asphalt penetration 60/70, which is widely used for road construction in Indonesia, is the cause of road deterioration due to high traffic loads and climate changes. Therefore, the addition of polymer additives can be one of the solutions to improve the quality and service period of asphalt. The aim of this research is to study the effect of cup lump natural rubber content on the characteristics of asphalt-rubber products. The mixture of asphalt-rubber is made by melt the cup lump at 130°C, then mixed it with asphalt at 170°C. The result showed that the addition of cup lump would decreased the penetration and weight loss properties, but increased the properties such as softening point, ductility, and Marshall stability of asphalt-rubber products. The optimal addition of cup lump was at 12% with penetration 52,4 dmm, softening point 63,5°C, ductility 142 cm, weight loss 0,068%, and Marshall stability 1904,36 kg.

## 1. Introduction

The land transport plays an important role in economic development of a city or region. In Indonesia, the growth rate of motor vehicles increased annually about 7,4%. However, if the growth was not accompanied by the length of the road, it will cause a traffic congestion. A traffic congestion for a long period gives big load into a road construction, it will cause cracks or rutting. In addition, other factors of road deterioration in Indonesia are climate change, water, humidity, and quality of asphalt that mostly bad [1]. In 2017, it was reported that length of paved roads reached 321.093 km where 16,72% damaged and 17,36% heavily damaged. In fact, refinement will increase the maintenance cost of paved roads annually [2]. Therefore, preventive action is by re-analyze road construction and improve quality of asphalt through modification. Modification can be done by addition of an additive into asphalt. It must be able to produce high stability and softening points, increase flexibility, increase durability, and increase binding capacity of asphalt to aggregate.

One of the additive is Cup Lump (CL) natural rubber. In fact, Indonesia is second ranked as rubber producers country in the world behind Thailand, as its production reached 3,774 million tons at the end of 2018 [3]. Rubber farmers are the main contributor (about 85%) of Indonesia's natural rubber production, which is a CL type. About 85% of Indonesia's natural rubber were exported to several countries such as United States, China, Japan, and etc [4]. The high export rate caused rubber price to

* To whom any correspondence should be addressed

be unstable. According to data January 2017 till January 2019, the rubber prices were decreased to 48,1% [5]. Thus, to maintain stability of rubber prices is by increasing consumption at the domestic level.

Some researches have been done with rubber as additive for the asphalt. CR from wasted-tires, Blok Skim Rubber (BSR) and Crumb Rubber Standard Indonesian Rubber 20 (SIR 20), if they were used directly as additives for asphalt, it will require very long mixing time and higher temperature. The addition of rubber in solid form is less desirable because the molecules chain of the rubber is very long. Therefore the rubber processing before it was mixed is very important. Wen [8] added latex with Dry Rubber Content (DRC) > 60% into conventional asphalt. The problem faced in using latex as additive is the formation of hot asphalt splashes due to the high water content in latex. Besides using rubber and latex as an asphalt additive, Cup Lump (CL) natural rubber can also be used as an asphalt additive. The addition of CL changes the asphalt rheology by increasing softening point but decreases penetration and ductility [9]. The use of cup lumps is more desirable than latex due to the short mixing time and lower mixing temperature.

In general, these studies show that modification of asphalt using various types of natural rubber have a great impact on improving the quality of asphalt. The improvement specifically occur on durability, resistance to temperature, resistance to rutting, and prevention to short aging. This study was conducted to study the effect of cup lump natural rubber as an additive on the characteristics of Marshall specifically roughness, stiffness, temperature sensitivity, and durability for asphalt to resist traffic load.

## 2. Materials and Methods

### *2.1 Materials*

The asphalt used in this research was asphalt Penetration 60/70 that obtained from PT. Pertamina (Persero) Indonesia, with specification as shown within Table 1. The type of natural rubber used in this research was Cup Lump obtained from locals' rubber plantation in Kampar, Riau Province, Indonesia. While the aggregate was processed aggregates from PT. Virajaya Riau Putra, Riau Province, Indonesia. Its specification shown within Table 2.

**Table 1**. Specification of Asphalt Penetration 60/70

| Properties | Test Standard | Test Result |
|---|---|---|
| Penetration at 25°C (dmm) | ASTM D5 | 70,2 |
| Softening point (°C) | ASTM D36 | 48 |
| Weight Loss by TFOT (%) | ASTM D6 | 0,365 |
| Penetration After TFOT (dmm) | ASTM D5 | 64,7 |
| Ductility (cm) | ASTM D113 | 110 |
| Marshall Stability (kg) | ASTM D6927 | 1179,25 |

**Table 2**. Composition of Aggregate on Marshall Stability Test

| Aggregate | Composition (%) |
|---|---|
| Coarse Aggregate | 14 |
| Medium Aggregate | 30 |
| Fine Aggregate | 45 |
| Filler Cement | 5 |

### *2.2 Preparation of Modified Asphalt Samples*

Cup Lump (CL) was first cut into small pieces into a cube size 1 cm x 1 cm x 1 cm. Then CL was melted at 130°C for 15 minutes. The CL was mixed together with asphalt under melted condition at 170°C, where CL content were varied at 4%, 6%, 8%, 10%, and 12% w/w. The materials were mixed by mixer with 100 rpm of speed for 30 minutes.

*2.3 Characterization of Modified Asphalt Samples*

Characterization of asphalt-rubber samples were consisted of Penetration Test (ASTM D5), Softening Point (ASTM D36), Weight Loss (ASTM D6 / D6M), Penetration After TFOT (ASTM D5), Ductility (ASTM D113), and Marshall Stability (ASTM D6927). Marshall Stability test were carried out using composition of aggregate as shown in Table 2 with optimum asphalt content 6% for all samples. Then, sample that gave the best Marshall test result were followed by Void Filled Aggregate (VFA) Test, Void In Mixture (VIM) Test, and Void Mineral Aggregate (VMA) Test.

## 3. Results and Discussion

*3.1 Effect of Compo Ratio on Penetration of Modified Asphalt*

The addition of CL as an additive into conventional asphalt produces Cup Lump Modified Asphalt (CMA) with different rheological properties compared to conventional asphalt, which is penetration properties. The result shows that the penetration decreased 16,38% with the penetration value 58,7 mm on additions of 4% CL. An excessive of CL ratio caused the penetration value to continue decrease till it reached 52,4 dmm on additions of 12% CL or decreased 25,35% of conventional asphalt. Figure 1 shows the result of CMA penetration test.

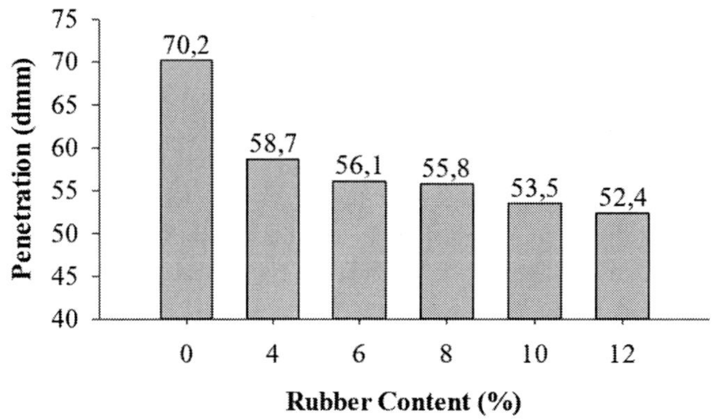


**Figure 1.** Effect of CL Ratio on Penetration of Modified Asphalt

The penetration decreased as the increasing of CL content, it is due to the increasing of asphaltene content and decreasing of maltene content. The ratio of asphaltene and maltene have a great impact on rheology of asphalt. The increase of asphaltene content hardened the bitumen, increased the softening point, and lower the penetration value which indicated higher viscosity [10]. The CL used in this study has a Dry Rubber Content (DRC) $\geq$ 55%. CL molecules are made up of straight polyisoprene hydrocarbon chains. Its physical appearance is like a flocculant sponge. Heating at temperatures below 100°C causes CL which is elastic turn to be stiff and slightly hard, whereas when heated above 100°C, CL will become flexible, soft, and transparent due to depolymerization [11].

The modifying of asphalt with CL used a mixing temperature of 170°C, a high temperature causes the water bound in the asphalt to be released in the form of water vapor. The release of large water vapor causes cavities between polymer molecules caused by CL structures such as sponges [11]. The basic components of asphalt such as saturate and aromatics will fill the cavity that was previously filled with water that has been released during the mixing process. This causes the asphaltene component to increase which results in strong interactions between the asphaltene molecules.

Rubber that expands for a long time and high temperatures causes the rubber to depolymerize [12]. Depolymerized CL will evenly distributed in the asphalt so that the absorption of light asphalt fractions increases. The increased asphaltene and resin ratio due to the light fraction absorbed by CL has a considerable impact on the performance of the resin. Resin in asphalt is polar and by increasing resin content have an impact on increasing adhesion, ductility, flexibility and plasticity [13]. Increasing the asphalt adhesion is related to the increase in adhesion and cohesion properties of asphalt. Increasing asphalt adhesion causes the asphalt to become harder with indications of low asphalt penetration [6].

*3.2 Effect of Compo Ratio on Softening Point of Modified Asphalt*

The result shows that penetration of CMA decreased as the increasing of CL content, simultaneously increased the softening point of modified asphalt. The softening point increased as the increasing of asphaltene content and decreasing of maltene content linearly [13]. Modified asphalt with the addition of CL gave resistance to higher temperature. The addition of 12% CL on CMA increased its softening point value by 32,3% from conventional asphalt, with softening point 63,5°C. The effect of CL addition on asphalt can be seen in Figure 2.

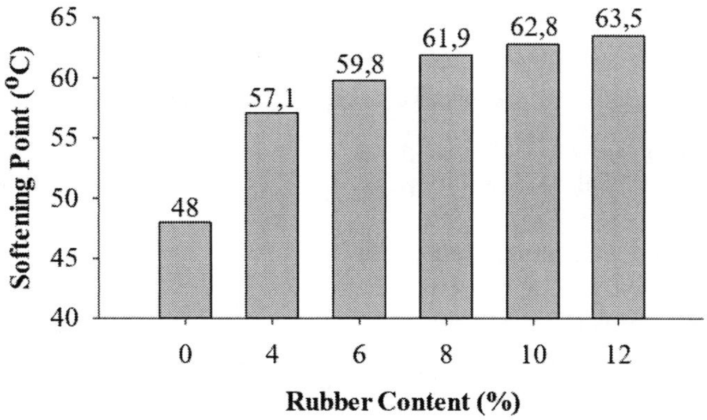


**Figure 2.** Effect of CL Ratio on Softening Point of Modified Asphalt

The increased of softening point directly proportional to the increased of asphalt viscosity and cohesion [10,11]. It is indicate that an increased of asphalt stiffness as an effect of increased molecular weight. The increase of molecular weight creates more obstacles on the asphalt when softening point test is carried out [14]. Besides CL content, other factor that affects the softening point is mixing temperature. The softening point will be optimal at mixing on 180°C, but the rubber will be degraded if it exceeds the temperature [15]. At high mixing temperature, depolymerization process will take place quickly and CL will evenly distributed in the asphalt [6,8]. The addition of 4% CL has a significant effect on the increase of softening point, with an increase 18,95% and softening point 57,1°C. The addition of small amount of CL in the asphalt has a great impact on rheology against stiffness and thermal resistance.

*3.3 Effect of Compo Ratio on Ductility of Modified Asphalt*

The result shows that a slight addition of CL has a great impact on penetration and softening point value. One of the reasons of penetration decreased and increased on softening point because of the high cohesion bond of the asphalt. The increased of cohesion bond of the asphalt is characterized by an increase in ductility. The effect of CL addition on ductility can be seen in Figure 3.

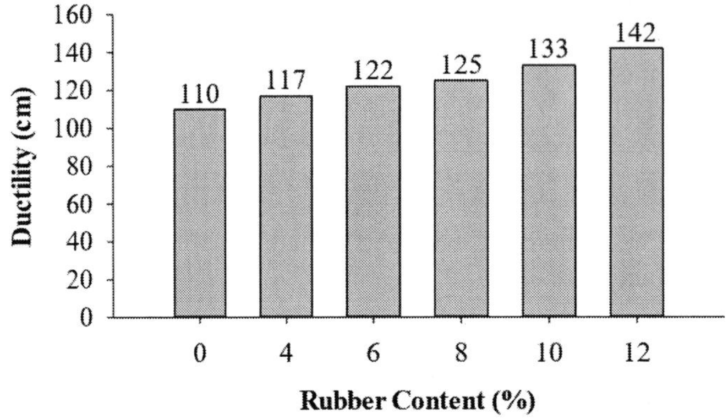


**Figure 3.** Effect of Compo Ratio on Ductility of Modified Asphalt

Fig. 3 shows the results of ductility values, it shows that the addition of 4% CL increased ductility 6,36% from conventional asphalt. The addition of 12% CL gave highest ductility to 142 cm or 29% from conventional asphalt. Ductility was strongly influenced by cohesion bond of asphalt [10], the fraction in asphalt that has great impact on cohesion is resin. Ductility of asphalt increased as the increasing of resin content, but will decrease if there is an increase of asphaltene and oil content. A high ductility indicated that the asphalt were elastic, it is related to the smaller the size of rubber particles in a mixture that can increase elasticity of rubber [11]. On this research the ductility were increased, it is due to a high mixing temperature so that CL were evenly distributed in the asphalt. Gagle [16] stated that the addition of Styrene and Butadiene could improve ductility at high mixing temperature. On the other hand, natural rubber particles that evenly distributed at high temperature can reduced the increased of asphaltene content, simultaneously improved cohesion and adhesion properties of asphalt without reduce ductility. The increased of ductility made asphalt more flexible and improved the susceptibility to cracks on low temperature. The addition of rubber also improved the tensile strength, toughness, and thermal resistance.

*3.4 Effect of Compo Ratio on Weight Loss and Penetration of Modified Asphalt After TFOT*

Weight loss referred referred in this test is the difference of weight sample before and after Thin Film Oven Test (TFOT). Weight loss occur because volatilization of aromatic components such as *benzene* and *xylene* [11]. The amount of weight loss indicated sensitivity of asphalt to temperature. TFOT testing aims to determine effect of volatile compounds on aging and sensitivity of asphalt to high temperature. The maximum weight loss of asphalt permitted were only 1% of asphalt weight [9].

Fig. 4 (a) shows the addition of CL reduced the weight loss of asphalt component. Weight loss was reduced from 63% to 81,36% of conventional asphalt at 4% - 12% of CL content. The CL was acted as membrane for light fractions in asphalt [9]. Aromatic and saturated components were decreased as the length of mixing time due to absorption of rubber, however, asphaltene component and resin were unchanged [11]. The greater the loss of weight, the more hardening or aging is found. Weight loss (volatilization) and oxidation are two main causes of the aging of asphalt. Short-term aging was caused by the loss of volatile and oxidizing components during the construction period. While the long-term aging was caused by traffic loads, temperature changes, sunlight, oxidation, etc. [17]. The aging of asphalt caused asphalt harden, so the asphalt was easier to cracked. The weight loss of asphalt caused decrease of penetration value and increase of softening point.

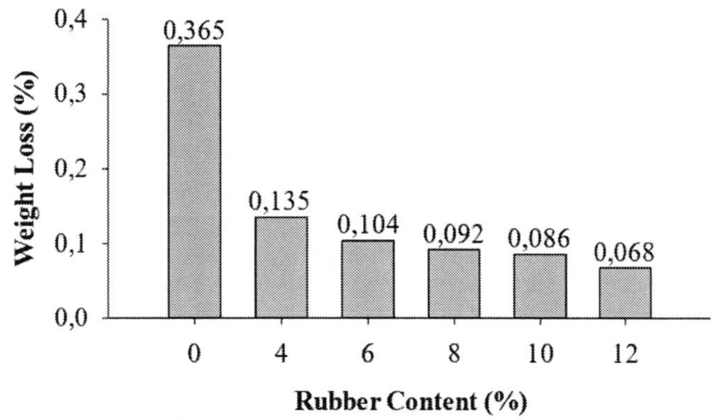


(A)

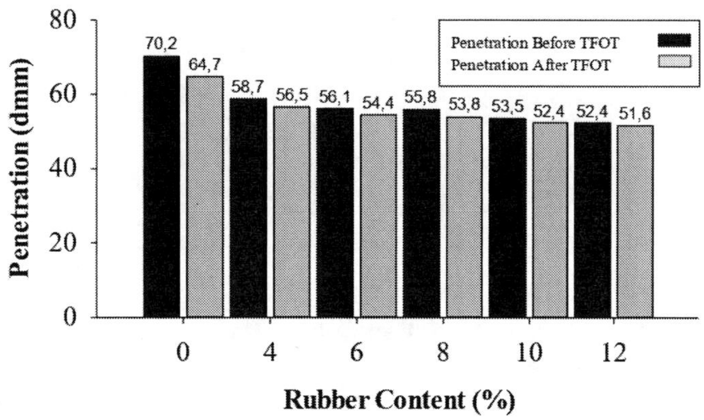


(B)

**Figure 4.** Effect of Compo Ratio on (a) Weight Loss and (b) Penetration of Modified Asphalt After TFOT

Figure 4 (b) shows the decreased of penetration value about 0,8% to 2,2% after TFOT, the lowest of penetration value were 51,6 dmm at the addition of 12% CL. The weight loss after TFOT reduced the penetration due to the increased of asphaltene and resin content. The increased of asphaltene ratio caused asphalt to become harder and has a low penetration value. Meanwhile low penetration asphalt tends to had weight loss and penetration after TFOT the lowest [11]. It is suitable for use in hot climates because low penetration asphalt tends to had a higher softening point [18]. Asphalt with low penetration tends to have rutting resistance and fatigue during service period [19].

### *3.5 Effect of Compo Ratio on Marshall Stability of Modified Asphalt*

Application of modified asphalt must possess ability to withstand traffic loads without permanent deformation [18]. In addition, the stability of asphalt mixture must have rutting resistance and better performance on high and low temperatures. The advantages of adding rubber into asphalt in its application are better driving quality, moisture resistant, higher durability, reduced sensitivity to temperature, and low noise levels. Figure 5 shows the highest stability Marshall was 1904,36 kg or up to 61,5% at the addition of 12% CL. The addition of 4% CL did not have significant effect on the increased of stability Marshall. The result shows that optimal CL addition is between 6% - 12% with an increase in stability 20.6% - 61.5%. This study used optimum asphalt content 6% of the total mixed weight applied to all samples. With asphalt content that are not optimum in the aggregate mixture, the greater possibility of rutting and deformation will be [20]. The addition of CL changed rheology of asphalt in binding the aggregate to be stronger due to increased cohesion and adhesion [11].

The addition of 12% CL is the most optimum to be applied as road construction because it has high stability and softening point. It has Void Filled Aggregate (VFA) up to 51,2%. A small VFA value indicates the asphalt was unable to fill the void between aggregate grains. However, even though the resulted VFA value is small, the cohesion and adhesion bond between asphalt and aggregate remain strong so that they have a high stability value. The VFA value is inversely proportional to the value of Void In Mixture (VIM) and the value of Void in Mineral Aggregate (VMA). The asphalt mixture with CMA had VIM value of 11,9% and VMA value of 24,37% which indicated poor asphalt tightness. The asphalt mixture that was not impermeable will easily oxidized which leads to short aging. The value of VIM and VMA indicated that the asphalt is pivot and susceptible to erosion of the asphalt film covering the aggregate by water, as well as the risk of deformation due to increasing the traffic loads.

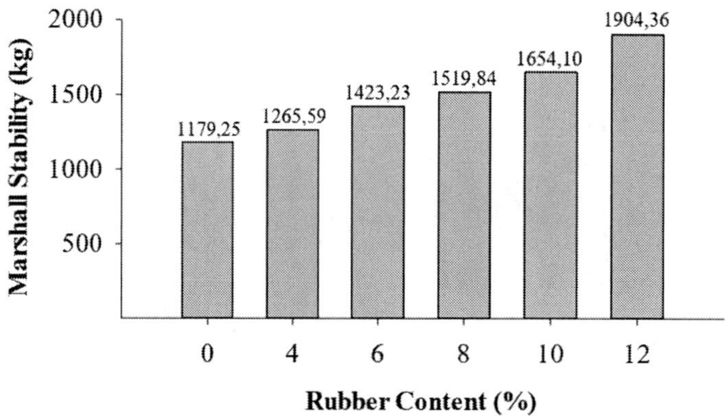


**Figure 5.** Effect of Compo Ratio on Marshall Stability of Modified Asphalt

The stability of the asphalt-rubber mixture can be influenced by several factors that make a significant difference to the results of other studies. These are several factors that have a great impact to the stability of the asphalt mixture are the optimum asphalt content in the mixture, gradation, and composition of aggregate in the mixture, the penetration of asphalt – rubber that used, mixed viscosity, mixing temperature, and compaction. The increased of Marshall stability gave better durability for asphalt to resist traffic load, increase the rutting resistance of asphalt, reduce deformation of asphalt at high temperature, increase fatigue resistance and anti aging performance.

## 4. Conclusions

Cup Lump (CL) natural rubber, which is used as an asphalt additive, is proven to be able to improve rheology of conventional asphalt from the aspect of resistance to traffic loads and high temperature. The result showed that along with the increase of CL content in asphalt, the penetration of CMA would decrease and the softening point would increase. The optimal addition of CL was at 12% with penetration 52,4 dmm, softening point 63,5°C, ductility 142 cm, weight loss 0,068%, and Marshall stability 1904,36 kg. Weight loss 0,068% indicated the asphalt was more resistant to short aging during service period. The high Marshall stability gave better durability for asphalt to resist traffic load, increase the rutting resistance of asphalt, reduce deformation of asphalt at high temperature, increase fatigue resistance and anti aging performance.

## 5. References

[1] Pais, J.C., Amorim, S.I. and Minhoto, M.J 2013 *J. Trans. Eng.* **139** (9) 873
[2] The Central Bureau of *Statistics*. Indonesian land transportation statistics. https://www.bps.go.id/publication/2018/11/27/43cba6b697f03cc2b272dfb7/statistik-transportasi-darat-2017.html. 2018 (accessed 06 April 2019)
[3] GAPKINDO. https://www.gapkindo.org/analisis-pasar/442-desember-2018/. 2018 (accessed 06 April 2019)
[4] The Central Bureau of *Statistics (*BPS). https://www.bps.go.id/statictable/2019/02/27/2030/ekspor-karet-remah-menurut-negara-tujuan-utama-2012-2017.html/. 2019 (accessed 06 April 2019)
[5] Tradingeconomics Statistic. rubber commodity price statistics. https://tradingeconomics.com/commodity/rubber/. 2019 (accessed 06 April 2019)
[6] Singh, B., Kumar, L., Gupta, M., Chauhan, M. and Chauhan, G.S 2013 *J. Appl. Polym. Sci.* **129** (5) 282
[7] Azahar, N.M., Hassan, N.A., Putrajaya, R., Hainin, M.R., Puan, O.C., Shukry, N.A.M. and Hezmi, M.A *IOP Conference Series: Earth and Environmental Science* **220** (1)
[8] Ramadhan, A., Cifriadi, A., Prastanto, H. and Kinasih, N.A 2019 *Majalah Kulit, Karet, dan Plastik* **34** (2) 69
[9] Wen, Y., Wang, Y., Zhao, K. and Sumalee, A 2017 *Int. J. Pavement Eng.* **18** (6) 547
[10] Read, J. and Whiteoak, D 2003 *The shell bitumen handbook*
[11] Peralta, E.J.F 2009 *Study of the Interaction between Bitumen and Rubber* (Doctoral dissertation)
[12] Jensen, W. and Abdelrahman, M 2006 *Crumb rubber in performance-graded asphalt binder*
[13] Oyekunle, L.O 2006 *Oil & Gas Science and Technology-Revue de l'IFP* **61** (3) 433.
[14] Fang, C., Qiao, X., Yu, R., Yu, X., Liu, J., Yu, J. and Xia, R 2016 *J. of Appl. Polym. Sci* **133** (27)
[15] Micaelo, R., Santos, A. and Duarte, C 2012 *Proceedings of 5 th Eurasphalt & Eurobitume Congress. Istanbul*
[16] Gagle, D.W. and Draper, H.L 1978 *U.S. Patent 4,130,516*
[17] Liu, G., Yang, T., Li, J., Jia, Y., Zhao, Y. and Zhang, J 2018 *Constr. Build. Mat* **168** 501
[18] Speight, J.G 2015 *Asphalt materials science and technology*
[19] Limbong, R.H., Hadiwardoyo, S.P., Sumabrata, R.J. and Ariyapijati, R.H 2017 *Key Eng.Mat.***753** 321
[20] Siswanto, H., Supriyanto, B., Pranoto, Chandra, P.R. and Hakim, A.R 2017 *Conference Proceedings* **1887**

IOP Conf. Series: Materials Science and Engineering **845** (2020) 012051 doi:10.1088/1757-899X/845/1/012051

# Magnetic iron oxide particles ($Fe_3O_4$) fabricated by ball milling for improving the environmental quality

**A. Erwin [1*], S. Salomo[1], P. Adhy[2], N. Utari [3], W. Ayu [3], Y. Wita[3] and S. Nani [3]**
[1]Physics Department Faculty of Mathematics and Natural Sciences Riau University
[2]Department of Mechanical Engineering Faculty of Engineering Riau University
[3]Magetic Laboratory, Physics Department Faculty of Mathematics and Natural Sciences Riau University, Indonesia 28293.

E-mail: erwin_amiruddin@yahoo.com

**Abstract.** Nowadays magnetic iron oxide particles ($Fe_3O_4$) have been extensively used as nanomaterials due to their broad range of applications. One of their applications is to improve the environmental quality by elimination of harmful pollutants in wastewater. Therefore, preparation method of magnetite ($Fe_3O_4$) nanoparticles is one of the most important keys to its adsorption efficiency. This study aims to prepare magnetite nanoparticles from natural iron sand of Arta Beach, Pariaman, West Sumatera using ball milling technique. This technique is simple and easy to carry out. The balls used in this milling are the iron balls with the diameter of 1.5 cm. The phase composition, elemental composition, microstructure and magnetic properties of milled powder of magnetic particles were studied by X-ray diffraction (XRD), X-ray fluorescence (XRF), scanning electron microscopy (SEM) and vibrating sample magnetometer (VSM) respectively. The results showed that the milling time is a key parameter of the magnetite nanoparticles preparation. By increasing the milling time, then high purity magnetite samples can be obtained. Moreover, the magnetic particle size of the sample decreases as the milling time increases. The SEM micrographs showed that the sample has a broad distribution of particle diameters, ranging from around 1 up to 107 μm and 0.5 to 36 μm after 30 and 90 hours of milling respectively.

## 1. Introduction

Iron sand is kind of mineral that is available in some main Indonesian islands coast such as West Sumatra coast especially Arta Beach, Pariaman, West Sumatera Province. It is commonly known that iron sand consisted of magnetite ($Fe_3O_4$), maghemite ($\gamma$-$Fe_2O_3$) and hematite ($\alpha$-$Fe_2O_3$) phases. These particles in nanometer size show an interesting property such as superparamagnetic in which the value of magnetic moment is high and low value of coercivity. Since the magnetic properties of magnetite nanoparticles can be controlled using external magnetic field, then their application became broad such as environmental materials [1-3], magnetic recording media [4-7], magnetic sensor [8], toners and inks for photocopy machine [9] and magnetic resonance imaging [10,11], biomedicine and bioengineering drug delivery target [12,13]. In environmental point of view, magnetite particles can remove the pollutant in contaminated water based on magnetic extraction technique [14]. One of the

[*] To whom any correspondence should be addressed.

most important parameters for controlling the magnetic properties of magnetite nano particles is the size of the particles. In reality, preparation of magnetite nanoparticles with controllable size and size distribution is a great challenge for researchers [15].

Today there are several methods for preparation of magnetite nanoparticles such as chemical, [16] biological [17] and physical method [18]. One of the physical methods in preparation of magnetite nanoparticles is ball milling method [17]. Ball milling method is a simple and efficient an as well as low cost method in crunching the iron sand concentrate in a rotating tube using metallic or ceramic balls. In this method, ceramic balls that have higher mechanical energy are used to crunch the iron sand concentrate from iron sand separator. The size of magnetic particles obtained through this method depends on the time, speed and weight of the ceramic balls [18]. Previous researchers [16,17] have used this method to synthesize magnetic nanoparticles from iron salt precursor. They found that the size of magnetic nanoparticles is around 10-15 nm. However, they also found that magnetic nanoparticles that synthesize using ball milling resulted higher magnetic saturation (14emu/gr) compared to that of using chemical method that gives magnetic saturation value of 3 emu/gr. In this article, we report the preparation of magnetite particles using ball milling method for improving the environmental quality.

## 2. Experimental Method

Natural iron sand samples were collected from Arta Beach, Pariaman, West Sumatera. The samples were dried under sun ray until it completely dry before synthesizing them using iron sand separator The product of iron sand separator called concentrate 1 was milled using ball milling as a function of time. The separation between magnetic and non-magnetic particles, then Neodymium iron boron magnet (NdFeB) was employed. The elemental identification of samples before and after milling was recorded using X- Ray Fluorescence Spectrometer (XRF). Magnetic phases of synthesized samples were studied using X-Ray Diffractometer (XRD) technique equipped with Cu $K_\alpha$ radiation of wavelength of $\lambda$=0.15406 nm. Morphological studies were performed using scanning electron microscope (SEM). The magnetic properties of synthesized samples were studied using vibration sample magnetometer (VSM).

## 3. Results and Discussion

### *3.1 Elemental composition*

Elemental composition of an iron sand before and after ball milling process was determined using X-ray fluorescence (XRF). This composition is presented in Table 1.

Table 1 shows that the elemental composition was affected by processing time. It shows that the Fe contents were increased very significantly after being processed by the ball milling for 90 hours. Some other elements for examples Si, Al, K, and Ca were decreased, however, the other elements such as Ti, V and Mn were increased. This indicates that the ball milling process crunch the iron sand grains into smaller parts so that the non-magnetic and magnetic grain were separated during ball milling process. Moreover, from the table it can be seen that especially Fe and Si concentration improved after being processed by ball milling. Therefore, through ball milling process, some nonmagnetic elements were eliminated from the iron sand.

### *3.2 X-ray Diffraction Analysis*

Magnetic phases for a synthesized iron sand after 30 and 90 hours ball milling times were characterized using X-Ray Diffractometer Phillips that equipped with Cu tube producing x-ray radiation with wavelength of 0. 15406 nm. In this measurement, the diffraction angle was selected in interval of 10° to 70° with the step of 0.02°. X-ray diffraction pattern was obtained by applying high voltage source that is around 40 kV and 30 mA. X- ray diffraction patterns of a synthesized iron sand after 30 and 90 hours' ball milling times are shown in Figure 3.

**Table1**. Elemental composition of iron sand before and after ball milling process

| Element | Concentration (%) | |
|---|---|---|
| | Before Milling | After 90 hours milling |
| Mg | 3.452 | 3.345 |
| Al | 10.706 | 8.118 |
| Si | 61.182 | 11.246 |
| P | 0.908 | 0.591 |
| Cl | 2.415 | 0.582 |
| K | 7.245 | 0.537 |
| Ca | 5.835 | 1.265 |
| Ti | 0.472 | 5.836 |
| V | 0.006 | 0.403 |
| Mn | 0.136 | 0.321 |
| Fe | 6.118 | 66.238 |
| others | 1.525 | 1.518 |

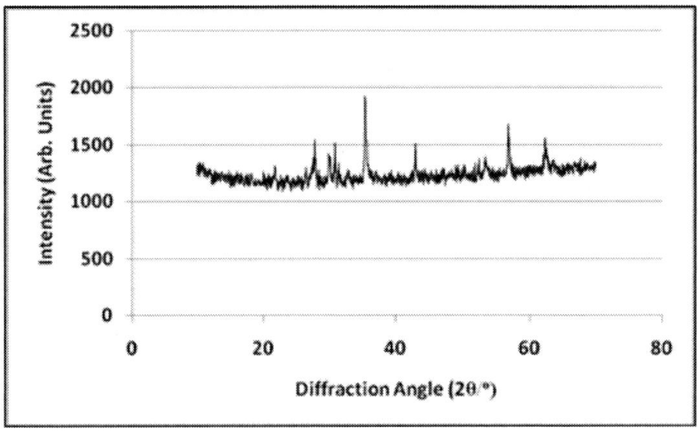


(a)

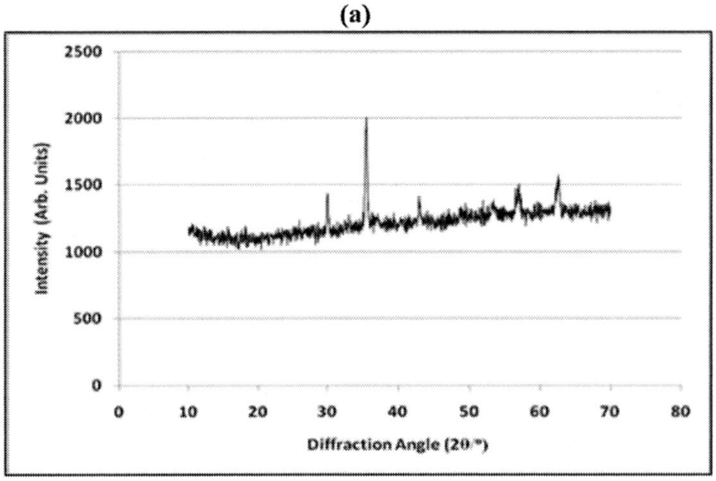


(b)

**Figure 1**. X-ray diffraction patterns for a synthesized iron sand after (a) 30 and (b) 90 hours' ball milling times

Fig. 1 shows two XRD patterns for a synthesized iron sand after 30 and 90 hours ball milling times. In general, it was found that by increasing the ball milling time, then high magnetite purity can be obtained. These XRD results clearly indicate variation of the phase of magnetic iron oxide for these milling times. It can be easily found that the diffraction peaks on a synthesized iron sand after 90 hours milling at 2θ value of 30.04°, 32.47°, 35.47°, 43.00°, 53.45°, 57.02° and 62.70° were completely matched the reflections of (220), (311), (400), (422), (511), (440), respectively. It is clear that these crystal planes are remained unchanged as the milling time increases. Moreover, the XRD pattern become broadened and the diffraction peak intensities increased as the particles size decreased with increasing ball milling time as confirmed with scanning electron micrographs Fig 2. From XRD profile, it is clear that no XRD peaks for iron oxide were observed for high ball milling time namely 90 hours. The magnetite iron oxide ($Fe_3O_4$) phase increased as the ball milling time increased from 30 to 90 hours. Since no obviously peaks observed with increasing ball milling time, then this indicated that FeO phase does not exist for high ball milling time. The comparisons between magnetic phase ($Fe_3O_4$ and $Fe_2O_3$) as observed from XRD peaks are presented in Table 2 and 3.

**Table 2.** Diffraction parameters for iron sand milled for 30 hours

| Diffraction Angle (2θ) | d spacing (Å) | Reflection Plane hkl (for $Fe_3O_4$) | Magnetic Phase | FWHM Left (°2Th.) |
|---|---|---|---|---|
| 21.9228 | 4.05441 | | $Fe_3O_4$ | 0.1535 |
| 26.6172 | 3.34905 | | $Fe_3O_4$ | 0.1535 |
| 27.9454 | 3.19282 | | $Fe_2O_3$ | 0.0768 |
| 30.0392 | 2.97489 | 220 | $Fe_3O_4$ | 0.2047 |
| 30.9676 | 2.88777 | | $Fe_2O_3$ | 0.0768 |
| 32.8377 | 2.72746 | | $Fe_2O_3$ | 0.4093 |
| 35.3469 | 2.53939 | 331 | $Fe_3O_4$ | 0.1791 |
| 42.9686 | 2.10497 | 400 | $Fe_3O_4$ | 0.0768 |
| 53.3876 | 1.71616 | 422 | $Fe_3O_4$ | 0.4093 |
| 56.7803 | 1.62141 | 511 | $Fe_2O_3$ | 0.0768 |
| 62.3647 | 1.48774 | 440 | $Fe_3O_4$ | 0.0936 |

**Table 3.** Diffraction parameters for iron sand milled for 90 hours

| Diffraction Angle (2θ) | d spacing (Å) | Reflection Plane hkl | Magnetic Phase | FWHM Left (°2Th.) |
|---|---|---|---|---|
| 30.0424 | 2.97457 | 220 | $Fe_3O_4$ | 0.2558 |
| 32.4741 | 2.75717 | | $Fe_2O_3$ | 0.1791 |
| 35.4722 | 2.53072 | 331 | $Fe_3O_4$ | 0.2047 |
| 43.0041 | 2.10331 | 400 | $Fe_3O_4$ | 0.2047 |
| 53.4531 | 1.71421 | 422 | $Fe_3O_4$ | 0.6142 |
| 57.0175 | 1.61523 | 511 | $Fe_3O_4$ | 0.7164 |
| 62.7019 | 1.48178 | 440 | $Fe_3O_4$ | 0.3072 |

### *3.3 Microstructural Properties*

The scanning electron microscopy (SEM) images of 30 and 90 hours ball milled samples with 400x magnification are shown in Fig. 2. This analysis was carried out in order to investigate the shape and size of the as-synthesized magnetite particles. Scanning electron microscopy (SEM) images presented n Figure 2 (a) and (b) show the magnetic particle size and the morphology obtained from ball milling with time of 30 and 90 hours. These images show the samples contain particles of different sizes and shapes. SEM images of the sample of 30 hours milled consisted of particles covering the range 1–107 μm and more than 30% of these particles are large approximately 107 μm. The shape of these particles is clearly not systematic shapes. Moreover, SEM image of the samples milled for 90 hours (Figure 2b) shows particles of smaller size of 0.5- 36μm. From Figure 2 it can be estimated that the average particles size decreases with increasing ball milling time. This result is in agreement with that obtained by previous study [22]. This is a notable result, obtained using this simple method for preparing magnetite particles which might be useful for environmental material applications.

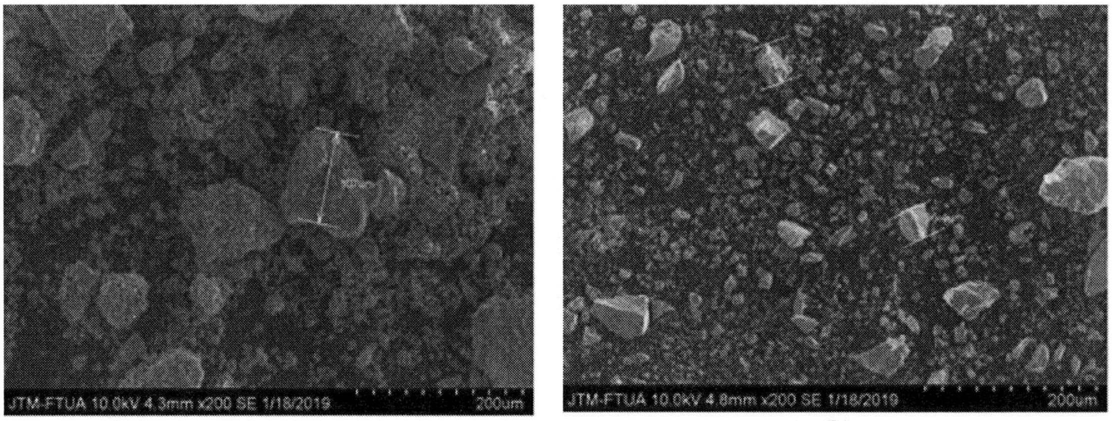


**(a)** **(b)**

**Figure 2.** SEM images of (a) 30 and (b) 90 hours ball milled samples

*3.4 Magnetic Properties*

Magnetic properties of the sample were measured by vibration sample magnetometer (VSM) available at Department of Physics National Atomic Energy Agency (BATAN) Serpong, Tangerang, Banten. Indonesia. Figure 5 below shows the variations of magnetization values and applied magnetic field for the as synthesized iron sand after 30 and 90 hours ball milling times.

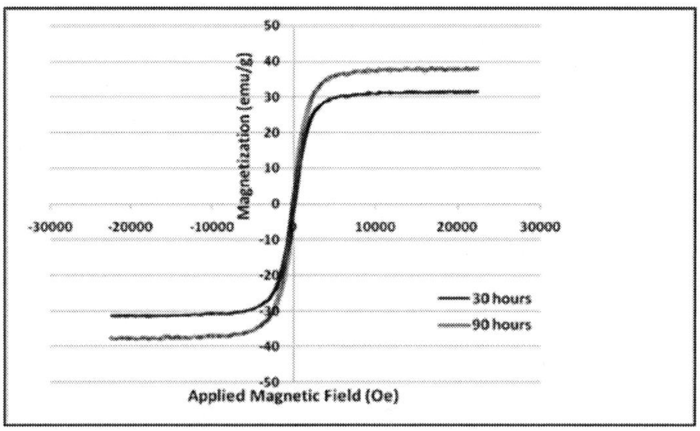


**Figure 3.** Loop hysteresis for the as synthesized iron sand after (a) 30 and (b) 90 hours ball milling times

The hysteresis loops shown in Fig. 3 are the loop hysteresis of iron oxide prepared by ball milling with time of 30 and 90 hours. The applied magnetic field in this measurement was chosen to be 3000 Oe. These two samples show a nonlinear relationship in magnetization as a function of applied magnetic field. It is clear from Figure 2 that magnetic remanence is almost absent in both sample. It shows almost no remnant magnetization for zero applied magnetic field. This results show superparamagnetic behaviour of particles. One important parameter of the hysteresis loops shown in Fig 3, is small value of coercivity. It can be noticed from Fig. 3 that the coercivity value of the samples has been found to remain constant (173 Oe) with the increase in milling time. Moreover, it can be seen that as the milling time was increased from 30 hours to 90 hours, the magnetisation values of the samples increase from 31.48 emu/g to 37.80 emu/g. The increase in magnetization values for longer milling time is being thought due to the presence of an additional magnetite iron oxide ($Fe_3O_4$) phase and reduction of nonmagnetic elements as confirmed by XRF results in Table 1. The loop hysteresis of the as synthesized sample of 90 hours shows magnetization values smaller than the bulk values [28].

## 4. Conclusion

Magnetite ($Fe_3O_4$) particles with minimum size of 0.5 μm have been successfully prepared by varying the milling time. The variations of milling time in this study play an important role in preparing magnetite iron oxide particles in which the size of magnetite particles decreases as milling time increases. This method is simple and low cost, however, it takes longer time to obtain a desired result. Moreover, the milled particles show high percentage of magnetite phase for longer milling time as shown by XRD results. The result also showed that the size of particles has become smaller as milling time increased as revealed by SEM images and broadening of the XRD peaks. Furthermore, it was also found that the magnetization of as synthesized sample increases as milling time increase.

## 5. References

[1] Wang, Y., Du, W., Xu, Y 2009 *J. Am. Chem. Soc* **25** 2895–2899
[2] M.Hua, S. Zhang, B. Pan, W. Zhang, L. Lv, and Q. Zhang 2012 *Journal of Hazardous Materials* **212** 317-331
[3] J.W. Farrell, J. Fortner, S. Worketal 2014 *Environmental Engineering Science* **31** 393-402
[4] Srivastava, A., Ojha, A., Chaubey, S., Singh, J., Sharma, P 2010 *Journal of Alloys and Compounds* vol. **500**, p. 206
[5] C.T. Black, C.B. Murray, R.L. Sandstrom, S. Sun 2000 *Science* **290**, 1131.
[6] Erwin A, Prayitno A 2017 *ARPN Journal of Engineering and Applied Sciences* Vol. **12**, no. 12, June 2017, pp 3832.
[7] Erwin 2017 *The 6th International Conference on Theoretical and Applied Physics* AIP Conf. Proc. 1801, 020003-1–020003-6.
[8] H. Zeng, J. Li, J.P. Liu, Z.L. Wang, S.H. Sun 2002 *Nature* **420**, 395.
[9] M. Ataeefard, E. Ghasemi, and M. Ebadi, 2014. *The Scientific World Journal* vol. **2014** Article ID706367 ,7 pages,
[10] Zhang, S., Qi, Y.Y., Yang, H., Gong, M.F., Zhang, D. and Zou, L.G 2013 *Journal of Nanoparticle Research* **15**,2023.
[11] L.X. Tifenauer, A. Tschirky, G. Kuhne, R.Y. Andres 1996 *Magn. Reson.Imaging* **14**, 391.
[12] Mahmoudi, M., Sant, S., Wang, B., Laurent, S., Sen, T 2011 Advanced *Drug Delivery Reviews* vol. **63**, p. 24
[13] Morteza Mahmoudi, Shilpa Sant, Ben Wang, Sophie Laurent, Tapas Sen. 2011 Advanced *Drug Delivery Reviews* **(6)3** pp 24–46
[14] Elliott, D. W., and W.-X. Zhang 2001 *Environmental Science and Tech* **35(24)** ,4922-4926.
[15] L Ing H.L., Pilko S., Wu J.H., Jung M.H., Min J.H., Lee J.H., An B.H., Kim Y.K 2007 *J. Magn. Mater* **310**, 815.
[16] Chin, A. B. and Yaacob, I. I 2007 *J Mater Process Tech* **191,** 235-237.
[17] Lisy, M. R., Hartung, A., Lang, C., Schuler, D., Richter, W., Reichenbach, J. R., Kaiser, W. A. and Hilger, I 2007 *Invest Radiol* **42,** 235-241.
[18] Wang, Z. M., Liu, Y. and Zeng, X. Y 2006 *Powder Technol* **161,** 65-68.
[19] David Marihot Silaban, Erwin, Yanuar dan U. Malik 2017 Prosiding *Fisika Nasional Universitas Riau* pp. 142-145

[20] Goya, G. F 2004 *Solid State Communications* ***130*(12)**, 783–787.
[21] Almásy, L., Creanga, D., Nadejde, C., Rosta, L., Pomjakushina, E., & Ursache-Oprisan, M. 2015. Journal *of the Serbian Chemical Society* ***80*(3)**, 367–376. http://doi.org/10.2298/JSC140313053A.
[22] Davit Marihot Silaban, Erwin, Yanuar, Adhy Prayitno 2017 Program Book: *The 3rd International Conference on Science Technology and Interdisciplinary Research 2017*, pp 50
[23] Erwin Amiruddin and Adhy Prayitno 2019 *MATEC Web of Conferences* **276**, 06014, https://doi.org/10.1051/matecconf /201927606014.
[24] J.F. de Carvalho, S.N. de Medeiros, M.A. Morales, A.L. Dantas, A.S. Carrico 2013 *J. Applied Surface Science* **275**, pp. 84-87
[25] G.F. Goya, 2004. *Journal of Materials Science*, **39**, pp. 5045–5049.
[26] M. Mutlu Can, S. Ozcan, A. Ceylan, T. Firat, 2010 *Materials Science and Engineering: B* 172 pp. 72–75
[27] Lu, A., Hui, E. L. Salabas, and F. Schüth 2007 *Angewandte Chemie International Edition* **46(8)**, 1222-1244.
[28] R.M. Cornell, U. Schwertmann 2003 The *Iron Oxides*, 2nd ed., Wiley-VCH, Weinheim, Germany,

# Landfill mining dominated by organic solid waste: a review on its benefits, potential and challenges to recovery landfills in growing cities in Indonesia

**Zaulfikar[1,2*], Sudarno[1,3] and Budiyono[1,4]**

[1]Doctoral Program Environmental Sciences of Diponegoro University, Semarang, Indonesia
[2]Industrial Engineering Department of Teuku Umar University, Meulaboh, West Aceh, Indonesia
[3]Environmental Engineering Department of Diponegoro University, Semarang, Indonesia
[4]Chemical Engineering Department of Diponegoro University, Semarang, Indonesia
*E-mail: zaulfik@yahoo.com

**Abstract.** One of serious impacts of increasing human population in cities is the availability of land, strategic land in particular becomes rare and as a result, its value increases rapidly. Another impact is the amount of solid waste produced which also continues to increase sharply. Although the perception and knowledge of the community regarding solid waste is improving, it does not necessarily change their paradigm, mindsets and behaviors in minimizing solid waste, the landfill is still the final solution to solid waste. In general, the carrying capacity and supporting capacity of landfills in the cities have reached the maximum limit and must be closed immediately. Initially the locations of the landfills were far from city centres, but now generally they are in densely populated urban areas. Procurement of new landfill is constrained by regulations, costs, technical and social issues. NIMBY syndrome (not in my back yard) is the culmination of public rejection, and has even become a vertical conflict between the community and the government. The study of landfill mining, particularly the study of benefits, potential and challenges in choosing landfill mining, which is dominated by organic solid waste, is an interesting study to explore. The use of compost landfill mining as an organic fertilizer for plants, or as biocover is a logical argument for recover or extend the service life of old landfill.

## 1. Introduction

Rapid population growth has a serious impact on the amount of solid waste produced, especially in cities. The increasing amount of solid waste becomes more serious due to the changes in society's lifestyles and consumption patterns as a result of an increase in urban income. Although the perceptions and knowledge of the urban community about solid waste are increasingly positive, it does not necessarily change their paradigm, mindsets and behaviours in reducing solid waste, and the landfill is still the final solution to solid waste [1–3].

* To whom any correspondence should be addressed.

The general characteristics of solid waste in developed countries are inorganic and dry solid waste, such as metal-containing waste - e.g. spring beds, chairs, car seats, etc - which can go through the combustion process easily. Thus, landfill mining in developed countries generally focuses on metal mining from incinerator ash deposits.

Meanwhile, in developing countries such as Indonesia, urban household solid waste is dominated by organic solid waste - which is up to 70% of the total solid waste entering landfills - and categorized as wet solid waste with a water content up to 80% [1]. This makes solid waste management with the combustion process impossible. However, the organic materials can decompose naturally and produce methane, carbon and water.

The main objective of this paper is to examine the unique landfill mining concept which is different from the concept of landfill mining in developed countries, i.e. the landfill mining which is dominated by organic solid waste, so that the results obtained are also unique and different, such as decomposed organic material mass in the form of fine soil-like material, called compost landfill mining.

## 2. Definition of Landfill Mining

From the economic aspect, it is stated that the cost of solid waste management with the spreading system in landfills is lower than solid waste management with incinerators [4]. However, population density factors need to be considered - in areas with low population and land availability, a landfill is still a better choice [5]. Whereas in densely populated areas in which land prices continue to increase rapidly, obtaining sites for new landfills that comply with regulations will be difficult and generally fail [6,7].

From an environmental perspective, disposal of solid waste into landfills, both by burying and leaving them without management, is a serious mistake, because the solid waste still contains energy and materials which can be utilized [8,9]. One way to get back the materials is by digging up the waste. The activity is known as Landfill Mining (LFM), which is a process for extracting other solid materials or natural resources from solid waste materials that has previously been disposed of by burying it in soil [10].

Landfill Mining is a way to recover resources buried in landfills and minimize environmental burdens caused by emissions from landfills [11]. Landfill mining can extend landfill service life; repair and recover land conditions; and restore the adverse environmental effects of leachate pollution and gas emissions produced by landfill [10,12,13]. Another definition of landfill mining is a process of extracting solid waste from old landfills with the aim of treating the unprocessed solid waste in order to be used as a resource, for example as secondary raw material or as a raw material for substitute fuels "Refused Derived Fuel" [14].

## 3. Benefits of Landfill Mining

Limitations of urban land require good management of the landfill, so that the service life of the landfills can be extended. Landfill mining is one of alternatives to achieve the goal. The process includes extracting, sorting and reusing materials which are still useful from landfills. Thus, landfill mining has the potential to recover and reuse landfills.

Furthermore, landfill mining can also reduce and restore the adverse effects of the environment that can threaten the health of local communities. These adverse effects include ground water pollution by leachate into the soil. Meanwhile, air pollution can occur from carbon gas and methane gas emissions [12].

In addition to its economic benefits and environmental recovery, other indirect benefits of landfill mining activities include being able to avoid obstacles (e.g. legal, environmental impact, community acceptance) and large costs through low costs of land liability and cover [15]. Therefore, landfill mining can be adopted as a viable way of ecological recovery of landfills [16,17].

Some literature states that the benefits, costs and environmental impacts of landfill mining activities depend much on several factors, such as age of land, characteristics of excavated materials, landfill depth, waste decomposition level of decomposition, content of hazardous materials, waste management methods, economic/social developments of local communities, climate, humidity and landfill location

[10,11,15,18]. Therefore, in order to find out the full potential of a landfill, it is important to conduct both qualitative and quantitative analysis of the characteristics of the excavation materials beforehand [19,20].

Other indirect benefits that can also be obtained from the selection of land mining are the restoration of the aesthetic values in the region, including recovery from the impression of being dirty, slum, bad smell, dust, flies, rats and stray dogs. The recovery of aesthetic values has a real impact on the landfill recovery and increases the economic values of the landfill and its surrounding areas. This is because the existence of landfill has a negative impact on the price of nearby houses and buildings [21,22].

The impact of the decline in environmental quality also has a close relationship with the social conditions of the community. Therefore, the presence of landfills is often a cause of social conflict with the surrounding community. This social conflict has become a common phenomenon in big cities in Indonesia, and has even led to a vertical conflict between the community and the local government. Strong protests from the community have given birth to NIMBY syndrome (not in my back yard) [22].

Both of newspaper and television media have often reported public protests about the existence of landfills, including:

i. In 2003; Community protests to the Surabaya government against the closure of the Sukolilo landfill, because in the 2000 protest it was agreed that the deadline for closing the landfill would have been in 2002 [23].
ii. In 2009 and 2016; Community protests against Galuga landfill in Bogor, West Java [24–26].
iii. In 2010; due to the trauma of the 2005 landfill landslide tragedy which claimed 157 deaths and removed 2 villages from the map, the community around Leuwi Gajah landfill in West Java rejected the plan to reactivate the site [27,28].
iv. In 2017; Public protest against the end of the operation period of the Pasir Bajing landfill in Garut, West Java [29].
v. In 2018; Conflict between the government of Bekasi city and the DKI Jakarta government regarding compensation payment for the community around the Bantar Gebang landfill in Bekasi, West Java, for its decreasing environmental quality [30].

## 4. Landfill Mining Potential

### *4.1. Raw Materials Recycle*

The composition of solid waste in Indonesia generally consists of 70% of organic solid waste, and the rest is inorganic solid waste. These inorganic solid waste have the potential to be reused or recycled into new products or other products. This is because most of the inorganic solid wastes still have economic value or calorific value as an energy source. Some of them are paper waste, plastic, glass, metal and batteries [31].

Most of formal sectors in Indonesia have not been able to carry out the process of sorting solid waste, so the activity is generally carried out by the informal sector, namely scavengers. Scavengers are people who treat rubbish as something of value, a source from which high-value materials can be obtained [32].

In general, landfills in Indonesia have not been properly managed, so they are categorized as being dirty and smelly, and a source of diseases. Scavengers and people who live around the landfills are highly susceptible to diseases and exposure to toxic and hazardous gas emissions. Therefore, there must be a paradigm shift from collecting used goods manually by scavengers to the paradigm of using a environmental quality and high market demand for recycled raw materials [20].

### *4.2. A Source of Organic Fertilizer*

Organic solid waste is rubbish whose constituent materials come from living things produced by nature or by human activities. Household solid waste is one example of many sources of solid wastes produced by human activities. The examples of organic household solid waste includes food scraps, vegetables, fruits, leaves, meat, vegetable/fruit peel, etc. The long term impact of the accumulation of organic solid waste is the decomposition of these organic solid wastes into fine materials resembling soil, called compost.

Composting is a fermentation process or also known as the process of organic material degradation caused by oxidation. The decomposition will produce carbon dioxide and water, so that the remaining biomass becomes a component that is simpler, stable and resembles humus. The basic principle of composting is to reduce the comparative value of carbon and nitrogen (C/N) elements from organic solid waste to near to the value of the C/N ratio of land, which is less than 15 [33].

One of the physical characteristics of good compost is one that has a texture and smell like soil. Good compost can improve texture and maintain soil moisture. Good compost can also increase the content of organic matter and soil nutrients. Therefore, good compost can be used as fertilizer for agriculture, landscaping, nurseries, garden soil conditioning, golf courses and reforestation of critical lands. So, the market potential for demand for compost is very large. This condition provides a large space for landfill mining activities. In addition to its economic potential and recovery of environmental quality, landfill mining activities can also create jobs [34].

### *4.3. As Biocover*
Soil-like fine materials obtained from the landfill excavation process which contain heavy metals or other hazardous toxic materials cannot be used as organic fertilizer for plants in the food chain. Instead, they can be used for other plants that are not for consumption. The soil can also be used as a landfill top cover (biocover). Thus, landfill mining not only can recover the lands of old closed landfills, but also can increase the service life of actively operating landfills [10,12].

To reduce methane gas emissions into the air, the landfill area can be covered by using soil and compacted until the thickness of soil of 20-30 cm. Reusing compost landfill mining as landfill top covers (biocover) not only reduces methane emissions from landfills, but also enhances the methane decomposition process to become carbon dioxide, water and biomass. This is because compost landfill mining contains various groups of microorganisms, especially methanotropic bacteria which can oxidize methane biologically, so that the use of compost landfill mining as biocover is better than using ordinary soil [35,36].

## 5. The Challenges on Selecting Landfill Mining
### *5.1. Highly Costs*
In general, the cost of landfill mining is greater than income, so that landfill mining activities are not feasible without additional government mandates and funds [37,38]. However, for the municipal governments, landfill mining is important in order to obtain additional lands in densely populated areas. Therefore, the issue of additional funding needs is not a serious consideration. This is due to the value of the recovered lands which exceeds the value of the landfill contents [15,39].

### *5.2. Heavy Metal Content and Hazardous Toxic Material*
In the past, landfill management used an open dumping system without facilities to control environmental quality. The most serious impact of this practice is ground water contamination by leachate. Any landfill will produce leachate, that is liquid produced by dissolving organic and inorganic compounds as the result of physical, chemical and biological processes of solid waste decomposition; The source of the liquid can come from the solid waste itself and rainwater in landfills.

Shallow groundwater is found in the depth up to 20-40 meters underground, for example wells or shallow bore wells which are used as clean water sources by people who do not have access to water from the PDAM (water company). If the water source is polluted, the water quality will decrease and have a serious health impact on the community.

Jaya [40] have reported that up to a distance of 400 meters from landfills, the quality of community water wells is categorized as heavily polluted. From the report presented in Table 1, it can be seen that the heavy metal contents of cadmium (Cd) and lead (Pb) up to a distance of 100 meters away from the landfill are 0.019 mg/l and 0.981 mg/l, respectively. These results are higher than the specified quality standard: 0.01 mg/l for cadmium (Cd) and 0.03 mg/l for lead (Pb). At a distance of 400 meters away from the landfill, although cadmium content has dropped to 0.003 mg/l (below the quality standard), lead content reduces to only 0.602 mg/l (still above the quality standard).

**Table 1.** The results of measuring the quality of wells around Suwung landfill in Bali

| Parameters | Units | Class I Water Quality Standard (Bali Governor's Regulations No. 8/2007) | Results (Based on distance of samples from the landfill site) | |
|---|---|---|---|---|
| | | | 100 meter | 400 meter |
| Cadmium (Cd) | mg/l | 0,01 | 0,019 | 0,003 |
| Lead (Pb) | mg/l | 0,03 | 0,981 | 0,602 |

The content of heavy metals is also found at 4 landfills in West Java as presented in Table 2 [41]. These heavy metals include copper (Cu), chromium (Cr), zinc (Zn), lead (Pb) and cadmium (Cd). Heavy metal content is also found in the Sukolilo landfill in Surabaya. The content are 0.087 mg/l for Cd, 0.593 mg/l for Pb and 1.067 mg/l for Cr, respectively [42].

**Table 2.** Contents of some heavy metals at 4 waste landfills in West Java

| Parameters | Jelekong Landfill | Pasir Impun Landfill | Leuwi Gajah Landfill | Sarimukti Landfill |
|---|---|---|---|---|
| Cu (ppm) | 7,00725 | 12,1166 | 12,3735 | 12,2879 |
| Cr (ppm) | 1,65529 | 3,128 | 3,7935 | 4,5102 |
| Zn (ppm) | 9,74202 | 4,4761 | 4,2712 | 4,6744 |
| Pb (ppm) | 4,76249 | 14,1736 | 15,2153 | 14,9236 |
| Cd (ppm) | 0 | 0,2755 | 0,4057 | 0,3201 |

**Figure 1.** Construction of playground/city recreation park on the closed Sukolilo landfill in Surabaya. (Source: Observation results on October 1, 2017)

**Figure 2.** Construction of residential homes on the closed Sukolilo landfill in Surabaya. (Source: Observation results on October 1, 2017)

Based on reports of the contents of heavy metals and other hazardous toxic substances potentially found in landfills in developing cities in Indonesia, the utilization of the former landfills as city forests is a wise decision made by municipal governments. This is considering that soil and groundwater contamination by leachate can last for a long time and can reach 30-50 years [43].

However, the use of closed landfills as city playgrounds, recreational park or even for residential developments is a questionable decision (Figure 1 and Figure 2). As the health and safety aspects of the land users are still in doubt, the study of all aspects (which will take a long time and require large funds) must be done first. This is based on the consideration that the landfill management in the past has not specifically separated the solid waste: domestic, medical and industrial waste.

### *5.3. Researched on Potential of Landfill Mining*

Landfilled wastes buried for a long time can be excavated and used as compost for plants or can also be reused as biocover for actively operating landfills. The excavation of the landfill is termed landfill mining. The series of landfill mining activities include excavation, drying, sorting organic and inorganic waste, enumeration, screening and weighing to determine the composition of the extracted materials.

Studies to examine the potential of landfill mining in Indonesia have been carried out by several researchers. For example, a research at Sukolilo landfill in Surabaya found that the composition of the landfill materials is as follows: soil 34.26%, plastic 38.10% and others 2-10% as wood, cloth, rubber and metal [44].

Another research was conducted at the Ngipik landfill in Gresik district. Based on the mass balance of excavated materials, it was reported that 97.57% of the materials obtained could be reused and the remaining 2.43% in the form of rubber, glass and others were materials that could not be utilized. Some of the materials that can still be reused include 25.91% compost, 19.37% urug material and the rest are plastic, wood and fabric, which can be used as raw materials for refused derived fuel (RDF) after requiring further processing. Detailed data from the excavation results are presented in Table 3 [45].

**Table 3.** Waste Composition

| Components | Compositions (%) | Utilization |
|---|---|---|
| Fine Material | 25.91 | Compost |
| Rough Material | 19.37 | Additional soil |
| Plastic | 41.40 | Alternative fuel / RDF |
| Wood | 5.31 | Alternative fuel / RDF |
| Linen | 5.49 | Alternative fuel / RDF |
| Iron/metal | 0.09 | Recycle |
| Rubber | 1.40 | Re-landfill |
| Glass | 1.01 | Re-landfill |

### *5.4. Waste Management Problem in Indonesia*

In Indonesian law (UU No. 18, 2008) of Solid Waste Management states that it local governments are obliged to close every landfill which uses an open dumping system no later than 5 years after the law came into force. It means that by 2014 there is no open dumping landfill in Indonesia. Nevertheless, in reality, the local governments are not able to obey the law [46].

Lack of accurate data on solid waste, lack of facilities in landfill sites, lack of efforts to compost and recycle waste are evidence that the technical aspects, costs and human resources are serious obstacles in local solid waste management. This fact is in line with the Executive Summary data of the 2013 Sanitary Landfill Policy Study of the Coordinating Ministry for Economic Affairs of the Republic of Indonesia who stated that the technical, financing and quality aspects of human resources are the main obstacles in implementing a landfill using a sanitary landfill system [47].

Solid waste problems, especially in developing cities, are increasingly complicated by the high rate of urbanization, which directly affects the population and the amount of solid waste produced. Although the perceptions and knowledge of the urban community about solid waste are increasingly positive, it does not necessarily change their paradigm, mindsets and behaviors in minimising solid waste significantly; landfills are still final solution for solid waste, which in turn present serious problem for local governments [1–3].

### *5.5. Regulations*

Article 28H paragraph 1 of the UUD 1945 states that everyone has a right to good and healthy environment. Furthermore, Article 33 paragraph 3 states that the earth, water and natural resources contained therein are controlled by the state and used as much as possible for the prosperity of the people [48]. The prosperity is not only for the present generation, but also for future generations. To guarantee sustainable prosperity, the Government and the House of Representatives agreed to issue Republic of Indonesia Law Number 32 of 2009 concerning Environmental Protection and Management. The mandate is clearly stated in article 1 paragraph 3: sustainable development is a conscious and planned effort which combines environmental, social and economic aspects into development strategies in order to ensure the environmental integrity as well as the safety, capability, welfare, and quality of life of present and future generations [49].

These three aspects are also applied in solid waste management. This means that solid waste management activities must ensure the integrity of the environment and carrying capacity of the environment. Solid waste management must also profit and provide economic benefits for the community and the region. Therefore, solid waste management activities must also consider the socio-cultural aspects of the local community so that community involvement, participation and empowerment can be fully achieved.

Increasing population and changes in communities' consumption patterns are among the main causes of the current increase in the amount of solid waste. In general, throughout the territory of the Republic of Indonesia, solid waste management is also not in accordance with the methods and techniques of environmentally-friendly solid waste management, thus having negative impacts on the health of the local community and the surrounding environment. Therefore, the Indonesia Government considers that solid waste problem is a national problem; its management needs to be carried out comprehensively and integratedly so that it can provide both economic and health benefits for the local community, and safety for the environment. In order for solid waste management to be carried out proportionally, effectively and efficiently, legal management, responsibility and authority are needed in solid waste management, either for the central government, regional governments, community or the business world. Based on these considerations, the Government and the House of Representatives of the Republic of Indonesia agreed to pass Law No. 18 of 2008 concerning Solid Waste Management.

Law No. 18 of 2008 [50] explains in details and in full: the definitions of terms of solid waste and solid waste management, the scope of solid waste management, principles and objectives of solid waste management, government duties and authorities (for central government, provincial governments and regional governments), as well as the rights and obligations of each individual regarding solid waste. Other important points in the Act include a special chapter on the implementation of solid waste management (household solid waste and household-like solid waste), solid waste reduction regulations, solid waste handling regulations, and specific waste management rules. Other important material contained in the Law includes the role of the community in solid waste management, prohibitions relating to the solid waste, supervision, administrative sanctions, criminal provisions and transitional provisions. There are some interesting points stated in the transitional provisions chapter, including the affirmation of each regional government to make a plan to close the open dumping system landfill no later than 1 (one) year after the enactment of Law Number 18 of 2008. The next assertion is that the regional government must close any landfills which uses an open dumping system within 5 (five) years from the enactment of Law Number 18 of 2008. The affirmation is also addressed to managers of residential, commercial, industrial, special, public, social and other facilities which do not have solid waste sorting facilities before the enactment of Law Number 18 of 2008 to build or provide solid waste sorting facilities within 1 (one) year.

To regulate more clearly the specific provisions of several articles in Law Number 18 Year 2008, the government stipulates Government Regulation Number 81 of 2012 concerning Management of Household Solid Waste and Household-like Solid Waste. The Government Regulation includes a regulation on solid waste management in question, with the aim of maintaining public health, preservation and functioning of the environment, and making the solid waste as a resource. The regulation on the solid waste management includes policies, strategies, implementation, guidance, information systems, community roles, compensation, and the development and application of technology [51].

There are some very interesting things from the direction of policy, strategy and implementation of solid waste management referred to in the Government Regulation, including the reduction and handling of solid waste. Solid waste reduction includes limiting solid waste generation, recycling and reusing. Meanwhile, solid waste handling includes sorting, collecting, transporting, processing and final processing of solid waste. The final solid waste processing referred to in the Government Regulation is directed to the district/city governments to use controlled landfill methods, or sanitary landfill methods, or environmentally-friendly technology.

The Government Regulation still requires further guidance so that it can be implemented by the regional governments. Law No. 18 of 2008 mandates the relevant ministers to prepare further instructions. For example, the Regulation of the Minister of Public Works of the Republic of Indonesia Number 03 of 2013 concerning the Implementation of Solid Waste Infrastructure and Facilities in Handling Household Solid Waste and Similar Household Solid Waste [52]; and the Regulation of the Minister of Environment and Forestry of the Republic of Indonesia Number 59 of 2016 concerning Standard Quality of Leachate for Businesses and/or Landfills [53].

To achieve the quality set by the government, the National Standardization Agency stipulates the Indonesian National Standard (SNI) for every activity and goods produced and marketed within the territory of the Unitary State of the Republic of Indonesia. Some of the standards relating to solid waste include SNI 19-3964-1994 concerning Methods of Collecting and Measuring Samples of Solid Waste Generation and Composition of Urban Solid Waste [54]; SNI 19-2454-2002 concerning Procedures for Operational Management of Urban Solid Waste [55]; and SNI 3242:2008 concerning Solid Waste Management in Residential Settlements [56].

The above explanation illustrates that regulations regarding solid waste management in Indonesia are adequate for a developing country. However, changes in behaviours, perceptions and paradigms about solid waste and its management expected from the community, economic players and the government still require time. The inability of most local governments to close open dumping landfills is an evidence that costs and human resources are still the main obstacles in solid waste management. Their inability to procure new lands for controlled landfills or sanitary landfills is a challenge and provides potential for landfill mining activities in order to increase landfill service life or recovery of open dumping landfills.

## 6. Conclusion

### *6.1. Conclusion*

i. Naturally, any organic material can decompose to produce gas methane, carbon, water and soil-like fine material, called compost. Thus, the recovery of landfills dominated by organic solid waste can be done using landfill mining methods, from which compost rich in nutrients can be used as organic fertiliser for plants.
ii. To find out the potential in each landfill, the physical, chemical and biological characteristics of the excavated materials should be analysed beforehand.
iii. The analysis results of the chemical characteristics of the landfill mining compost will determine its proper use: organic fertilizer for plants or biocover for actively operating landfills.

### *6.2. Suggestion*

i. In order to reduce negative impacts on the environment and also to avoid lawsuits from landfill mining activities, the environmental impact analysis (EIA) must be prepared first.
ii. Recovery of leachate ponds is not included in this review study. Therefore, before landfill mining activities are carried out, in-depth studies on the recovery of leachate ponds and studies on the environmental impact analysis (EIA) must also be carried out beforehand.

**Acknowledgements**
The authors gratefully acknowledge the financial support of the Indonesia Endowment Fund for Education and the technical support of the Ministry of Research, Technology and Higher Education of the Republic Indonesia.

## References

[1] Fehr M 2006 *Environmentalist* **26(1)** 21
[2] Mahyudin RP 2014 *EnviroScienteae* **10(1)** 33
[3] Akhtar H and Soetjipto HP 2014 *J. Humans Environ.* **21(3)** 386
[4] Thorneloe SA, Weitz K and Jambeck J 2007 *Waste Manag.* **27(8)** 1006
[5] Barrett A and Lawlor J 1997 *J. Environ. Plan. Manag.* **40(1)** 19
[6] Chang NB and Davila E 2007 *Waste Manag.* **27(6)** 820
[7] Massarutto A 2012 *Econ. Policy Energy Environ.* 81
[8] Cobb CE and Ruckstuhl K 1988 *Proc. of the National Waste Processing Conf.* Detroit MI USA 145
[9] Hogland W, Marques M and Nimmermark S 2004 *J. Mater. Cycles Waste Manag.* **6(2)** 119
[10] Krook J, Svensson N and Eklund M 2012 *Waste Manag.* **32(3)** 513

[11] Hogland W 2002 *Environ. Sci. Pollut. Res.* **9(1)** 49

[12] Frändegård P, Krook J, Svensson N and Eklund M 2013 *J. Clean. Prod.* **55** 24

[13] Fisher H and Findlay DM 1995 *Wolrd Wastes 38.* **38(7)** 11

[14] Ortner ME, Knapp J and Bockreis A 2014 *Proc. of Institution of Civil Engineers: Waste and Resour. Manag.* ICE Publishing 167 51

[15] Zee DJ van der, Achterkamp MC and Visser BJ de 2004 *Waste Manag.* **24(8)** 795

[16] Krook J 2010 *J. Clean. Prod.* **18** 1772

[17] Krook J and Baas L 2013 *J. Clean. Prod.* **55** 1

[18] Rosendal R 2015 *Conf.: Proc Sardinia 2015, 15th Int. Waste Manag. and Landfill Symp.* Cagliari Italia 5

[19] Prechthai T, Padmasri M and Visvanathan C 2008 *Resour. Conserv. Recycl.* **53(1-2)** 70

[20] Burlakovs J, Kriipsalu M, Klavins M, Bhatnagar A, Vincevica-Gaile Z, Stenis J, Jani Y, Mykhaylenko V, Denafas G, Turkadze T, Hogland M, Rudovica V, Kaczala F, Rosendal RM and Hogland W 2016 *Resour. Conserv. Recycl.* **123** 73

[21] Bouvier R, Halstead JM, Conway KS and Manalo AB 2000 *J. Reg. Anal. Policy.* **30(2)** 23

[22] Hadi SP 2001 *Dimensi Lingkungan Perencanaan Pembangunan (Dimensions of the Development Planning Environment)* Gadjah Mada University Press Yogyakarta Indonesia

[23] Wibowo KS 2003 *TEMPO.CO*

[24] Purnama D 2009 *TEMPO.CO*

[25] Sopia S and Yuwanto E 2016 *REPUBLIKA.co.id*

[26] Sudarno A 2016 *Liputan6.com*

[27] San 2010 *TRIBUNNEWS.COM*

[28] Kompas Editor 2010 *kompas.com*

[29] Bams 2017 *focusjabar.co.id*

[30] Ul Haq MF and Ferdinan 2018 *detiknews.com*

[31] Leverenz H and Kreith F 2002 *Handbook of Solid Waste Management* vol 9 ed G Tchobanoglous and F Kreith McGraw-Hill Companies New York USA

[32] Hogland W and Marques M 2000 *Solid Waste Manag.* 313

[33] Sudrajat HR 2006 *Mengelola Sampah Kota (Managing Municipal Solid Waste)* Niaga Swadaya Jakarta Indonesia

[34] Jones PT, Geysen D, Tielemans Y, Passel SV, Pontikes Y, Blanpain B, Quaghebeur M and Hoekstra N 2013 *J. Clean. Prod.* **55** 45

[35] Barlaz MA, Green RB, Chanton JP, Goldsmith CD and Hater GR 2004 *Environ. Sci. Technol.* **38(18)** 4891

[36] Albanna M, Fernandes L and Mostafa W 2007 *J. Environ. Eng. Sci.* **6(2)** 191

[37] Hull RM, Krogmann U and Strom PF 2005 *J. Environ. Eng.* **131(3)** 478

[38] Vossen W van 2013 *Istanbul Int. Solid Waste, Water and Wastewater Congress* Turkey 347

[39] Passel S van, Dubois M, Eyckmans J, Gheldere S de, Ang F, Jones PT and Acker K van 2013 *J. Clean. Prod.* **55** 92

[40] Jaya IWAES, Suarna IW and Aryanta IWR 2016 *Ecotrophic* **10(1)** 62

[41] Kurniasari O, Damanhuri E, Padmi T and Kardena E 2014 *J. Bumi Lestari (J. Sustain. earth).* **14(1)** 46

[42] Silas J, Ratna SH, Setyawan W, Dewanti W, Hartatik and Firmaningtyas S 2014 *Simp. Nas. RAPI XIII* FT UMS A1

[43] Tchobanoglous G, Theisen H and Vigil S 1993 *Integrated Solid Waste Management: Engineering Principles and Management Issues* McGraw-Hill Companies New York USA

[44] Wharmadewanti 2003 *Evaluation of degraded waste from a landfill related to the concept of landfill management, a case study in Keputih Landfill Surabaya* Thesis Institut Teknologi Bandung

[45] Mahmudin I and Trihadiningrum Y 2013 *Prosid. Sem. Nas. Manaj. Tekn. XIX* Prog. Studi MMT-ITS Surabaya 1 D-2-1

[46] Sidik MA 2011 *J. Teknol. Lingkung.* **12**(**3**) 319

[47] Coordinating Ministry for Economic Affairs 2013 *Executive Summary: Studies on Sanitary Landfill Policy in Indonesia* Jakarta Indonesia

[48] Undang Undang Dasar Negara Republik Indonesia Tahun 1945 Jakarta Indonesia

[49] Undang Undang Republik Indonesia *Nomor 32 Tahun 2009* Jakarta Indonesia

[50] Undang Undang Republik Indonesia *Nomor 18 Tahun 2008* Jakarta Indonesia

[51] Peraturan Pemerintah *Nomor 81* Tahun *2012* Jakarta Indonesia

[52] Peraturan Menteri Pekerjaan Umum *Nomor 03 Tahun 2013* Jakarta Indonesia

[53] Peraturan Menteri Lingkungan Hidup dan Kehutanan *Nomor P.59 Tahun 2016* Jakarta Indonesia

[54] Badan Standardisasi Nasional (National Standardization Agency) *SNI 19-3964-1994*

[55] Badan Standardisasi Nasional (National Standardization Agency) *SNI 19-2454-2002*

[56] Badan Standardisasi Nasional (National Standardization Agency) *SNI 3242:2008*

# AUTHOR INDEX